ECOLOGICAL WATER QUALITY – WATER TREATMENT AND REUSE

Edited by **Kostas Voudouris**
and **Dimitra Voutsa**

INTECHOPEN.COM

Ecological Water Quality – Water Treatment and Reuse

Edited by Kostas Voudouris and Dimitra Voutsa

Published by InTech

Janeza Trdine 9, 51000 Rijeka, Croatia

Edition 2014

Publishing Process Manager Marija Radja

Technical Editor Teodora Smiljanic

Cover InTech Design Team

Additional hard copies can be obtained from orders@intechopen.com

Ecological Water Quality – Water Treatment and Reuse,
Edited by Kostas Voudouris and Dimitra Voutsa
p. cm.
ISBN 978-953-51-0508-4

INTECH

open science | open minds

Contents

Preface

Human activities may seriously affect the quality of aquatic ecosystems. Pathogen organisms, nutrients, heavy metals, toxic elements, pesticides, pharmaceuticals and various other organic micropollutants enter to aquatic environment through a range of point and diffuse sources. The presence of these compounds has adverse impacts on aquatic biota. It is well recognised that the distribution and the abundance of various species in aquatic systems are directly related to the water quality and hydrological conditions.

The achievement of good chemical and ecological status of waters are the main targets of Water Framework Directive (2000/60/EC) that has established the framework for actions in the field of water policy for the protection of inland surface waters, groundwaters, transitional waters and coastal waters. The assessment of good chemical status is based on the monitoring of priority pollutants that have to meet the proposed quality environmental standards. The assessment of ecological status is based on various biological elements such as composition and abundance of phytoplankton, aquatic flora, benthic invertebrate fauna and fish fauna. Moreover, biological diversity is among the criteria for assessing the good environmental status of marine waters as described in Marine Strategy Framework Directive (2008/56/EC).

The pollution of aquatic environment also reduces possible uses of water, especially those that require high quality standards i.e. for drinking purposes. A wide range of treatment technologies, from advanced techniques up to low cost systems, are available in order to remove possible pollutants from water cycle. The choice of suitable methods depends on the physicochemical behaviour of pollutants, the required quality standards, the cost, and the available infrastructure. In any case, sustainable choices of water use that prevent water quality problems aiming at the protection of available water resources and the enhancement of the aquatic ecosystems should be our main target.

This book entitled *"**Ecological water quality –Water treatment and reuse**"* attempts to cover various issues of water quality in the fields of Hydroecology and Hydrobiology and present various Water Treatment Technologies. Particularly, this book is divided into two sections:

1) Water quality and aquatic ecosystems

The first ten (10) chapters focus on the biological aspects of water quality using bio-indices and biosensors.

2) Water treatment technologies

This section includes fifteen (15) chapters related to the water treatment technologies in order to improve the water quality.

We would like to express our thanks to the authors contributed to this volume, to the reviewers for their valuable assistance as well as to the organizers and the staff of the INTECH Open Access Publisher, especially **Marija Radja**, for their efforts to publish this series of books on Water Quality.

Kostas Voudouris
Laboratory of Engineering Geology & Hydrogeology,
Department of Geology, Aristotle University of Thessaloniki,
Greece

Dimitra Voutsa
Laboratory of Environmental Pollution Control,
Department of Chemistry, Aristotle University of Thessaloniki,
Greece

List of reviewers

Bobori Dimitra,
Lab. of Ichthyology, School of Biology, Aristotle University of Thessaloniki
Genitsaris Savvas,
Dep. of Botany, School of Biology, Aristotle University of Thessaloniki
Georgiou Pantazis,
School of Agriculture, Aristotle University of Thessaloniki
Kaklis Akis,
Dr. of Hydrogeology, Aristotle University of Thessaloniki
Karayanni Hera,
Dep. of Biological Applications and Technology, University of Ioannina
Katsiapi Maria,
Dep. of Botany, School of Biology, Aristotle University of Thessaloniki
Kormas Kostas,
Dep. of Ichthyology and Aquatic Environment, University of Thessaly
Lazaridou Maria,
Professor of Biology, Aristotle University of Thessaloniki
Mattas Christos,
Dr. of Hydrogeology, Aristotle University of Thessaloniki

Melfos Basil,
Lecturer of Geology, Aristotle University of Thessaloniki
Michaloudi Evagelia,
Lab. of Ichthyology, School of Biology, Aristotle University of Thessaloniki
Moustaka-Gouni Maria,
Dept. of Botany, School of Biology, Aristotle University of Thessaloniki
Polemio Maurizio,
Istituto di Ricerca per la Protezione Idrogeologica, Bari, Italy
Theodosiou Nikolaos,
Assistant Professor, Civil Engineering, Aristotle University of Thessaloniki
Voudouris Kostas,
Ass. Professor of Hydrogeology, Aristotle University of Thessaloniki, Greece
Voutsa Dimitra,
Ass. Professor, Dep. of Chemistry, Aristotle University of Thessaloniki, Greece

Section 1

Water Quality and Aquatic Ecosystems

1

Evaluation of Ecological Quality Status with the Trophic Index (TRIX) Values in the Coastal Waters of the Gulfs of Erdek and Bandırma in the Marmara Sea

Neslihan Balkis[1], Benin Toklu-Aliçli[1] and Muharrem Balci[2]
[1]Istanbul University, Faculty of Science, Department of Biology,
[2]Istanbul University, Institute of Science,
Vezneciler-Istanbul,
Turkey

1. Introduction

In developing countries, more than 90 percent of wastewater and 70 percent of industrial wastes are discharged into coastal waters without being treated (Creel, 2003). The entry of wastes into marine environment not only changes water quality parameters but also affects benthic organisms, causes habitat change and increases the risk of eutrophication and, thereby, causes the area to become susceptible. The Urban Wastewater Treatment Directive (UWTD; EC, 1991) defines eutrophication as the "enrichment of water by nutrients, especially compounds of nitrogen and/or phosphorus, causing an accelerated growth of algae and higher forms of plant life to produce an undesirable disturbance to the balance of organisms present in the water and to the quality of the water concerned". Karydis (2009) characterized "oligotrophic" waters as nutrient poor with low productivity, "eutrophic" waters as nutrient rich with high algal biomass and "mesotrophic" waters as moderate conditions. Hypoxia or even anoxia is the last stage of eutrophication (Gray, 1992) and this phase is often characterized as "dystrophic" (Karydis, 2009). In addition, eutrophication of coastal waters has been considered as one of the major threats to the health of marine ecosystems in the last few decades (Andersen et al., 2004; Yang et al., 2008). The risk of eutrophication may increase or decrease depending on the speed and direction of flow and wind. It can occur as a result of natural processes, for example, where there is upwelling of nutrient rich deep water to nutrient poor but light rich surface water of the photic zone of the water column (Jørgensen & Richardson, 1996). Cultural eutrophication arising from anthropogenic activities is particularly evident in marine areas with limited water exchange, and in lagoons, bays and harbours (Crouzet et al., 1999).

Various factors may increase the supply of organic matter to coastal systems, but the most common is clearly nutrient enrichment. The major causes of nutrient enrichment in coastal areas are associated directly or indirectly with meeting the requirements and demands of human nutrition and diet. The deposition of reactive nitrogen emitted to the atmosphere as a consequence of fossil fuel combustion is also an important anthropogenic factor (Nixon, 1995). Nutrients are the essential chemical components of life in marine environment. Phosphorus and nitrogen are incorporated into living tissues, and silicate is necessary for the formation of the skeletons of diatoms and radiolaria (Baştürk et al., 1986). In the sea, most of the nutrients are present in sufficient concentration, and lack of some of them limits the growth of phytoplankton (Pojed & Kveder, 1977). While some nutrient enrichment may be beneficial, excessive enrichment may result in large algal blooms and seaweed growths, oxygen depletion and the production of hydrogen sulphide, which is toxic to marine life and can cause high mortality, red tide events, decreasing fishery yields, and nonreversible changes in ecosystem health (Daoji & Daler, 2004).

Trophic conditions of European coastal waters vary considerably from region to region and within regions. A trophic index (TRIX) characterizing eutrophic levels, was introduced by Vollenweider et al. (1998). The European Environmental Agency has evaluated this index and suggested that TRIX scales at regional level should be developed. TRIX values are very sensitive and any slight change of oxygen, chlorophyll *a*, dissolved inorganic nitrogen and total phosphorus concentrations results in changed index values (Boikova et al., 2008). This simple index seems to help synthesize key eutrophication variables into a simple numeric expression to make information comparable over a wide range of trophic situations (Anonymous, 2001).

Bays and gulfs are very important for fishing since they provide habitats for sheltering and reproduction for most living species. They are influenced by environmental conditions, especially pollution, very rapidly, which causes negative changes in their structures. Bays and gulfs are quieter compared to open seas and have a semi-closed structure, which increases the frequency of such events as eutrophication and red-tide events. The influence level of pollution on living organisms is directly related with the changes in species diversity, and the effects of pollution on a specific environment can be determined by monitoring the natural process. However, in order to achieve this, the most important requisite is to determine the ecological status of the area(s) that will be studied before pollution (Koray & Kesici, 1994). Seasonal changes and global warming considerably affects the biological structure of seas (Goffart et al., 2002). These effects in the marine environment come into being with different phenomena. For instance, mucilage formation in seas is the aggregation of organic substances that are produced by various marine organisms under special seasonal and trophic conditions (Innamorati et al., 2001; Mecozzi et al., 2001). In Turkish territorial waters, mucilage formation was observed firstly in the Gulf of İzmit in the Marmara Sea in October 2007 and mainly fisheries and tourism have been damaged seriously (Tüfekçi et al., 2010). Then, mucilage formations were reported on the shores of Büyükada Island (Balkıs et al., 2011) and in the Gulf of Erdek (Tüfekçi et al., 2010). This study is important because there is not a sufficient amount of comprehensive research conducted on the subject in the Gulfs of Erdek and Bandırma. Besides, the two gulfs are important for fishing and they are under the threat of heavy pollution.

In recent years, the scientific and technological advances have shown that studying sea and oceans, which cover 70% of the earth, is considerably important. Today in order to meet the increasing need for food, the studies on food sources in our seas have gained speed. This study aims to determine the ecological quality of coastal waters in the Gulfs of Erdek and Bandırma, and firstly the two gulfs will be compared in terms of environmental factors.

2. Material and methods

2.1 Study areas

The Marmara Sea forms "the Turkish Straits System" together with the Bosphorus and the Dardanelles. It has a surface area of 11,500 km^2 and the maximum depth is 1390 m. It is a small basin located between Asia and Europe. It is connected on the northeast with the Black Sea through the Bosphorus and on the southwest with the Aegean Sea through the Dardanelles (Ünlüata et al., 1990; Yüce & Türker, 1991; Beşiktepe et al., 1994). The brackish Black Sea waters with a salinity of about 17.6 ppt flow through the Bosphorus towards the Marmara Sea at the surface while the waters of Mediterranean origin with a salinity of about 38 ppt flow through the Dardanelles towards the Marmara Sea in a lower layer. There is an intermediate salinity mass with 25 m depth between these two water masses due to the difference in their densities (Ullyott & Pektaş, 1952; Yüce & Türker, 1991). The bottom water with a high salinity value includes a low amount of dissolved oxygen, and the water exchange and oceanographic conditions in the Marmara Sea are controlled by the two straits. The density stratification in the halocline impedes oxygen transfer from the surface layer that is rich in oxygen to the lower layer. Besides, biogenic particles in the bottom water increase oxygen consumption, which decreases the dissolved oxygen content of the lower water layer (Yüce & Türker, 1991; Beşiktepe et al., 2000).

The annual volume influx from the Black Sea to the Marmara Sea is nearly twice the salty water outflux from the Marmara basin to the Black Sea via the Bosphorus undercurrent (Ünlüata et al., 1990; Beşiktepe et al., 1994). Cyclonic alongshore currents in the Black Sea carry the polluted surface waters of the northwestern shelf as far as the Bosphorus region, with modified hydrochemical properties (Polat & Tuğrul, 1995). In addition, the salinity and nutrient contents of inflow slightly increase at the southern exit of the Bosphorus due to vertical mixing of the counter flows during the year. Concomitant photosynthetic processes in the Marmara upper layer, however, lead to consumption of biologically available nutrients and thus to a net export of particulate nutrients to the lower layer (Tuğrul et al., 2002). Primary production in the Sea is limited with halocline layer including the interface between 15 and 25 m (Polat & Tuğrul, 1995).

The oceanographic characteristics of the Gulfs of Erdek and Bandırma are similar to the Marmara Sea and the water column has a two-layer structure. The Gulf of Erdek has lower population density and industrial activities compared to the Gulf of Bandırma. The Biga River and the Gönen River flow into the Gulf of Erdek. The load of both rivers are affected by the mineral deposits and agricultural and food industries in their basin and the domestic wastes from the Boroughs of Biga and Gönen.

The Gulf of Bandırma is affected by industrial pollution and is more densely inhabited. The studies showed that the surface waters of the Gulf of Bandırma and the region that is on the northeast of the Kapıdağ Peninsula include more phosphate compared to the other parts of the Marmara Sea. This increase is caused by the domestic wastes and especially the fertilizer plant located in the Gulf. The Borough of Bandırma is rich in regards to nutrients both surface and ground waters. Most of the surface waters in Bandırma flow into the Susurluk River through Lake Manyas and the Kara River and reach the Marmara Sea. The most important harbor on the south of the Marmara Sea is located in this gulf. Although the intensive production of white meat and fertilizer raises the importance of the borough, it at the same time affects the Gulf of Bandırma negatively (Özelli & Özbaysal, 2001; Koç, 2002).

2.2 Sampling and primary analysis

Samples were collected from different depths of the water column (0.5, 15, 30 m) at three stations in each gulf, in total at six stations, seasonally (November, February, May, and August) for two years between November 2006 and August 2008 (Fig.1). The maximum depth at the stations is approximately 50 m. Photic zone, where photosynthesis occurs, was used as base in the selection of sampling depth. Water transparency was usually measured with the Secchi disk. A 3 l Ruttner water sampler with a thermometer was used for water analyses at each sampling point. The salinity was determined by the Mohr-Knudsen method (Ivanoff, 1972) and the dissolved oxygen (DO) by the Winkler method (Winkler, 1888).

Water samples for the determination of nutrients were collected in 100-mL polyethylene bottles and stored at -20 °C until the analysis in the laboratory. Nitrate+nitrite-N concentrations were measured by the cadmium reduction method using a "Skalar" autoanalyzer (APHA, 1999). Phosphate-P, Silicate-Si and chlorophyll *a* were analyzed by the methods described by Parsons et al. (1984). Chlorophyll *a* was measured after filtering 1 liter of the sample through Whatman GF/F filters. One milliliter of a 1% suspension of $MgCO_3$ was added to the sample prior to filtration. Samples were stored in a freezer, and pigments were extracted in a 90% acetone solution and measured with a spectrophotometer.

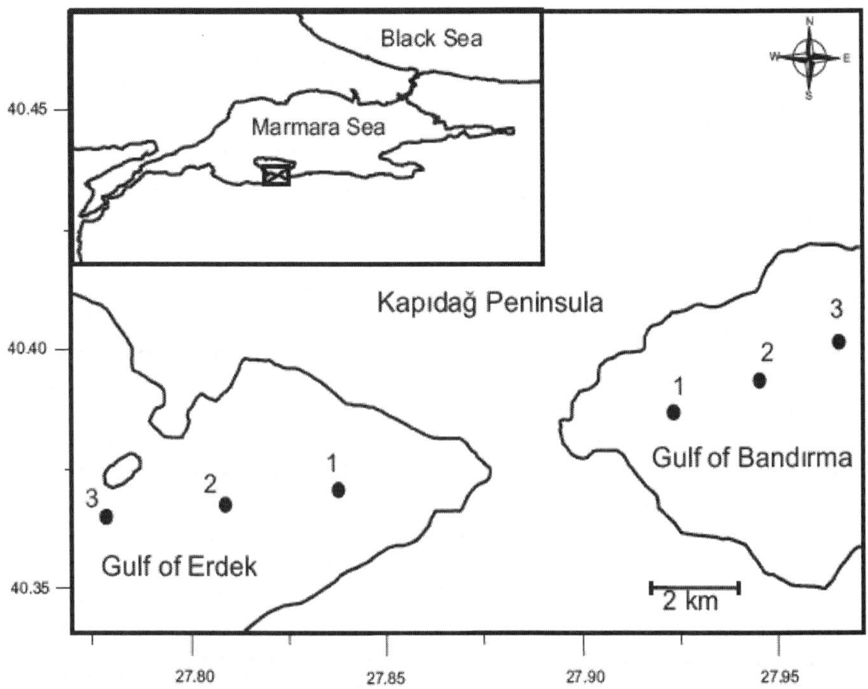

Fig. 1. Research stations in the Gulfs of Bandırma and Erdek

2.3 Data analysis

Trophic Index (TRIX) values were calculated in order to determine the eutrophication level of the sampling area and the quality of waters (Vollenweider et al., 1998). The index is given by:

$$TRIX= [\log_{10}(Chl\ a \cdot D\%O \cdot N \cdot P)+1.5]/1.2$$

Chl *a*= Chlorophyll *a* (µg L^{-1}), D%O= Oxygen as an absolute deviation (%) from saturation, N= Dissolved inorganic nitrogen N-NO_3+NO_2 (µg-at L^{-1}), P= Total phosphorus P-PO_4 (µg-at L^{-1}). Ammonium values were not used in nutrient ratios and calculation of TRIX, because NH_4-N values were not measured in this paper. TRIX was scaled from 0 to 10, covering a range of four trophic states (0-4 high quality and low trophic level; 4-5 good quality and moderate trophic level; 5-6 moderate quality and high trophic level and 6-10 degraded and very high trophic level) (Giovanardi & Vollenweider., 2004; Penna et al., 2004).

Spearman's rank correlation coefficient (Siegel, 1956) was used to detect any correlation among biotic (chlorophyll *a*) and abiotic variables (temperature, salinity, DO and nutrients), and Bray-Curtis similarity index in Primer v6 software, based on log(x+1) transformation was calculated to detect the similarity between sampling stations (Clark & Warwick, 2001).

3. Results

The vertical distribution of temperature, salinity and dissolved oxygen in the coastal waters of the Gulfs of Bandırma and Erdek is shown in Figs. 2, 3. During the study, temperature, salinity and dissolved oxygen levels of the seawater ranged between 6.5 and 26 °C, 21.4 and 38.6 ppt, and 3.5 and 15.62 mg L^{-1} in the gulfs, respectively. Also, chlorophyll *a* values ranged between 0.1 and 14.79 µg L^{-1} (Figs. 2, 3).

In the Gulf of Bandırma, the highest temperature value (26 °C) was measured at the depth of 0.5 m at all stations (Fig. 2). Homogenous distribution of water temperature was observed at Station 1 in February 2007. Also, sudden changes were more pronounced after the depth of 15 m at all stations. In this Gulf, the highest salinity value (38.5 ppt) was determined at the depth of 30 m at Station 3 in November 2006. While upper layer salinity values were low, sudden increases were observed after the depth of 15 m. Dissolved oxygen content of the gulf was lower in the deeper layer compared to the upper. The highest DO value (15.62 mg L^{-1}) was measured at the depth of 15 m at Station 2 in November 2006, and the lowest (3.5 mg L^{-1}) at the depth of 30 m at Station 1 in May 2008. For chlorophyll *a* concentration, the highest value (14.79 µg L^{-1}) was determined at the depth of 0.5 m at Station 2 in August 2008 and higher chlorophyll *a* value was observed in the upper water column. The lowest value was detected at the depth of 30 m (0.21 µg L^{-1}) at Station 2 in May 2008. Water transparency was 4.5 m (February 2008) - 16 m (November 2006) in the Gulf of Bandırma.

In the Gulf of Erdek, the highest temperature value (25.5 °C) was recorded in the surface water at all stations in August 2007. Sudden temperature changes were detected after the depth of 15 m at all stations as in the Gulf of Bandırma. Salinity values ranged from 22.4 ppt (st.2, 0.5 m, May 2007) to 38.6 ppt (st.3, 30 m, November 2006). While a sudden increase was detected in salinity values after the depth of 15 m at Station 2 and Station 3, the values increased in some seasons and decrease in others at Station 1, which is a coastal station. As in the Gulf of Bandırma, oxygen values were determined to be higher in the upper layer and showed decrease towards the deeper layers. The lowest value (3.78 mg L^{-1}) was detected at the depth of 30 m at Station 3 in August 2008. During most of the sampling period, chlorophyll *a* values were higher in the euphotic zone and showed decrease in correlation with the increase of depth. Chlorophyll *a* values were between 0.10 µg L^{-1}(st.3, 30 m, May 2008) and 2.83 µg L^{-1} (st.3, 15 m, February 2008). Water transparency was 6 m (February 2008) - 15 m (February 2007) in this gulf.

Nutrient concentrations and TRIX index values are shown in Figs. 4, 5. The amounts of nitrate+nitrite-N (0.07-5.83 µg-at N L^{-1}), phosphate-P (0.09-8.6 µg-at P L^{-1}) and silicate-Si (0.05-21.62 µg-at Si L^{-1}) concentrations were measured. The consumption of nitrogen and silica in the upper layer was determined to be higher in both of the gulfs. However, phosphorus values were quite high in the upper layer compared to the lower especially in August 2008. A similar situation was observed at Station 1 in February 2008. There were low levels of dissolved oxygen in the deeper layers, which were rich in nutrients.

Mean ratios of nutrients and chlorophyll *a* at the sampling stations are presented in Table 1. The lowest and highest mean ratios of N/P were 0.04 (0.5 m depth, in May 2007) and 4.73 (30 m depth, in May 2007) in the Gulf of Bandırma and 0.35 (0.5 m depth, in August 2007) and 7.08 (30 m depth, in May 2008) in the Gulf of Erdek, respectively. Also these ratios were recorded lower than the Redfield ratio (16/1). This result indicates that N is limiting nutrient for both gulfs. A considerable increase was observed in both N/P and Si/P ratios especially from the upper layer to the lower in both gulfs. Besides, these values were higher in the Gulf of Erdek compared to the Gulf of Bandırma. In both gulfs, N/Si ratio was lower than 1 during all sampling periods. P/Chl *a* ratio was low in the upper layer due to the increase of chlorophyll *a* value depending on phytoplankton activity and the use of phosphorus by these organisms.

	N/P			N/Si			Si/P			P/Chl *a*		
Period	0.5m	15m	30m	0.5m	15m	30m	0.5m	15m	30m	0.5m	15m	30m
Gulf of Bandırma												
Nov.2006	0.34	0.35	3.42	0.21	0.26	0.29	1.64	1.49	11.22	1.12	0.87	2.88
Feb.2007	0.35	0.13	3.82	0.19	0.07	0.52	1.80	1.81	7.76	0.66	0.70	3.28
May.2007	0.04	0.33	4.73	0.17	0.26	0.34	0.28	1.30	13.30	1.18	1.38	2.99
Aug.2007	0.65	0.23	3.18	0.45	0.21	0.27	4.20	1.81	11.39	0.98	0.80	1.89
Nov.2007	0.68	1.26	2.40	0.30	0.29	0.55	1.98	3.71	4.52	0.37	0.55	1.78
Feb.2008	0.57	0.47	3.08	0.19	0.24	0.70	3.29	1.90	4.39	0.57	0.19	0.49
May.2008	0.48	3.28	4.37	0.77	0.80	0.75	0.60	3.66	5.77	3.00	1.68	3.84
Aug.2008	0.08	0.08	2.37	6.25	2.38	1.35	0.02	0.17	3.19	0.64	0.58	0.72
Gulf of Erdek												
Nov.2006	0.68	2.32	4.70	0.12	0.30	0.36	6.05	7.33	12.88	0.25	1.18	1.91
Feb.2007	0.91	1.00	3.24	0.17	0.10	0.14	5.54	16.05	23.62	0.31	0.68	4.39
May.2007	0.85	0.55	3.70	0.13	0.07	0.21	6.45	13.11	17.78	0.19	0.25	2.06
Aug.2007	0.35	0.77	1.82	0.06	0.14	0.10	5.69	5.30	13.67	0.61	0.52	0.37
Nov.2007	0.72	1.47	0.97	0.44	0.66	0.49	1.42	1.98	2.62	0.61	1.02	6.42
Feb.2008	1.91	1.70	3.22	0.15	0.13	0.43	12.88	13.51	8.97	0.06	0.05	0.13
May.2008	1.80	2.33	7.08	0.92	0.54	0.75	1.93	4.60	9.42	0.35	0.52	4.93
Aug.2008	1.12	0.71	5.09	0.55	0.44	0.54	2.07	1.44	9.30	0.19	0.55	0.75

Table 1. The atomic ratios of nutrients and chlorophyll *a* at the sampling stations.

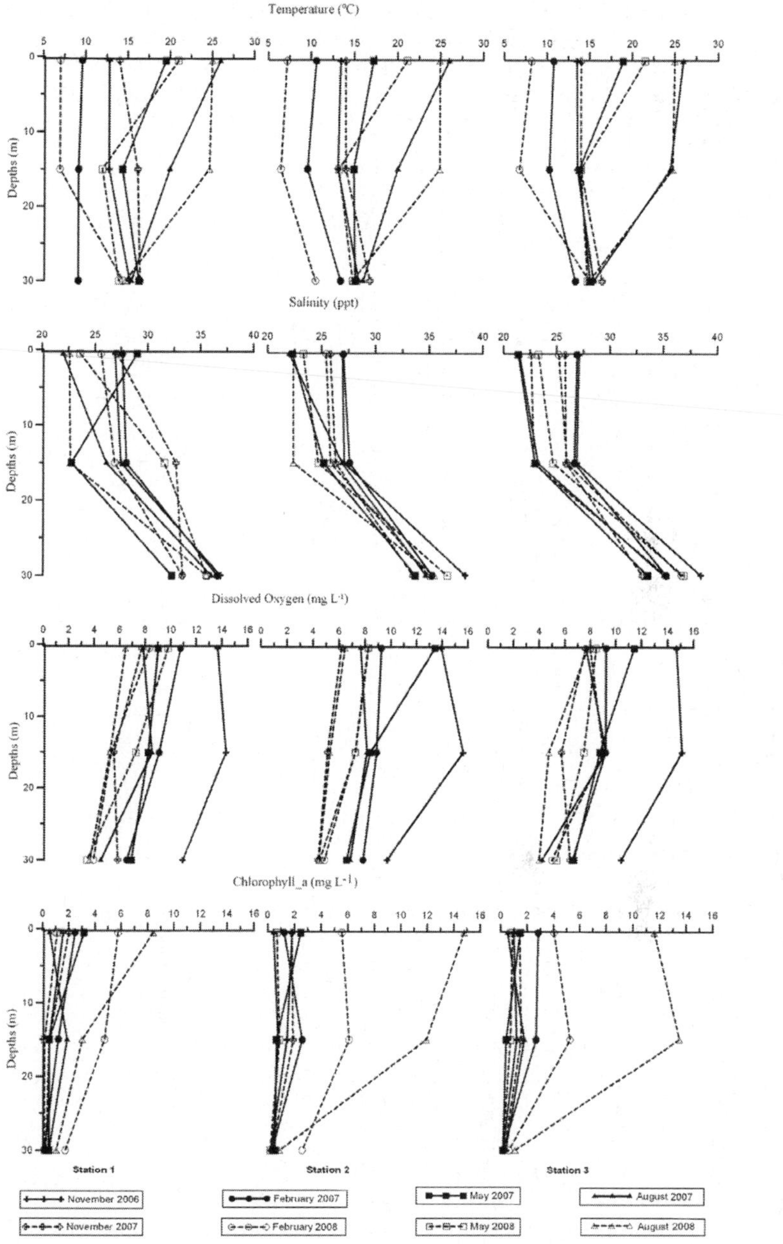

Fig. 2. Vertical variations of temperature (°C), salinity (ppt), dissolved oxygen (DO, mg L^{-1}) and chlorophyll *a* (μg L^{-1}) along the water column in the Gulf of Bandırma.

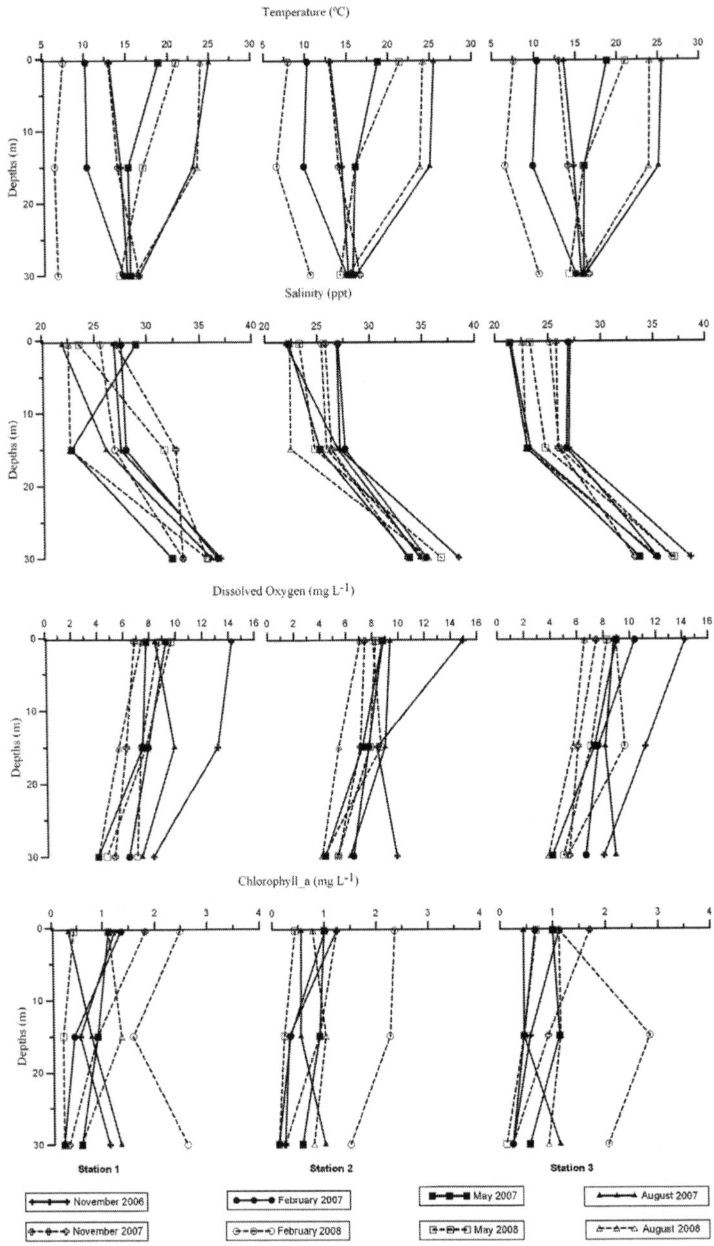

Fig. 3. Vertical variations of temperature (°C), salinity (ppt), dissolved oxygen (DO, mg L^{-1}) and chlorophyll *a* (μg L^{-1}) along the water column in the Gulf of Erdek.

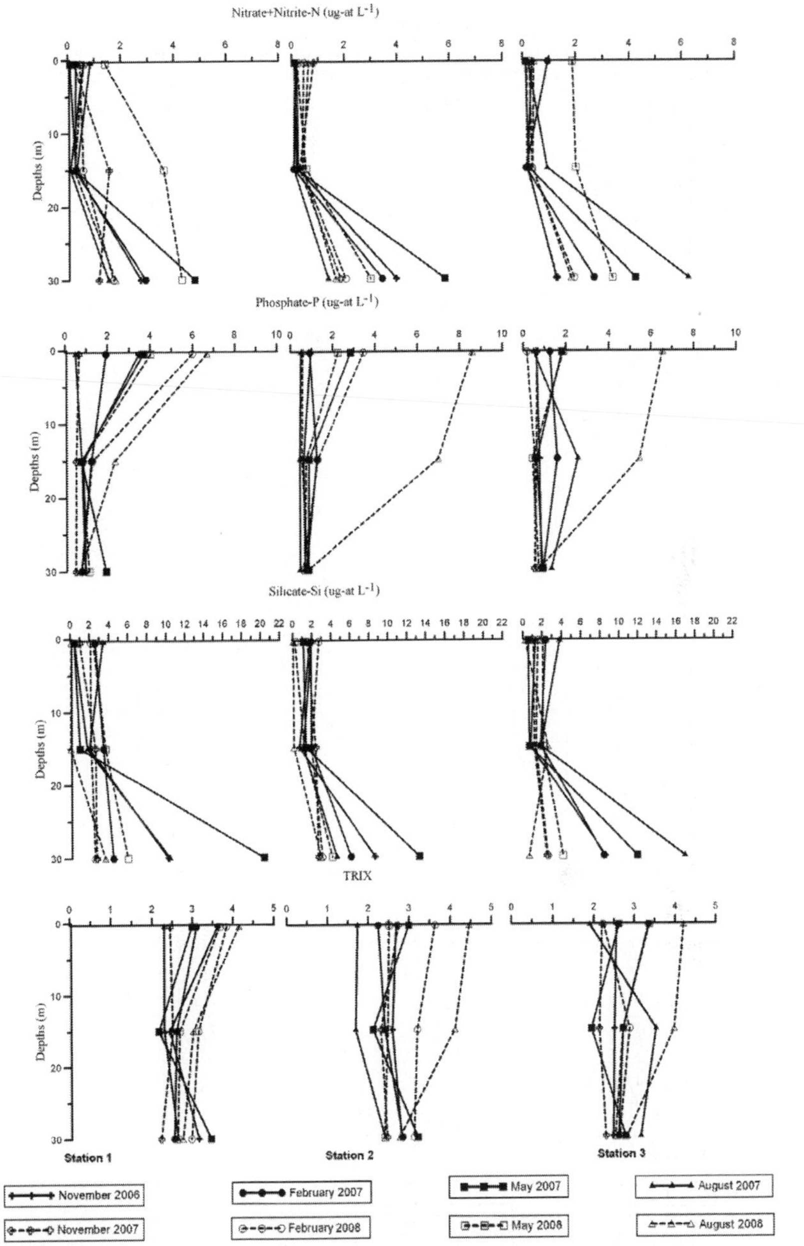

Fig. 4. Vertical variations of nutrient and TRIX index values along the water column in the Gulf of Bandırma.

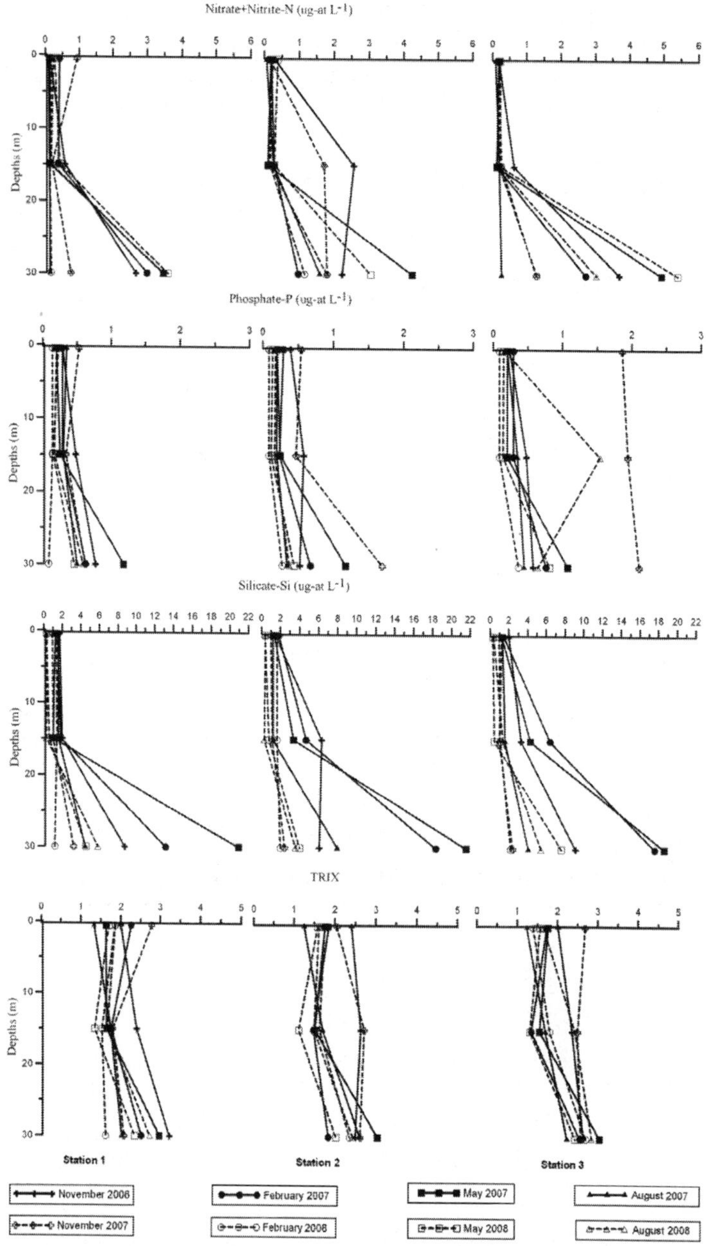

Fig. 5. Vertical variations of nutrient and TRIX index values along the water column in the Gulf of Erdek.

While the Trophic Index (TRIX) value for the Gulf of Erdek was determined to range between 1.12 and 3.23, it ranged between 1.68 and 4.46 for the Gulf of Bandırma (Figs. 4, 5). In addition, an increase was observed in TRIX values with the increase in concentrations of phosphorus and chlorophyll *a* in the last season (August 2008) of the second sampling period in the Gulf of Bandırma.

The results of Spearman's rank order correlation were employed to explain the relationship among the ecological parameters in the gulfs (Table 2, 3). The nutrients were negatively correlated to dissolved oxygen, but positively to salinity except phosphorus in the Gulf of Bandırma. Also, chlorophyll *a* was negatively correlated with N and Si in the gulfs, however it was positively correlated with P in the Gulf of Bandırma and negatively in the Gulf of Erdek.

	Dissolved oxygen	Temperature	Salinity	Chlorophyll *a*	Nitrogen	Phosphorus	Silica
Temperature	-,190						
Salinity	-.274**	-.279**					
Chlorophyll *a*	,139	-,132	-.545**				
Nitrogen	-.482**	,020	.604**	-.521**			
Phosphorus	,170	,106	-.238*	.443**	-,055		
Silica	-.247*	-,175	.706**	-.579**	.699**	-,106	
TRIX	,022	-,043	-,022	.331**	.355**	.640**	,118

Table 2. Spearman's rank-correlation matrix (r_s) to correlate among ecological variables in the Gulf of Bandırma (** *P*<0.01, * *P*<0.05, n=72)

	Dissolved oxygen	Temperature	Salinity	Chlorophyll *a*	Nitrogen	Phosphorus	Silica
Temperature	-,071						
Salinity	-.390**	-.360**					
Chlorophyll *a*	,151	-.335**	-.280**				
Nitrogen	-.455**	-,117	.675**	-.448**			
Phosphorus	-.397**	,005	.625**	-.345**	.593**		
Silica	-.266*	-.216*	.751**	-.336**	.525**	.618**	
TRIX	-.365**	-.206*	.653**	,000	.745**	.807**	.551**

Table 3. Spearman's rank-correlation matrix (r_s) to correlate among ecological variables in the Gulf of Erdek (** *P*<0.01, * *P*<0.05, n=72)

The Bray-Curtis similarity index did not show significant differences at the sampling stations according to ecological parameters. Sampling stations in the gulfs were approximately 91% similar to each other (Fig. 6).

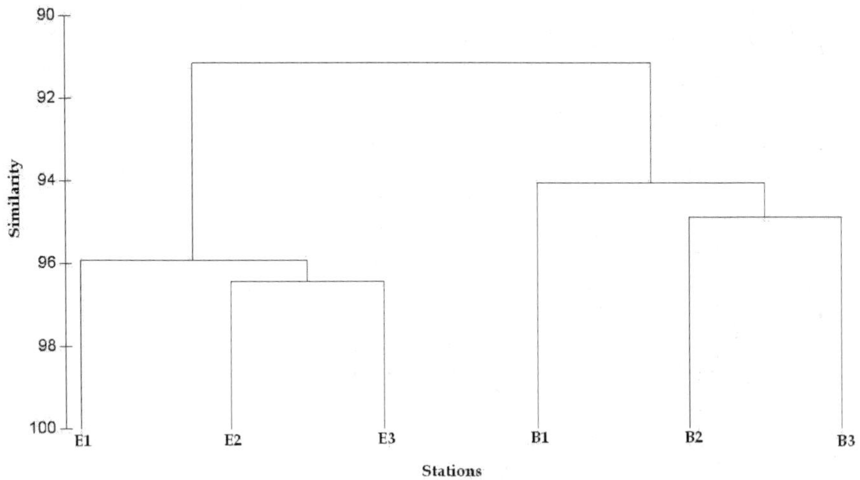

Fig. 6. Bray-Curtis similarity dendogram of the sampling stations in the Gulfs of Bandırma and Erdek.

4. Discussion

The chemical oceanography of the Marmara Sea is remarkably affected by the Black Sea and the Aegean Sea, and the basin includes two different masses. In this study, the highest temperature values were measured at the depth of 0.5 m in August 2007 in both gulfs with similar characteristics. Especially, the variations of temperatures in surface water showed typically seasonal trend and this was caused by the effect of light and the contact of this layer with the atmosphere. While lower values were measured in the lower layer water compared to the upper in spring and summer, temperature values increased in correlation with the increase in depth in autumn and winter. The increase in temperature at the depth of 30 m during cold periods indicates the effect of the Mediterranean waters. Especially, a sharp decrease was detected in water temperature after the depth of 15 m in August in 2007 and 2008 during all sampling periods in the both gulfs. It was found that less saline water from the Black Sea via the Bosphorus was effective at the depths that were close to the surface and salinity was noted to increase from the surface to the bottom, reaching its highest value at the depth of 30 m due to the Mediterranean current. After 15 m, a sudden increase in salinity was remarkable, which indicates the presence of a halocline layer. Besides, in gulfs salinity values changed seasonally throughout the water column at Station 1, which is an inner one while seasonal changes were more stable until the depth of 15 m at Station 2 and 3 and the values showed sudden increases after this depth.

Rather high oxygen concentrations were observed in the upper water, probably coinciding with the maximum of the photosynthetic activity. A decrease was observed in dissolved oxygen values generally from the surface to the bottom along the water column. This was due to excessive oxygen consumption during the decomposition of detritus, which was produced as a consequence of primary production in the upper layer and biochemical reactions occurring in the deeper layer. Excessive bacterial and animal activity due to increased phytoplankton biomass and high organic loads in eutrophic systems can lead to oxygen depletion (Karydis, 2009). This decrease in the water column was in accordance with the result of the previous review study (Yılmaz, 2002). The most remarkable period in terms of seasonal changes in oxygen was November 2006 in both gulfs. Considering the other sampling periods, the highest dissolved oxygen values were recorded at all depths in this sampling period. The fate and behavior of DO is of critical importance to marine organisms in determining the severity of adverse impacts (Best et al., 2007). When the DO falls below 5 mg L^{-1}, sensitive species of fish and invertebrates can be negatively impacted, and at the DO levels below 2.5 mg L^{-1} most fish are negatively impacted (Frodge et al., 1990). Best et al. (2007) provided DO thresholds in accordance with 5 ecological categories in the European Water Framework Directive (≥5.7 mg L^{-1} High; ≥4.0<5.7 mg L^{-1} Good; ≥2.4<4.0 mg L^{-1} Moderate; ≥1.6<2.4 mg L^{-1} Poor; <1.6 mg L^{-1} Bad). In our study, high and good quality status were detected in both gulfs and moderate quality status were only detected at the depth of 30 m at Station 1 in May and August 2008 in the Gulf of Bandırma and at Station 3 in August 2008 in the Gulf of Erdek. Especially, salinity at 30 m, which increase with the effect of the Mediterranean waters, showed a negative correlation with DO and water quality is moderate at this depth according the DO levels. The main physical factors affecting the concentration of oxygen in the marine environment are temperature and salinity: DO solubility decreasing with increasing temperature and salinity (Best et al., 2007). Negative correlation was found between DO and salinity in the present study ($p<0.01$). Although pollution has clearly increased in last 30 years in the upper water of the Marmara Sea, dissolved oxygen values in the deeper layer have not changed compared to the values measured in 1970s (Tuğrul et al., 2000).

Chlarophyll *a* production and nutrient availability are closely associated with eutrophication (Nixon, 1995; Kitsiou & Karydis, 2001). Chlorophyll *a* distribution depends on hydro-chemical conditions, namely nutrient availability, temperature changes, light conditions, water turbulence etc. (Lakkis et al., 2003; Nikolaidis et al., 2006 a, b). In this study, the highest chlorophyll *a* values were generally determined in winter period in both gulfs and P/Chl *a* ratio was low in the upper layer due to the increase of chlorophyll *a* value depending on phytoplankton activity and the use of phosphorus by these organisms; however, chlorophyll *a* showed excessive increase only in summer (August 2008) in the Gulf of Bandırma. Especially a serious environmental problem was observed in 2008 in the whole Marmara Sea. In recent studies, it was stated that mucilage formation, which was observed mainly due to excretory activity of some diatoms together with bacteria, the dinoflagellate *Gonyaulax fragilis*, the presence of sharp picnocline and thermocline caused by the two-layered water system of the Marmara Sea in 2008; besides, which the weather conditions and the status of currents during that time effected this formation (Tüfekçi et al., 2010; Balkıs et al., 2011). During the mucilage formation chlorophyll *a* values changed between 0.1-22 µg L^{-1} in these studies. According to the study by Ignatiades (2005), the limits of average concentration in chlorophyll *a* are <0.5 µg L^{-1} for oligotrophic, 0.5-1.0 µg L^{-1} for mesotrophic and >1.0 µg L^{-1} for eutrophic waters. According to chlorophyll *a* results obtained from both gulfs, the gulfs showed mesotrophic conditions in some periods and eutrophic, hyper-eutrophic (during the mucilage formation event) conditions in others.

Fiocca et al. (1996) reported that the availability of dissolved inorganic nitrogen and dissolved inorganic phosphorus leads to a seasonal change in N/P ratio, high in winter and low in summer. In the euphotic zone, nutrients, especially nitrate+nitrite-N and silicate-Si, are practically depleted by the phytoplankton uptake. In the Marmara Sea, the highest abundance of phytoplankton was recorded in the surface water (0.5-5 m) in recent comprehensive studies (Balkıs, 2003; Deniz & Taş, 2009; Tüfekçi et al., 2010; Taş et al., 2011). The highest nutrient values were recorded generally in the bottom layer where there was aggregation. During the study, positive correlations ($p<0.01$) of nitrogen and silicate were detected with salinity, which increased with depth. The strong positive correlation ($p<0.01$) between nitrogen, phosphate and silicate in the Gulf of Erdek and between nitrogen and silicate in the Gulf of Bandırma might indicate that these nutrients come from the same sources into the water column. Especially, the increase in the amount of nitrogen and silica at the depth of 30 m is remarkable since these elements are produced by the bacterial decomposition of the organic substances aggregating at the bottom.

Smith (1984) mentioned that nitrogen budget in the ocean was associated with air-water interaction such as N_2 fixation, losses of fixed nitrogen, or sediment back to gaseous form but that it depends on availability of phosphorus because phosphorus is not exchanged between the ocean and atmosphere as nitrogen. In both gulfs, there was negative correlation ($p<0.01$) between chlorophyll *a* and nutrients except for the positive correlation ($p<0.01$) between chlorophyll *a* and phosphorus in the Gulf of Bandırma (Tables 2, 3). It is known that eutrophication variables such as nutrient and chlorophyll *a* do not seem to follow a linear relationship (Karydis, 2009). Interestingly, there was a negative correlation ($p<0.05$) between salinity and phosphate in the Gulf of Bandırma and positive correlation ($p<0.01$) in the Gulf of Erdek. The inverse relationship between phosphate and salinity occurs with increasing of evaporation, which depends on temperature while remain nitrogen is at low level concentrations in the gulf (Smith, 1984). Actually, the positive correlation ($p<0.01$) between phosphate and salinity, which increased with depth indicates the source of phosphate in the water column in the Gulf of Erdek. Smith (1984) argued that the sediments could be reliable records for net nitrogen and phosphorus accumulation in the bays, which have very slow water turnover. Generally, nutrients are depleted by phytoplankton at the points where the light reaches while it increases in direct proportion to depth. The negatively correlated relationships between chlorophyll *a* and nutrients might indicate that nutrients are controlled by primary producers in the short term (Pérez-Ruzafa et al., 2005). In terms of phosphorus inputs, the significant positive relationships between phosphorus and chlorophyll *a* might indicate that the Gulf of Bandırma has not reached the saturation level, which has an enhanced effect on primary productivity. Jaanus (2003) asserted that phosphorus was more important on primary production in eutrophied coastal areas rather than nitrogen when a positive correlation between phosphorus and chlorophyll *a* was detected. On the other hand, the significant negative relationship between nutrients and chlorophyll *a* probably indicates that these nutrients have excessive concentrations, which have a restricted effect on primary production with regard to phosphorus in the Gulf of Bandırma.

Many authors are of the opinion that it is useful to look at the N/Si/P ratios in various parts of the ocean, and that only certain ratios are favorable for bioproductivity. It is known that P stress is common in freshwater systems, whereas N stress is found in marine systems (Ryther & Dunstan, 1971). The nitrogen limitation of phytoplankton growth is common in coastal systems (Nixon, 1986). Redfield et al. (1963) mentioned a C/N/P ratio of 106/16/1 ratio among the elements of sea water. If N/P ratios are below the Redfield value of 16/1, N is limiting nutrient (Stefanson et al., 1963). In the coastal waters of the Gulfs of Bandırma and Erdek, the atomic ratio of N/P was lower than the Redfield ratio of 16, and N was limiting nutrient. The increase in the N/P ratio was remarkable in lower layers in comparison to the surface water, which showed that phosphorus increased in proportion to nitrogen in the surface water while nitrogen increased in proportion to phosphorus in lower layers. In addition, the reason for lower values of N/P ratio in summer months is the limiting effect of nitrogen (Marcovecchio et al., 2006) and in the eutrophic areas, nitrogen might be a significant growth-limiting factor under the conditions, which have high total phosphorus concentration and low total N/P ratio (Attayde & Bozelli, 1999). Diatom growth in marine waters is likely to be limited by dissolved silica when the N/Si ratios are above 1/1 (Roberts et al., 2003). During the study period, the N/Si ratios were low (<1), and this indicates that silicate is not a limiting factor, especially for the growth of diatoms. In the studies conducted in the Marmara Sea, it was reported that dinoflagellates and diatoms constitute most of the plankton population, which supports our finding (Balkıs, 2003; Deniz & Taş, 2009; Tüfekçi et al., 2010; Balkıs et al., 2011). It is known that especially diatoms show an excessive increase in summer and spring (Balkıs, 2003; Deniz & Taş, 2009).

The Secchi disk depth for oligotrophic waters varies from 20 to 40 m (Ignatiades et al., 1995). Secchi disk depth detected between 10 and 20 m characterizes mesotrophic conditions whereas it is less than 10 m for eutrophic waters (Ignatiades et al., 1995). The lowest-highest Secchi range recorded in this study is 4.5-16 m. The values measured in 2008 are lower than those recorded in the previous year. Especially in winter when environmental inputs are intense lower values were recorded. In addition, intense vertical mixtures also caused turbidity in this period.

Trophic condition of vast marine areas, like the Mediterranean, varies considerably from region to region and within regions (Vollenweider et al., 1996). Vollenweider et al. (1998) calculated TRIX mean values as 3.37-5.60 for the Adriatic Sea. Moncheva et al. (2001) detected that TRIX index values varied from 5.0 to 6.0 in the Thermaikos Gulf of the northern Aegean Sea, and the lowest values were recorded in summer. In addition, the index was recorded between 1.9-4.7 in Kalamitsi on the east coasts of the central Ionian Sea (Nikolaidis et al., 2008), 6.90-7.70 in southern Black Sea (Baytut et al., 2010), 0.86-2.98 in the Edremit Bay of the Aegean Sea (Balkıs & Balcı, 2010). Calculated TRIX values were slightly lower than expected because NH_4 was not measured in this study and used in the original formula. Low TRIX values, as defined in this paper, indicate poorly productive waters corresponding to high water quality in the Gulfs of Erdek and Bandırma. On the other hand, according to chlorophyll *a* results the environment generally showed mesotrophic-eutrophic conditions. The chlorophyll *a* scale used above (Ignatiades, 2005) is the one suggested for the Aegean Sea. The Marmara Sea, different from the Aegean Sea, is an inland sea and has a two-layered water system. The calculated Trix values without NH_4 could caused to obtained low values. Especially the upper layer waters are under the effect of the waters of Black Sea, which has intense river inputs. Therefore, in order to determine the water quality of the environment not only physical and chemical studies but also biological studies that will show the organism communities in the environment and their abundance should be conducted. This study showed the current state of water quality in both gulfs and can be used as a main source in determining possible changes in these gulfs in future. Moreover, effective wastewater management including nutrient control may be required for an effective pollution prevention program in these regions.

5. Acknowledgements

This study was supported by the Research Fund of Istanbul University, project number 541.

6. References

American Public Health Association (APHA) (1999). *Standard Methods for the Examination of Water and Waste Water*, 20th ed. Washington DC, USA.

Andersen, J.H., Conley, D.J. & Hedal, S. (2004). Palaeoecology, reference conditions and classification of ecological status: the EU Water Framework Directive in practice. *Marine Pollution Bulletin*, 49, 283-290.

Attayde, J. L. & Bozelli, R. L. (1999). Environmental heterogeneity patterns and predictive models of chlorophyll *a* in a Brazilian coastal lagoon. Hydrobiologia, 390 (1-3), 129-139.

Balkıs, N. (2003). Seasonal variations in the phytoplankton and nutrient dynamics in the neritic water of Büyükçekmece Bay, Sea of Marmara. *Journal of Plankton Research*, 25(7), 703-717.

Balkıs, N. & Balcı, M. (2010). Seasonal variations of nutrients and chlorophyll *a* in the coastal waters of the Edremit Bay and The trophic index (TRIX) values of the environment. *Fresenius Environmental Bulletin*, 19(7), 1328-1336.

Balkıs, N., Atabay, H., Türetgen, I., Albayrak, S., Balkıs, H. & Tüfekçi, V. (2011). Role of single-celled organisms in mucilage formation on the shores of Büyükada Island (the Marmara Sea). *Journal of the Marine Biological Association of the United Kingdom*, 91 (4), 771-781.

Baştürk, Ö., Saydam, A.C., Salihoğlu, İ. & Yılmaz, A. (1986). Oceanography of the Turkish Straits', *First Annual Report*, Vol. III: health of Turkish Straits, II. Chemical and environmental aspects of the Sea of Marmara. Institute of marine Sciences, METU, Erdemli-İçel, Turkey.

Baytut, O., Gonulol, A. & Koray, T. (2010). Temporal variations of phytoplankton in relation to eutrophication in Samsun Bay, Southern Black Sea. *Turkish Journal of Fisheries and Aquatic Sciences*, 10, 363-372.

Best, M.A., Wither, A.W. & Coates, S. (2007). Dissolved oxygen as a physico-chemical supporting element in the water framework directive. *Marine Pollution Bulletin.*, 55, 53-64.

Beşiktepe, Ş.T., Özsoy, E., Abdullatif, M.A. & Oğuz, T. (2000). Marmara Denizi'nin Hidrografisi ve Dolaşımı, *Marmara Denizi 2000 Sempozyumu Bildiriler Kitabı*, 11-12 Kasım 2000 Ataköy Marina, İstanbul, Yayın No: 5 Türk Deniz Araştırmaları Vakfı, İstanbul: 314-326.

Beşiktepe, Ş.T., Sur, H.İ., Özsoy, E., Abdullatif, M.A., Oğuz, T. & Ünlüata, Ü. (1994). The circulation and hydrography of the Marmara Sea, *Progress in Oceanography*, 34, 285-334.

Boikova, E., Botva, U. & Licite, V. (2008). Implementation of trophic status index in brackish water quality assessment of Baltic coastal waters. *Proceedings of the Latvian Academy of Sciences*, Section B, 62(3), 115-119.

Clarke, K.R. & Warwick, R.M. (2001). Change in marine communities: an approach to statistical analysis and interpretation, second ed., PRIMER-E: Plymouth.

Creel, L. (2003). Ripple effects: Population and coastal regions. (policy brief). Washington, DC: Population Reference Bureau.

Crouzet, P., Nixon, S., Laffon, L., Bøgestrand, J., Lallana, C., Izzo, G., Bak, J. & Lack, T.J. (1999). Status and impact on marine and coastal waters, In: *Nutrients in European ecosystems,* Thyssen, N., (Ed.), pp. 81-93, Office for Official Publications of the European Communities, EEA, Copenhagen.

Daoji, L. & Daler, D. (2004). Ocean pollution from land-based sorces: East China Sea, China. *Ambio,* 33, 107-113.

Deniz, N & Taş, S. (2009). Seasonal variations in the phytoplankton community in the north-eastern Sea of Marmara and a species list. *Journal of the Marine Biological Association of the United Kingdom,* 89(2), 269-276.

EC. 1991. Urban Wastewater Treatment Directive 91/271/EEC. Official Journal of the European Communities, L135/40–52, 30 May 1991.

Fiocca, F., Luglie, A. & Sechi, N. (1996). The phytoplankton of S'Ena Arrubia Lagoon (centre-western Sardinia) between 1990 and 1995. *Giornale Botanico Italiano,* 130, 1016-1031.

Frodge, J.D., Thamas, G.L. & Pauley, G.B. (1990). Effects of canopy formation by floating and submergent aquatic macrophytes on the water quality of two shallow Pacific Northwest lakes. *Aquatic Botany,* 38, 231-248.

Giovanardi, F. & Vollenweider, R.A. (2004). Trophic conditions of marine coastal waters: experience in applying the Trophic index TRIX to two areas of the Adriatic and Tyrrhenian seas. *Journal of Limnology,* 63, 199-218.

Goffart, A., Hecq, J.H. & Legendre, L. (2002). Changes in the development of the winter-spring phytoplankton bloom in the Bay of Calvi (NW Mediterranean) over the last two decades: a response to changing climate? *Marine Ecology Progress Series,* 236, 45-60.

Gray, S.J. (1992). Eutrophication in the sea. In: *Marine eutrophication and pollution dynamics,* Colombo, G. and Viviani, R. (Eds), Olsen&Olsen, Fredensborg, 394p.

Ignatiades, L. (2005). Scaling the trophic status of the Aegean Sea, eastern mediterranean. *Journal of Sea Research.,* 54, 51-57.

Ignatiades, L., Georgooulos D. & Karydis, M. (1995). Description of a phytoplanktonic community of the oligotrophic waters of SE Aegean Sea (Mediterranean), *P.S.Z.I. Marine Ecology,* 16, 13-26.

Innamorati, M., Nuccio, C., Massi, L., Mori, G. & Melley, A. (2001). Mucilages and climatic changes in the Tyrrhenian Sea. *Aquatic Conversation: Marine and Freshwater Ecosytems,* 11, 289-298.

Ivanoff, A. (1972). *Introduction al'océanographie,* Tome I. Librairie Vuibert, Paris.

Jaanus, A. (2003). Water environment of Haapsalu Bay in retrospect (1975-2000). Proceedings of the Estonian Academy of Sciences. Ecology, 52 (2), 91-111.

Jorgensen, B.B., Richardson, K. (Eds). (1996). *Eutrophication in coastal marine ecosystems.* Coastal and Estuarine Studies 52. American Geophysical Union, Washington, DC.

Karydis, M. (2009). Eutrophication assessment of coastal waters based on indicators: a literature review. *Global NEST Journal,* 11(4), 373-390.

Kitsiou, D. & Karydis, M. (2001). Marine eutrophication: a proposed data analysis procedure for assessing spatial trends, *Environmental Monitoring and Assessment*, 68, 297-312.

Koç, T. (2002). Bandırma ilçesinde tavukçuluğun çevresel etkisi, *Ekoloji*,11 (43), 11-16.

Koray, T. & Kesici, U. Y. (1994). Bodrum Körfezi'nin (Ege Denizi) fitoplankton ve protozooplankton tür kompozisyonu. *Ege Üniversitesi Fen Fakültesi Dergisi*, Seri B, Ek 16/1, 971-980.

Lakkis, S., Jonsson, L., Zodiatis, G. & Soloviev, D. (2003). Remote sensing data analysis in the Levantine basin: SST and chlorophyll *a* distribution. In: Oceanography of Eastern mediterranean and Black Sea. Similarities and differences of two interconnected basins (Ed. A. Yilmaz). 14-18 October, 2002, TUBITAK Publishers, Ankara, pp. 266-273.

Marcovecchio, J., Freije, H., De Marco, S., Gavio, A., Ferrer, L., Andrade, S., Beltrame, O. & Asteasuain, R. (2006). Seasonality of hydrographic variables in a coastal lagoon: Mar Chiquita, Argentina, Aquatic Conserv: Marine Freshwater Ecosystem, 16, 335-347.

Mecozzi, M., Acquistucci, R., Di Noto, V., Pietrantonio, E., Amirici, M. & Cardarilli, D. (2001). Characterization of mucilage aggregates in Adriatic and Tyrrhenian Sea: structure similarities between mucilage samples and the insoluble fractions of marine humic substance, *Chemosphere*, 44, 709-720.

Moncheva, S., Gotsis-Skretas, O., Pagou, K. & Krastev, A. (2001). Phytoplankton blooms in Black Sea and Mediterranean coastal ecosystems subjected to anthropogenic eutrophication: similarities and differences. *Estuarine Coastal Shelf Science*, 53, 281-295.

Nikolaidis, G., Moschandreou, K. & Patoucheas, D.P. (2008). Application of a trophic index (TRIX) for water quality assessment at Kalamitsi coasts (Ionian Sea) after the operation of the wastewater treatment plant. *Fresenius Environmental Bulletin*, 17(11), 1938-1944.

Nikolaidis, G., Patoucheas, D.P. & Moschandreou, K. (2006a). Estimating breakpoints of chl-a in relation with nutrients from Thermaikos Gulf (Greece) using piecewise linear regression. *Fresenius Environmental Bulletin*, 15 (9b), 1189 - 1192.

Nikolaidis, G., Moschandreou, K., Koukaras, K., Aligizaki, K., Kalopesa, E. & Heracleous, A. (2006b). The use of chlorophyll *a* for trophic state assessment of water masses in the inner part of Thermaikos Bay (Nw Aegean Sea). *Fresenius Environmental Bulletin*, 15 (9b), 1193 - 1198.

Nixon, S.W. (1986). Nutrient dynamics and the productivity of marine costal waters, In: *Coastal eutrophication* (Eds. R.D. Halwagy, B. Clayton, M. Behbehani), pp. 97-115, The Alden Press, Oxford.

Nixon, S.W. (1995). Coastal marine eutrophication: a definition, social causes, and future concerns. *Ophelia*, 41, 199-219.

Özelli C. & Özbaysal M.K. (2001). Kalkınmada öncelikli yöreler, Balıesir il profile, TC. Sanayi Bakanlığı KOSGEB.

Parsons, T.R., Maita, Y. & Lalli, C.M. (1984). A manual of chemical and biological methods for seawater analysis. Pergamon Press, U. K.

Penna, N., Capellacci, S. & Ricci, F. (2004). The influence of the Po River discharge on phytoplankton bloom dynamics along the coastline of Pesaro (Italy) in the Adriatic Sea. *Marine Pollution Bulletin*, 48, 321-326.

Perez-Ruzafa, A., Fernandez, A.I., Marcos, C., Gilabert, J., Quispe, J.I. & Garcia-Charton, J.A. (2005). Spatial and temporal variations of hydrological conditions, nutrients and chlorophyll *a* in a Mediterranean coastal lagoon (May Menor, Spain). Hydrobiologia, 550, 11-27.

Pojed, I. & Kveder, S. (1977). Investigation of nutrient limitation of phytoplankton production in the Northern Adriatic by enrichment experiments. *Thalassia Jugoslavica*, 13, 13-24.

Polat, Ç.S. & Tuğrul, S. (1995). Nutrient and organic carbon exchanges between the Black and Marmara seas through the Bosphorus strait, *Continental Shelf Research*, 15(9), 1115-1132.

Redfield, A.C., Ketchum, B.K. & Richards, F.A. (1963). The influence of organisms on the composition of sea water. In: *The sea*. Vol.2. (Ed. M.N. Hill), , pp. 26-77, Wiley, New York.

Roberts, E.C., Davidson, K. & Gilpin L.C. (2003). Response of temperate microplankton communities to N:Si ratio perturbation. Journal of Plankton Research, 25, 1485-1495.

Ryther, J.H. & Dunstan, W.M. (1971). Nitrogen, phosphorus and eutrophication in the coastal marine environment. Science, 171, 1008-1013.

Siegel, S. (1956). Non parametric statistics for the behavioral sciences. McGraw-Hill, New York.

Smith, S.V. (1984). Phosphorus versus nitrogen limitation in the marine environment. Limnology and Oceanography, 29, 1149-1160.

Stefanson, U. & Richards, F.A. (1963). Processes contributing to the nutrient distribution of Colombia River and the Strait of Juan de Fuca. Limnology and Oceanography, 8, 394-410.

Taş, S., Okuş, E., Ünlü, S. & Altıok, H. (2011). A study on phytoplankton following 'Volgoneft-248' oil spill on the north-eastern coast of the Sea of Marmara. *Journal of the Marine Biological Association of the United Kingdom*, 91(3), 715-725.

Tuğrul, S., Beşiktepe, Ş. & Salihoğlu, İ. (2002). Nutrient exchange fluxes between the Aegean and Black Seas through the Marmara Sea, *Mediterranean Marine Science*, 3 (1), 2002, 33-42.

Tuğrul, S., Çoban-Yıldız, Y., Ediger, D. & Yılmaz, A. (2000). Karadeniz, Marmara ve Akdeniz'de partikül organic madde dağılımı ve kompozisyonu, I. Ulusal Deniz Bilimleri Konferansı, 30 Mayıs-2 Haziran 2000, Kültür ve Kongre Merkezi, ODTÜ, Ankara.

Tüfekçi, V., Balkıs, N., Polat-Beken, Ç., Ediger, D. & Mantıkçı, M. (2010). Phytoplankton composition and environmental conditions of a mucilage event in the Sea of Marmara. *Turkish Journal of Biology*, 34, 199-210.

Ullyott, P. & Pektaş, H. (1952). Çanakkale Boğazındaki yıllık temperatür ve tuzluluk değişmeleri hakkında ilk araştırmalar, *Hidrobiologi Mecmuası, İstanbul Üniversitesi Fen Fakültesi Hidrobiyoloji Araştırma Enstitüsü Yayınlarından*, Seri A, cilt 1, sayı, 1, 19-33.

Ünlüata, U., Oğuz, T., Latif, M.A. & Özsoy, E. (1990). On the Physical Oceanography of the Turkish Straits, In: The Physical Oceanography of Sea Straits, L.J. Pratt (Ed.), NATO/ASI Series, 318, 26-60, Kluwer Academic Publishers, Dordrecht.

Vollenweider, R.A., Rinaldi, A., Viviani, R. & Todini, E. (1996). *Assessment of the State of Eutrophication in the Mediterranean Sea. MAP Technical Report Series* 106. UNEP/FAO/WHO., Athens.

Vollenweider, R.A., Giovanardi, F., Montanari, G. & Rinaldi, A. (1998). Characterization of the trophic conditions of marine coastal waters with special reference to the NW Adriatic Sea: Proposal for a trophic scale, turbidity and generalized water quality index. *Environmetrics*, 9, 329-357.

Winkler, L.W. (1888). The determination of dissolved oxygen in water. *Berichte der Deutschen Chemischen Gesellschaft*, 21, 2843-2855.

Yang, X., Wu, X., Hao, H. & He, Z. (2008). Mechanisms and assessment of water eutrophication. *Journal of Zhejiang University Science B*, 9(3), 197-209.

Yılmaz, A. (2002). Biogeochemistry of the seas surrounding Turkey: Cycling and distributions. *Turkish Journal of Engineering and Environmental Sciences*, 26, 219-235.

Yüce, H. & Türker, A. (1991). Marmara Denizi'nin fiziksel oşinografik özellikleri ve Akdeniz suyunun Karadeniz'e girişi, *Uluslararası Çevre Sorunları Sempozyumu Tebliğleri, İstanbul Marmara Rotary Kulübü, İstanbul*, 284-303.

2

Ecological Water Quality and Management at a River Basin Level: A Case Study from River Basin Kosynthos in June 2011

Ch. Ntislidou, A. Basdeki, Ch. Papacharalampou, K. Albanakis, M. Lazaridou and K. Voudouris*

Interdisciplinary Postgraduate Study Program "Ecological Water Quality and Management at a River Basin Level" Departments of Biology, Geology & Civil Engineering, Aristotle University of Thessaloniki, Thessaloniki, Greece

1. Introduction

The European Parliament and Council decided a policy on the protection, an appropriate treatment and management of water field leading on the Water Framework Directive 2000/60/EC (WFD, European Commission, 2000) in October 2000. The WFD obliges Member States to achieve the objective of at least a good ecological quality status before 2015 and requires them to assess it by using biological elements, supported by hydromorphological and physico-chemical ones. The assessment must be done at a basin level and authorities are obliged to follow efficient monitoring programs in order to design integraded basin management plans. Efforts are being made to adapt national programmes for the WFD requirements (Birk & Hering, 2006). In most European countries, river monitoring programmes are based on benthic macroinvertebrate communities (Sánchez-Montoya et al., 2010).

The WFD (EC, 2000) suggests a hierarchical approach to the identification of surface water bodies (Vincent et al., 2002) and the characterization of water body types is based on regionalization (Cohen et al., 1998). The directive proposes two systems, A and B, for characterizing water bodies according to the different variables considered (EC, 2000). The WFD allows the use of both systems, but considers system A as the reference system. If system B is used by Member States, it must achieve at least the same degree of differentiation. System A considers the following obligatory ranged descriptors: eco-region, altitude, geology and size, whereas system B considers five obligatory descriptors (altitude, latitude, longitude, geology and size) and fifteen optional ones.

A prerequisite for a successful implementation of the WFD in European waters is the intercalibration of the national methods for each biological quality element on which the

*Corresponding Author

classification of ecological status is based (Simboura and Reizopoulou, 2008). According to the Mediterranean intercalibration exercise (MED-GIG) (Casazza et al., 2003, 2004, five river types are proposed, based on the catchment area, the altitude, the geological background and the flow regime of the rivers. Greece participates in this exercise and belongs in the Mediterranean geographical intercalibration group (MED-GIG) (Casazza et al., 2003, 2004).

The pressures and impacts play a key role in the likelihood that a water body will fail to meet the set objectives. IMPRESS analysis (CIS Working Group 2.1: IMPRESS, 2003) assesses the impact and evaluates the likelihood of failing to meet the directive's environmental objectives. Additionally, the Driving force-Pressure-State-Impact-Response (DPSIR) framework represents the relations between socio-economic driving forces and impact on the natural environment (Kristensen, 2004) and the SWOT analysis helps the understanding of the Strength-Weakness-Opportunities-Threats.

This chapter deals with the ecological water quality of the Kosynthos river basin based on (a) the distinction of the water bodies by applying System B and taking into consideration the pressures, (b) the calculation of an approximate water balance according to the activities developed in the river basin, (c) the assessment of the ecological water quality, using benthic macroinvertebrates, (d) the implementation of Impress analysis DPSIR and SWOT analyses.

2. Study area

The Kosynthos River is located in the north-eastern part of Greece, flows through the prefectures of Xanthi and Rhodopi and discharges into the Vistonis lagoon (Figure 1) as a result of the diversion of its lowland part in 1958. Kosynthos' length is approximately 52 Km (Pisinaras et al., 2007). In the present study, 8 sites were selected in Kosynthos river basin (Figure 1) during the period June 2011, depending on the different pressures that presented in the area. Four sites belonged to the mountainous area and the rest sites to the low-land one. The Kosynthos river basin belongs to the water district of Thrace (12th water district), covering an area of 460 Km^2. The region consists of forest and semi-natural areas (69.6%), rural areas (27.7%), artificial surfaces (2.5%) and wetlands (0.3%) (Corine Land Cover 2000). It is considered to be a mountainous basin (Gikas et al, 2006) of steep slopes and its average elevation is about 702 m. In total, the 7.3% of the basin is protected by the Ramsar Convention or belongs to the EU Natura 2000 sites.

Geologically speaking, the study area belongs entirely to Rhodope massif (Figure 2) consisting of old metamorphic rocks (gneisses, marbles, schists), observed mainly in the northern part of the basin. Moreover, igneous rocks (granites, granodiorites) have intruded the Rhodope massif through magmatic events during Tertiary and outcrop in the central part of the basin. Because of the granite intrusion in the calcareous rocks and the contact metamorphosis, a sulfur deposit is created, consisting mainly of pyrites. Quaternary and Pleistocene mixed sediments cover the south-eastern part of the catchment. The boundary between the highland area and the lowland is characterized by a sharp change of slope.

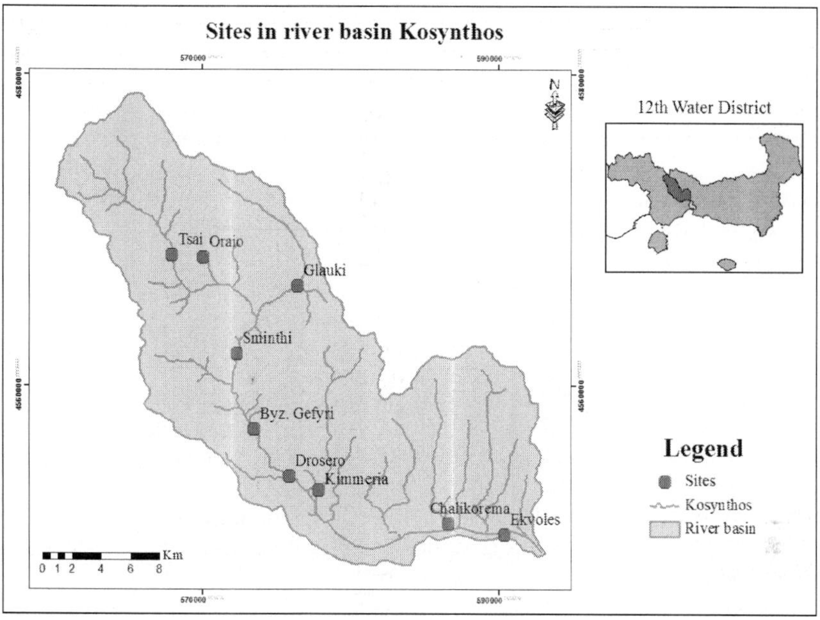

Fig. 1. Map of Kosynthos river basin showing the sampling sites.

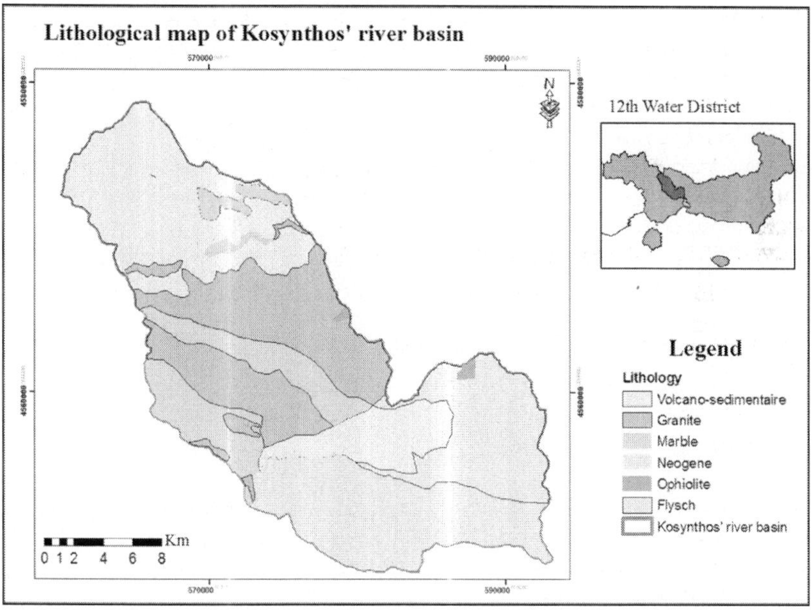

Fig. 2. Lithological map of Kosynthos' river basin.

From a hydrogeological point of view, two main aquifers are developed within the aforementioned geological formations: 1) an unconfined aquifer in the Quaternary deposits of lowlands and 2) a karst aquifer in marbles of the northern part of the basin (Diamantis, 1985). Karst aquifer system often discharges groundwater through springs in the hilly part of the basin, where permeable marbles are in contact with impermeable basement rocks. Previous studies (Hrissanthou et al., 2010; Gikas et al., 2006) show significant sediment transportation to Vistonis lagoon from Kosynthos river because of intense erosion. However, no deltaic deposits are observed in the outfall of Kosynthos, while an inner delta is created right before the stream's diversion (Figure 3). The steep topography combined with the inclination of the diverted section prevents the transportation of coarse sediments, allowing only fine-grained fractions to Vistonis lagoon.

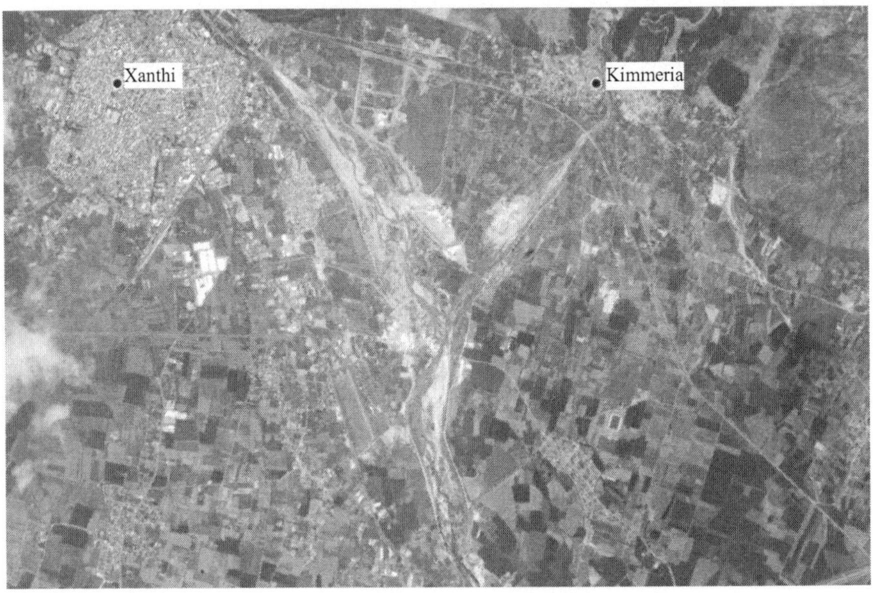

Fig. 3. The inner delta of Kosynthos River, right before the diverted part (Google Earth).

3. Material and methods

3.1 Typology

In this study system B was selected because the basin of Axios River (a transboundary Greek-FYROM river) belongs to two different ecoregions according to System A. In order to distinguish the water bodies of the Kosynthos river basin, apart from the obligatory descriptors the slope, from the optional ones, was selected and a new category in the basin descriptor was added (0-10 Km^2). The rivers were characterized according to the MED-GIG intercalibration exercise (Van de Bund et al, 2009).

3.2 Approximate water balance

The estimation of the approximate water balance of Kosynthos catchment is based on monthly rainfall and temperature data of 7 weather stations (Genisea, Iasmos, Xanthi, Semeli, Gerakas, Thermes, Dimario) distributed equally across and beyond the basin, for the period 1964-1999 and GIS technique (Voudouris, 2007). As part of the estimation process, components of the hydrological cycle (precipitation *P*, actual evapotranspiration *E*, infiltration *I* and surface runoff *R*), instream flow, available water capacity and water needs (demand for urban, farming, irrigation and industrial water) of the river basin are calculated.

3.3 Quality elements

Dissolved oxygen (DO mg/l), water temperature (WTemp, °C), pH and conductivity (μS/cm) were measured in situ with probes (EOT 200 W.T.W./Oxygen Electrode, pH-220, CD-4302, respectively). TSS (mg/l), nutrients (N-NH_4 and P-PO_4, mg/l) and oxygen demand (BOD_5, mg/l) were estimated following A.P.H.A. (1985). Flow was quantified with a flow meter (type FP101) and stream discharge (m^3/s) was calculated for each site. The percentage composition of the substrate was visually estimated according to Wentworth (1922) scale. The Habitat Modification Scores (HMS) was calculated to assess the extent of human alterations at each site (Raven et al., 1998).

Benthic macroinvertebrates were collected using a standard pond net (ISO 7828:1985, EN27828:1994) with the semi-quantitative 3-minute kick and sweep method according to Armitage et al (1983) and Wright (2000) proportionally to the approximate coverage of the occurring habitats (Chatzinikolaou et al., 2006). The animals were preserved in 4% formaldehyde.

In the laboratory, they were sorted and identified to family level. To assess the ecological quality of each site the Hellenic Evaluation System (HES) (Artemiadou & Lazaridou, 2005) and the European polymetric index STAR ICMi (European Commission 2008/915/EC) were applied to the benthic macroinvertebrate samples.

3.4 Statistical analysis

For the statistical analyses all data were log (x+1) transformed except for pH and temperature which were standardized. Parameter expressed as percentages (substrate) was arcsine transformed (Zar, 1996). The hierarchical clustering analysis, based on Bray-Curtis index (Clarke and Warwick, 1994) was applied to the samples of benthic macroinvertebrates for grouping them.

Similarity percentages analysis (SIMPER analysis) (Clarke & Warwick, 1994) was used to distinguish the macroinvertebrate taxa contributing to similarity and dissimilarity between the groups. Redundancy Analysis (RDA) was performed in order to detect covariance between environmental variables and abundances of taxa (Ter Braak, 1988). Correlated variables were excluded with the use of the inflation factor (<20) and the Monte Carlo permutations test ($p<0.05$).

3.5 Impress analysis/DPSIR and SWOT analysis

Impress analysis estimates the impacts taking into account the morphological alterations and the pollution pressures. The morphological alterations estimated through the calculation of a Habitat Modification Score (HMS) (Raven et al., 1998) which is based on the artificial modifications. The pollution pressures are treated differently for point and non point sources. As point sources of pollution are considered the urban wastewater and septic tanks, producing BOD, N and P combinations, which are calculated according to the emission factors (Fribourg-Blanc and Courbet, 2004) whereas livestock according to Ioannou et al., (2009) and Andreadakis et al., (2007) calculated the pollutants.

The human population and the species numbers of breeding animals derived from the Greek National Statistical Service. Industries data, the point sources of pollution, are not available from National Services. Non point sources of pollution, being the land uses, are determined using the Corine Land Cover 2000 and their pollutants are calculated according to the immission factors of WL-Delft et al, (2005). The morphological alterations of pressures were significant if the agricultural land cover was more than 40% (LAWA, 2002) and urban land cover more than 2.5% (Environment Agency, 2005) of the total extent of the river basin.

The pressures from pollution sources would be significant if the total immissions exceeded the proposed limits for irrigation (Decision 4813/98) and for fish life (European Commission 2006/44/EC). All limit scores were adjusted to the river basin, taking into consideration the river flow, estimated as 5.8 m^3/s (Gikas et al., 2006). Multiplied by the estimated river flow, the limit scores were adjusted to the river basin.

The impact assessment, the evaluation of likelihood of failing to meet the environmental objectives and the risk management used the methodology proposed by Castro et al., (2005). Finally, the conceptual model DPSIR (at a river basin level) and SWOT analysis (at the level of Municipalities Mykis and Dimokritos) were applied.

4. Results

4.1 Typology

In accordance with the hierarchical approach, the river flowing in the basin is separated in two main water bodies, due to the canalization of the low-land part of Kosynthos in 1958. Therefore, the diverted part is characterized as heavily modified water body (HMWB), while the rest of the river is characterized as natural water body (NWB). The classification of the river by types leads in 17 types in the catchment area, of which 15 in the drainage network (Figure 4). Finally, the subdivision of a water body of one type into smaller water bodies according to the existing pressures, results in 44 water bodies, from which in 9 sampling of biological, hydromorphological and physico-chemical parameters were executed in June 2011. Based on the European common intercalibration river types (Van de Bund et al., 2009), two types (RM1 and RM2) appear in the river basin.

4.2 Approximate water balance

The climate is semi-humid with water excess and deficiency during winter and summer respectively (Angelopoulos & Moutsiakis, 2011). The annual rainfall (P) is influenced by the elevation (H) of the region (P=0.92H+625, R^2=0.95). The mean annual precipitation in the basin for the period 1964-1999 is 1085.6 mm (Figure 5). Based on Turc method, the coefficient of the actual evapotranspiration was estimated to be 71% of the mean annual precipitation. The remaining amount is allocated to surface runoff (15.3%) and infiltration (13.4%).

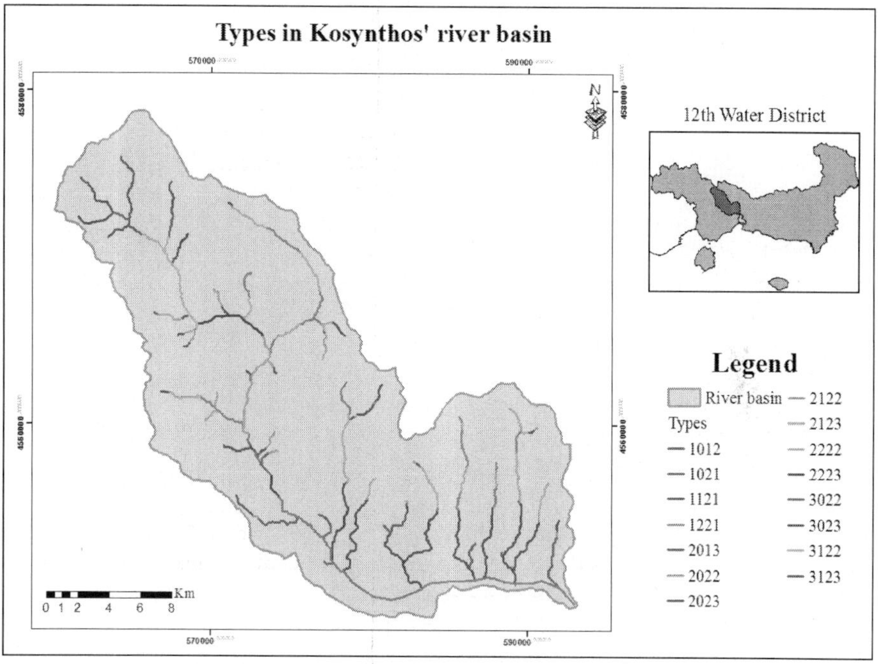

Fig. 4. Types in Kosynthos' river basin according to system B.

A great amount of water infiltrates in marbles and alluvial deposits and then a part of this amount discharged through springs. Instream flow ('environmental flow') is a term that refers to the water required to protect the structure and function of aquatic ecosystems at some agreed level. (Zhang et al, 2006). In accordance with the legislation (M.D. 49828/2008), instream flow equals to 30% of surface runoff and the rest 70% is estimated as available water potential. Assuming that 50% of the infiltration also involves in the available water, the total water potential of Kosynthos basin is calculated to be 86.9x10^6 m^3/yr for the period 1964-1999.

The relevant agents considered for the calculation of water demand are the municipalities that configure the Kosynthos river basin (Municipalities of Myki, Xanthi, Dimokritos, Iasmos). The needs for urban and farming water are calculated equal to 1.7x10^6 m^3 each for irrigation water 62x10^6 m^3 and for industrial water 6.3x10^6 m^3 (demographic and population data 1991-2001. N.S.A.G.). It should be although mentioned that any analysis of water resource management suffers the same handicap with regard to the availability of complete and homogenous information. particularly on municipality level (Torregrosa et al., 2010). Comparing the amount of water potential of Kosynthos catchment for the period 1964-1999 with the water demands, the approximate water balance for the same period is characterized as positive; the water potential is greater that the water demands.

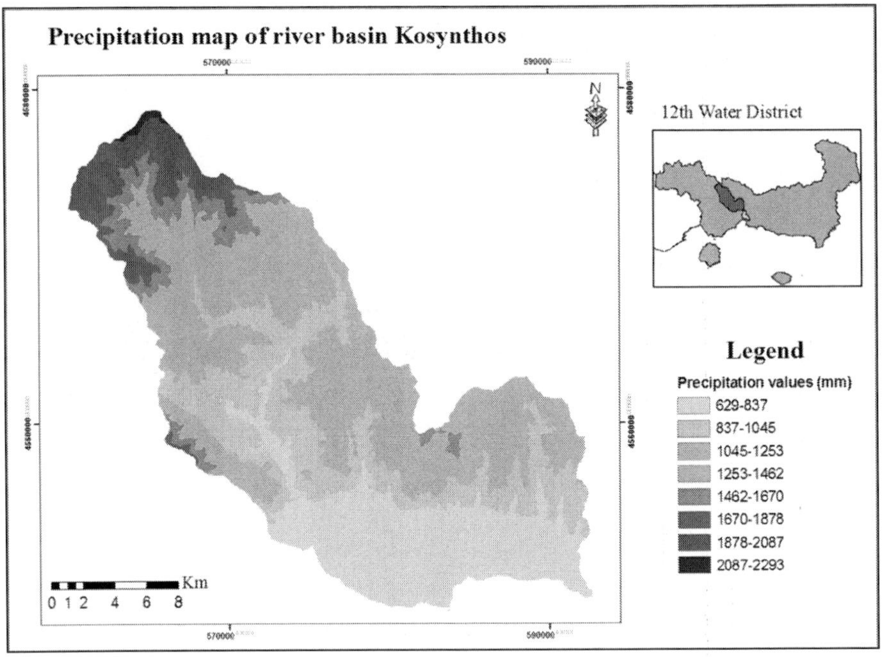

Fig. 5. Precipitation map in Kosynthos river basin.

	P	=	E	+	I	+	R
Water amount (10^6 m^3)	499.4		356.3		66.7		76.4
Precipitation (mm)	1085.6		774		145.5		165.5
Percentage (%)	100		71.3		13.4		15.3

Table 1. Approximate water balance for the Kosynthos river basin (1964-1999).

4.3 Quality elements

The results of the physico-chemical parameters of the river water are presented in Table 2. Ammonium concentration was found to exceed the boundaries of Cyprinid life in all sites, except the site Oraio which exceeded the boundary of portable water and the site Tsai which exceeded boundary of Salmonid life. Also, T.S.S. concentration exceeded the boundary of portable water in sites Kimmeria. Chalikorema and Ekvoles. The substrate composition is represented in Figure 2. The sites Oraio, Byz. Gefyri and Kimmeria are mostly consisted of fine substrate. According to the index HMS most of the sites are characterized as "Predominantly unmodified" (HMS score 3-8, Figure 6).

Sites	D.O. (mg/l)	WTemp (⁰C)	pH	Conductivity (μS/cm)	T.S.S. (mg/l)	P-PO_4 (mg/l)	N-NO_3 (mg/l)	N-NH_4 (mg/l)	Discharge (m^3/sec)	B.O.D.$_5$ (mg/l)
KYA portable water Y2/2600/2001			**6.5-9.5**	**2.5**		**2.143**	**11.29**	**0.318**		
Boundaries of Directive 2006/44/EC										
(Salmonid)	**6**		**6.0-9.0**		**25**			**0.031**		**3**
(Cyprinid)	**4**		**6.0-9.0**		**25**			**0.155**		**6**
Tsai	10.17	15.7	7.52	0.076	1.2	0.022	0.139	**0.141**	0.307	1.97
Oraio	7.29	18.4	7.68	0.312	20.4	0.146	0.715	**0.393**	<0.001	1.45
Glauki	8.38	22.1	8.05	0.258	3.2	0.120	0.424	**0.215**	0.033	1.5
Sminthi	9.22	19.1	7.7	0.17	8.2	0.030	0.309	**0.282**	0.583	1.76
Byz. Gefyri	8.19	26.6	7.71	0.26	4.4	0.041	0.190	**0.207**	0.441	0.94
Drosero	8.68	28.2	8.16	0.25	7.2	0.030	0.190	**0.214**	0.231	0.64
Kimmeria	6.59	33.2	8.01	0.431	36.2	0.024	1.539	**0.228**	<0.001	1.13
Chalikorema	8.25	20.5	6.83	0.501	41.6	0.045	1.248	**0.308**	<0.001	1.33
Ekvoles	8.14	29.1	7.42	0.42	47.6	0.044	0.396	**0.270**	0.306	2.32

Table 2. Physicochemical parameters of the studied sites in the Kosynthos river basin during the period June 2011(with black letters mentioned the concentrations which exceeded the proposed limits).

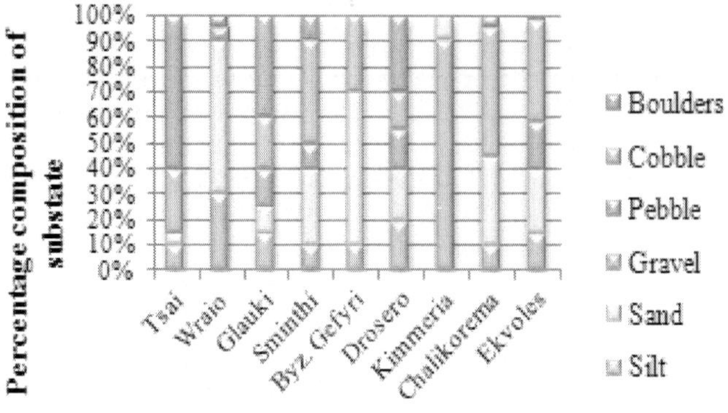

A

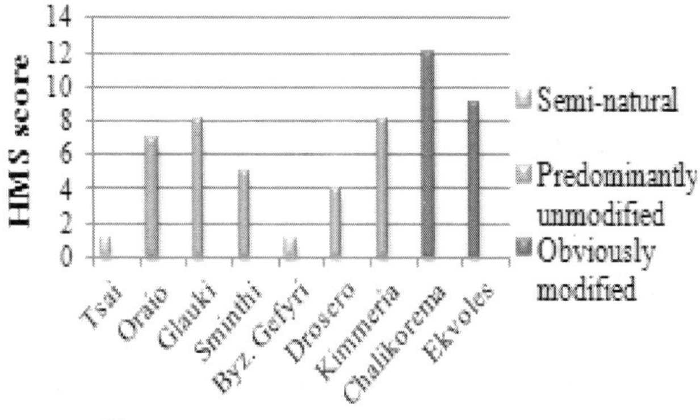

B

Fig. 6. (A) Percentage composition of substrate, and (B) HMS Score at the studied sites of the Kosynthos river basin in period 2011.

In this study 22.005 benthic macroinvertebrates were identified belonging to 48 different taxa. Abundances were found to be higher in the site Oraio and the site Chalikorema had the lowest. The ecological quality of the sites Tsai, Sminthi and Kimmeria, according to the Hellenic Evaluation Score (HES), was characterized as good, sites Oraio Byz. Gefyri, Chalikorema and Ekvoles as moderate and Glauki and Drosero as poor (Figure 7). By the European polymetric index STAR ICMi it was found the same quality, except for the site Kimmeria which was characterized less than good (Table 3). This difference is related to the fact that the HES index takes into account more sensitive taxa.

Sites	Type	Interpretation of HES	Interpretation of STAR ICMi
Tsai	R-M1	Good	Good
Oraio	R-M1	Moderate	< Good
Glauki	R-M1	Poor	< Good
Sminthi	R-M2	Good	Good
Byz. Gefyri	R-M2	Moderate	< Good
Drosero	R-M2	Poor	< Good
Kimmeria	R-M2	Good	< Good
Chalikorema	R-M2	Moderate	< Good
Ekvoles	R-M2	Moderate	< Good

Table 3. The ecological water quality of the studied sites at the river basin Kosynthos in June 2011.

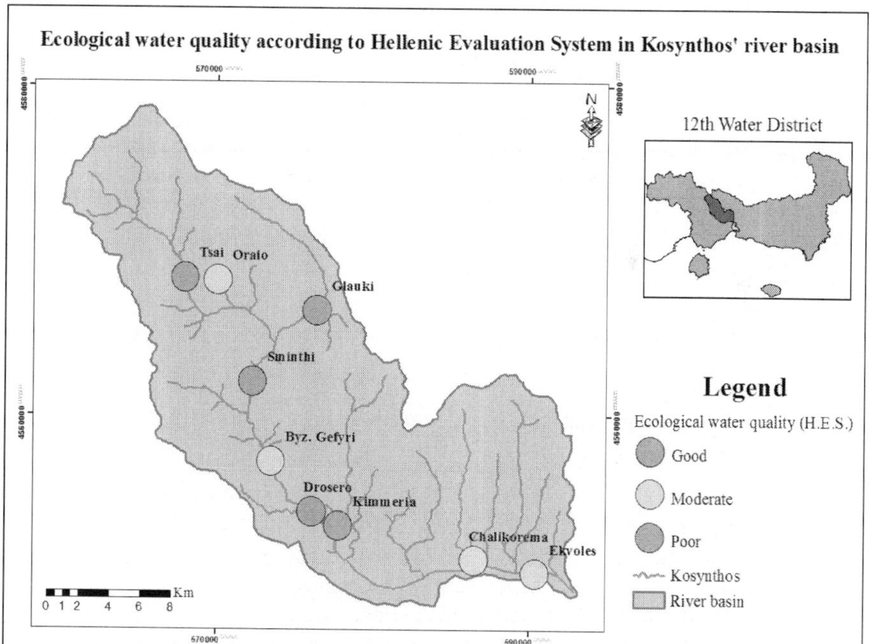

Fig. 7. Ecological water quality according to Hellenic Evaluation System of the studied sites at the river basin Kosynthos in June 2011.

4.4 Statistical analysis

The hierarchical clustering analysis, based on Bray-Curtis index, clustered the benthic macroinvertebrates of the different sites into three clusters (Figure 8). The groups clustered modified sites with an excess of human activities (Ekvoles & Xalikorema) (Group A), the inland delta sites (Drosero & Kimmeria) (Group B) and the high altitude sites (the rest of the stations) (Group C).

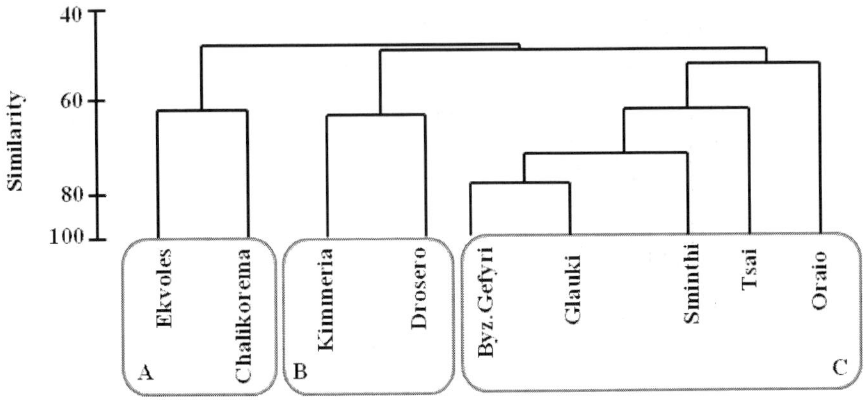

Fig. 8. Hierarchical clustering analysis, based on Bray-Curtis index of the studied sites at the Kosynthos river basin in June 2011.

Simper Analysis showed that the dissimilarity between the groups was around 51%. The families Gammaridae and Simulidae were the key taxa for the differences between the clusters (Figure 9). According to CCA the eigenvalues of the first two axes accounted for 73.8% of the variance. P-PO_4 was the variable best correlated with the first axis, whereas the second axis was best correlated with discharge (Figure 10).

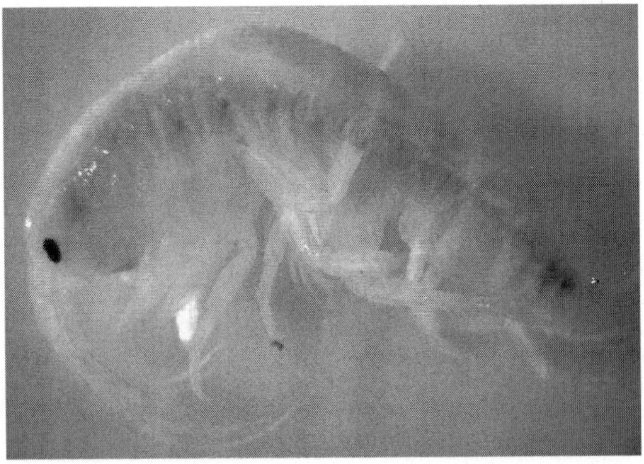

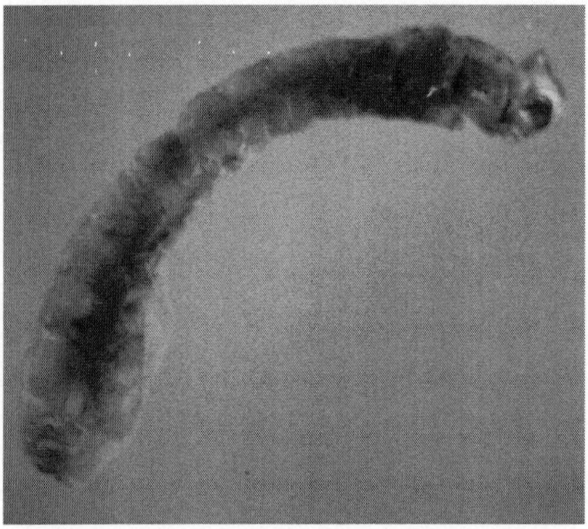

Fig. 9. Benthic macroinvertebrates (A) Gammaridae and (B) Simulidaae which are responsible for the differences between the groups (Photos: Patsia, 2009).

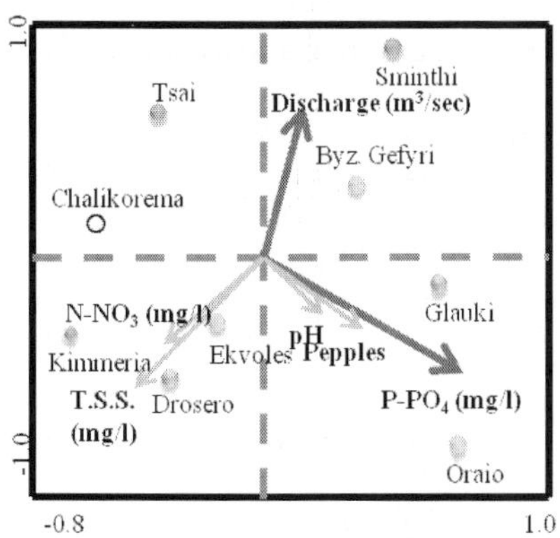

Parameters	ENVI AX1	ENVI AX2
pH	0.319	-0.285
TSS (mg/l)	-0.356	-0.468
N-NO$_3$ (mg/l)	-0.359	-0.356
P-PO$_4$ (mg/l)	0.736	-0.506
Discharge (m^3/sec)	0.144	0.673
Pebbles	0.268	-0.248

AXES	I	II
Eigenvalues	0.421	0.221
Species-environment correlations	0.991	0.987
Cumulative percentage variance		
a) of species data	42.1	64.1
b) of species-environment relation	48.4	73.8
Test of significance of I axis. P-value	0.11	
T Test of significance of all axes. P-value	0.03	

Fig. 10. Canonical correspondence analysis diagram with environmental variables and nine (9) sites at Kosynthos river basin in June 2011.

4.5 Pressures from pollution sources and morphological alteration pressures

The total emissions, immissions loads produced within the river basin Kosynthos and the environmental quality standards for irrigation and fish life are presented in Table 4. Only BOD exceeded the limits for the salmonid life standard. It is evident that livestock breeding is the most polluting activity (Figure 11). Agriculture is the second diffuse pollution source of total nitrogen (30%) and nitrogen immissions. The morphological alterations for the urban land cover were 1.5% and the agricultural land cover was 27.7% lower than the proposed levels so not significant.

	Emissions	Immissions	Irrigation standards	Fish life standards	
				Salmonid	Cyprinid
BOD (Kg/day)	11.142	2.744	12.562	1.507	3.500
Total N (Kg/day)	4.051	1.814	25.375	2.900	3.200
Total P (Kg/day)	538	85	-	100	201

Table 4. Comparison of emission and immission loads with maximum permitted immission loads.

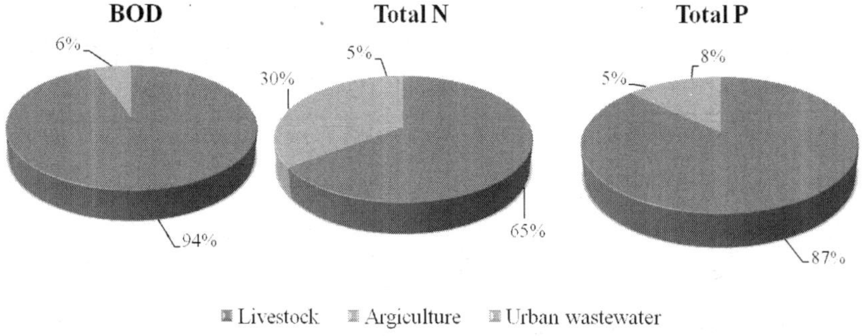

Fig. 11. BOD, total P and total N immissions that each activity produces.

4.6 Impact assessment

The impacts from the morphological alterations are probable, because the mean score of HMS is 5.5. Also, the impacts from the pollution pressures are probable, because the mean biological quality is inferior to good quality and because the nitrogen of $N\text{-}NH_4$ exceeds the limit for potable water in the site Oraio. Hence, the impacts from the morphological alterations and pollution pressures are probable. The likelihood of failing to meet the environmental objectives for the morphological alterations is medium because the impacts are probable and there are no significant pressures (urban land cover 1.5% and agricultural land cover 27.7%). Additionally, for the pollution pressures, the likelihood of failing to meet the environmental objectives is medium because the impacts are probable and there are no data for significant pressures (lack of the inputs of industrial pollutants). So in the Kosynthos river basin, an operational monitoring of the risk management for both the morphological and pollution pressures is proposed.

4.7 DPSIR and SWOT analysis

According to the DPSIR framework there is a chain of causal links starting with 'driving forces' (D) (human and economic activities) through 'pressures' (P) (emissions, waste) to 'state' (S) (physical, chemical and biological) and 'impacts' (I) on ecosystems, human health and functions, and eventually leading to political 'responses' (R) (prioritization, target setting, indicators). Consequently, all the above were examined in the Kosynthos river basin (Table 5).

D	P	S	I	R
Urban growth	1. Sewage 2. Urban waste 3. Morphological alterations	1. High concentration in $N\text{-}NH_4$ organic pollution and medium water quality 2. Reject of urban solid waste 3. Alteration of river channel and bank (bridges, passage of vehicles). HMS score 5.5	1. Degradation of water quality 2. Alteration of natural landscape and degradation of water quality 3. Interruption of continuity of riparian area and degradation of riparian habitat	1. Creation of installation of treatment of sewages 2 & 3. Collection, recycling and environmental sensitization
Farming activity	Untreated forage sewages	High content in $N\text{-}NH_4$ $N\text{-}NO_2$, $P\text{-}PO_4$ total organic pollution and medium water quality	Degradation of water quality	Creating organized bands & modern livestock units, wastewater treatment and their use as fertilizer.
Rural activity	Use of pesticides and chemical fertilizers	High concentration in $N\text{-}NO_2$ and medium water quality	Degradation of water quality	Sensitization of citizens and farmers
Anthropogenic activities	1 Drilling and over-exploitation of aquifer systems 2. Morphological alterations 3. Deviation and regulation of river watercourse with terraces 4. Forest clearing and sand extraction.	1. Falling water table 2. Degradation of riparian vegetation. low QBR scores in the lowlands 3 & 4. Modification of habitat	1. Depletion of aquifer 2, 3 & 4. Alteration of riparian habitats, increased erosion and input of nutrients.	1.Drilling at regions with highest potential, rational water use and water pricing 2, 3 & 4. Re-establishment of riparian vegetation with native species

Table 5. DPSIR analysis in the Kosynthos river basin

For the sustainable development of the study area, SWOT analysis was applied in Municipality Mykis, which is in the mountainous part of the basin and in Municipality Dimokritos, which is in the lowland part of the basin, in order to estimate the Strengths, Weaknesses, Opportunities and Treats. Based on the SWOT analysis, which is a useful tool for local authorities and decision makers (Diamantopoulou & Voudouris, 2008), some recommendations are proposed to maximize the existing opportunities (Table 6) for achievement of good quality status in Kosynhtos river basin.

A		B	
Strength Protected areas with biodiversity Byzantine city of 6 AD	Weakness Use of septic tanks Absence of treatment in industries Incomplete maintenance of irrigation supply network Over-tapping ground water	Strength Protected areas with biodiversity Archaeological, historical & folklore interest Riparian forest	Weakness Use of septic tanks Intensive livestock farming Illegal disposal of debris Illegal logging from riparian forest
Opportunities Growth of ecotourism in protected areas Exploitation of traditional buildings	Threats Absence of urban and industrial waste water treatment Incomplete maintenance of irrigation supply network Absence of administrative policy	Opportunities Growth of ecotourism in protected areas and in the riparian forest Exploitation of traditional buildings Investments	Threats Absence of urban waste water treatment Veterinary surgeon units Absence of administrative policy for the riparian forest

Table 6. SWOT analysis in Municipalities: (A) Dimokritos and (B) Mykis.

5. Discussion

In this study System B was selected because of the flexibility in the choice of abiotic parameters and better distinction in relation to the animals than the System A (Dodkins et al., 2005). According to Kanli (2009) the descriptor "Altitude" significantly affects the structure of communities of benthic macroinvertebrates in relation to the other descriptors used in the typology. Also, according to Rundle et al. (1993) and Brewin et al. (1995) "Basin size" is the second most important descriptor that affects the structure of biocommunities after the altitude. In this case, the hierarchical clustering analysis, based on Bray-Curtis index, showed that the descriptor of "Altitude" was the most important descriptor for the separation of benthic macroinvertebrates. For Mediterranean types of RM there was no apparent difference between the stations on the distribution of benthic macroinvertebrates (most of them were R-M2).

The approximate water balance for the period 1964-1999 is characterized as positive, since the water potential in the basin is sufficient to meet the needs arising from activities. The intense infiltration due to the karstic marbles of the Rhodope Mass. and the hydraulic conditions developed in the mountainous area by the presence of impermeable formations does not allow high surface runoff. Moreover, the largest city in the basin (Xanthi) is not watered from this basin.

The concentration of total suspended solids is affected by the dissolution of mineral matter and the intense evaporation (Voudouris, 2009). In this study, the highest TSS concentration measured in the lowland sites (47.6 mg/l) due to the large sediment transportation, mainly fine-grained material derived from the intense erosion, weathering and dissolution of lithological formations because of steep slopes.

The physico-chemical and biological characteristics are modified from the discharge and are related to the ability dissolution of pollutants (Prat et al., 2002). According to Hubbard et al. (2011) the importance of intense flooding in rivers demonstrates the inverse relationship between supply and nutrient concentration. In this study, in the site Kimmeria was found the lowest discharge (0.6 l/s) and low concentration of P-PO_4 and N-NH_4. This occurs because the actual band width is greater than the measured during the sampling period. Instead the highest concentration of N-NO_3 may be due to the influx of water from underground sources upstream of the site. Finally, in the site Oraio it was measured the second smallest discharge (0.8 l/s) and the highest values of nutrients, because the active band width is small and leads to accumulation of nutrients.

The ecological water quality of the site Tsai is connected to the absence of pressures. In the sites Oraio, Glauki and Byz. Gefyri, the ecological water quality is characterized as poor due to the present livestock feeding and the septic tanks. Also, the sites Chalikorema and Ekvoles were characterized as moderate because of the intensive agricultural land use, livestock feeding and septic tanks. Finally, the ecological quality in the site Sminthi and Chalikorema is good, because of the self-purification of the system and the presence of water sources respectively.

Impress Analysis showed that the immissions loads in the basin of Kosynthos is lower than the proposed irrigation limits (Decision 4813/98) issued for another region. It is suggested that an adoption of a similar Decision for Xanthi is important, since it is a rural and agricultural basin with intense activity in the lowland section. Also, the immissions loads did not exceed the limits for the cyprinid life standard, although the total organic load exceeded the limits for salmonid life standard. As livestock breeding appears to be the most polluting activity there is a certain amount of uncertainty involved due to lack of data concerning the location of breeding farms, their grazing fields, their antipollution technologies and the disposal processes of pollutants into the environment (Ioannou, 2009). The latter is mainly due to intense livestock activity observed in the municipality Dimokritos (40% total organic load from the entire river basin). Consequently, a risk management operational monitoring is proposed for both morphological and pollution pressures in the Kosynthos river basin, in order to achieve good quality status in 2015.

6. Conclusions

In conclusion, in the Kosynthos river basin 15 river types are present in the hydrographic network, according to the System B of the WFD. When taking into account the existing pressures in the basin, 44 water bodies are detected. The approximate water balance for the period 1964-1999 is characterized as positive. Among the nine stations selected for sampling benthic macroinvertebrates and according to Hellenic evaluation system of the ecological quality in three stations (Tsai, Sminthi, Kimmeria) water quality was estimated as good, in four stations (Oraio, Gefyri, Chalikorema, Ekvoles) medium and in two (Glauki, Drosero) as poor. Finally, by applying the Impress Analysis, operational monitoring was recommended.

7. References

A.P.H.A. (1985). 'Standard Methods for the Examination of Water and Wastewater'. 16th edition American Public Association.

Andreadakis A., E. Gavalakis, L. Kaliakatsos, C. Noutsopoulos and A. Tzimas (2007). 'The implementation of the Water Framework Directive (WFD) at the river basin of Anthemountas with emphasis on the pressures and impacts analysis'. Proc. Int. Conf. Conference 9th Environmental Science and Technology. Rhodes. Greece.

Angelopoulos C. and E. Moutsikis (2011). 'Hydrological conditions of the Kotyli springs (N. Greece) based on geological and hydrochemical data'. 9th International Hydrogeological Congress. Kalavryta, Greece. 2011.

Armitage P.D., D. Moss, J.F. Wright and M. T. Furse (1983). 'The performance of a new biological water quality score system based on macroinvertebrates over a wide range of unpolluted running-water sites'. Water Research, Vol. 17, pp. 333–347.

Artemiadou V. and M. Lazaridou (2005). 'Evaluation Score and Interpretation Index for the ecological quality of running waters in Central and Northern Hellas'. Environmental Monitoring and Assessment, Vol. 110, pp. 1–40.

Brewin P., T.M.L Newman and S.J. Ormerod (1995). 'Patterns of macroinvertebrate distribution in relation to altitude, habitat structure and land use in streams of the Nepalese Himalaya'. Arch. Hydrobiol., Vol. 135, pp. 79-100.

Birk S. and D. Hering (2006). 'Direct comparison of assessment methods using benthic macroinvertebrates: a contribution to the EU Water Framework Directive intercalibration exercise'. Hydrobiologia, Vol. 566, pp. 401-415.

Casazza G., C. Lopez y Royo and C Silvestri (2004). 'Implementation of the Water Framework Directive for coastal waters in the Mediterranean ecoregion: the importance of biological elements and of ecoregional co-shared application'. Biologia Marina Mediterranea, Vol. 11, pp. 12-24.

Casazza G., C. Silvestri and E. Spada (2003). 'Implementation of the Water Framework Directive for coastal waters in the Mediterranean eco-region'. 6th International Conference on Mediterranean Coastal Environment (MEDCOAST) Eds Ozhan. MEDCOAST Publication, pp. 1157-1168

Castro C.D., AP Infante, J.R. Rodriguez, R.S.X. Lerma and F.J.S. Martinez (2005). 'Manual para la identificacion de las presiones y analisis del impacto en aguas superficiales'. Centro de Publicaciones.

Chatzinikolaou Y., V. Dakos and M. Lazaridou (2006). 'Longitudinal impacts of anthropogenic pressures on benthic macroinvertebrate assemblages in a large transboundary Mediterranean river during the low flow period'. Acta hydrochim. Hydrobiologia. Vol. 34. pp. 453-463.

CIS Working group 2.1: IMPRESS (2003). 'Guidance on the Analysis of Pressures and Impacts'. Office for Official Publications of the European Communities.

Clarke K.R. and M.R. Warwick (1994). 'Change in Marine Communities: An Approach to Statistical Analysis and Interpretation'. Plymouth Marine Laboratory.

Cohen P., H. Andrianamahefa and J.G. Wasson (1998). 'Towards a regionalization of aquatic habitat: distribution of mesohabitats at the scale of a large basin'. Regul Rivers Res Manage. Vol. 14. pp. 391-404

Diamantis J. (1985). 'Hydrogeological study of the basin of Vistonida lake.' PhD thesis. Dep. of Civil Engineering, Polytechnic School. Democritus University of Thrace (in Greek).

Diamantopoulou P., K. Voudouris (2008). 'Optimization of water resources management using SWOT analysis: The case of Zakynthos island, Ionian Sea, Greece'. Environmental Geology 54, pp. 197-211.

Decision 4813/98 (1998). 'Determination of usage of surface waters and special terms for waste disposal in natural receivers at Trikala Prefecture' FEK-575B/11-6-1998.

Dodkins I., B. Rippey, T.J. Harrington, C. Bradley, B.N. Chathain, M. Kelly, M. Quinn, McGarrigle, S. Hodge and D. Trigg (2005). 'Developing an optimal river typology for biological elements within the Water Framework Directive'. Water Research, Vol. 39. pp. 3479–3486.

Environment Agency (2005). 'Technical Assessment Method for Morphological Alterations in Rivers' Water Framework Directive Programme - Environment Agency.

European Commission (2000). 'Directive 2000/60/EC of the European Parliament of the Council of 23rd October 2000 establishing a framework for community action in the field of water policy'. Official Journal of the European Communities.

European Commission (2006). 'Directive 2006/44/EC of the European Parliament of the Council of 6th September 2006 on the quality of fresh waters needing protection or improvement in order to support fish life' Official Journal of the European Communities.

European Commission (2008). 'Commission decision of 30 October 2008 establishing. pursuant to Directive 2000/60/EC of the European Parliament and of the Council. the values of the Member State monitoring system classifications as a result of the intercalibration exercise (notified under document number C (2008) 6016) (Text with EEA relevance) (2008/915/EC)' Official Journal of the European Communities.

Fribourg Blanc B. and C. Courbet (2004). 'Calcul des émissions de polluants organiques et de fertilisants dans l'eau. Second lot: collecte et mise en forme des informations de base nécessaires au calcul des émissions urbaines et industrielles répertoriées, dispersées et surfaciques'. Office International de l'Eau.

Gikas G., T. Yiannakopoulou and V.A. Tsihrintzis (2006). 'Modeling of non-point source pollution in a Mediterranean drainage basin'. Environmental Modeling and Assessment, Vol. 11, pp. 219-233.

Hrissanthou V., P. Deplimani and G. Xeidakis (2010). 'Estimate of sediment inflow into Vistonis Lake. Greece'. International Journal of Sediment Research, Vol. 25, pp. 161-174.

Ioannou A., Y. Chatzinikolaou and M. Lazaridou (2009). 'A preliminary pressure –impact analysis applied in the Pinios River Basin (Thessaly. Central Greece)'. Water and Environment Journal, Vol. 23, pp. 200-209.

Kanli L. (2009). 'Comparison of river typology systems in Greece' MSc thesis. School of Biology. Aristotle University of Thessaloniki. Thessaloniki, Greece.

Kristensen P. (2004). 'The DPSIR Framework'. UN Workshop on a comprehensive assessment of the vulnerability of water resources to environmental change in Africa using river basin approach. UNEP Headquarters. Kenya.

LAWA (2002). 'German Guidance Document for the implementation of the EC Water Framework Directive' Länderarbeitsgemeinschaft Wasser. Mainz, Germany

Ministerial Degree 49828/2008 concerning the 'Adoption of Special Framework for Spatial Planning and Sustainable Development for Renewable Energy and the strategic environmental assessment'.

Ministry of Industry Energy and Technology (1987). 'Weather-raingauge stations'. National Department of Water and Natural Resources. Athens.

National Statistical Agency of Greece Demographic and population data 1991-2001.

Patsia A. (2009). 'Photographic key characters of the Hellenic Evaluation System taxa of benthic macroinvertebrates. Case studies: Strymonas and Nestos rivers' MSc thesis. School of Biology. Aristotle University of Thessaloniki. Thessaloniki, Greece.

Petalas C., F. Pliakas, J. Diamantis and A. Kallioras (2004). 'Study of the distribution of precipitation and surface runoff in the Region of Eastern Macedonia and Thrace for the period 1964-1998'. Bulletin of the Greek Geological Company. Proceedings of the 10th International Congress. Thessaloniki.

Pisinaras V., C. Petalas, A. Gemitzi and V.A. Tsihrintzis (2007). 'Water quantity and quality monitoring of Kosynthos River. North-Eastern Greece'. Global NEST Journal, Vol.9, pp. 259-268.

Raven P., N. Holmes, F. Dawson and M. Everard (1998). 'Quality assessment using River Habitat Survey data'. Aquatic conservation: Marine and Freshwater Ecosystems, Vol. 8, pp. 477-499.

Rundle S.D., A. Jenkins and S.J. Ormerod (1993). 'Macroinvertebrate communities in streams in the Himalaya. Nepal'. Freshwater Biology, Vol. 30, pp. 169–180.

Sánchez-Montoya M.M., M.R. Vidal- Abarca and M.L. Suárez (2010). 'Comparing the sensitivity of diverse macroinvertebrate metrics to a multiple stressor gradient in Mediterranean streams and its influence on the assessment of ecological status'. Ecological Indicators, Vol. 10, pp. 896-904.

Simboura N. and S. Reizopoulou (2008). 'An intercalibration of classification metrics of benthic macroinvertebrates in coastal and transitional ecosystems of the Eastern Mediterranean ecoregion (Greece)'. Marine Pollution Bulletin, Vol. 56, pp. 116-126

Ter Braak C. J. F. and P. Šmilauer (1998). 'CANOCO Reference Manual and User's Guide to CANOCO for Windows. Software for Canonical Community Ordination (version 4)'. Centre for Biometry.

Torregrosa T., M. Sevilla, B. Montano and V. Lopez-Vico (2010). 'The integrated Management of Water Resources in Marina Baja (Alicante. Spain). A Simultaneous Equation Model'. Water Resources Management, Vol. 24, pp. 3799-3815.

Van de Bund WJ (2009). 'Water Framework Directive intercalibration technical report. Part 1: rivers'. JRC Scientific and Technical Reports.

Vincent C., H. Heinrich, A. Edwards. K. Nygaard and J. Haythronthwaite (2002). 'Guidance on typology: Reference conditions and classification systems for transitional and coastal waters'. European Commission.

Voudouris K. (2007). 'An application of SWOT analysis and GIS for the optimization of water resources management in Korinthia prefecture. Greece'. Proc. of International Conference "Water resources management: New approaches and technologies". European Water Resources Association (EWRA). Chania, Crete. 14-16 June 2007, pp. 307-315 (Eds Karatzas G., Papadopoulou M., Tsagarakis K.).

Voudouris K. (2009). 'Environmental Hydrogeology' Tziolas Publications. Thessaloniki, Greece (in Greek).

Wentworth C. (1922). 'A scale of grade and class terms for clastic sentiments'. Journal of Geology, Vol. 30, pp. 377–392.

WL-Delft, S.A. ENVECO and D. Argyropoulos (2005). 'Water Framework Directive Summary Report for articles 5 & 6'. Natural Resources and Environment of Cyprus.

Wright J.F. (2000). 'An introduction to RIVPACS'. In Wright J.F.D., D.W. Sutcliffe & M. Furse (Eds). Assessing the Biological Quality of Freshwaters. RIVPACS and Other Techniques. FBA. Ambleside.

Zar J.H. (1996). 'Biostatistical Analysis' 2nd eds. Prentice-Hall International Inc.

Zhang Y., Z. Yang and X. Wang (2006). 'Methodology to determine regional water demand for instream flow and its application in the Yellow River Basin'. Journal of Environmental Sciences, Vol.18. pp. 1031-1039.

3

An Ecotoxicological Approach to Evaluate the Environmental Quality of Inland Waters

M. Guida[4], O. De Castro[1], S. Leva[4], L. Copia[4], G.D'Acunzi[3], F. Landi[3], M. Inglese[2] and R.A. Nastro[5]
[1]*Department of Biological Sciences-University "FedericoII" of Naples,*
[2]*Laboratory of Environmental and Food Analyses Studies "Inglese & co. ltd",*
[3]*Salerno District Administration Bureau,*
[4]*Department of Structural and Functional Biology,University "Federico II"of Naples,*
[5]*Department of Sciences for the Environment, University Parthenope of Naples,*
Italy

1. Introduction

Water is traditionally considered a renewable resource as the quantity theoretically available depends essentially on meteoric water contributions. Modern theories consider river ecosystem as a central element of the environment but, actually, water ecosystems are more and more polluted because of agricultural and industrial activities. In Italy, where there is a widespread urbanization, most part of river ecosystems are exposed to a severe risk of damage, with consequent loss of biodiversity. The institution of national and regional parks makes possible the preservation of significant natural value areas. In 1974, the UNESCO established a "World Network of Biosphere Reserves" (WNBR) aimed to preserve areas, where there is a close relation between man and nature, through environmental preservation and sustainable development (UNESCO,2010). Nevertheless, a constant monitoring along the whole watercourse (from spring to mouth) is needed as, even in a protected area, human activities could cause, directly or indirectly, damages to valuable ecosystems.

1.1 The ecotoxicology

In recent years, ecotoxicology started a new approach to the environmental analysis as it covers:

- chemistry, i.e. the fate of chemicals in the environment
- environmental toxicology, dealing with the evaluation of toxic effects of a pollutant at different levels of biological integration
- ecology, which provides indications on regulation of both structure and function of ecosystems and, at the same time, on interactions between biotic and abiotic components (De Castro et al., 2007).

Ecotoxicology is based on the use of bio-indicators belonging to different levels of an ecosystem trophyc chain. Mathematical models are also used to foresee the environmental fate of chemicals and their effects on exposed organisms (man included) and ecosystems. An organism can be considered a" bio-indicator" when, in presence of pollutants, shows detectable variations from its natural state. Well defined responses to different concentrations of pollutants are also needed. Moreover, a good bio-indicator should be sensitive to pollutants and have a wide distribution in the investigated environment, low mobility, long life cycle and genetic uniformity. Actually, *Daphnia magna, Lepidium sativum, Cucumis sativus, Sorghum saccharatum, Pseudokirchneriella subcapitata, Vibrio fischeri* are widely used to evaluate water and soil quality as well as the toxicity of chemicals, wastes, pharmaceutical products which have to be processed in a wastewater plant or directly dumped in the environment. The main symptoms or endpoints of an ecotoxicological test could be:

- change in community structure;
- morphological changes;
- change in vitality;
- damage to genes.

Actually, a contemporary utilization of different bio-indicators in order to evaluate the ecotoxicity of a matrix (wastewater, contaminated soil, pharmaceutical by-products, etc.) let the researchers able to gain useful data about the possible toxic effects on an ecosystem (Guida et al., 2006). In the next lines, informations about biology of some test-organisms adopted in ecotoxicological analyses and about the main indices and parameters used in the evaluation of a river ecosystem quality are resumed.

1.1.1 *Daphnia magna*

Daphnia magna is a freshwater crustacean belonging to the class Brachiopoda, order Cladocera, phylum Arthropoda. It has a small size (not more than 5 mm in length) with an oval body compressed laterally. It is dorsally characterized by a welded bivalve structure (called "carapace"), which encloses the entire animal except for the head. Its body has a single compound eye, sessile, strongly pigmented; a small ocellus and two pairs of antennae. One pair of antennae is greedy and very developed, having essentially a swimming function. The transparency of carapace allows an observer to notice some internal organs: heart, located dorsally in the post-cephalic region, middle intestine (clearly visible when it contains food) and ovaries, that are located laterally. At a temperature of 20°C, the life cycle of *D. magna* is 60-100 days. Under favourable environmental conditions, the population is exclusively composed of female animals, with a parthenogenetic reproduction. This species is very sensitive to many pollutants able to cause variations to aquatic ecosystem, since it is considered the "perfect" bioindicator and test-organisms in ecotoxicological essays (Guida et al.,2004).

1.1.2 *Vibrio fisheri*

Vibrio fischeri is a rod-shaped Proteobacterium, gram negative, characterized by polar flagella. It is widespread in marine environments, living as symbiotic of various marine animals, such as the bobtail squid. *V. fischeri* is most often found as a symbiont of *Euprymna scolopes*, a small shallow water squid found on the shores of Hawaii.

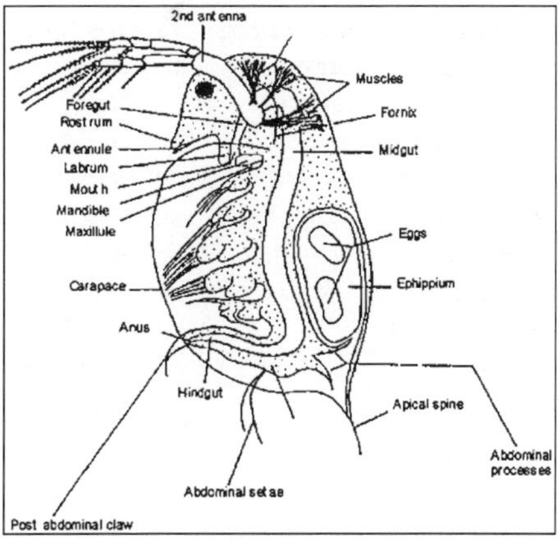

Fig. 1. A female individual of *Daphnia magna*: internal and external anatomy. (FAO, 1996)

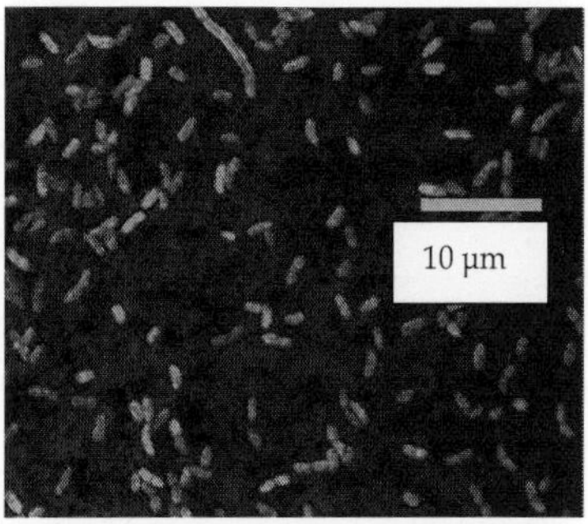

Fig. 2. The bioluminescence of *Vibrio fisheri* (1000X). Image taken by E. Nelson and L. Sycuro (http://microbewiki.kenyon.edu; last accessed: September 2011).

The bioluminescent bacterium *Vibrio fischeri* and juveniles of *Euprymna scolopes* specifically recognize and respond to each other during the formation of a persistent colonization within the host's nascent light-emitting organ. The biolumiscence depends on metabolic activity of bacteria so, damaged microrganisms less in biolumiscence. This feature has been used to study toxicity in marine samples.

1.1.3 *Pseudokirchneriella subcapitata*

Pseudokirchneriella subcapitata (previously called *Selenastrum capricornutum*) is a single-celled freshwater algae, belonging to the *Chlorococcales* family. It is a representative species of oligotrophic and eutrophic aquatic system. The cells, sickle-shaped, have a volume of 40-60 µm^3, size of 6-7 µm, and its life cycle is very short. This alga shows a good level of sensitivity to toxicants and it is commonly used to perform multispecies tests (Chen C.Y. et al., 2007).

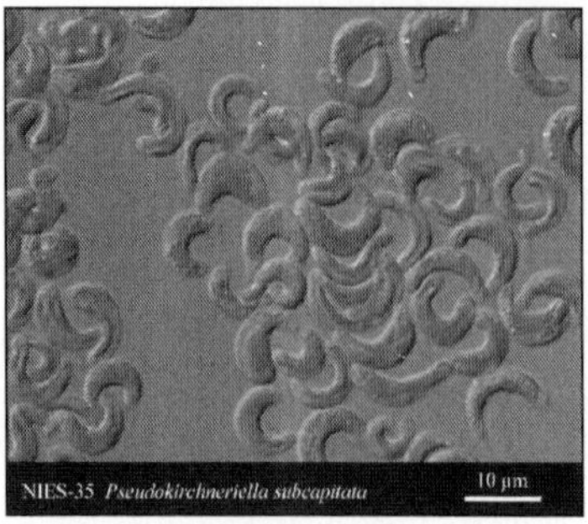

Fig. 3. Image at light microscope of *P.subcapitata* (1000x). (http://www.shigen.nig.ac.jp; last accessed: September 2011)

1.1.4 Phytotoxicity essays

The toxicity of a pollutant on a soil can be assessed by specific plants which, in natural ecosystems, can be considered good bioindicators. Ecotoxicological bioassays on plants consists generally in simple methods. It is possible to assess different endpoints for each essay such as seeds survival, germination rate, growth speed in the light and in the dark. Different plants could be used in ecotoxicological tests depending on the investigated matrix. Nevertheless, plants very commonly used in such essays in Italy and in some other Countries are *Lepidium sativum* and *Cucumis sativus* (Youn-Joo An; 2004).

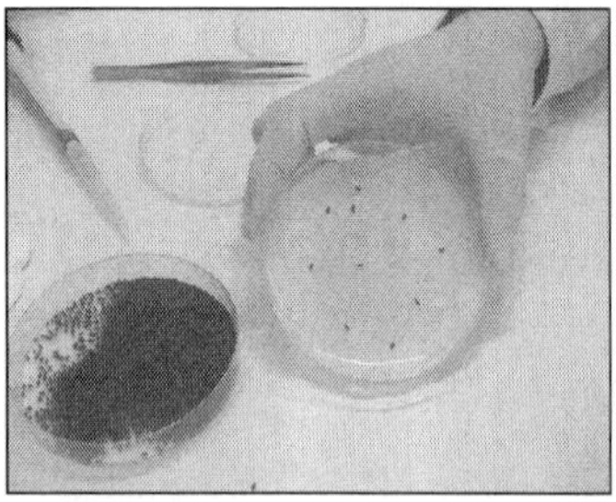

Fig. 4. Phytotoxicity test on *L. sativum*.

1.2 The evaluation of watercourse ecological quality: Extended Biotic Index (EBI) and Fluvial Functional Index (FFI)

If ecotoxicology allows scientists to gain informations about the potential toxicity of a single pollutant or a mixture of chemicals (like industrial wastewater) through the implementations of laboratorial tests, a complete study of a water ecosystems requires the collection of data directly on site. So, analyses of microfauna (small invertebrates colonizing water ecosystems) and ichtyofauna, the measure of parameters like water Dissolved Oxygen (DO), Turbidity, pH, Conductivity, Temperature, besides of quality of riverside flora and evaluation of watercourse erosion are needed to evaluate the environmental quality of a water ecosystem like a river. The collection and characterization of microfauna is at the basis of the Extended Biotic index (EBI), whose value is used to classify the quality of freshwater courses in Italy (Ghetti P., 1997). Further biotic indices based on the quali-quantitative analysis of invertebrates community are also used with the same aim (Kalyoncu H. and Zeybek M., 2011). Once calculated the EBI value (ranging from 0 to 12), a class of environmental quality is attributed to each stream or river or part of it. There are 5 classes available, corresponding to a well defined colour (Table 1). Colours are used to report the classification of environmental quality on a map of the watercourse. Like EBI, Fluvial Functioning Index (FFI) values correspond to well defined quality classes (Negri P.et al., 2011). Recently, National Environmental Protection Agency of Italy suggested FFI for a correct evaluation of the health of river ecosystems (APAT, 2007). It must be underlined that FFI categories are as recommended by Water Framework Directive 2000/60 (Negri P. et al., 2011).

Classes	E.B.I. (Values)	Outcome	Colour
Class I	10-11-12...	Watercourse not significantly polluted or not polluted at all	Blue
Class II	8-9	Watercourse moderately polluted or altered	Green
Class III	6-7	Watercourse polluted or altered	Yellow
Class IV	4-5	Watercourse highly polluted or altered	Brown
Class V	1-2-3	Watercourse extraordinarily polluted of altered	Red

Table 1. Classes of environmental quality corresponding to EBI values.

The above mentioned indices are based on a simply concept: a watercourse is a dynamic system, formed by different habitats which continuously follow each other from the source down to the mouth and which interconnect with the surrounding terrestrial ecosystems. It is important to notice how, along the course of a river, the environmental conditions (like morphological, hydrodynamic, physical and chemical parameters) change and, with respect to these, the biological populations vary too. That's why the state of a river ecosystem must be evaluated along the whole course in pre-determinate stations. The study of the whole territory crossed by the river and the identification of possible critical situation is of a great importance as it could affect heavily the results of a monitoring activity. The hydro-geologic conformation of the river is also important as it could affect the accessibility to sampling stations. So, an adequate preliminary study, a good experience in the choice of sampling sites, besides of a suitable equipment, are necessary.

1.3 Other indices

According to EU Regulation 2000/60/CE published on 10/23/2000 (the Water Framework Directive), the quality of a watercourse has to be evaluated through both Ecological and Chemical States indices. From the elaboration of these indices, the Ecological State of Watercourse (ESW Index) could be calculated. Another index widely used is the Evaluation of Macro-descriptors Pollution Level (EMPL) which is calculated from the following parameters: Chemical Oxygen Demand (COD), Biochemical Oxygen Demand at 5 days (BOD_5), NO_3^-, NH_4^+,Total Phosphorous, *Escherichia coli*, Oxygen saturation rate.

1.4 Evaluation of freshwaters microbial quality

Water quality evaluation can't exempt from a microbiological characterization of water itself. The presence of some bacterial species like *Escherichia coli* gives important informations about the grade of a watercourse pollution (Edberg S.C. et al., 2000). Different microbial parameters were considered in the present survey as more parameters led more informations about the organic pollution of a river ecosystem. *E. coli* and fecal coliforms are indicators of recent fecal pollution. Streptococci and Clostridia, instead, are abundant in case of a past fecal pollution. *Staphylococcus aureus, Pseudomonas aeruginosa, Aeromonas hydrophila* are potential pathogens of both fishes and men (Pathak S.P. et al., 2008; Health Canada, 2011; Fazli M., 2009).

1.5 Phytoplankton

Phytoplankton includes different algal species belonging to different taxa (*Bacillariophyceae, Dinophyceae, Chrysophyceae, Cryptophyceae, Dictyochophyceae, Prymnesiophyceae, Raphydophyceae and Euglenophyceae, Prasinophyceae and Chlorophyceae*). The analysis of the phytoplanctonic component of a water ecosystem, gives important informations about the amount of nutrients (mainly nitrates and phosphates), the presence of toxics and some chemical-physical factors as water temperature and turbidity. Moreover, phytoplankton populations and corporations oscillate widely even in short time or in small space according to the environmental conditions (as nutrients, dissolved oxygen, turbidity, temperature, etc.). That's why phytoplankton is considered, since long time, an important indicator of the trophic level of an aquatic ecosystem (freshwater and marine waters) (Greene J.C et al., 1975; Mahoney J.B., 1983).

2. Aim of the study

Ecotoxicological tests are usually carried out to assess the potential toxicity of wastes or chemicals before their introduction in the environment. In the course of two years, we tried to apply an ecotoxicological approach to evaluate the environmental quality of Tanagro and Bussento rivers flowing through a WNBR area (UNESCO, 2010), in Southern Italy. We used ecotoxicological tests findings to identify possible critic situations (due to a possible chemical pollution) along both river courses. But, since a river is a complex system, a polyphasic approach in a correct environmental quality evaluation was needed. In fact, the real challenge researchers have to face consists in consider all possible elements of river ecosystem (water quality, fauna and flora composition, erosion phenomena, etc.) in order to have a realistic picture of the ecological state of a watercourse. So, besides of ecotoxicological essays, we collected data about chemical, microbiological and physical characteristics of water while ecological quality of watercourses was evaluated by the use of EBI. Moreover, further important data were gained through the interpolation of different parameters, calculating, this way, ecological indices like Evaluation of Macro-descriptors Pollution Level (EMPL) and Ecological State of Watercourses (ESW). At last, we tried to obtain a complete picture of the river environmental quality. The dependability of ecotoxicological tests for river quater quality evaluation was assessed.

3. Material and methods

Standardized procedures were applied in sampling procedures and water analyses. A complete list of all parameters measured and indices calculated is reported in Table 2. The choice of sampling sites was carried out according to EU Regulation 2000/60/CE recommendations, taking in account the introduction of waste waters through spillways . On the whole, 14 sampling points (stations) were chosen for each river. Collected data were stored in a database and georefentiated by an ArcGis software (data not shown). Frequency of sampling and analytic procedures are reported in Table 3.

Chemical parameters	*Microbiological parameters*	*Ecological and Ecotoxicological parameters*
Temperature (°C)	Aerobic Colony Count at 22°C	Daphnia magna acute toxicity essay
Dissolved Oxygen (DO)	Aerobic Colony Count at 37°C	Pseudokirchneriella subcapitata acute toxicity test
Ph	Total Coliforms	Phytotoxicity test
Specific Conductivity (SC)	Fecal Coliforms	Inihibition test on algeae
NH_4^+	Escherichia coli	Fitoplankton
NO_2^-	Streptococci	Extended Biotic Index (EBI)
NO_3^-	Clostridia	Fluvial Functional Index (FFI)
PO_4^-	Staphylococci	
BOD_5	Stafilococcus aureus	
COD	Pseudomonas aeruginosa	
	Aeromonas hydrophila	

Table 2. Complete list of analyses carried out on each water sample analyzed.

Biological Parameters	*Sampling frequency*	*Chemical-physical parameters*	*Sampling frequency*
Ecotoxicological essays	5 times/year	Water temperature	5 times/year
Microbiological analyses	5 times/year	Dissolved Oxygen	5 times/year
Fitoplancton	3 times/year	Salinity	5 times/year
EBI	2 times/year	pH	5 times/year
FFI	1 time/year	Nutrients (NO_3^-,NO_2^-)	5Times/year

Table 3. Frequency of sampling for each parameter in each station.

3.1 Chemical analyses

The chemical analyses were carried out according to Standard Methods (2006).

3.2 Microbiological analyses

All microbiological analyses were carried out according to ISO methods.

3.3 Phytoplankton

The EN 15204:2006 and the EN ISO 5667-1 and EN ISO 5667-3 were applied during this survey.

3.4 *Daphnia magna* – Acute bioassay

Daphnia magna acute toxicity test were carried out according to ISO 6341:1999. According to this method, ten newborns no older than 24 hours must be exposed to a sample. The newborns of *D. magna* are transferred in each container filled with 50 ml of sample: this operation must be carried out paying attention to don't damage the daphnides. Moreover, they must not be fed during all test procedure. After 24 hours, the number of immobile crustaceans (or of the ones showing at least a change in their usual way of swimming) is calculated. If the essay is carried out considering different sample concentrations, it is possible to calculate the EC_{50} , which gives, for a well defined toxic, the concentration value inhibiting the 50% of test organisms.

3.5 *Vibrio fisheri* – Acute bioassay

APAT IRSA-CNR 2003 n. 8030 is the method chosen to carry out acute toxic bioassays with *V. fischeri*. The method gives an evaluation of acute toxicity of freshwater and marine samples through the evaluation of *V. fischeri,* strain NRRL-B-11177 bioluminescence inhibition. Luminescence can be measured after 5, 15 and 30 minutes of exposition to a sample by the use of a luminometer. Sample toxicity is measured as EC50, which represents the sample concentration in correspondence of which there is a decrease of 50% of the light emitted by bacteria. *V. fisheri* toxicity test demonstrated a good correlation with tests carried out on other aquatic organisms like *D. magna, Artemia salina, Chlorella* sp., *Tetrahymena pyriformis* (Kaiser K.L., 1998). Its reliability for the evaluation of soil and colourful samples had been also demonstrated (Lappalainen J. et al., 2001).

3.6 Chronic bioassay

ISO 8692:2004 is the method applied for *Pseudokirchneriella subcapitata* bioessay. The bioassay analyzes the toxic effect of a sample by measuring the inhibition of algal growth. Selected cultures are exposed during their exponential phase of growth, at well defined concentrations of a sample for 72 hours. After the exposition to the sample, algae density is measured reading absorbance of the culture at 663 nm.

3.7 Phytotoxicity

Phytotoxicity test assesses the potential toxicity of a sample measuring the inhibition of germination and /or root elongation of seeds under controlled conditions. Negative controls are prepared. Seeds of two dicotyledons (*L. sativum* and *C. sativus*) and a monocotyledon (*S. saccharatum*) are exposed to a water sample and incubated in the dark at 25 ± 2 °C for 72 hours. Then, germinated seeds are counted and root length measured using a calibre. The effect on both germination and radical elongation is expressed as percentage germination index (GI%) (UNICHIM 1651:2003). Such tests are widely used to assess the ecotoxicological effects of soils and waters contaminated with organic molecules and/or heavy metals (An Y. et al., 2004).

4. Results

4.1 Ecotoxicological results

Ecotoxicological tests didn't show any important inhibitory effect resulting in a good quality of both river waters. So, *D. magna* didn't suffer any significant toxic effect: the percentage of immobility didn't ever overcome, on the average, the 20% of individuals. Sampling sites number 7 and 13 of Tanagro river showed a modest effect on *P.subcapitata* (an increase in cell reproduction rate of about 10% on average). As *to L. sativum, C. sativus* and *S. saccharatum* no toxic effects were detected in both river samples. In some cases, especially for *L. sativum* and *S. saccharatum* some kind of bio-stimulation was observed (Fig.5 and 6). The results of phytotoxicity tests didn't show any important inhibition of seeds germination or

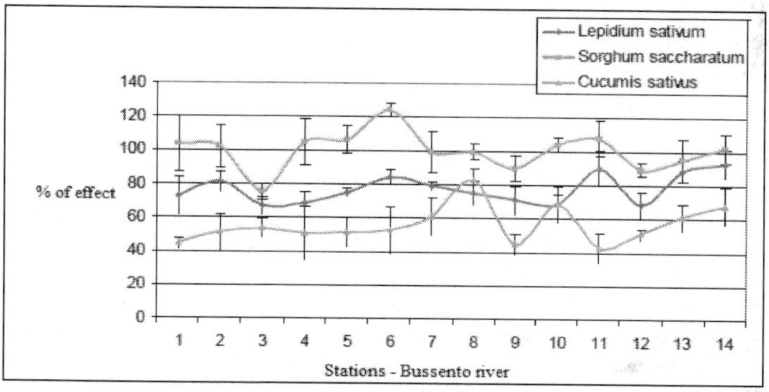

Fig. 5. Bussento river phytotoxicity tests on *L. sativum, S. saccharatum, C. Sativus.*

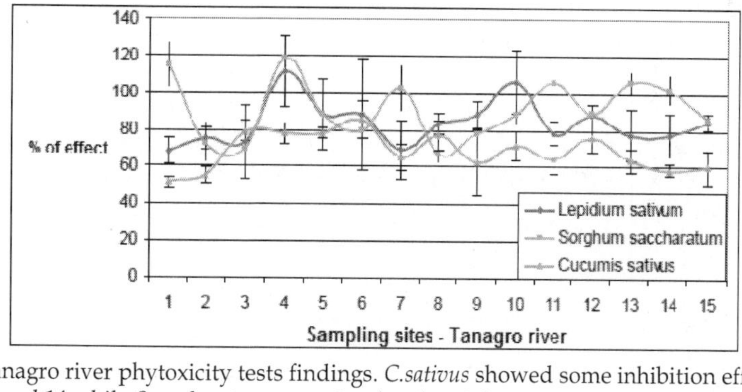

Fig. 6. Tanagro river phytoxicity tests findings. *C.sativus* showed some inhibition effects at station 1 and 14 while *S.saccharatum* was stimulated mainly in stations 1 and 4.

root elongation on the most part of the samples, even if *C. sativus* showed a less growth rate than the other two plants. This could be explained on the base of different physiology of the species used in the test. All samples were tested also on *V. fisheri* and they never showed any inhibitory effect on bacteria. It is important to notice that, in no case and for no test organism, it was possible to calculate EC50 value because of the really low samples toxicity.

4.2 Chemical parameters and EMPL

In all stations, COD values were, on the average, always less than 20 mgO_2/mL, while BOD_5 never overcome 10 mgO_2/mL in Bussento river (Fig.7). As to Tanagro river, an increase of COD and BOD_5 was detected at station number 3, whose values overcame 20 mgO_2/L for COD and 10 mgO_2/L for BOD_5 (Fig.8). Moreover, chemical analyses showed some significant variations of nitrogenous compounds (as NO_3^-), not only among the stations but

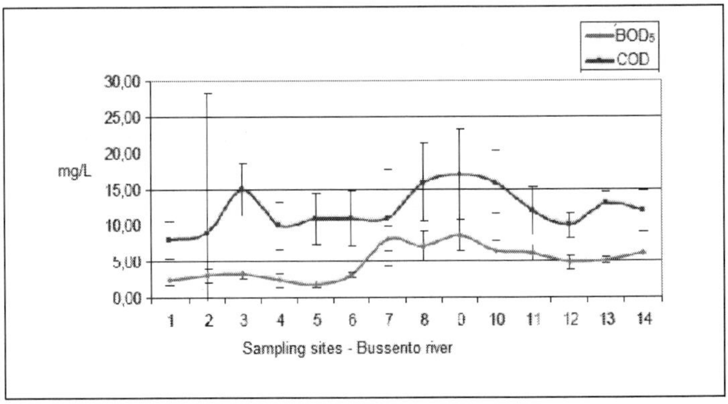

Fig. 7. BOD_5 and COD values in Bussento river. Average values and standard error are reported.

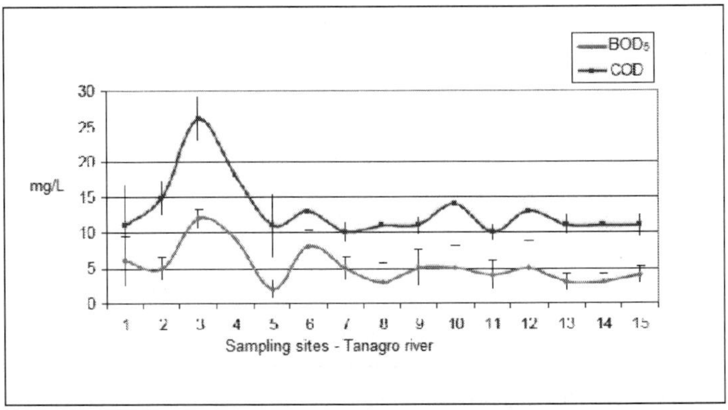

Fig. 8. Tanagro river: COD and BOD_5 average values and standard errors.

even for a single sampling site in the course of the time, causing an increase in variability (Fig. 9 and 10). So, the greatest variations were detected in correspondence of the stations number 6 and 7 of Bussento river and 3 and 4 for Tanagro ones. Nevertheless, as *E. coli* concentration never overcame the 5000 CFU/100mL limit, the increase of nitrates could be due to agricultural rather than to wastewater intakes. Both rivers, in fact, flow through a not highly urbanized land characterized by an agriculture-based economy.

If NO_2^- values were low in all samples, NH_4^+ concentration showed some different values between the two watercourses, as shown in figures 9 and 10. On the average, NH_4^+ was higher in Tanagro rather than in Bussento water. In any case, NH_4^+ concentration decreased from station number 10 till the river month. A similar tendency was detected for PO_4^{2-}, suggesting an agricultural origin of both nutrients.

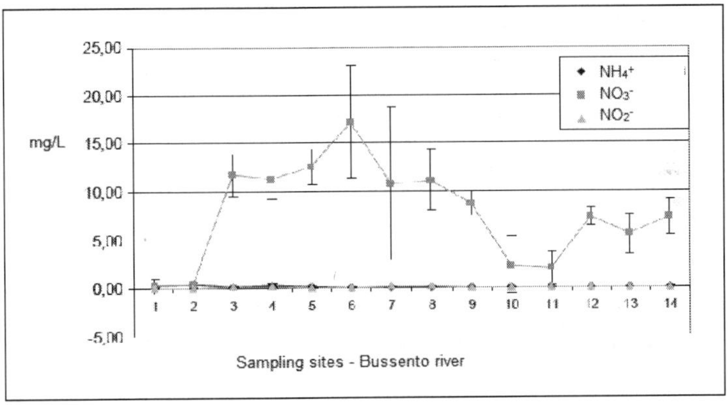

Fig. 9. Nitrogenous compounds concentrations in Bussento water samples. In the graphic, average values and standard errors are reported.

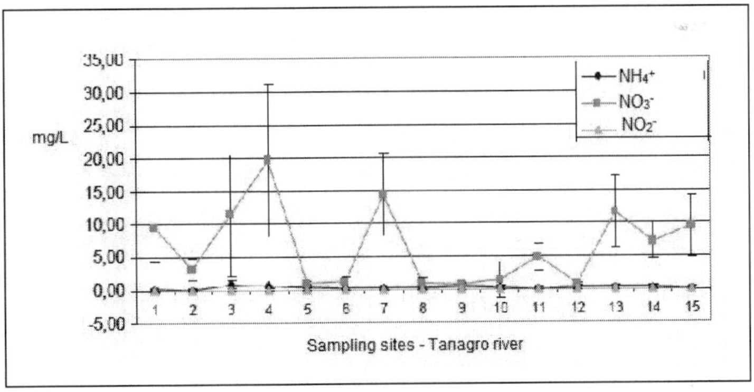

Fig. 10. Nitrogenous compounds concentrations (average ± standard error) along Tanagro river course.

No water acidification was detected in both rivers as pH values showed little variation and, in any case, they ranged between 7.5-8.5 values. The results of chemical analyses, together with *E. coli* concentrations, carried out on both rivers were compared to values shown in table 4, in order to calculate the EMPL index for both watercourses. From collected data, all stations were classified as belonging to the 2nd and 3rd levels, showing a moderate pollution. Furthermore, Tanagro river showed a better water quality than Bussento one as 73% of collected samples belonged to level 2 versus a 53% of Bussento ones. On the whole, 66% of all samples belonged to level 2.

Parameters	Level 1	Level 2	Level 3	Level 4	Level 5
100-DO (% sat.)	≤|10|	≤|20|	≤|30|	≤|50|	≥|50|
BOD_5 (O_2/mg/L)	<2.5	≤4	≤8	≤15	>15
COD (O_2/mg/L)	<5	≤10	≤15	≤25	>25
NH_4^+ (N mg/L)	<0.03	≤0.1	≤0.5	≤1.5	>1.5
NO_3^- (N mg/L)	<0.30	≤1.5	≤5	≤10	>10
Total-P (P mg/L)	<0.07	≤0.15	≤0.30	≤0.6	>0.6
***Escherichia coli* (CFU/mL)**	<100	≤1000	≤5000	≤20000	>20000
Score referable to each parameter (75° percentile of the sampling period)	80	40	20	10	5
Evaluation of Macro-descriptors Pollution Level	480-560	240-475	120-235	60-115	<60

Table 4. Reference values of EMPL index.

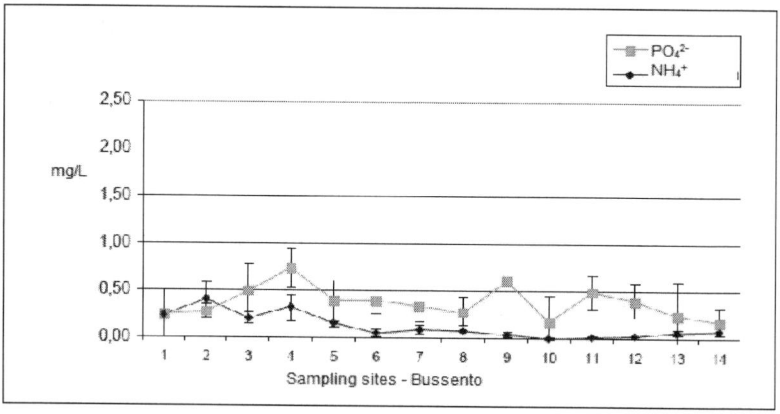

Fig. 11. PO_4^{2-} and NH_4^+ concentrations in Bussento waters.

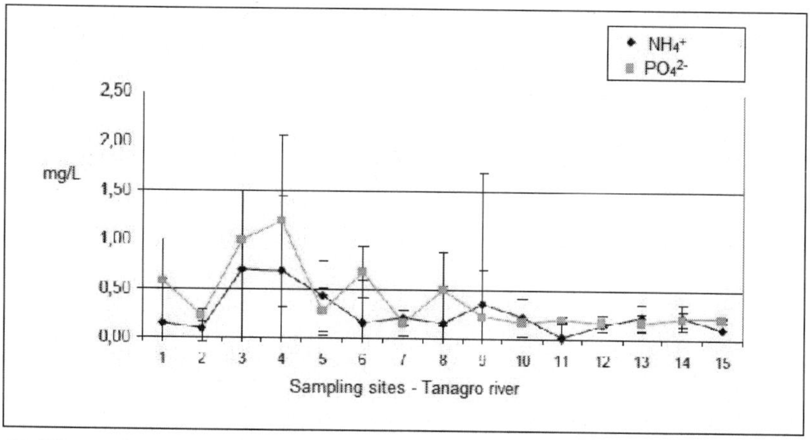

Fig. 12. PO_4^{2-} and NH_4^+ values along Tanagro river stations.

4.3 Extended biotic index

The analysis of macro-invertebrates population were carried out two times a year (on summer and on winter) in correspondence of the same stations used to take samples for chemical-physical and microbiological analyses. Temperature, DO, pH and water conductivity were measured contemporaneously as they affect macro-invertebrates distribution heavily and a correct EBI evaluation can not exempt from a determination of the above mentioned parameters. Even if it is possible to notice a similar trend in the EBI variation along both watercourses (from spring to mouth), the second sampling campaign (on summer) showed lower values than the first one (carried out on winter). Apart from few sites showing a real deterioration of river ecosystem (especially in Tanagro river), EBI average values didn't change significantly during the sampling activity. Most part of the collected samples were classified as belonging to the second class (corresponding to moderate pollution) or to the third one (altered ecosystem).

4.4 The Ecological State of Watercourses (ESW)

The EBI values together with EMBL ones were used to gain another index, the Ecologic State of Watercourses (ESW) (Table 5). Our results showed, on the whole, a good or sufficient ecological quality. Only 36% of Bussento river sampling sites reached the 2nd ESW class of quality, suggesting a certain vulnerability of the freshwater ecosystem itself

As the remaining sites resulted just in a 3rd class. Tanagro river was characterized by a better environmental state as 46% of stations were classified as belonging to a 2nd class. Nevertheless, our results shed light on a critical environmental state regarding the station 3 of Tanagro river (4th class of quality). Even if most part of ESW values were determined by both EBI and EMBL values, it is interesting to notice how, in some cases, ESW values were affected in most part by the EBI values rather than EMBL ones as reported in Table 6.

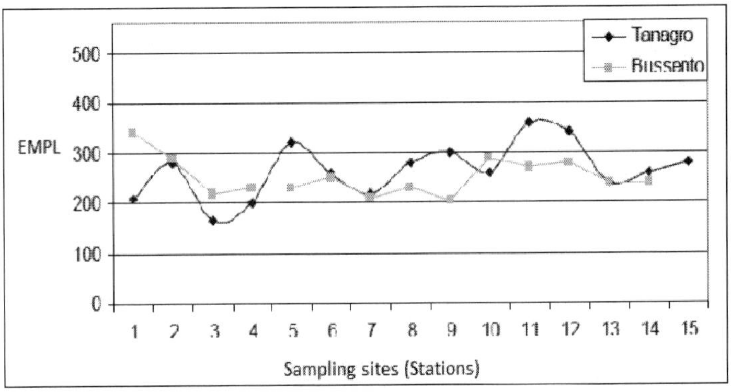

Fig. 13. EMPL values of Tanagro and Bussento river. There are not great variations in value.

The biological index not only is complementary to chemical characterization of freshwaters but it could even indicate, in advance, possible environmental criticisms.

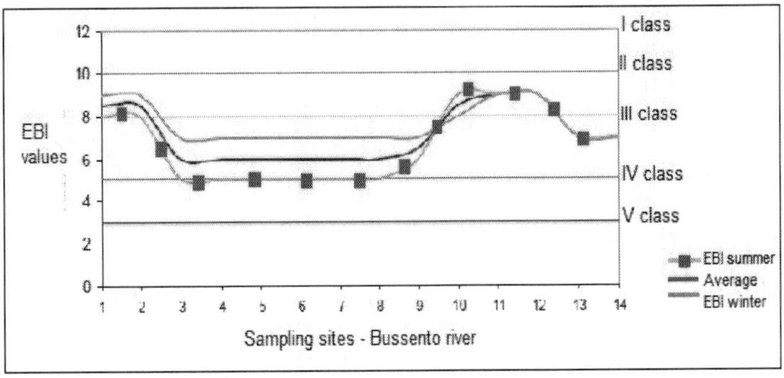

Fig. 14. EBI values of Bussento river stations. The values range from the 2nd to the 4 class of quality.

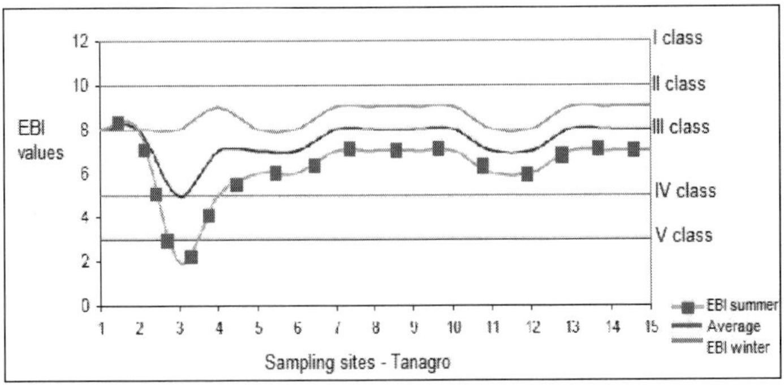

Fig. 15. Tanagro river EBI values in winter and in summer. In station n°3 the lowest value corresponding to a 5th class.

ESW	Class I	Class II	Class III	Class IV	Class V
EBI EMBL	≥10 480-560	8-9 240-475	6-7 120-235	4-5 60-115	1,2,3 <60
Overcome	High	Good	Sufficient	Poor	Very Poor
Conventional colour	Blu	Green	Yellow	Orange	Poor

Table 5. ESW class of quality and the correspondent EBI and EMBL values are reported.

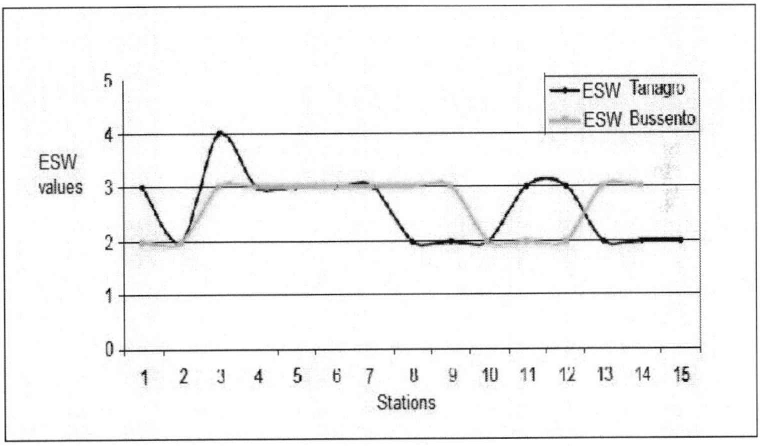

Fig. 16. ESW values: environmental quality reached the worst value just in station number 3 of Tanagro river.

	EMBL+EBI	EMBL	EBI
Tanagro	54%	13%	33%
Bussento	79%	21%	0%

Table 6. Percentage of samples whose ESW value was affected by EMBL+EBI or by just EMBL or EBI values.

FFI was applied in order to consider both hydro-morphological and biological factors in the evaluation of river courses environmental quality. Our findings, compared to some reference values, showed an environmental quality ranging from good-moderate to high (Table 7). The 57% of Tanagro river course fell into the high and good categories while Bussento river just 35%.

	High	Good	Good-moderate
Tanagro	7%	50%	43%
Bussento	21%	14%	65%

Table 7. FFI values for Tanagro and Bussento rivers.

4.5 Phytoplankton

Phytoplankton analyses didn't show any particular dystrophy, except for three stations of Bussento river, where high values of phytoplankton density were found.

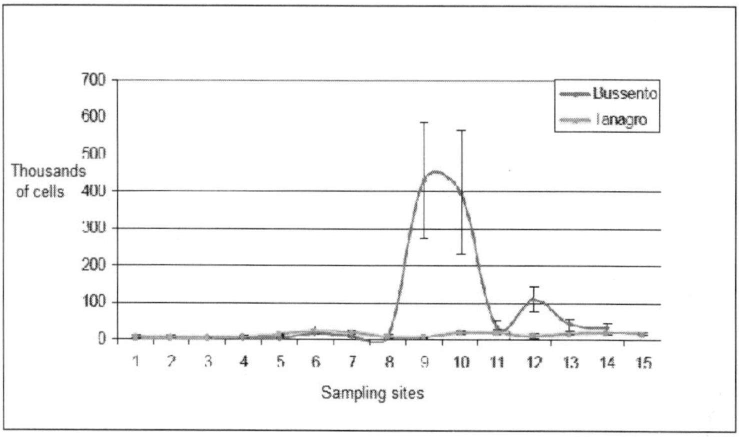

Fig. 17. Phytoplankton concentrations along the watercourses stations.

4.6 Microbiological results

Even if microbiological analyses are not strictly required to evaluate the environmental quality of a watercourse, the presence of some bacteria could be useful in order to recognize anthropogenic impacts due to wastewater intakes. As to *E. coli* concentration, Bussento water showed in all sites an amount ranging between 100 and 1000 CFU/100ml. On the other hand, most part of Tanagro samples (56%) showed a concentration ranging from 100 to 1000 CFU/100 ml, a 35% less than 100 CFU/100 ml and just a 9% of the stations was characterized by an amount overcoming 1000 CFU/100ml. These data were substantially confirmed by total and fecal coliforms, streptococci amounts. As to the other microbial parameters, no clostridia were found in both rivers water samples while *S.aureus* (typical of human and mammalian skin and mucosa) was seldom isolated. *Aeromonas hydrophyla* and *Pseudomonas aeruginosa*, instead, were widely present in water samples as they are part of environmental microflora. In Figures 18 and 19 the results concerning *A. hydrophyla* and *P. aeruginosa* are reported. From a comparison between both *A. hydrophyla, P.aeruginosa* and *E.coli* amounts along the rivers, there is no evidence of any correlation between such bacterial strains amounts.

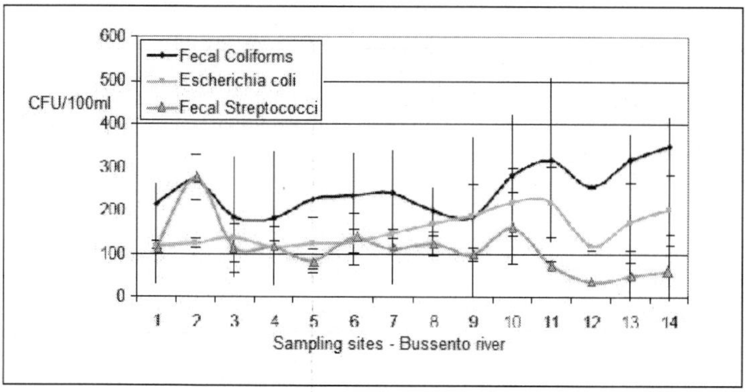

Fig. 18. Bussento river: microbiological values overcame the 100 CFU/100mL values except for streptococci in the last four stations.

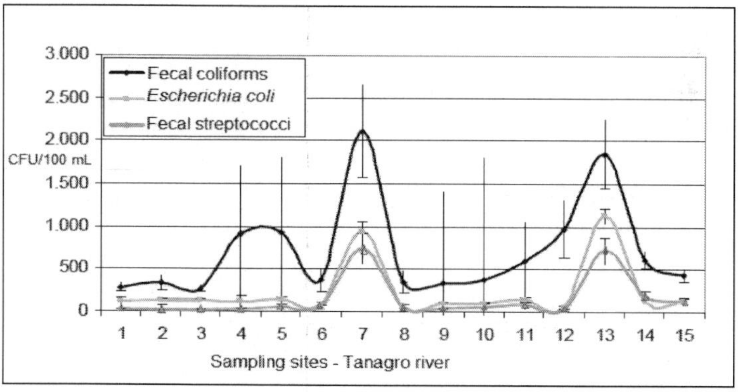

Fig. 19. *E.coli*, fecal coliforms and streptococci in Tanagro river waters. Stations 7 and 13 showed some high values.

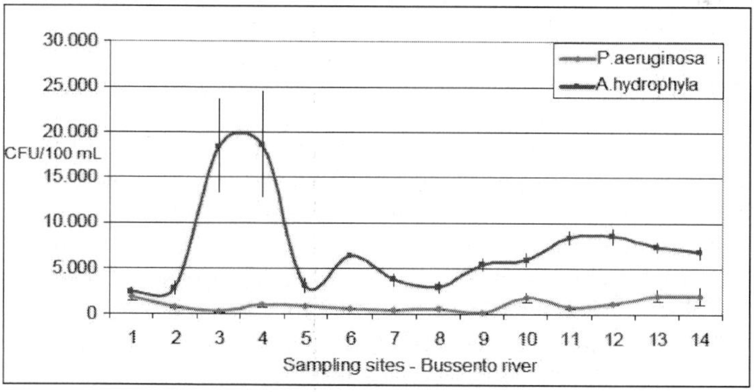

Fig. 20. *P. aeruginosa* and *A. hydrophila* concentrations in Bussento waters.

5. Conclusions

The environmental quality of both rivers resulted, on the whole, in a good or, at least, a sufficient quality, comparable, at least, to the upper course of Sele river flowing in the same area (Rizzo D. et al., 2009). These data are important as in a so highly urbanized district like the Campania one, most part of watercourses are heavily polluted and compromised by human activities. Sarno river, for example, is one of the most polluted river in Europe (Arienzo M. et al.,2000; De Pippo T. et al.,2006) and it flows in a high anthropized area. In this context of environmental degradation, the protection of high naturalistic value areas is a must and our data confirmed the efficacy of natural reserves and of a sustainable development policy. Under a strictly technical point of view, a polyphasic approach provides detailed informations about environmental quality of a river ecosystem. Nevertheless, it must be underlined that some experience in data analyses and competence in different fields are needed in order to give a right interpretation of findings yelded by so different analyses. In fact, it has to be noticed, for example, that phytotoxicity tests outcomes weren't fully overlapping with other ecotoxicological tests and of some difficult interpretation. While the *D.magna* and *P.subcapitata* tests gave no evidence of any significant toxic effect of water samples, *L.sativum, C.sativus, S.saccharatum* showed a modest inhibition of germination after 5 days (between 22% and 30% of seeds for each species didn't germinate) in correspondence to the station number 3 of Tanagro river, in accordance to the EBI, ESW, EMPL whose values were corresponding to a poor or sufficient environmental quality. It is clear that ecotoxicological tests are not enough to evaluate the environmental quality of a complex ecosystem like a river but they showed to be a useful tool in the evaluation of river environmental quality.

6. Acknowledgments

We thank Dr. Daniela Santafede for her contribution to the elaboration of chemical and microbiological data.

7. References

An Y. , Kim Y., Kwon T., Jeong S. (2004) Combined effect of copper, cadmium, and lead upon Cucumis sativus growth and bioaccumulation. Science of The Total Environment Volume 326, Issues 1-3, pp. 85-93 .

APAT (2007). Indice di Funzionalità Fluviale - Nuova versione del metodo revisionata ed aggiornata. http://info.apat.it/pubblicazioni

Arienzo M., Adamo P., Bianco M.R., Violante P. (2000). Impact of Land Use and Urban Runoff on the Contamination of the Sarno River Basin in Southwestern Italy. Water, Air, & Soil Pollution Volume 131, Numbers 1-4, 349-366

Autorità di Bacino Regionale Friuli Venezia Giulia (ABR.FVG). River Continuum Concept. http://www.abr.fvg.it/studies/ecologia-acque/water-ecology/river-continuum-concept Visited on January 2012.

Chen CY, Ko CW, Lee PI.(2007). Toxicity of substituted anilines to Pseudokirchneriella subcapitata and quantitative structure-activity relationship analysis for polar narcotics. Environ Toxicol Chem. Jun;26(6):1158-64.

De Castro O., Gianguzzi L., Colombo P., De Luca P., Marino G., Guida M.: (2007) Multivariate Analysis of Sites Using Water Invertebrates and Land use as Indicators of the Quality of Biotopes of Mediterranean Relic Plant (Petagnaea gussonei, Apiaceae) Environmental Bioindicators, 2, pp. 161 - 171.

De Pippo T., Donadio C., Guida M., Petrosino C. (2006). The case of Sarno River (Southern Italy): effects of geomorphology on the environmental impacts. Environ Sci Pollut Res Int. 2006 May;13(3):184-91.

Directive 2000/60/EC of the European Parliament and of the Council establishing a framework for the Community action in the field of water policy". Official Journal (OJ L 327). December 2000.

Edberg SC, Rice EW, Karlin RJ, Allen MJ. (2000). Escherichia coli: the best biological drinking water indicator for public health protection. Symp Ser Soc Appl Microbiol. 2000;(29):106S-116S.

FAO (1996). Manual on the Production and Use of Live Food for Aquaculture. FAO Fisheries Technical Paper 361, pp. 283-288.

Fazli M., Bjarnsholt T., Kirketerp-Møller K., Jørgensen B., Schou Andersen A., Krogfelt K.A., Givskov M., and Tolker-Nielsen T.(2009) Nonrandom Distribution of *Pseudomonas aeruginosa* and *Staphylococcus aureus* in Chronic Wounds. J Clin Microbiol.47(12): 4084–4089.

Ghetti PF (1997). Application Manual: Extended Biotic Index - Macroinvertebates in quality control of running water environments, in Italian [I Macroinvertebrati nel controllo della qualita di ambienti di acque correnti. Indise Biotico Esteso (I.B.E). Manuale di applicazione. Provincia Autonoma di Trento]. Trento Italy, 222 pp.

Greene J.C., Miller W.E., Shiroyama T. and Maloney T.E.(1975). Utilization of algal assays to assess the effects of municipal, industrial, and agricultural wastewater effluents upon phytoplankton production in the Snake River system. Water, Air, & Soil Pollution Volume 4, Numbers 3-4, 415-434.

Guida M., Mattei M., Melluso G., Pagano G., Meriç S.(2004) *Daphnia magna* and *Selenastrum capricornutum* in evaluating the toxicity of alum and polymer used in coagulation-flocculation" Fresenius Environmental Bulletin; vol 13- N° 11b 1244-7.

Guida M., Melluso G., Mattei M., Inglese M., Pagano G., Belgiorno V., Meriç S. (2006): A multi-battery lethal and sub-lethal toxicity investigation on organo-pesticides. 18-22 September Istanbul- Turkey. IWA DipCon 10th International Specialised Conference on Diffuse Pollution and Sustainable Basin Management.

Health Canada. Environmental and Workplace Health: Bacterial Waterborne Pathogens - Current and Emerging Organisms of Concern. www.hc-sc.gc.ca (visited on September 2011).

Kaiser K.L. (1998). Correlations of *Vibrio fischeri* bacteria test data with bioassay data for other organisms. Environ Health Perspect. 106 (Suppl 2): 583–591.

Kalyoncu H. and Zeybek M. (2011) An application of different biotic and diversity indices for assessing water quality: A case study in the Rivers Çukurca and Isparta (Turkey).African Journal of Agricultural Research Vol. 6(1), pp. 19-27.

Lappalainen J., Juvonen R., Nurmi J., Karp M. (2001) Automated color correction method for Vibrio fischeri toxicity test. Comparison of standard and kinetic assays Chemosphere Volume 45, Issues 4-5, Pages 635-64.

Mahoney J. B. (1983) Algal assay of relative abundance of phytoplankton nutrients in northeast United States coastal and shelf waters.

Negri P., Siligardi M.,Fuganti A.,Francescon M., Monauni C.,Pozzi S. The use of the Functioning Fluvial Index for river management. evis.sggw.waw.pl/wethydro/contents/.../107-115_PaoloNegri_e.pdf (visited on September 2011).

Pathak S.P. , Bhattacherjee J.W., Kalra N., Chandra S. (2008). Seasonal distribution of Aeromonas hydrophila in river water and isolation from river fish. Journal of Appl. Microbiol. Vol. 65 Issue 4, pp. 347 - 352.

Rizzo D., Gentile A., Sica M, Caracciolo L.,Mautone M.,Rastrelli L.,Saviello G. Caratterizzazione Qualitativa e Tipizzazione del Fiume Sele Individuazione e definizione del Corpo Idrico di Riferimento Sito-Specifico. Atti del Convegno Funzionalità fluviale:strumento di gestione e pianificazione. Trento, 19-20 November 2009.

Youn-Joo An (2004) Soil ecotoxicity assessment using cadmium sensitive plants. Environmental Pollution Volume 127, Issue 1, Pages 21-26.

UNESCO - MAB Secretariat (2010). World Network of Biosphere Reserves. http://www.unesco.org/mab/doc/brs/BRList2010.pdf visited on January 2012.

Emerging (Bio)Sensing Technology for Assessing and Monitoring Freshwater Contamination – Methods and Applications

Raquel B. Queirós[1,2], J.P. Noronha[3],
P.V.S. Marques[1] and M. Goreti F. Sales[2]
[1]INESC TEC (formerly INESC Porto) and Faculty of Sciences, University of Porto,
[2]BioMark Sensor Research and ISEP/IPP - School of Engineering, Polytechnic Institute of Porto,
[3]REQUIMTE/CQFB and Faculty of Sciences and Technology of University Nova de Lisboa,
Portugal

1. Introduction

Water is life and its preservation is not only a moral obligation but also a legal requirement. By 2030, global demands will exceed more than 40 % the existing resources and more than a third of the world's population will have to deal with water shortages (European Environmental Agency [EEA], 2010). Climate change effects on water resources will not help. Efforts are being made throughout Europe towards a reduced and efficient water use and prevention of any further deterioration of the quality of water (Eurostat, European Comission [EC], 2010). The Water Framework Directive (EC, 2000) lays down provisions for monitoring, assessing and classifying water quality. Supporting this, the Drinking Water sets standards for 48 microbiological and chemical parameters that must be monitored and tested regularly (EC, 1998). The Bathing Water Directive also sets concentration limits for microbiological pollutants in inland and coastal bathing waters (EC, 2006), addressing risks from algae and cyanobacteria contamination and faecal contamination, requiring immediate action, including the provision of information to the public, to prevent exposure. With these directives, among others, the European Union [EU] expects to offer its citizens, by 2015, fresh and coastal waters of good quality.

Freshwater quality is generally monitored with regard to chemical and microbiological parameters. Most regulated chemical parameters are monitored by more or less expensive but reliable techniques and some of which allow on-site analysis. The ease of access to such analytical procedures is reflected on the large amount of chemical data given by Water Information System for Europe [WISE] (EEA, 2011). Microbial parametric data are, on the contrary, more difficult to obtain. Microorganisms may be defined as those organisms that are not readily visible to the naked eye, requiring a microscope for detailed observation. These have a size range (maximum linear dimension) up to 200 μm, and vary from viruses, through bacteria and archea, to micro-algae, fungi and protozoa (Sigee, 2005).

There are many different microorganisms that can pose serious risks to the environment and public health (Figure 1). In general, waterborne pathogens cause 10–20 million deaths and 200 million non-fatal infections each year (Leonard et al., 2003). They contribute to harmful effects either by a direct or indirect way: through the direct contact with the pathogen (when they are ingested by susceptible men or other living organisms) or by their metabolites like the toxic products excreted to water bodies. Some of these contaminants of microbiological origin are subject to limiting values in waters.

Legal requirements indicate cultured-based methods to monitor cell activity or highly sophisticated and expensive instrumental-based methods for metabolite detection/quantification. In general, these methods are cumbersome and take too long to produce the desired response; within that time the contamination can move/spread out, while users of recreational waters and possible consumers are at risk of contracting serious infections. Regardless of their sensitivity and selectivity these methods also are unsuitable for routine and on-site applications (An & Carmichael, 1994; Health Protection Agency, 2005; International Organization for Standardization [ISO], 2005).

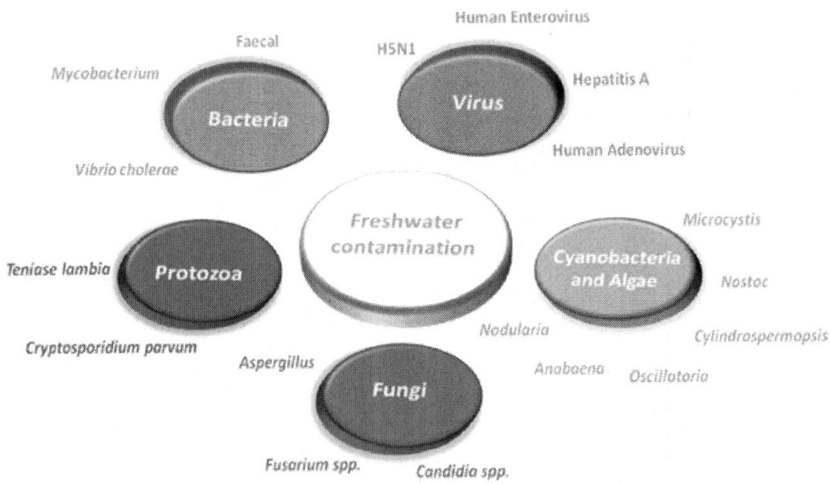

Fig. 1. Microorganisms contaminating freshwater.

Novel technologies for assessing and monitoring microbial water contamination are now becoming available. Biosensors are here of main interest (Figure 2). They have a capture-probe (or biorecognition element) on a standard transduction surface allowing on-site inexpensive determination. The biorecognition element (enzymes, cofactors, cells, antibodies, nucleic acids, or structured polymer) interacts with the target analyte, producing a physical-chemical change that is converted by the transducer into a detectable signal. This signal can be measured optically, mechanically, magnetically, thermally or electrically (Nayak et al., 2009; Su et al., 2011; Turner & Piletsky, 2005; Vo-Dinh, 2007).

Optical or radiant transducers can be classified according to mode or scattering. Classification by mode includes absorption, emission or combination thereof (absorbance or transmission in Ultra-Violet [UV], Visible [Vis], or Infra-Red [IR] region of the spectrum, Attenuated Total Reflectance [ATR], Evanescent Field, Surface Plasmon Resonance [SPR], Luminescence, and Photoemission). Classification by scattering consists on phase change, polarization, absorption and opto-thermal effect (Raman, Ellipsometry, SPR and Photo-Acoustic effect). Mechanical transducers are often called frequency transducers and include Surface Acoustic Wave [SAW], piezoelectric oscillators, and Quartz Crystal Microbalance [QCM]. Magnetic transducers are Nuclear Magnetic Resonance [NMR] and mass spectrometry detection systems. Thermal transducers include calorimetric systems. Electrical transducers can be called electrochemical transducers when the electrical measure is evaluated in a solution. The classification of electric transducers by mode can include voltage (Potenciometric) or current transducers (Amperometric), current-voltage transducers (Voltammetric, Field Effect Transistors [FETs], Metal Oxide Semiconductor [MOS], charge transfer and resistance transducers (coulometric, chemiresistors, ion mobility, and mass spectrometry) and dielectricity transducers (Capacity systems) (Spichiger-Keller, 1998).

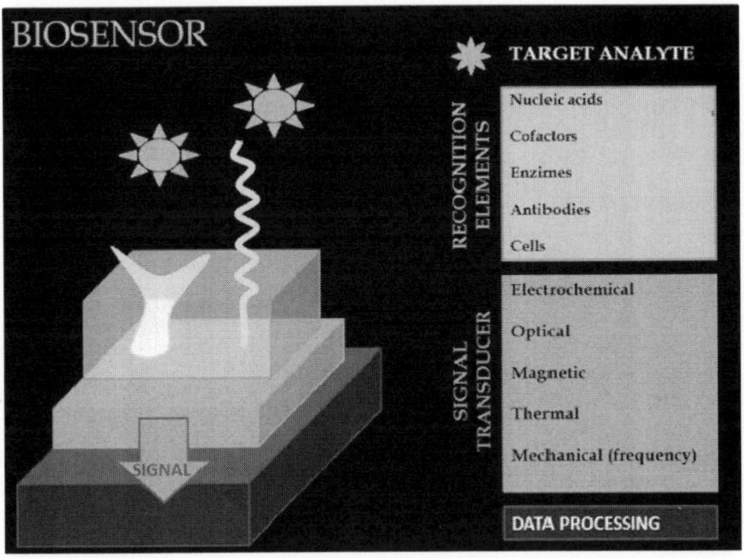

Fig. 2. General biosensor scheme.

Biosensors must meet essential requirements for a successful application. First, the output signal must be of relevance. The biosensor must be accurate and sensitive, providing zero "false-positives" and few "false negatives" and should be reproducible. Moreover the biosensor must be specific, discriminate between the target organism and other organisms. The response time of the biosensor must be inferior to the analytical reference methods, ideally providing a "real-time" response. Also, it must be physically robust and must be able to resist temperature, pH, and ionic strength changes as well as be insensitive to other environmental interferences. The assay should require minimal operator skills for routine detection. One of the most important criteria is the cost of either manufacture, running and life cost. Finally, the biosensing assay must be accurate against standard techniques (Leonard et al. 2003).

Biosensors have been applied by now to the most important forms of microbial contamination: virus, protozoa, bacteria and their toxins, harmful algae and fungi. This chapter will review emerging biosensing technology, methods and applications for assessing and monitoring microbial water contamination. It will focus the development of cost-effective on-site methods, based on biosensing devices that show the potential to complement both laboratory-based and field analytical methods. It will be structured according to biorecognition elements used in the biosensing system, and quote the microorganisms, metabolites or species that are a potential risk to humans; legal requirements; and emerging issues in water and infectious disease.

2. Virus biosensors

The most important waterborne viruses belong to Enteric viruses [EV]. Enterovirus, Hepatitis A virus, Noroviruses, Coxsackie viruses A and B and Rotaviruses are present in gastrointestinal tract of contaminated individuals and are eliminated through feces in larger quantities. Due to its incomplete removal from sewage (even after chlorination) EV are likely to be transmitted to humans from contaminated drinking water (Tavares, 2005). These pathogens can cause a variety of symptoms such as, non-bacterial gastroenteritis, infectious hepatitis, myocarditis and aseptic meningitis. EV can remain viable several months in water in adverse conditions (Tavares et al., 2005; Gilgen et al., 1997).

The most common assays for virus detection use antibody-based methods such as Enzyme-Linked ImmunoSorbant Assay [ELISA] (Bao et al. 2001), fluorescent antibody assays (Barenfanger et al., 2000) and serologic testing (O'Shea et al., 2005). PCR assays are frequently used as complement technique to improve the detection of low concentration of virus (Henkel et al., 1997), despite being a costly and laborious technique. Several authors (Brassard et al., 2005; Gilgen et al. 1997; Soule et al., 2000) presented Reverse Transcriptase - Polymerase Chain Reaction [RT-PCR] as sensitive and efficient method for detection of few EV. Since these diagnostic methods are generally unwieldy and often have limited sensitivity, a variety of new virus detection methods, including microcantilevers (Ilic et al., 2004), evanescent wave biosensors (Donaldson et al., 2004), immunosorbant electron microscopy (Zheng et al., 1999) and atomic force microscopy (Kuznetsov et al., 2005), have emerged to conquer these limitations.

Avian influenza virus [AIV], also known as H5N1, is considered by World Health Organization [WHO] an emerging issue and a potential risk of infection and recommends the monitoring of this virus in waters (WHO & WASH Inter Agency Group, 2007). The natural reservoirs of H5N1 virus are wild waterfowls; these birds may excret large quantities of the virus in their feces as well as in the saliva and nasal secretions. Infected migratory waterfowl may then enter water environments where the birds gather and transmit the low pathogenicity virus to other migratory waterfowl and to domestic birds. During replication, the virus can mutate and highly pathogenic avian influenza [HPAI] strains may occur. The main concern with respect to the HPAI H5N1 virus is that it may change into a form that is highly infectious for humans and that spreads easily from person to person (WHO, 2007).

AIV can persist for extended periods of time. H5N1 has been isolated from unconcentrated waters from lakes in Canada, United States and Hong Kong. There are no quantitative data available on levels of H5N1 virus in lake water where waterfowl gather, although its detection in unconcentrated waters and in small sample volumes suggest relatively high levels. Besides direct deposition of faeces into lake waters by migratory waterfowl, it has been suggested that faecal waste from duck and chicken farms may spread to bodies of water via wind, surface runoff or possibly enter groundwater through disposal and composting of waste on poultry farms (WHO, 2007).

Some authors reported a few new biosensors approaches for the rapid detection of H5N1 virus using impedance measurement of immuno-reaction coupled with red blood cells amplification (Lassiter et al., 2009; Lum & Li, 2010), interferometric immunobiosensor (Xu et al., 2007) and electrochemical-aptamer based biosensor (Liu et al., 2011). Lum & Li developed an impedimetric biosensor for the detection of AIV. The biosensor is composed by immunomagnetic nanoparticles [NPs], a microfluidic chip and an interdigitated microelectrode for impedance measurement. A polyclonal antibody against N1 subtype was immobilized on the surface of the microelectrode to specifically bind H5N1 virus. Therefore, chicken red blood cells were used as biolabels to attach H5N1 captured on the microelectrode to amplify impedimetric signals. A second antibody, also used against N1, offered greater specificity and reliability than the previous one. The results of this study showed that the biosensor was able to detect AIV in less than 2 hours (Lum & Li, 2010).

An optical interferometric waveguide immunoassay for direct and label-less detection of avian influenza virus is described by Xu et al., 2007. The assay is based on refraction index changes that occur upon binding of virus particles to antigen specific antibodies on the waveguide surface. Three virus subtypes were tested using both monoclonal and polyclonal capture antibodies. The detection limits were as low as 0.0005 HA units mL^{-1} (Xu et al., 2007). Liu et al. developed an electrochemical method for the detection of avian influenza virus using and aptamar of DeoxyriboNucleic Acid [DNA] as recognition element immobilized on a hybrid nanomaterial-modified electrode. The modified electrode was assembled with multi-walled carbon nanotubes [MWNT], polypyrrole nanowires [PPNWs] and AuNPs. The detection limit found was 4.3×10^{-13} M. These studies showed that the new hybrid nanomaterial MWNT/PPNWs/AgNPs and the DNA aptamer could be used to fabricate an electrochemical biosensor for gene sequence detection (Liu et al., 2011).

A spectroscopic assay based on Surface Enhanced Raman Scattering [SERS] using silver nanorod array substrates allowing the rapid detection of several human respiratory viruses with a high degree of sensitivity and specificity was developed by Shanmukh et al.. This novel SERS assay can detect spectral differences between viruses, viral strains and viruses with gene deletions in biological media. The method provides rapid detection and characterization of viruses generating reproducible spectra without viral manipulation (Shanmukh et al., 2006).

Persistent infection with hepatitis B virus [HBV] is a major health problem worldwide and may lead to chronic hepatitis, cirrhosis and primary liver cancer (Ye et al., 2003). The detection of HBV DNA in the serum of patient is becoming an important tool in the diagnosis of HBV infection. Two electrochemical detection methods based on DNA hydridization for detection of Hepatitis B viruses were reported by (Erdem et al., 2000 and Ye et al., 2003).

Varíola Virus, also call Smallpox virus [SpV], was irradicated in 1977 but with the increase of unprotected population this virus became an ideal warfare agent. Donaldson and co-workers developed a sensitive and rapid real time immunoassay to detect SpV (Donaldson et al., 2004). It consisted in a polystyrene optical waveguide coated with streptavidin where anti-Vaccinia antibody was attached as a biorecognition element. The detection was performed by a commercial single wavelength fluorometer designed for evanescent wave fluoroimmunoassays. The biosensor was able to detect a minimum of 2.5×10^5 pfu (pock-forming units)/ml of vaccinia virus in seeded throat culture swab specimens (Donaldson, 2004).

But other viruses have been detected rapidly by using biosensing technology. For example, immunosorbent electron microscopy was used to quantify recombinant baculovirus-generated bluetongue virus [BTV] core-like particles in either purified preparations or lysates of recombinant baculovirus-infected cells. The capture was the anti-BTV VP7 monoclonal antibody [MAb]. This technique is simple, rapid and accurate for the quantification of virus produced in large scale in vitro systems (Zheng et al., 1999).

Another example is the use of Atomic Force Microscopy [AFM], under dynamic conditions, for imaging single-stranded genomic RiboNucleic Acid [RNA] from four icosahedral viruses, in which the RNA was observed to unfold (Kuznetsov et al., 2005).

3. Bacteria biosensors

Bacteria are the major pathogen responsible for water-borne disease. Cholera, typhoid fever, bacillary dysentery, leptospirosis and gastroenteriris are some examples of waterborne diseases caused by *Vibrio cholera, Salmonella typhi, Shigella spp., Leptospira spp.,* and Enteropathogenic *Escherichia coli* [EPEC] (Ashbolt, 2004).

Table 1 resumes the latest screening methods for waterborne bacteria; capture and detection methods, limit of detection [LOD] and range of detection are compared (Baudart & Lebaron, 2010; Bharadwaj et al., 2011; Bruno et al., 2010; Chen et al., 2008; Duplan et al., 2011; Fu et al., 2010; Geng et al., 2011; Guven et al., 2011; Huang et al., 2008, 2011; Karsunke et al., 2009; Kwon et al., 2010; Li et al., 2011A; Luo et al., 2010; Miranda-Castro et al., 2009; Park et al. 2008; Sun et al., 2009; Wang et al., 2009; Wilbeboer et al., 2010; Wolter et al., 2008; Yoon et al., 2009; Yu et al., 2009; Xue et al., 2009).

Pathogens transmitted via water are mostly of faecal source (Ashbolt, 2004). Pathogenic *E. coli* strains are responsible for infection of the enteric, urinary, pulmonary and nervous systems. *E. coli* infection is usually transmitted through consumption of contaminated water or food. The routine detection methods for these microorganisms are based on Colony Forming Units [CFUs] count requiring selective culture, or biochemical and serological characterizations. Although bacterial detection by these methods is sensitive and selective, days are needed to get a result. Besides, these methods are costly and time consuming. Because of its great importance as faecal contaminant indicator in waters, the development of biosensors to detect and quantify *E. coli* has been extensively studied and there are a very large number of new methods and improvements to reference methods (Choi et al., 2007; Deobagkar et al., 2005; Gau et al., 2001; Yáñez et al., 2006; Liu & Li, 2002; Simpson & Lim 2005; Tang et al., 2006; Yoo et al., 2007; Yu et al., 2009).

Bacteria	Strain	Methods		LOD	Detection range	References
		Capture	Transduction			
E. coli	O157:H7	Immunological	Electrochemical	10^{2} CFU mL^{-1}	10^{2} - 10^{5} CFU mL^{-1}	Yu et al., 2009
E. coli	O157:H7	Nucleic acid	Electrochemical	0.5 nmol L^{-1}	-	Wang et al., 2009
E. coli	ATCC 8739	Nucleic acid	Optical	30 CFU mL^{-1}	30 - 3x10^{4} CFU mL^{-1}	Bruno et al., 2010
E. coli	unspecified	Nucleic acid	Electrochemical	50 cells mL^{-1}	1.0x10^{2} - 2.0x10^{3} cells mL^{-1}	Geng et al., 2011
E. coli	K12	Immunological	Optical	10^{4} CFU mL^{-1}	10^{2} - 10^{6} CFU mL^{-1}	Duplan et al., 2011
E. coli	O157:H7	Immunological	Electrical	61 CFU mL^{-1}	0 - 10^{4} CFU mL^{-1}	Luo et al., 2010
E. coli	O157:H7	Nucleic acid	Mechanical	1.2x10^{2} CFU mL^{-1}	10^{2} - 10^{6} CFU mL^{-1}	Chen et al., 2008
E. coli	ATCC 15597	Immunological	Optical	10 CFU mL^{-1}	10 - 10^{7} CFU mL-1	Yoon et al., 2009
E. coli	unspecified	Immunological	Optical	8 CFU mL^{-1}	10^{1} - 10^{4} CFU mL-1	Guven et al., 2011
E. coli	several	Nucleic acid	Optical	15 culturable *E. coli* (100 mL)$^{-1}$	3 5 and 14 5 viable E. coli (100 mL)$^{-1}$	Baudart & Lebaron, 2010
E. coli	DSM 30083	Enzimatic	Optical	7 CFU mL^{-1}	100 – 10^{8} CFU mL^{-1}	Wildeboer et al., 2010
E. coli	ACTT 25922	Nucleic acid	Mechanical	8 Cells (800 mL)$^{-1}$	-	Sun et al., 2009
E. coli	K12	Immunological	Optical	100 CFU mL^{-1} (33% Ab coverage) 10 CFU mL^{-1} (50% and 100 % Ab coverage)	-	Kwon et al., 2010
E. coli	BD2399	Immunological	Mechanical	10^{5} CFU mL^{-1}	5x10^{4} - 5x10^{9} CFU mL^{-1}	Fu et al., 2010
E. coli	DH5α	Nucleic acid	Electrochemical	5 CFU mL-1	1x10^{3} - 5x10^{5} CFU mL^{-1}	Li et al., 2011
E. coli	ATCC 35218	Immunological	Optical	< 1000 CFU mL-1	10^{1} – 10^{8} CFU mL^{-1}	Bharadwaj et al., 2011
E. coli	unspecified	TGA	Optical	10^{2} CFU mL^{-1}	10^{2} – 10^{7} CFU mL^{-1}	Xue et al., 2009
E. coli	O157:H7	Immunological	Optical	1.8x10^{3} CFU mL^{-1}	1.8x10^{3} – 1.8x10^{8} CFU mL^{-1}	Park et al., 2008
E. coli	O157:H7	Immunological	Optical	1x10^{4} cells mL^{-1} (single) 3x10^{3} cells mL^{-1} (multi)	1x10^{4} - 1x10^{6} cells mL^{-1} (single) 3x10^{4} - 3x10^{6} cells mL^{-1} (multi)	Wolter et al., 2008
E. coli	O157:H7	Immunological	Optical	1.8x10^{4} cells mL^{-1}	8.5x10^{4}-3.7x10^{6} cells mL^{-1}	Karsunke et al., 2009
E. coli	O157:H7	PU	Mechanical	2x10^{2} cells mL$^{?1}$	2x10^{2} - 3x10^{6} cells ml$^{?2}$	Huang et al., 2008
E. coli	K12 ER2925	Immunological	Electrical	10 cfu mL^{-1}	0 - 10^{5} cfu mL^{-1}	Huang et al., 2010
Vibrio Chorela	O1	Immunological	Optical	1x10^{3} CFU mL^{-1}	1x10^{3} - 1x10^{7} CFU mL^{-1}	Sungkanak et al., 2010
Vibrio Chorela	O1; O139	Immunological	Optical	10^{5} CFU mL^{-1} (O1) (WOE) 10^{7} CFU mL^{-1} (O139) (WOE) 10^{2} CFU mL^{-1} (O1; O139) (WE)	10^{1} -10^{8} CFU mL^{-1}	Yu et al., 2011
Vibrio Chorela	several	Nucleic acid	Optical	10 CFU mL^{-1}	10 - 10^{6} CFU mL^{-1}	Zhou et al., 2011
Leptospira interrogans	several	Nucleic acid	Optical	10 CFU mL^{-1}	10 - 10^{6} CFU mL^{-1}	Zhou et al., 2011
Salmonella spp.	several	Nucleic acid	Optical	1 CFU mL^{-1}	10 - 10^{6} CFU mL^{-1}	Zhou et al., 2011
Salmonella typhimurium	ATCC 14028	Immunological	Optical	3 × 10^{6} cells mL-1 (single/multi)	3 × 10^{6} - 3 × 10^{8} cells mL^{-1} (single) 3 × 10^{6} - 1 × 10^{9} cells mL^{-1} (multi)	Wolter et al., 2008
Salmonella typhimurium	ATCC 14028	Immunological	Optical	2.0×10^{7} cells mL^{-1}	5x10^{7}-1.1x10^{9} cells mL^{-1}	Karsunke et al., 2009
Salmonella typhimurium	unspecified	Immunological	Mechanical	5×10^{3} CFU mL^{-1}	5×10^{3} - 5×10^{8} CFU mL^{-1}	Guntupalli et al., 2007
Mycobacterium sp.	H37rv	Nucleic acid	Electrochemical	1.25 ng mL^{-1}	1.25 - 50 ng mL^{-1}	Thiruppathiraja et al., 2011

LOD – Limit Of Detection; WE – With Enrichment; WOE – WithOut Enrichment. ¿ CFU – Colony Form Unit; Ab – Antibody.

Table 1. Latest screening techniques for bacteria detection in waters.

The immunological methods are the most widely used as recognition methods (Bharadwaj et al., 2011; Duplan et al., 2001; Fu et al., 2010; Guven et al., 2011; Huang et al., 2011; Karsunke et al., 2009; Know et al., 2010; Luo et al., 2010; Park et al., 2008; Wolter et al., 2008; Yoon et al., 2009; Yu et al. 2009), but nucleic acid capture probe are starting to gain some importance in the field (Baudart & Lebaron, 2010; Bruno et al., 2010; Chen et al., 2008; Geng et al., 2011; Li et al., 2011; Sun et al., 2009; Wang et al., 2009). The use of aptamers instead of antibodies [Abs] as capture probes are increasing due to the advantages they present against Abs. Despite its high specificity and affinity, aptamers offer higher chemically stability and can be selected in vitro for a specific target, ranging from small molecules to large proteins and even cells. Moreover, once selected, aptamers can be synthesized with high reproducibility and purity from commercial sources. Furthermore, aptamers often undergo significant conformational changes upon target binding. This offers great flexibility in the design of novel biosensors with high detection sensitivity and selectivity (Song et al., 2008).

Bruno et al. described a Fluorescence Resonance Energy Transfer [FRET]-Aptamers assay applied to *E. coli* detection. In this assay, 25 reverse and 25 forward aptamer candidate sequences against *E. coli* were tested and compared in order to select the most sensitive to SPR and competitive-FRET analysis (Bruno et al., 2010). An iron oxide [Fe_2O_3] gold [Au] core/shell nanoparticle-based electrochemical DNA biosensor was developed for the amperometric detection of *E. coli* by Li et. al., 2011A The assay doesn't require any amplification step and the lowest detection value found was 5 CFU mL^{-1} of *E. coli* after 4.0 h of incubation (Li et al., 2011A). Another aptamer-based biosensor using magnetic particles was reported by Geng et al., 2011. This work included cobalt NPs and an enrichment process was necessary to find 10 *E. coli* cells mL^{-1} in real water samples by differential pulse voltammetry.

A flow piezoelectric biosensor based on synthesized thiolated probe specific to *E. coli* O157:H7 eaeA gene was immobilized onto the piezoelectric biosensor surface. Then, DNA hybridization was induced by exposing the immobilized probe to the *E. coli* O157:H7 eaeA gene fragment, resulting in a mass change and a consequent frequency shift of the piezoelectric biosensor. A second thiolated probe complementary to the target sequence was conjugated to the AuNPs to amplify the frequency change of the piezoelectric biosensor. The products amplified from concentrations of 1.2×10^{2} CFU mL^{-1} of *E. coli* O157:H7 were detectable by the piezoelectric biosensor (Chen et al., 2008).

Sun et al. described a nano silver Indium Tin Oxide [ITO]-coated piezoelectric quartz crystal [PQC] electrode using DNA hybridization to detect *E. coli* cells. Neutravidin and a biotinylated probe were loaded at nano silver ITO-coated PQC and binding ratio was assessed. The binding ratio between the neutravidin and biotinylated DNA probe is increased from 1.00:1.76 of normal PQC to 1.00:3.01 using the nano-silver[Ag]-modified electrode, leading to an increase of more than 71% of the binding capacity of neutravidin to biotinylated DNA probes and an enhancement of 3.3 times for binding complementary DNA onto the nano-Ag-modified neutravidin/biotinylated DNA PQC biosensor. Under the optimized conditions, the detection limit was 0.4 ng/L for DNA Polymerase Chain Reaction [PCR] products or 8 *E. coli* cells in 800 mL in order to detect a single *E. coli* (Sun et al., 2009).

Magnetoelastic biosensors are based on the principle of change in the frequency of magnetoelastic materials (Nayak et al., 2009). Fu et al. developed a magnetostrictive microcantilever using physical absorption as detection system and an Ab against *E. coli* immobilized onto the surface of the microcantilever to form a biosensor. It was found a detection limit of 10^{5} CFU mL^{-1} for a microcantilever with the size of 1.5mm×0.8mm×35μm (Fu et al., 2010).

Immuno-optical biosensors are largely the most detection systems used and many different approaches are developed. A poly(ethylene glycol) [PEG] hydrogel based microchip with patterned nanoporous aluminum oxide membrane [AOM] for bacteria fast patterning and detection with low frequency impedance spectrum was reported by Yu et al. The PEG hydrogel micropatterns on the saline-modified nanoporous alumina surface created controlled spatial distribution of hydrophobic and hydrophilic regions. These microwell arrays were composed of hydrophilic PEG sidewalls and hydrophobic silane-modified AOM bottom. Abs against *E. coli* were added and washed to form the patterns in the microwells. Then, the target bacteria was successfully patterned and captured inside the microwells. The microchip was able to detect bacteria concentrations of 10^2 CFU mL^{-1} (Yu et al., 2009).

Luo et al., 2010 developed a nitrocellulose nanofibers surface functionalized with anti-E. coli Abs on the top of silver electrodes Another Ab were coupled with conductive magnetic NPs and incubated with the test sample for target conjugation. After, the purified sample with the conductive label was dispensed on the application pad (a conductometric lateral flow biosensor). After capillary flow equilibrium, the direct-charge transfer between the electrodes was proportional to captured sandwich complex, which could be used to determine the pathogen concentration. The detection time of the biosensor was 8 min, and the detection limit 61 CFU mL^{-1} (Luo et al., 2010).

A microfluidic device with a portable spectrometer and UV to identify the signal intensity at 375 nm was developed by Know et al., 2010 and Yoon et al., 2009. The device fabricated by Kwon et al. was constructed in acrylic using an industrial-grade milling machine eliminating the need for photolithography and internal or external pumping. An automatic sampling system was built using drip emitters, such that the system could be connected to a pressurized water pipe for real-time detection of *E. coli* (Sungkanak et al., 2010). The microdevice developed by Yoon was fabricated by standard soft lithography with the poly(dimethyl siloxane) [PDMS]. A hole was made through the PDMS to make a view cell. Two cover glass slides were bonded to the top and bottom slides of a view cell using oxygen plasma asher. Two inlets and one outlet were then connected to Teflon tubing. A syringe pump was used to inject anti-E. coli conjugated latex particles and E. coli target solutions into the microchannel device (Yoon et al., 2009). Both devices presented 10 CFU mL^{-1} as detection limit.

A label-free technique based on evanescent wave absorbance changes at 280 nm from a U-bent optical fiber sensor was reported by Bharadwaj et al. Bending a decladded fiber into a U-shaped structure enhanced the penetration depth of evanescent waves and hence the sensitivity of the probe. A portable optical set-up with a UV Light-Emitting Diode [LED], a spectrometer and U-bent optical fiber probe of 200μm diameter, 0.75mm bend radius and effective probe length of 1 cm demonstrated an ability to detect less than 1000 CFU mL^{-1} (Bharadwaj et al., 2011).

A method combining immunomagnetic separation and SERS was developed by Guven et al., 2011. Au-coated magnetic spherical NPs were prepared by immobilizing biotin-labeled anti-*E. coli* Abs onto avidin-coated magnetic NPs and used in the separation and concentration of the *E. coli* cells. Raman labels have been constructed using rod shaped AuNPs coated with 5,5-dithiobis-(2-nitrobenzoic acid) [DTNB] and subsequently with a molecular recognizer. Then DTNB-labeled gold nanorods interacted with the gold-coated magnetic spherical nanoparticle-Ab-*E. coli* complex. The limit of detection value of this method was 8 CFU mL^{-1}.

The enzyme β-D-glucuronidase [GUD] is a specific marker for *E. coli* and 4-methylumbelliferone-β-D-glucuronide [MUG] a sensitive substrate for determining the presence of *E. coli* in a sample. Wildeboer et al., 2010, described a novel hand-held fluorimeter to directly analyse real samples for the presence of *E. coli*. The miniaturized fluorescence detector reduced the incubation time to 30 min and detect *E. coli* as low as 7 CFU mL^{-1} in river water samples (Wildeboer et al., 2010).

The use of carbon allotropes like graphene is of a great potential in biosensing due to their extraordinary electrical, physical and optical properties. Huang et al. reported a graphene based biosensor to electrically detect *E. coli* bacteria with high sensitivity and specificity. The device has a graphene film immobilized with Abs against *E. coli* and a passivation layer. After exposure to *E. coli* bacteria, the graphene device conductance increased significantly. The lowest concentration detected was 10 CFU mL^{-1}. The sensor also detected glucose induced metabolic activities of the bound *E. coli* bacteria in real time (Huang et al., 2011).

Some authors used synthetic materials as recognition materials (Huang et al., 2008; Xue et al., 2009). Xue et al. described a thioglycol acid [TGA] coupled with water-soluble quantum dots as a fluorescence marker for *E. coli* total count. TGA covalently bound to membrane protein of bacteria cells and after 30 min. fluorescence signal are clean seen at a fluorescent microscope (Xue et al., 2009). Huang et al. also reported a synthetic biosensor with polyurethane [PU] as capture probe of *E. coli* cells and a remote-query magnetoelastic system. The resonance frequency of a liquid immersed magnetoelastic sensor, measured through magnetic field telemetry, changed mainly in response to bacteria adhesion to the sensor and the liquid properties of the culture medium. During its growth and reproduction, *E. coli* consumed nutrients from a liquid culture medium that decreased the solution viscosity, and, in turn, changed the resonance frequency of the medium-immersed magnetoelastic sensor (Huang et al., 2008).

Cholera is described by WHO as an acute intestinal infection caused by ingestion of food or water contaminated with *Vibrio cholerae*. It has a short incubation period, from less than one day to five days, and produces an enterotoxin that causes a painless, watery diarrhea that can quickly lead to severe dehydration and death if treatment is not promptly given. Its severity may be confirmed by looking at Haiti recent outbreaks detected after the earthquake. Within 4 days of detecting an unusually high numbers of patients with acute watery diarrhea and dehydration, in some cases leading to death, the National Public Health Laboratory in Haiti isolated *Vibrio cholerae* serogroup O1, from patients in the affected areas of the earthquake. By March 2011, 4,672 people have died and 252,640 cases had been reported (Centers for Diseases Control and prevention [CDC], 2010). Now, cholera outbreaks are being reported since March 2011 along the Congo River, affecting the Democratic Republic of Congo and the Republic of Congo (WHO, 2011)…

Recently, an ultrasensitive microcantilever-based biosensor with dynamic force microscopy for the detection of *Vibrio cholera* O1 with a detection limit of 1×10^3 CFU mL^{-1} and a mass sensitivity, $\Delta m/\Delta F$, of 146.5 pg/Hz was reported by Sungkanak et al. 2010. A dry-reagent gold nanoparticle-based lateral flow biosensor for the simultaneous detection of *Vibrio cholerae* serogroups O1 and O139, based on immunochromatographic principle, was also developed by (Yu et al., 2011).

Legionella pneumophila [L. pneumophila], which is associated with Pontiac fever and legionnaires disease, is commonly found in air conditioning and refrigerating towers, but can also be transported in drinking and freshwaters (Ashbolt, 2004). The detection of *L. pneumophila* in water samples using standard microbiological culture techniques is a delayed process because the bacterium is slow-growing and nutritionally fastidious, to the point that other species may compete with *Legionella* even when antibiotic are supplemented (Cooper et al., 2009).

Miranda-Castro et al. described an electrochemical sensing method for semiquantitative evaluation of *L. pneumophila* (Miranda-Castro et al., 2009). The DNA fragments were amplified by PCR and hybridized to a biotin-labeled reporter sequence and then to a thiolated stem-loop structure immobilized onto gold electrodes as a reporter molecule with 1-naphthyl phosphate as a substrate. 1-Naphthol enzymatically generated was determined by differential pulse voltammetry [DPV].

An immnuosensor using side-polished fiber based on SPR with halogens and LED light source (850 nm LED) for the detection of *L. pneumophila* was reported by (Lin et al., 2007). The SPR curve on the optical spectrum described on the optical spectrum analyzer [OSA] demonstrated a width wavelength of sensing range of SPR effects and sensitive responses. The detection limit was 10^1 CFU mL^{-1} and the detection range of 10^1 to 10^3 CFU mL^{-1} (Lin et al., 2007).

Optical Waveguide Lightmode Spectroscopy [OWLS] technique uses the evanescent field of a He-Ne laser which is coupled into a planar waveguide via an optical grating (Vörös et al., 2002). An OWLS real-time analytical system for the detection of *L. pneumophila* was described by Cooper et al.. An aqueous suspension of *L. pneumophila* was passed across the surface of waveguides functionalized with a specific anti-*Legionella* antibody. The binding between the bacterial cells and the Ab specific for that cell resulted in an increase in the refractive indices of the transverse electric and transverse magnetic photoelectric currents. The assay is presented as a rapid (25 min.) and sensitive (1.3×10^4 CFU mL^{-1}) detection method for *L. pneumophila* contamination in water samples (Cooper et al., 2009).

Regularly, water is contaminated with more than one pathogen, so the development of multi-analyte arrays are quiet relevant. Wolter et al. and Karsunke et al. reported a chemiluminescence microarray for the simultaneous detection of *E. coli, Salmonella typhimurium* and *L. pneumophila* in waters (Karsunke et al., 2009; Wolter et al., 2008).

The immuno-microarray described by Wolter et al. were produced on PEG-modified glass substrates by means of a contact arrayer. The chemiluminescence reaction was accomplished by a streptavidin-horseradish peroxidase catalyzed reaction of luminol and hydrogen peroxide, and was recorded by a sensitive charge coupled device [CCD] camera. The detection limits, in multianalyte experiments, achieved in 13 minutes, were 3×10^6 cells mL^{-1}, 1×10^5 cells mL^{-1}, and 3×10^3 cells mL^{-1} for *S. typhimurium, L. pneumophila,* and *E. coli* O157:H7, respectively (Wolter et al., 2008).

Zhou et al. reported the development of a DNA microarray for detection and identification of *L. pneumophila* and ten other pathogens like *Salmonella spp., Leptospira interrogans* and *Vibrio Cholera,* in drinking water. The combined two-step enrichment procedure allows detection sensitivity of 0.1 ng DNA or 10^4 CFU mL^{-1} achieved for pure cultures of each target organism (Zhou et al., 2011).

Salmonella is related with water and food-borne severe diarrhea. It contaminates more often dairy products from poultry but because these pathogens can be present in animal faeces water resources can as well be contaminated (Karsunke et al., 2009; Nayak et al., 2009; Wolter et al., 2008; Zhou et al., 2011).

Guntupalli et al. developed a magnetoelastic biosensor for *S. typhimurium* in a mixed microbial population. A varying magnetic field was applied to magnetoelastic particles coated with an Au/Chromium [Cr] thin film and a specific Ab in contact with the analyte and the resulting signal was measured. A detection limit of 5×10^3 CFU mL^{-1} and the sensitivity of 139 Hz $decade^{-1}$ were observed for a 2×0.4×0.015 mm sensor (Guntupalli et al., 2007).

About 90 species of Mycobacteria have been found, 20 of which are known to cause disease in humans. Among these, *M. avium, M. intracellulare, M. chelonae, M. kansasii, M. marinum, M. fortuituma and M. ulcerans* are considered potential human pathogens. These are called Non-tuberculosis mycobacteria [NTM] and have been identified in numerous environmental sources, including water. NTM species have been isolated from water sources as well as waste water, surface water, recreational water, ground water and tap water. In this field, there are some recent advances in the development of fast-screening method to detect Mycobacteria. Hunter et al. published a research report about treatment, distribution, standard methods isolation, comparison of culture media and analysis in water (Hunter et al., 2001). Most recent developments in these area report PCR-based detection methods (Jing et al., 2007; Tobler et al., 2006), and demand new emerging biosensing technology.

Recent efforts for detecting whole bacterial communities have been made in order to detect possible bacterial pathogens which are not included in the standard monitoring processes (Hong et al., 2010; Kormas et al., 2010; Poitelon et al., 2010; Revetta et al., 2010). The analysis of bacterial 16S rRNA gene diversity in drinking water distribution systems [WDS] can indicate the presence of Proteobacteria even after the chlorine disinfection treatment (Revetta, 2010).

Poitelon and coworkers presented the variations of 16S rDNA phylotypes prior and after chlorination treatments in 2 WDS in France. The 16S rDNA sequences were grouped into operational taxonomic units [OTUs] standards and acquired for each sample. Significant differences were found in terms of structure and composition of the bacterial community before and after the chlorination disinfection. The predominant bacteria found were alpha- and betaproteobacteria (Poitelon et al., 2010). Kormas, 2010, also presented a similar study in the WDS of Trikala city, Greece. This work showed a possible microbiological risk, not from microorganisms that are routinely monitoring, but from Mycobacteria-like Bacteria.

Overall, the wiseness of the ecology of the bacterial communities in the WDS is important to be acquainted with the chlorination resistance of pathogen in these microbiological communities (Kormas et al., 2010).

4. Cyanobacteria and algae biosensors

The toxic metabolites produced by *Cyanobacteria* in surface waters are another potential threat to drinking water. This has been recognized recently after detecting cyanobacterial toxins [cyanotoxins] in a great number of water samples, from nearly every region on earth. Outbreaks of human poisoning attributed to toxic *Cyanobacteria* have been reported, following exposure of individuals to contaminated water, while drinking, swimming and canoeing (WHO, 1999).

Cyanotoxins [CNs] belong to rather diverse groups of chemical substances each of which shows specific toxic mechanisms in vertebrates. Some CNs are strong neurotoxins (anatoxin-a, saxitoxins), others are primarily toxic to the liver (microcystins, nodularin and cylindrospermopsin), and yet others (such as the lipopolysaccharides) appear to cause gastroenteritis (WHO, 1999). Microcystins [MCs] are geographically the most widely CNs distributed in freshwaters. MCs are peptide toxins produced by *Cyanobacteria* populations in high proportions, a typical symptom of eutrophication. Microcystin-LR [MC-LR] is the most typical element of this group and seems to display severe hepatotoxic and carcinogenic activity. MC-LR concentration in waters for human consumption is regulated by environmental Protection Agency [EPA], EU and WHO, who recommended a limit of 1 μg L^{-1} MC-LR in drinking waters (EC, 2000, 2006; WHO, 1999). Current standard methods to monitor MC-LR require sophisticated and expensive procedures and specific laboratorial conditions that take long time to reach the intend result. Within that time the contamination can move/spread out, while users of recreational waters and possible consumers are at risk of contracting serious infections.

A huge number of novel methods and techniques have been developed recently for this purpose. They are presented in table 2 (Almeida et al., 2006; Allum et al., 2008; Campàs et al., 2007; Campàs & Marty, 2007; Dawan et al., 2011; Ding & Mutharasan, 2010; Gregora & Marsálek, 2005; Hu et al., 2008; Lindner et al., 2009; Long et al., 2009, 2010; Loypraseta et al., 2008; Pyo et al., 2005, 2006; Pyo & Jin, 2007; Queirós et al., 2011; Sheng et al., 2007; Wang et al., 2009; Xia et al., 2010, 2011; Zhang et al., 2007). Optical immuno-based techniques are by large the most widely techniques that have been developed for the detection of MCs (Campàs et al., 2007; Hu et al., 2008; Long et al., 2010; Loypraseta et al., 2008; Pyo et al., 2005, 2006; Pyo & Jin, 2007; Sheng et al., 2007; Xia et al., 2010).

The chemiluminescence sensing technique used in microbial biosensors relies on the generation of electromagnetic radiation as light by the release of energy from a chemical reaction using synthetic compounds (highly oxidized species) which respond to the target analyte in a dose-dependent manner (Hu et al., 2008; Lindner et al., 2009; Su et al., 2011). Lindner et al. reported a rapid immunoassay for sensitive detection of MC-LR using a portable chemiluminescence multichannel immunosensor. The sensor device is based on a capillary ELISA technique in combination with a miniaturized fluidics system and uses chemiluminescence as the detection principle. Minimum concentrations of 0.2 μg L^{-1} MC-LR could be measured in spiked buffer as well as in spiked water samples (Wang et al., 2009). Hu et al. also presented a chemiluminescence immunosensor based on gold nanoparticles. The immunoassay included three main steps: indirect competitive immunoreaction, oxidative dissolution of AuNPs, and indirect determination for MCs with Au^{3+}-catalysed luminol chemiluminesent system. The method has an extensive working range of 0.05–10 μg L^{-1}, and a limit of detection of 0.024 μg L^{-1} (Hu et al., 2008).

Evanescent wave fiber optic immunosensors have been developed to determine numerous target compounds based on the principle of immunoreaction and total internal reflect fluorescent. They also show potential advantages such as miniaturization, sensitivity, cost-effectiveness, and capability of real-time rapid measurements (Long et al., 2009). A portable trace organic pollutant analyzer based on the principle of immunoassay and total internal reflection fluorescence was developed by Long et al.. The reusable fiber optic probe surface was produced by covalently immobilizing a MC-LR-ovalbumin conjugate onto a self-assembled thiol-silane monolayer of fiber optic probe through a heterobifunctional reagent. The recovery of MC-LR added to water samples at different concentrations ranged from 80 to 110% with relative standard deviation values less than 5% (Long et al., 2009).

Pyo et al. developed a few number of optical immuno-based assays to detect MC-LR: a gold colloidal immunochromatographic strip, a fluorescence immunochromatographic strip and cartridge and a PDMS microchip using a liquid-cord waveguide as optical detectors and a MAb against MC-LR as recognition element. The detection limits found were 0.05 μg L^{-1}, 0.15 μg L^{-1} and 0.05 μg L^{-1}, respectively (Pyo et al., 2005, 2006; Pyo & Jin, 2007).

Optical sensing techniques, and specially integrated in immunoarrays, are particularly powerful tools in high throughput screening to monitoring multiple analytes simultaneously (Long et al., 2010; Su et al., 2011). Long et al. developed an optical fiber-based immunoarray biosensor for the detection of multiple small analytes (MC-LR and trinitroluene [TNT]). These compounds can be detected simultaneously and specifically within an analysis time of about 10 min for each assay cycle. The limit of detection for MC-LR was 0.04 μg L^{-1}. This compact and portable immunoarray shows good regeneration performance and binding properties, robustness of the sensor surface and accuracy in the measurement of small analytes which can be considered an excellent multiple assay platform for clinical and environmental samples (Long et al., 2010).

A suitable disposable type biosensor for on-site monitoring of MC-LR in environmental waters was described by Zhang et al.. They reported a competitive binding non-separation electrochemical enzyme immunoassay, using a double-sided microporous gold electrode in cartridge-type cells. Mean recovery of MC-LR added to tap water was 93.5%, with a coefficient of variation of 6.6% (Zhang et al., 2007).

Electrochemical detection methods are also commonly used (Campàs et al., 2007; Campàs & Marty, 2007; Dawan et al., 2011; Loypraseta et al., 2008; Pyo & Jin, 2007, Wang et al., 2009; Zhang et al., 2007).

Campàs et al. developed a highly sensitive using both MAb and Polyclonal Ab [PAb] against MC-LR. This amperometric immunosensor was compared with colorimetric immunoassay, PP inhibition assay and High Performance Liquid Chromatography [HPLC]. The amperometric immunosensor simplifies the analysis, and offer faster and cheaper procedures than others assays.

Wang et al. described a simple, rapid, sensitive, and versatile ELISA-SWNT-paper based sensor for detection of MC-LR in waters. This paper described the use of paper saturated with SWNT embebed in a poly(sodium 4-styrenesulfonate) solution where the Abs are after immobilized. The chronoamperometric detection system was tested and performed a LOD of 0.6 μg L^{-1} with a linear range up to 10 μg L^{-1} (Wang et al., 2009).

Methods		Microcystins	LOD (µg L)^{1}	Detection range (µg^{-1}L^{-1}	References
Capture	Transduction				
Immunological	Electrochemical	-LR	0.6	up to 10	Wang. et al., 2009
Immunological	Optical	-LR	0.03	0.1-10	Long et al., 2009
Immunological	Optical	-LR -RR	down to 0.1	0.001-30	Sheng et al., 2007
Immunological	Piezoelectric	-LR	0.1 (river) 1.00E^{3} (tap)	1-100	Ding & Mutharasan, 2010
Immunological	Optical	Any	0.024	0.05–10	Hu et al., 2008
Molecular Imprinting	Optical	-LR	0.3	0.3-1.4	Queirós et al., 2011
Enzimatic	Optical	-LR	1.00E-02	0-100	Almeida et al., 2006
Immunological	Optical	-LR	1.50E-01	0.150-1.6	Pyo & Jin, 2007
Immunological	Electrochemical	-LR	0.10 (MAb) 1.73 (PAb)	10^{-4}-10^{2} (MAb) $10^{?1}$-10^{2} (PAb)	Campàs & Marty, 2007
Immunological	Mechanical	-LR	1	1-1000	Xia et al., 2011
Enzimatic	Optical	-LR	1	0-1000	Allum et al., 2008
Immunological	Optical	-LR	0.2	0-10	Lindner et al., 2009
Enzimatic	Electrochemical Optical	Any	37 (Electrochemical) 2 (Optical)	37-188 (Electrochemical) - (Optical)	Campàs et al., 2007
Immunological	Electrochemical	-LR	0.1	0.01-3.16	Zhang et al., 2007
Immunological	Electrochemical	-LR	7.00E-06	10E^{6}-1	Loypraserta et al., 2008
Immunological	Optical	-LR	1	1-100	Xia et al., 2010
Immunological	Optical	-LR	0.05	0.05-1.2	Pyo et al., 2005
Immunological	Optical	-LR	0.05	0-1.6	Pyo et al., 2006
Immunological	Optical	-LR	0.04	0-1000	Long et al., 2010
Immunological	Electrochemical	-LR	1.00E-06	1.00E^{6}-1.00E^{4}	Dawan et al., 2011

LOD – Limit Of Detection; WE – With Enrichment; WOE – WithOut Enrichment.

Table 2. Emerging techniques for the detection of microcystins in waters.

The use of electrochemical capacitance systems as detection methods in biosensors allows obtaining exceptionally low limits of detection. Loypraserta et al. and Dawan et al. described a label-free immunosensor based on a modified gold electrode incorporated with AgNPs to enhance the capacitive response to MC-LR has been developed. Anti-MC-LR was immobilized on AgNPs bound to a self-assembled thiourea monolayer, and compared with a bare electrode without the modified AgNPs (Dawan et al., 2011; Loypraseta et al., 2008). MC-LR could be determined with a detection limit of 7.0 and 1.0 µg L^{-1} for (Loypraseta et al., 2008) and (Dawan et al., 2011), respectively. Compared with the modified electrode without AgNPs, this assay presented higher sensitivity and lower limit of detection.

The Adda (3-amino-9-methoxy-2,6,8-trimethyl-10-phenyldeca-4,6-dienoic acid) group of MCs is responsible by the infiltration of toxins in the liver cells where MCs irreversibly bind to serine/threonine protein phosphatases type 2A [PP2A] and type 1 [PP1], inhibiting their enzymatic activity (Almeida et al., 2006; Campàs et al., 2007). These enzymes are involved on the dephosphorylation of proteins. Consequently, their inhibition results in hyperphosphorylation and reorganisation of the microfilaments, promoting tumours and liver cancer (Almeida et al., 2006; Campàs et al., 2007). Almeida et al., Allum et al. and Campàs et al. developed enzymatic-based capture probes based on PP1 and PP2A inhibition. Almeida et al. and Allum et al. presented a simple, rapid and reproducible PP1 inhibition colorimetric test, with the first one presenting a detection limit of 1.00E^{-2} μg L^{-1}. Allum et al. showed an optical fluorometric biosensor based on protein phosphate for the detection of MC-LR or diarrhetic shellfish toxins. The immobilised format described was used to evaluate the potential to translate protein phosphate into a prototype biosensor suitable as a regulatory assay allowing faster throughput than a solution assay and in particular, in assessments of sensitivity and reusability (Allum et al., 2008).

A chronoamperometric biosensor was developed by Campàs et al. 2007 also based on the inhibition of the PP2A. The enzyme was immobilized by the use of poly(vinyl alcohol) azide-unit pendant water-soluble photopolymer. The standard inhibition curve has provided a 50% inhibition coefficient [IC50] of 83 μg L^{-1}, which corresponds to a limit of detection of 37 μg L^{-1} (35% inhibition). Real samples were analysed using the developed amperometric biosensor and compared to those obtained by a conventional colorimetric protein phosphatase inhibition [PPI] assay and HPLC. The results demonstrated that the developed amperometric biosensor may be used as screening method for MCs detection (Campàs et al., 2007).

A dendritic surfactant for MC-LR detection by double amplification in a QCM biosensor was presented by Xia et al.. For primary amplification, an innovative interface on the QCM was obtained as a matrix by the vesicle layer formed by a synthetic dendritic surfactant. The vesicle matrix was functionalized by a MAb against MC-LR to detect the analyte. The results showed that a detection limit of 100 μg L^{-1} was achieved by the first amplification. A secondary amplification was implemented with anti-MC-LR AuNPs conjugates as probes, which lowered the detection limit for MC-LR to 1 μg L^{-1} (Xia et al., 2011).

A piezoelectric-excited millimeter-sized cantilever sensor was developed by Ding et al. for the sensitive detection of MC-LR in a flow format using MAb and PAb that bind specifically to MC-LR. Monoclonal antibody against MC-LR was immobilized on the sensor surface via amine coupling. As the toxin in the sample water bound to the antibody, resonant frequency decreased proportional to toxin concentration. Positive verification of MC-LR detection was confirmed by a sandwich binding on the sensor with a second antibody binding to MC-LR on the sensor which caused a further resonant frequency decrease (Ding &_Mutharasan, 2010).

Very recently, Queirós et al. described a completely new approach a Fabry-Pérot sensing probe based on an optical fibre tip coated with a MC-LR selective thin film. The membranes were developed by molecular imprinting of MC-LR in a sol-gel matrix that was applied over the tip of the fibre. The sensor showed low thermal effect, thus avoiding the need of temperature control in field applications. It presented a detection limit of 0.3 μg L^{-1} and shows excellent selectivity for MC-LR against other species co-existing with the analyte in environmental waters (Queirós et al., 2011).

Saxitoxins were originally isolated from shellfish where they are concentrated from marine dinoflagellates and are one of the causative agents of paralytic shellfish poisoning [PSP] and have caused deaths in humans (Haughey et al., 2011; WHO, 1999; Yakes et al., 2011).

Saxitoxins have been found in the cyanobacteria *Aphanizomenon flos-aquae, Anabaena circinalis, Lyngbya wollei* and *Cylindrospermopsis raciborskii*. Saxitoxins have been found in some countries in diverse cyanobacteria genera, such as *Aphanizomenon flos-aquae* strains NH-1 and NH-5 in North America, C1 and C2 toxins from *Anabaena circinalis* strains in Australia and *Cylindrospermopsis raciborskii* in Brazil (WHO, 1999).

Institute of Agri-Food and Land Use, the US Food and Drug Administration, and the Joint Institute for Food Safety and Applied Nutrition recently reported several SPR platforms biosensor based on inhibition assays to detect PSP toxins (Haughey et al., 2011; Yakes et al., 2011). A saxitoxin PAb (R895) and a MAb (GT13A) were tested and compared. MAb GT13A shows a higher sensitivity 77.8%-100% and was then immobilized into the SPR biosensor surface. The final system provides rapid substrate formation, about 18 h for saxitoxin conjugation with low reagent consumption, contains a reference channel for each assay, and is capable of triplicate measurements in a single run with detection limits well below the regulatory action level (Haughey et al., 2011; Yakes et al., 2011).

Other biosensors were developed for detection of other toxic algae, beyond cyanobacteria, present in environmental waters. Orozco et al. described an electrochemical DNA-based sensor device for detecting toxic algae (*Prymnesium parvum, and Gymnodinium catenatum*). A sandwich hybridization assay was developed with a thiol (biotin) labelled capture probe immobilized onto gold or carbon electrodes, the synthetic DNA was applied to the sensor and allowed to hybridize to the capture probe. A signal probe with HRP label was then applied, followed by an antibody to the HRP and a substrate. The electrical signal obtained from the redox reaction was proportional to the amount of DNA applied to the biosensor (Orozco et al., 2011; Orozco & Medlin, 2011).

Diercks-Horn et al. reported very recently the ALGADEC device which is a semi-automated ribosomal RNA [rRNA] biosensor for the detection of *Alexandrium minutum* toxic algae. The biosensor consisted of a multiprobe chip with an array of 16 gold electrodes for the detection of up to 14 target species. The multiprobe chip was placed inside an automated hybridization chamber, which was in turn placed inside a portable waterproof case with reservoirs for different reagents. The device processed automatically the main steps of the analysis and completed the electrochemical detection of toxic algae in less than 2 h (Diercks-Horn et al., 2011). Furthermore, Diercks et al. reported a colorimetric sandwich hybridization in a microtiter plate assay also for the detection of *Alexandrium minutum*. The system is an adaptation of a commercially available PCR ELISA Dig Detection Kit to be possible the rapid assessment of specificity of the two probes. The mean concentration of RNA per cell of was determined to be 0.028 ng±0.003 (Diercks et al., 2008).

Metfies et al. described an electrochemical DNA-biosensor for the detection of the toxic dinoflagellate *Alexandrium ostenfeldii* (Metfies et al., 2005). This device is similar to the one reported by Diercks-Horn et al. very recently.

5. Fungi biosensors

The knowledge-based concerning the occurrence of fungi in water is low. Fungal contamination of water has been reported for decades, although investigations have been inadequate compared with those of bacteria. Fungi are responsible by infections, allergic reactions and production of toxins [mycotoxins]. Mycotoxins contaminate food and drinks with harsh effects on human and animal health, including cancer, immunological effects and death (Russell et al., 2005). Water, either drinking or nondrinking, may be an effective medium for toxin and biological weapon dispersal. In nondrinking water, the toxin could be spread, from a shower and then inhaled. Workplaces such as farms or car washes where high volumes of water are employed could be susceptible to toxins or fungi. Furthermore, drinking water for animals may be at a considerably higher level of risk than that for humans, thereby increasing the threat to food sources (Russell et al., 2010).

Russell and Paterson proved the production of the mycotoxin zearalenone [ZEN] in drinking waters by *Fusarium graminearum* (Russell & Paterson, 2007). ZEN was purified with an immunoaffinity column and quantified by HPLC with fluorescence detection. The extracellular bear of ZEN was 15.0 ng L^{-1} (Russell et al., 2005; Russell & Paterson, 2007).

Aflatoxins, from Aspergillus flavus, were detected from a cold water storage tank and first reported as the first reported natural occurrence of any mycotoxin in waters (Paterson et al., 1997). Very recently Kattke et al. reported a FRET-based quantum dot immunoassay for rapid and sensitive detection of *Aspergillus amstelodami* (Kattke et al., 2011). The biosensor complex is formed when a quencher-labeled analyte is bound by the antigen-binding site of the quantum dot-conjugated antibody; when excited, the quantum dot will transfer its energy through FRET to the quencher molecules due to their close proximity. With the addition of the target analytes the quencher-labeled analytes dislocation causes disruption of FRET, which translates to increased quantum dot donor emission signal. The optimized displacement immunoassay detected *A. amstelodami* concentrations as low as 10^3 spores mL^{-1} in less than five minutes (Kattke et al., 2011).

Few biosensors for the rapid detection of different mycotoxins in food and beverages samples with special focus to electrochemical detection methods (Alonso-Lomillo et al., 2010; Dinçkaya et al., 2011; S. Li et al., 2011) and enzymes as capture probes (Alonso-Lomillo et al., 2010; S. Li et al., 2011) have been developed recently.

Li et al. presented an amperometric aflatoxin B1 biosensor developed by aflatoxin-oxidase [AFO], embedded in sol-gel, linked to MWCNTs-modified platinum [Pt] electrode (Li et al., 2011B). A DNA-based gold nanoparticles biosensor was developed for detection of aflatoxin M1. A self-assembled monolayer of cysteamine and gold nanoparticles on the SAM were prepared on gold electrodes, layer-by-layer. The assembly processes of cysteamine, gold nanoparticles, and ss-HSDNA were monitored with the help of electrochemical impedance spectroscopy [EIS] and cyclic voltammetry [CV] techniques. The biosensor provided a linear response to aflatoxin M1 over the concentration range of 1–14 ng mL^{-1} with a standard deviation of ±0.36 ng mL^{-1} (Dinçkaya et al., 2011).

Biosensors for the rapid detection of Ochratoxin [Ochra] were also described. Ochra is a group of mycotoxins produced as secondary metabolites by fungi which presents a serious hazard to human and animal health (Metfies et al., 2005; Russell et al., 2010; Yuan et al., 2009).

A sensitive enzyme-biosensor based on screen-printed electrodes was presented by Alonso-Lomillo et al., 2010. A competitive immunoassay with optical detection (SPR using AuNPs) was described by Yuan et al., in whom a mixed self-assembled monolayer was arranged with the immobilization of Ochra through its ovalbumin conjugated with a polyethylene glycol linker (Yuan et al., 2009).

Sapsford et al. reported an array biosensor for the detection of multiple mycotoxins such as ochratoxin A, fumonisin B, aflatoxin B1 and deoxynivalenol in food or beverages. The arrangement of the test was a competitive-based immunoassay, using monoclonal antibodies and the simultaneous detection was performed in less than 15 minutes (Sapsford et al., 2006).

6. Protozoa biosensors

Further on, it is estimated that 2.5 million cases of world annual illness outcome from food and water parasites. Four parasites, *Cryptosporidium parvum, Cyclosporacayetanensis, Giardia lamblia* and *Toxoplasma gondii,* account for the majority of cases, with 71% of waterborne diseases caused by *G. lamblia* and *Cryptosporidium* (Ashbolt, 2004; Rasooly & Herold, 2006; Smith & Nichols, 2010).

Frequent methods for protozoa detection are PCR-based assays (Loge et al., 2002; Toze, 1999) and more recently immunomagnetic separation and fluorescent assays (Ferrari et al., 2006; Mons et al., 2009). These techniques are used to evaluate the microbiological quality of water, and to assess the efficiency of pathogen removal in drinking and wastewater treatment plants (Girones et al., 2010).

Li et al., 2011B, reported a most likely biosensor approached for the detection of *G. lamblia* cysts based on the catalytic growth of AuNPs (X.X. Li et al., 2009). The assay can be described in 7 steps. First the sample is transferred and secondly concentrated; the third step is the incubation of the AuNPs immobilized with a MAb against *G. lamblia* as capture probe, the free gold probes are separated and the sample are resuspended and left growing; the seventh step is the detection by UV-VIS. Detection limit of *G. lamblia* cysts was determined as low as $1.088X10^{3}$ cells ml^{-1} (X.X. Li et al., 2009).

7. Conclusion

Although WHO considers H5N1 water contamination an emerging issue, with high potential risks of infection, it seems that most issued waterborne viruses include EV. The toxic metabolites produced by *Cyanobacteria* in surface waters are another potential threat to drinking water, being MC-LR widely studied and target of biosensing development. Despite the low occurrence of fungi in water, mycotoxins have been also reported and target of biosensing technology. *Bacteria* are definitely the major pathogenic specie responsible for water-borne disease. They are mostly from faecal source. *E. coli* has been assayed thoroughly by immunological and nucleic acid methods, coupled to electrochemical or optical detection methods. *Protozoa* can be 0.01 mm to 1.0 mm in *length* and are quite easy to identify, turning out biosensors less desired. They are much less reported and detected mostly by PCR-based assays, which are used to evaluate the microbiological quality of water.

The biosensor schemes for detection of pathogens in waters presented in this chapter are only a few of many emerging devices and assays that are being developed for several applications. Biosensors offer high sensitivity and specificity, allowing the detection of extremely low infection doses and multiple target analyte at once. The integration of several sensors in networks providing multi-analyte detection and their portability of these sensor networks are at the forefront. This allows the fast screening and the prevention of outbreaks and its consequences.

The future directions of the use of biosensors in environmental contaminant analysis may lead to the combination of these sensor networks with wireless signal transmitters for remote sensing at real-time monitoring. Low manufacturing and production costs are also attractive features that may lead the market expansion in a very short time. They are indeed promising tools for field detection and fast screening of pathogens in waters…

8. Nomenclature

Ab	Antibody
Abs	Antibodies
Adda	3-amino-9-methoxy-2,6,8-trimethyl-10-phenyldeca-4,6-dienoic acid
AFM	Atomic Force Microscopy
AFO	Aflatoxin-oxidase
Ag	Silver
AgNPs	Silver NanoParticles
AIV	Avian Influenza Virus
AOM	Aluminium Oxide Membrane
ATR	Attenuated Total Reflectance
Au	Gold
BTV	BlueTongue Virus
CCD	Charge Coupled Device
CFU	Colony Form Unit
CNs	Cyanotoxins
Cr	Chromium
CV	Cyclic Voltammetry
DNA	DeoxyriboNucleic Acid
DPV	Differential Pulse Voltammetry
DTNB	5,5-dithiobis-(2-nitrobenzoic acid)
E. Coli	*Escherichia coli*
EIS	Electrochemical Impedance Spectroscopy
ELISA	Enzyme-Linked ImmunoSorbant Assay
EPA	Environmental Protection Agency
EPEC	EnteroPathogenic Escherichia coli
EU	European Union
EV	Enteric Viruses
FETs	Field Effect Transistors
FRET	Fluorescence Resonance Energy Transfer
GUD	b-D-glucuronidase
HBV	Hepatitis B Virus

HPAI	Highly Pathogenic Avian Influenza
HPLC	High Performance Liquid Chromatography
HRP	HorseRadish Peroxidase
IC50	half maximal Inhibitory Concentration
IR	Infra-Red
ITO	Indium tin oxide
LED	Light-Emitting Diode
LOD	Limit Of Detection
MAb	Monoclonal Antibody
MC-LR	Microcystin-LR
MCs	Microcystins
MOS	Metal Oxide Semiconductor
MUB	4-methylumbelliferone-b-D-glucuronide
MWNT	Multi-Walled carbon NanoTubes
NMR	Nuclear Magnetic Ressonance
NPs	Nanoparticles
NTB	Non-Tuberculosis Mycobacteria
NWs	NanoWires
Ochra	Ochratoxin
OSA	Optical Spectrum Analyzer
OWLS	Optical Waveguide Lightmode Spectroscopy
PAb	Polyclonal Antibody
PCR	Polymerase Chain Reaction
PEG	Poly(ethylene glycol)
Pfu	Pock-forming units
PMDS	Polydimethylsiloxane
PP	PolyPyrrole
PP1	Protein Phosphatase 1
PP2A	Protein Phosphatase 2A
PPI	Protein Phosphatase Inhibition
PPNWs	PolyPyrrole NanoWires
PQC	Piezoelectric Quartz Crystal
PSP	Paralytic Shellfish Poisoning
Pt	Platinum
PU	Polyurethane
QCM	Quartz Crystal Microbalance
RNA	RiboNucleic Acid
rRNA	ribosomal RNA
RT-PCR	Reverse Transcriptase - Polymerase Chain Reaction
SAW	Surface Acoustic Wave
SERS	Surface Enhanced Raman Scattering
SPR	Surface Plasmon Resonance
SpV	Smallpox Virus
SWNT	Single-Wall NanoTube
TGA	Thioglycol Acid
TNT	Trinitroluene
UV	Ultra-Violet

Vis	Visible
WHO	World Health Organization
WISE	Water Information System for Europe
ZEN	Zearalenone

9. References

Allum, L.L., Mountfort, D.O., Gooneratne, R., Pasco, N., Goussaind G., & Halle, E.A.H. (2008). Assessment of protein phosphatase in a re-usable rapid assay format in detecting microcystins and okadaic acid as a precursor to biosensor development. *Toxicon 52 (7)*, pp. 745-753.

Almeida, V., Cogo, K., Tsai, S.M., & Moon, D.H. (2006). Colorimetric test for the monitoring of microcystins in Cyanobacterial culture and environmental samples from southeast - Brazil. *Brazilian Journal of Microbiology 37*, pp. 192-198.

Alonso-Lomillo, M.A., Domínguez-Renedo, O., Ferreira-Gonçalves L., & Arcos-Martíneza, M.J. (2010). Sensitive enzyme-biosensor based on screen-printed electrodes for Ochratoxin A. *Biosensors and Bioelectronics 25 (6)*, pp. 1333-1337.

An, J., & Carmichael, W.W. (1994). Use of colorimetric protein phosphatase inhition assay and enzime linked immunosorbent assay for the study of microcystins and nodularin. *Toxicon 32*, pp. 1495-1507.

Ashbolt, N.J. (2004). Microbial contamination of drinking water and disease outcomes in developing regions. *Toxicology 198*, pp. 229-238.

Bao, P.-D., Huang, T.-Q., Liu, X.-M., & Wu, T.-Q. (2001). Surface-enhanced Raman spectroscopy of insect nuclear polyhedrosis virus. *J. Raman Spectrosc. 32*, pp. 227-230.

Barenfanger, J., Drake, N., Leon, N., Mueller, T., & Troutt, T. J. (2000). Clinical and Financial Benefits of Rapid Detection of Respiratory Viruses: an Outcomes Study. *Clin. Microbiol. 38*, pp. 2824-2828.

Baudart, J., & Lebaron, P. (2010). Rapid detection of *Escherichia coli* in waters using fluorescent in situ hybridization, direct viable counting and solid phase cytometry. *Journal of Applied Microbiology 109(4)*, pp. 1253–1264.

Bharadwaj, R., Sai, V.V.R., Thakare, K., Dhawangale, A., Kundu, T., Titus, S., Verma, P.K., & Mukherji, S. (2011). Evanescent wave absorbance based fiber optic biosensor for label-free detection of *E. coli* at 280nm wavelength. *Biosensors and Bioelectronics 26*, pp. 3367–3370.

Brassard, J., Seyer, K., Houde, A., Simard, C., & Trottier, Y.-L. (2005). Concentration and detection of hepatitis A virus and rotavirus in spring water samples by reverse transcription-PCR. *Journal of Virological methods 123*, pp. 163-169.

Bruno, J.G., Carrillo, M.P., Phillips, T., & Andrews, C.J. (2010). A Novel Screening Method for Competitive FRET-Aptamers Applied to E. coli Assay Development. *J Fluoresc. 20*, pp. 1211–1223.

Campàs, M., & and Marty, J.-L. (2007). Highly sensitive amperometric immunosensors for microcystin detection in algae. *Biosensors and Bioelectronics 22*, pp. 1034–1040.

Campàs, M., Szydłowska, D., Trojanowicz, M., & Marty, J.-L. (2007). Enzyme inhibition-based biosensor for the electrochemical detection of microcystins in natural blooms of cyanobacteria. *Talanta 72 (1)*, pp. 179-186.

Centers for Diseases Control and Prevention [CDC]. (2010), Morbidity and Mortality Weekly Report, November 19th, Atlanta, USA.

Chen, S.-H., Wu, V.C.H., Chuang, Y.-C., & Lin, C.-S. (2008). Using oligonucleotide-functionalized Au nanoparticles to rapidly detect foodborne pathogens on a piezoelectric biosensor. *Journal of Microbiological Methods 73*, pp. 7–17.

Choi, J.W., Lee, W., Lee, D.B., Park, C.H., Kim, J.S., Jang, Y.H., & Kim, Y. (2007). Electrochemical Detection of Pathogen Infection Using Cell Chip. *Environ. Monit. Assess. 129*, pp. 37–42.

Cooper, I.R., Meikle, S.T., Standen, G., Hanlon, G.W., & Santin, M. (2009). The rapid and specific real-time detection of *Legionella pneumophila* in water samples using Optical Waveguide Lightmode Spectroscopy. *Journal of Microbiological Methods 78*, pp. 40–44.

Dawan, S., Kanatharana, P., Wongkittisuksa, B., Limbut, W., Numnuam, A., Limsakul, C., & Thavarungkul, P. (2011). Label-free capacitive immunosensors for ultra-trace detection based on the increase of immobilized antibodies on silver nanoparticles. *Analytica Chimica Acta 699 (2)*, pp. 232-241.

Deobagkar, D.D., Limaye, V., Sinha, S., & Yadava, R.D.S. (2005). Acoustic wave immunosensing of *Escherichia coli* in water. *Sensors and Actuators B 104*, pp. 85–89.

Diercks, S., Medlin, L.K., & Metfies, K. (2008). Colorimetric detection of the toxic dinoflagellate Alexandrium minutum using sandwich hybridization in a microtiter plate assay. *Harmful Algae 7*, pp. 137-145.

Diercks-Horn, S., Metfies, K., Jäckel, S., & Medlin, L.K. (2011). The ALGADEC device: A semi-automated rRNA biosensor for the detection of toxic algae. *Harmful Algae 10*, pp. 395-401.

Dinçkaya, E., Kınık, Ö., Sezgintürk, M.K., Altuğ, Ç., & Akkoca, A. (2011). Development of an impedimetric aflatoxin M1 biosensor based on a DNA probe and gold nanoparticles. *Biosensors and Bioelectronics 26 (9)*, pp. 3806-3811.

Ding Y., & Mutharasan R. (2010). Highly Sensitive and Rapid Detection of Microcystin-LR in Source and Finished Water Samples Using Cantilever Sensors. *Environmental Science and Technology 45(4)*, pp. 1490-1496.

Donaldson, K.A., Kramer, M.F., & Lim, D.V. (2004). A rapid detection method for Vaccinia virus, the surrogate for smallpox virus. *Biosensors &Bioelectronics 20*, pp. 322-327.

Duplan, V., Frost, E., & Dubowski, J.J. (2011). A photoluminescence-based quantum semiconductor biosensor for rapid in situ detection of Escherichia coli. *Sensors and Actuators B*, doi:10.1016/j.snb.2011.07.010.

European Commission [EC], Directive 2006/7/EC of the European Parliament and of the Council of 2006, February 15th, Concerning the management of bathing water quality and repealing Directive 76/160/EEC, In: *Official Journal of the European Communities, L64*, pp. 37-51.

EC, Council Directive 98/83/EC of 1998, November 3st. On the quality of water intended for human consumption, In: *Official Journal of the European Communities, L330*, pp. 32-54.

EC. Directive 2000/60/CE of 2000, October 23th. Water Frame Directive, In: *Official Journal of European Commission, L327/1.*

European Environment Agency [EEA]. (2010). The European environment, State and outlook 2010, Water resources: quantity and flows, In: *Publications Office of the European Union*, Luxembourg, (ISBN 978 92 9213 162 3).

EEA, Wise TCM hazardous substances. 2011, February 18th. Available from: <http://www.eea.europa.eu/themes/water/interactive/tcm-hs>

Erdem, A., Kerman, K., Meric, B., Akarca, U.S., & Ozsoz, M. (2000). Novel hybridization indicator methylene blue for the electrochemical detection of short DNA sequences related to the hepatitis B virus. *Analytica Chimica Acta 422*, pp. 139–149.

Eurostat, European Commission. (2010). Environmental statistics and accounts in Europe, In: *Eurostat Statistical Book.*

Ferrari, B.C., Stoner, K., & Bergquist, P.L. (2006). Applying fluorescence based technology to the recovery and isolation of Cryptosporidium and Giardia from industrial wastewater streams. *Water Research 40*, pp. 541–548.

Fu, L., Zhang, K., Li, S., Wang, Y., Huang, T.-S., Zhang, A., & Cheng, Z.-Y. (2010). In situ real-time detection of E. coli in water using antibody-coated magnetostrictive microcantilever. *Sensors and Actuators B 150*, pp. 220–225.

Gau J.-J., Lan E.H., Dunn B., Ho C.-M., & Woo J.C.S. (2001). A MEMS based amperometric detector for *E. Coli* bacteria using self-assembled monolayers. *Biosensors & Bioelectronics 16*, pp. 745–755.

Geng, P., Zhang, X., Teng, Y., Fu, Y., Xu, L., Xu, M., Jin, L., & Zhang, W. (2011). DNA sequence-specific electrochemical biosensor based on alginic acid-coated cobalt magnetic beads for the detection of E. coli. *Biosensors and Bioelectronics 26*, pp. 3325–3330.

Gilgen, M., Germann, D., Lüthy, J., & Hübner, Ph. (1997). Three-step isolation method for sensitive detection of enterovirus, rotavirus, hepatitis A virus, and small round structured viruses in water samples. *International Journal of Food Microbiology 37*, pp. 189–199.

Girones, R., Ferrus, M.A., Alonso, J.L., Rodriguez-Manzano, J., Calgua, B., Corrêa, A., Hundesa, A., Carratala, A., & Bofill-Mas, S. (2010). Molecular detection of pathogens in water - The pros and cons of molecular techniques. *Water Research 44*, pp. 4325-4339.

Gregora J., & Marsálek B. (2005). A Simple In Vivo Fluorescence Method for the Selective Detection and Quantification of Freshwater Cyanobacteria and Eukaryotic Algae. *Acta hydrochim. hydrobiol. 33*, pp. 142–148.

Guntupalli, R., Lakshmanan, R.S., Hu, J., Huang, T.S., Barbaree, J.M., Vodyanoy, V., & Chin, B.A. (2007). Rapid and sensitive magnetoelastic biosensors for the detection of *Salmonella typhimurium* in a mixed microbial population. *Journal of Microbiological Methods 70*, pp. 112–118.

Guven, B., Basaran-Akgul, N., Temur, E., Tamer, U., & Boyac, I.H. (2011). SERS-based sandwich immunoassay using antibody coated magnetic nanoparticles for Escherichia coli enumeration. *Analyst 136*, pp. 740–748.

Haughey, S.A., Campbell, K., Yakes, B.J., Prezioso, S.M., DeGrasse, S.L., Kawatsu, K., & Elliott C.T. (2011). Comparison of biosensor platforms for surface plasmon resonance based detection of paralytic shellfish toxins. *Talanta 85 (1)*, pp. 519-526.

Health Protection Agency. (2005). Enumeration of coliforms and Escherichia coli by Idexx (colilert 18) Quanti-trayTM. National Standard Method W 18. <http://www.hpastandardmethods.org.uk/pdf_sops.asp.>

Henkel, J.H., Aberle, S.W., Kundi, M., & Popow-Kraupp, T. (1997). Improved detection of respiratory syncytial virus in nasal aspirates by seminested RT-PCR. *J. Med. Virol. 53*, pp. 366-371.

Hong, P.-Y., Hwang, C., Ling, F., Andersen, G.L., LeChevallier, M.W., Liu, W.-T. (2010). Pyrosequencing analysis of bacterial biofilm communities in water meters of a drinking water distribution system. *Applied and Environmental Microbiology 76*, pp. 5631-5635.

Hu C., Gan N., He Z., & Song L. (2008). A novel chemiluminescent immunoassay for microcystin (MC) detection based on gold nanoparticles label and its application to MC analysis in aquatic environmental samples. *Intern. J. Environ. Anal. Chem. 88*, pp. 267–277.

Huang, S., Pang, P., Xiao, X., He, L., Cai, Q., & Grimes, C.A. (2008). A wireless, remote-query sensor for real-time detection of *Escherichia coli* O157:H7 concentrations. *Sensors and Actuators B 131*, pp. 489–495.

Huang, Y., Dong, X., Liu, Y., Lic, L.-J., & Chen, P. (2011). Graphene-based biosensors for detection of bacteria and their metabolic activities. *Journal of Materials Chemistry 21*, pp. 12358-12362.

Hunter P., Lee J., Nichols G., Rutter M., Surman S., Weldon L., Biegon D., Fazakerley T., Drobnewski F., & Morrell P. (2001). Fate of *Mycobacterium avium Complex* in Drinking Water Treatment and Distribution Systems. Research report DWI 70/2/122.

Ilic, B., Yang, Y., & Craighead, H. G. (2004). Virus detection using nanoelectromechanical devices. *Appl. Phys. Lett.85*, pp. 2604-2606.

International Organization for Standardization [ISO]. (2005). ISO 8199: 2005, Water quality - General Guidance on the enumeration of microorganisms by culture.

Jing G., Polaczyk A., Oerther D.B., & Papautsky I. (2007). Development of a microfluidic biosensor for detection of environmental mycobacteria. *Sensors and Actuators B 123*, pp. 614–621.

Karsunke X.Y.Z., Niessner R., & Seidel, M. (2009). Development of a multichannel flow-through chemiluminescence microarray chip for parallel calibration and detection of pathogenic bacteria. *Anal Bioanal Chem 395*, pp. 1623–1630.

Kattke, M.D., Gao, E.J., Sapsford, K.E., Stephenson, L.D., & Kumar, A. (2011). FRET-Based Quantum Dot Immunoassay for Rapid and Sensitive Detection of Aspergillus amstelodami. *Sensors 11*, pp. 6396-6410.

Kormas, K., Neofitou, C., Pachiadaki, M., & Koufostathi, E. (2010). Changes of the bacterial assemblages throughout an urban drinking water distribution system. *Environmental Monitoring Assess. 165*, pp. 27-38.

Kuznetsov, Y.G., Daijogo, S., Zhou, J., Semler, B.L., & McPherson, A. (2005). Atomic force microscopy analysis of icosahedral virus RNA. *J. Mol. Biol. 347*, pp. 41-52.

Kwon, H.-J., Dean, Z.S., Angus S.V., & Yoon, J.-Y. (2010). Lab-on-a-chip for Field *Escherichia coli* Assays: Long-term Stability of Reagents and Automatic Sampling System. *JALA - Journal of the Association for Laboratory Automation 15*, pp. 216-223.

Lassiter, K., Wang, R., Lin, J., Lum, J., Srinivasan, B., Lin, L., Lu, H., Hargis, B., Bottje, W., Tung, S., Berghman, L., & Li, Y. (2009). Comparison study of an impedance biosensor and rRT-PCR for detection of avian influenza H5N2 from infected chickens. *Program Book of PSA 2009 Annual Meeting*, Raleigh, NC, July, 2009.

Leonard, P., Hearty, S., Brennan, J., Dunne, L., Quinn, J., Chakraborty, T., & O'Kennedy, R. (2003). Advances in biosensors for detection of pathogens in food and water. *Enzyme and Microbial Technology 32*, pp. 3–13.

Li, K., Lai, Y., Zhang, W., & Jin, L. (2011A). Fe_2O_3@Au core/shell nanoparticle-based electrochemical DNA biosensor for *Escherichia coli* detection. *Talanta 84*, pp. 607–613.

Li, S., Chen, J., Cao, H., Yao, D., & Liu, D. (2011B). Amperometric biosensor for aflatoxin B1 based on aflatoxin-oxidase immobilized on multiwalled carbon nanotubes. *Food Control 22 (1)*, pp. 43-49.

Li, X.X., Cao, C., Han, S.J., & Sim, S.J. (2009). Detection of pathogen based on the catalytic growth of gold nanocrystals. *Water Research 43*, pp. 1425–1431.

Lin, H.-Y., Tsao, Y.-C., Tsai, W.-H., Yang, Y.-W., Yan, T.-R., & Sheu, B.-C. (2007). Development and application of side-polished fiber immunosensor based on surface plasmon resonance for the detection of *Legionella pneumophila* with halogens light and 850 nm-LED. *Sensors and Actuators A 138*, pp. 299–305.

Lindner, P., Molz, R., Yacoub-George, E., & Wolf, H. (2009). Rapid chemiluminescence biosensing of microcystin-LR. *Analytica Chimica Acta 636 (2)*, pp. 218-223.

Liu Y., & Li Y. (2002). Detection of *Escherichia coli* O157:H7 using immunomagnetic separation and absorbance measurement. *Journal of Microbiological Methods 51*, pp. 369- 377.

Liu, X., Cheng, Z., Fan, H., Ai, S., & Han R. (2011). Electrochemical detection of avian influenza virus H5N1 gene sequence using a DNA aptamer immobilized onto a hybrid nanomaterial-modified electrode. *Electrochimica Acta 56 (18)*, pp. 6266-6270.

Loge, F.J., Thompson, D.E., Call, D.R. (2002). PCR Detection of Specific Pathogens in Water: A Risk-Based Analysis. *Environ. Sci. Technol. 36*, pp. 2754-2759.

Long F., He M., Zhu A.N., & Shi H.C. (2009). Portable optical immunosensor for highly sensitive detection of microcystin-LR in water samples. *Biosensors and Bioelectronics 24*, pp. 2346–2351.

Long, F., He, M., Zhu, A., Song, B., Sheng, J., & Shi, H. (2010). Compact quantitative optic fiber-based immunoarray biosensor for rapid detection of small analytes. *Biosensors and Bioelectronics 26 (1)*, pp. 16-22.

Loypraserta, S., Thavarungkul, P., Asawatreratanakul, P., Wongkittisuksa, B., Limsakul, C., & Kanatharana, P. (2008). Label-free capacitive immunosensor for microcystin-LR using self-assembled thiourea monolayer incorporated with Ag nanoparticles on gold electrode. *Biosensors and Bioelectronics 24 (1)*, pp. 78-86.

Lum, J., & Li, Y. (2010). Rapid detection of avian influenza H5N1 virus using impedance measurement of immuno-reaction coupled with RBC amplification. Master Thesis, Pub. N.º 1485925, University of Arkansas.

Luo, Y., Nartker, S., Miller, H., Hochhalter, D., Wiederoder, M., Wiederoder, S., Setterington, E., Drzal, L.T., & Alocilja, E.C. (2010). Surface functionalization of electrospun nanofibers for detecting E. coli O157:H7 and BVDV cells in a direct-charge transfer biosensor. *Biosensors and Bioelectronics 26*, pp. 1612–1617.

Metfies, K., Huljic, S., Lange, M., & Medlin, L.K. (2005). Electrochemical detection of the toxic dinoflagellate Alexandrium ostenfeldii with a DNA-biosensor. *Biosensors and Bioelectronics 20*, pp. 1349–1357.

Miranda-Castro, R., de-los-Santos-Álvarez, N., Lobo-Castañón, M.J., Miranda-Ordieres, A.J., & Tuñón-Blanco, P. (2009). PCR-coupled electrochemical sensing of *Legionella pneumophila. Biosensors and Bioelectronics 24*, pp. 2390–2396.

Mons, C., Dumètre, A., Gosselin, S., Galliot, C., & Moulin, L. (2009). Monitoring of Cryptosporidium and Giardia river contamination in Paris area. *Water Research 43*, pp. 211–217.

Nayak, M., Kotian, A., Marathe, S., & Chakravortty, D. (2009). Detection of microorganisms using biosensors- A smarter way towards detection techniques. *Biosensors and bioelectronics 25*, pp. 661-667.

O'Shea, M. K., Ryan, M.A.K., Hawksworth, A.W., Alsip, B.J., & Gray, G.C. (2005). Symptomatic Respiratory Syncytial Virus Infection in Previously Healthy Young Adults Living in a Crowded Military Environment. *Clin. Infect. Dis. 41*, pp. 311-317.

Orozco, J., & Medlin, L.; 2011b. Electrochemical performance of a DNA-based sensor device for detecting toxic algae. *Sensors and Actuactors B 153*, pp. 71–77.

Orozco, J., Baudart, J., & Medlin, L.K. (2011). Evaluation of probe orientation and effect of the digoxigenin-enzymatic label in a sandwich hybridization format to develop toxic algae biosensors. *Harmful Algae 10 (5)*, pp. 489-494.

Park, E.J., Lee, J.-Y., Kim, J.H., Lee, C.J., Kim, H.S., & Min, N.K. (2010). Investigation of plasma-functionalized multiwalled carbon nanotube film and its application of DNA sensor for *Legionella pneumophila* detection. *Talanta 82*, pp. 904–911.

Park, S., Kim, H., Paek, S.-H., Hong, J.W., & Kim, Y.-K. (2008). Enzyme-linked immuno-strip biosensor to detect *Escherichia coli* O157:H7. *Ultramicroscopy 108*, pp. 1348– 1351.

Paterson R.R.M., Kelley J., & Gallagher M. (1997). Natural occurrence of aflatoxins and Aspergillus flavus (LINK) in water. *Lett. Appl. Microbiol. 25*, pp. 435–436.

Poitelon, J.-B., Joyeux, M., Welté, B., Duguet, J.-P., Prestel, E., DuBow, M. (2010). Variations of bacterial 16S rRNA phylotypes prior to and after chlorination for drinking water production from two surface water treatment plants. *J. Ind. Microbiol. Biotechnol.* 37, pp. 117-128.

Pyo, D., & Jin, J. (2007). Production and Degradation of Cyanobacterial Toxin in Water Reservoir, Lake Soyang. *Bull. Korean Chem. Soc. 28 (5)*, pp. 800-804.

Pyo, D., Choi, J., Hong, J., & Oo, H.H. (2006). Rapid Analytical Detection of Microcystins Using Gold Colloidal Immunochromatographic Strip. *Journal of Immunoassay & Immunochemistry 27*, pp. 291–302.

Pyo, D., Huang, Y., Kim, Y., & Hahn, J.H. (2005). Ultra-Sensitive Analysis of Microcystin-LR Using Microchip Based Detection System. *Bull. Korean Chem. Soc. 2005 26 (6)*, pp. 939-942.

Queirós, R.B., Silva, S.O., Noronha, J.P., Frazão, O., Jorge, P., Aguilar, G., Marques, P.V.S., & Sales, M.G.F. (2011). Microcystin-LR detection in water by the Fabry–Pérot interferometer using an optical fibre coated with a sol–gel imprinted sensing membrane. *Biosensors and Bioelectronics 26*, pp. 3932–3937.

Rasooly, A., & Herold, K.E. (2006) Biosensors for the Analysis of Food- and Waterborne Pathogens and Their Toxins. *Journal of AOAC International 89*, pp. 873-883.

Revetta, R., Pemberton, A., Lamendella, R., Iker, B., Santo Domingo, J.W. (2010). Identification of bacterial populations in drinking water using 16S rRNA-based sequence analyses. *Water Research 44*, pp. 1353-1360.

Russell R., Paterson M., & Lima N. (2005). Fungal Contamination of Drinking Water, In: *Water Encyclopedia*, John Wiley and Sons, Inc..

Russell, R., & Paterson, M. (2007). Zearalenone production and growth in drinking water inoculated with Fusarium graminearum. *Mycological Progress 6 (2)*, pp. 109-113.

Russell, R., Paterson, M., & Lima, N. (2010) The Weaponisation of Mycotoxins (Chapter 21). In: *Mycotoxins in Food, Feed and Bioweapons.* Springer-Verlag, Berlin, Heidelberg.

Sapsford, K.E., Ngundi, M.M., Moore, M.H., Lassman, M.E., Shriver-Lake, L.C., Taitt, C.R., & Ligler, F.S. (2006). Rapid detection of foodborne contaminants using an Array Biosensor. *Sensors and Actuators B 113,* pp. 599–607.

Shanmukh, S., Jones, L., Driskell, J., Zhao, Y., Dluhy, R., & Tripp, R.A. (2006). Rapid and Sensitive Detection of Respiratory Virus Molecular Signatures Using a Silver Nanorod Array SERS Substrate. *Nanoletters 6 (11),* pp. 2630-2636.

Sheng, J., He, M., Yu, S., Shi, H., & Qian, Y. (2007). Microcystin-LR detection based on indirect competitive enzyme-linked immunosorbent assay. *Front. Environ. Sci. Engin. China 1(3),* pp. 329–333.

Sigee, D. C. (2005). *Freshwater Microbiology.* John Wiley and Sons, LTD, Manchester, UK.

Simpson, J.M., & Lim, D.V. (2005). Rapid PCR confirmation of *E. coli* O157:H7 after evanescent wave fiber optic biosensor detection. *Biosensors and Bioelectronics 21,* pp. 881–887.

Smith, H.V., & Nichols, R.A.B. (2010). Cryptosporidium: Detection in water and food. *Experimental Parasitology 124,* pp. 61–79.

Song, S., Wang, L., Li, J., Zhao, J., & Fan, C. (2008). Aptamer-based biosensors. *Trends in Analytical Chemistry 27,* pp.108-117.

Soule, H., Genoulaz, O., Gratacap-Cavallier, B., Chevallier, P., Liu, J.-L., & Seigneurin J.-M. (2000). Ultrafiltration ans reverse transcription-polymerase chain reaction: an efficient process for polivirus, rotavirus and hepatitis A virus detection in water. *Water Research 34 (3),* pp. 1063-1067.

Spichiger-Keller U.E. (1998). Chemical and Biochemical Sensors (Chapter 2), In: *Chemical Sensors and Biosensors for Medical and Biological Applications,* Weinheim, WCH-Wiley.

Su, L., Jia, W., Hou, C., & Lei, Y. (2011). Microbial biosensors: a review. *Biosensors and bioelectronics 26,* pp. 1788-1799.

Sun, H., Choy, T.S., Zhu, D.R., Yam, W.C., & Fung, Y.S. (2009). Nano-silver-modified PQC/DNA biosensor for detecting *E. coli* in environmental water. *Biosensors and Bioelectronics 24,* pp. 1405–1410.

Sungkanak, U., Sappat, A., Wisitsoraat, A., Promptmas, C., & Tuantranont, A. (2010). Ultrasensitive detection of *Vibrio cholerae* O1 using microcantilever-based biosensor with dynamic force microscopy. *Biosensors and Bioelectronics 26,* pp. 784–789.

Tang, H., Zhang, W., Geng, P., Wang, Q., Jin, L., Wu, Z., & Lou, M. (2006). A new amperometric method for rapid detection of *Escherichia coli* density using a self-assembled monolayer-based bienzyme biosensor. *Analytica Chimica Acta 562,* pp. 190–196.

Tavares, T., Cardoso, D., & Diederichsen de Brito, W. (2005). Vírus entéricos veiculados por água: aspectos microbiológicos e de controle de qualidade da água. *Revisão 34,* pp. 85-104.

Thiruppathiraja, C., Kamatchiammal, S., Adaikkappan, P., Santhosh, D.J., & Alagar, M. (2011). Specific detection of *Mycobacterium sp.* genomic DNA using dual labeled gold nanoparticle based electrochemical biosensor. *Analytical Biochemistry 417,* pp. 73–79.

Tobler N.E., Pfunder M., Herzog K., Frey J.E., & Altwegg M. (2006). Rapid detection and species identification of *Mycobacterium spp.* using real-time PCR and DNA-Microarray. *Journal of Microbiological Methods 66*, pp. 116– 124.

Toze S. (1999). PCR and the Detection of Microbial pathogens in Water and Wastewater. *Water. Research 33*, pp. 3545-3556.

Turner, A.P.F., & Piletsky S. (2005). Biosensors and biommimetic sensors for the detection of drugs, toxins and biological agents., In: *NATO Security through Science Series*, Springer Ed., Volume 1, pp. 261-272.

Vo-Dinh, T. (2007). Biosensors and biochips, In: *Biomolecular sensing, processing and analysis*, Volume I, pp. 3-20.

Vörös, J., Ramsden, J.J., Csucs, G., Szendrö, I., Textor M., & Spencer, N.D. (2002). Optical Grating Coupler Biosensors. *Biomaterials 23 (17)*, pp. 3699-3710.

Wang L., Chen W., Xu D., Shim, B.S., Zhu Y., Sun F., Liu L., Peng C., Jin Z., Xu C., & Kotov N.A. (2009). Simple, rapid, sensitive, and versatile SWNT- Paper sensor for environmental toxin detection competitive ELISA. *Nano Letters 9(12)*, pp. 4147-4152.

Wang, L., Liu, Q., Hu, Z., Zhang, Y., Wu, C., Yang, M., & Wang, P. (2009). A novel electrochemical biosensor based on dynamic polymerase-extending hybridization for E. coli O157:H7 DNA detection. *Talanta 78*, pp. 647–652.

World Health Organization [WHO] and WASH Inter Agency Group. (2007) Questions & Answers on potential transmission of avian influenza (H5N1) through water, In: *Sanitation and Hygiene and ways to reduce the risks to human health.*

WHO. (1999). Toxic Cyanobacteria in Water: A guide to their public health consequences, monitoring and management. E & FN Spon, London.

WHO. (2006). Review of latest available evidence on potential transmission of avian influenza (H5N1) through water and sewage and ways to reduce the risks to human health, In: *Water, Sanitation and Health Public Health and Environment*, Geneva.

WHO. (2011). Global Alert and Response Programme. Cholera outbreaks: Cholera outbreaks in the Democratic Republic of Congo (DRC) and the Republic of Congo, July 22nd, 2011.

Wildeboer, D., Amirat, L., Price, R.G., & Abuknesha, R.A. (2010). Rapid detection of *Escherichia coli* in water using a hand-held fluorescence detector. *Water Research 44*, pp. 2621–2628.

Wolter, A., Niessner, R., & Seidel, M. (2008). Detection of *Escherichia coli* O157:H7, *Salmonella typhimurium*, and *Legionella pneumophila* in Water Using a Flow-Through Chemiluminescence Microarray Readout System. *Anal. Chem. 80*, pp. 5854–5863.

Xia, Y., Deng, J., & Jiang, L. (2010). Simple and highly sensitive detection of hepatotoxin microcystin-LR via colorimetric variation based on polydiacetylene vesicles. *Sensors and Actuators B: Chemical145 (2)*, pp. 713-719.

Xia, Y., Zhanga, J., & Jiang, L. (2011). A novel dendritic surfactant for enhanced microcystin-LR detection by double amplification in a quartz crystal microbalance biosensor. *Colloids and Surfaces B: Biointerfaces 86 (1)*, pp. 81-86.

Xu, J., Suarez, D., & Gottfried, D. (2007). Detection of Avian Influenza Virus Using An Interferometric Biosensor. *Analytical and Bioanalytical Chemistry 389*, pp. 1193-1199.

Xue, X., Pan, J., Xie, H., Wang, J., & Zhang, S. (2009). Fluorescence detection of total count of *Escherichia coli* and *Staphylococcus aureus* on water-soluble CdSe quantum dots coupled with bacteria. *Talanta 77*, pp. 1808–1813.

Yakes, B.J., Prezioso, S., Haughey, S.A., Campbell, K., Elliott, C.T., & DeGrasse, S.L. (2011) An improved immunoassay for detection of saxitoxin by surface plasmon resonance biosensors. *Sensors and Actuators B: Chemical 156 (2)*, pp. 805-811.

Yáñez M.A, Valor C., & Catalán V. (2006). A simple and cost-effective method for the quantification of total coliforms and *Escherichia coli* in potable water. *Journal of Microbiological Methods 65*, pp. 608–611.

Ye, Y.K., Zhao, J.H., Yan, F., Zhu, Y.L., & Ju, H.X. (2003). Electrochemical behavior and detection of hepatitis B virus DNA PCR production at gold electrode. *Biosensors and Bioelectronics 18*, pp. 1501-1508.

Yoo, S.K., Lee, J.H., Yun, S.-S., Gu, M.B., & Lee, J.H. (2007). Fabrication of a bio-MEMS based cell-chip for toxicity monitoring. *Biosensors and Bioelectronics 22*, pp. 1586–1592.

Yoon, J.-Y., Han, J.-H., Choi, C.Y., Bui, M., & Sinclair, R.G. (2009). Real-Time Detection of Escherichia coli in Water Pipe Using a Microfluidic Device with One-Step Latex Immunoagglutination Assay. *Transactions of the ASABE, 52(3)*, pp. 1031-1039.

Yu J., Liu Z., Liu Q., Yuen K.T., Mak A.F.T., Yang M., & Leung P. (2009). A polyethylene glycol (PEG) microfluidic chip with nanostructures for bacteria rapid patterning and detection. *Sensors and Actuators A 154*, pp. 288–294.

Yu, C.Y., Ang, G.Y., Chua, A.L., Tan, E.H., Lee, S.Y., Falero-Diaz, G., Otero, O., Rodríguez, I., Reyes, F., Acosta, A., Sarmiento, M.E., Ghosh, S., Ramamurthy, T., Yean, C.Y., Lalitha, P., & Ravichandran, M. (2011). Dry-reagent gold nanoparticle-based lateral flow biosensor for the simultaneous detection of *Vibrio cholerae* serogroups O1 and O139. *Journal of Microbiological Methods 86*, pp. 277–282.

Yuan, J., Deng, D., Lauren, D.R., Aguilar, M.-I., & Wu, Y. (2009). Surface plasmon resonance biosensor for the detection of ochratoxin A in cereals and beverages. *Analytica Chimica Acta 656 (1-2)*, pp. 63-71.

Zhang, F., Yang, S.H., Kang, T.Y., Cha, G.S., Nam, H., & Meyerhoff, M.E. (2007). A rapid competitive binding nonseparation electrochemical enzyme immunoassay (NEEIA) test strip for microcystin-LR (MCLR) determination. *Biosensors and Bioelectronics 22 (7)*, pp. 1419-1425.

Zheng, Y.Z., Hyatt, A., Wang, L.F., Eaton, B.T., Greenfield, P.F., & Reid, S.J. (1999). Quantification of recombinant core-like particles of bluetongue virus using immunosorbent electron microscopy. *Virol. Methods 80*, pp. 1-9.

Zhou, G., Wen, S., Liu, Y., Li, R., Zhong X., Feng, L., Wang, L., & Cao, B. (2011). Development of a DNA microarray for detection and identification of *Legionella pneumophila* and ten other pathogens in drinking water. *International Journal of Food Microbiology 145*, pp. 293–300.

5

Macroinvertebrates as Indicators of Water Quality in Running Waters: 10 Years of Research in Rivers with Different Degrees of Anthropogenic Impacts

Cesar João Benetti, Amaia Pérez-Bilbao and Josefina Garrido
University of Vigo
Spain

1. Introduction

The management of running waters is of great importance for the life of our society and one of the challenges to be met by future generations. The sustainable use of water resources for their exploitation in different aspects is essential. Also, the maintenance of good water quality, both sanitary and environmental, is essential, since it depends largely on the conservation of biodiversity (Fernández-Díaz, 2003).

Rivers are ecosystems of great ecological value with a rich fauna that consists of communities with a complex structure and high biological value. However, their special typology makes them fragile and vulnerable to environmental changes, especially those related to disturbances of anthropogenic origin, which often imply irreversible degradation of their biota (Beasley & Kneale, 2003; Dahl et al., 2004).

The vulnerability of these habitats is also evident in relation to the possible effects of climate change. Among the most affected ecosystems are rivers and streams. One of the predictable effects could be that some of these systems will be transformed from permanent to seasonal and some of them will even disappear. In consequence, the biodiversity of many of them will be reduced and their biogeochemical cycles will be altered (Jenkins et al., 1993).

One of the major impacts that affect rivers is the pollution of their waters by both domestic and industrial waste (Benetti & Garrido, 2010). Also agriculture, with intensive use of fertilizers and pesticides, has contributed significantly to eutrophication and contamination of aquatic ecosystems (García-Criado et al., 1999; Paz, 1993). Another important impact on running waters is the deliberate modification of streams by building dams and reservoirs which alter the ecological characteristics of their basins (Richter et al., 1997).

Fluvial ecosystems support very rich and diverse assemblages, with developed adaptations that allow them to prosper in these environments, and which, at the same time, make them very vulnerable to possible alterations in the habitat. In this sense, human activity often causes severe ecological damage to river systems. These disturbances produce alterations in the chemical composition of water and in the structure of the communities of organisms living in this environment (Oller & Goitia, 2005; Smolders et al., 1999, 2003).

Among the fauna of rivers that should be highlighted are macroinvertebrates. This group of great diversity and ecological importance consists of invertebrates of macroscopic size, normally more than 1mm, living permanently or during certain periods of their life cycle linked to the aquatic environment. They include insects, crustaceans, annelids, molluscs, leeches, etc.

Different groups of macroinvertebrates are excellent indicators of human impacts, especially contamination. Most of them have quite narrow ecological requirements and are very useful as bioindicators in determining the characteristics of aquatic environments (Benetti & Garrido, 2010; Fernández-Díaz et al., 2008; Pérez-Bilbao & Garrido, 2009), to identify the segments of a polluted river where self-purification of organic inputs is under process (Chatzinikolaou & Lazaridou 2007).

The Water Framework Directive (WFD 2000/60/EC) establishes common principles to coordinate the efforts of Member States to improve the protection of the European Community aquatic systems, to promote sustainable water use and to protect ecosystems. This directive is intended to prevent the deterioration of all types of water bodies and to ensure that these environments achieve good quality status. The WFD specifies several quality elements necessary for assessing the ecological state of a river. These elements are hydromorphological, physical, chemical and biological. For the latter the composition and abundance of benthic fauna, including invertebrates, are used. The presence/absence of certain taxa defines the quality state of a watercourse. This directive requires that member states of the European Union achieve good quality of all their water bodies by 2015.

So far, the European intercalibration process has produced class boundaries for four out of five types of Mediterranean rivers (R-M1, R-M2, R-M4 and R-M5) (European Commission, 2008) using benthic macroinvertebrates. The officially selected multimetric index for the intercalibration (European Commission, 2008) of the MedGIG rivers is the STAR_ICMi (Buffagni et al., 2006), which is also used by the Central European and Baltic GIG.

Biological monitoring has been established to control water quality. These studies are often based on the sampling of an area and the subsequent analysis of collected specimens that are suitable for monitoring the area and provide information on pollution trends. The structure of the community of one or more of these specimens (Cheimenopoulou et al., 2011) is used in classifying the watercourse ecological quality in a five-class system by using the ecological quality ratio between the observed to the reference conditions or biotic indices/scores.

Among these indices, diversity indices such as Shannon-Wiener or Margalef have been used or indices or scores based on aquatic macroinvertebrates. In Spain it is used the IBMWP score, which uses the presence of taxa and scores for their tolerance to pollution.

The purpose of this chapter is to study macroinvertebrate fauna as bioindicators of water quality in rivers. The questions are if invertebrates are good indicators of water quality in rivers and which are the effects of the impacts of human activities on invertebrate assemblages living in these habitats.

This chapter presents results from studies conducted over 10 years (1998-2008) in 10 rivers in the Autonomous Region of Galicia (North-western Spain) located in areas with different degrees of anthropogenic impacts. The selection of sampling sites was based on land uses near the river banks (woodlands, agriculture, transport system, urban areas, and industrial activities) in connection to some other habitat parameters. Several abiotic variables were also recorded at the same time as fauna was sampled. Benthic macroinvertebrates and their indices were used for the quality assessment. We also analyzed the biotic indices-environment and assemblage composition-environment relationships in order to study responses to structural characteristics of the habitat (natural or artificial) and variations in water quality parameters.

Fig. 1. Site LI3 of the Limia River (Ourense, NW Spain), before the central power station.

2. Water quality and anthropogenic impacts

Freshwater biodiversity provides a broad variety of valuable goods and services for human societies, some of them irreplaceable (Dudgeon et al., 2006), but human activities have always affected aquatic ecosystems. Rivers are highly vulnerable to change caused by anthropogenic impacts, and their flow is often manipulated to provide water for human use (Bredenhand & Samways, 2009). Globally, the biodiversity of freshwater ecosystems is rapidly deteriorating as a result of human activities (Dahl et al., 2004). According to

Dudgeon et al. (2006), there are five major threat categories to freshwater biodiversity: overexploitation, water pollution, flow modification, destruction or degradation of habitat, and invasion by exotic species.

It is possible that in future decades human pressure on water resources will further endangering aquatic biodiversity present in these systems (Strayer, 2006). Overexploitation of rivers and aquifers for irrigation is already a severe problem in many places, especially in the Mediterranean region. In most countries in the south of Europe, irrigation accounts for over 60% of water (Abellán et al., 2006). This activity can lead to drought and the disappearance of inland aquatic habitats (Abellán et al., 2006; Belmar et al., 2010) or changes in physical and chemical characteristics (Velasco et al., 2006).

Contamination due to different types of pollutants such as fertilizers, sewage, heavy metals or pesticides, is a serious problem worldwide. Increasing urbanization and industrialization generates different non-point sources of contamination, causing impairment of water quality of rivers (Beasley & Kneale, 2003). Many studies have dealt with the negative effect of different pollutants on aquatic biota, which results in biodiversity loss and poor water quality (Beasley & Kneale, 2003, 2004; Benetti & Garrido, 2010; Fernández-Díaz et al., 2008; Garrido et al., 1998; Harper & Peckarsky, 2005; Hirst et al., 2002; Lytle & Peckarsky, 2001; Smolders et al., 2003; Song et al., 2009).

Dam construction is one of the most important modifications that rivers are subjected to (Belmar et al., 2010). The general effect is the transformation of dynamic patterns into static, relatively stable ones with reduced flows (Baeza et al., 2003; Benejam et al., 2010; Stanford & Ward, 1979). Changes in marginal vegetation and in flow velocity may produce changes in the composition of aquatic assemblages, with the replacement of some species by others due to the destruction of microhabitats and the creation of new ones (Fulan et al., 2010; Lessard & Hayes, 2003; Sarr, 2011).

Widespread introduction and invasion of exotic species constitutes another human impact on freshwaters. This usually causes the extinction of indigenous species by competition or predation and biotic homogenization (Raehl, 2002). There are many examples of exotic species invasions, for instance the Nile perch, the American signal crayfish or the zebra mussel.

3. Biological indicators

In the past, water quality was assessed using only physicochemical parameters, but these variables only reflect punctual pollution. The use of biological indicators is more adequate to detect long-term changes in water quality, since aquatic organisms are adapted to specific environmental conditions. If these conditions change, some organisms can disappear (intolerant) and be replaced by others (tolerant). Therefore, variations in the composition and structure of aquatic organism assemblages in running waters can indicate possible pollution (Alba-Tercedor, 1996).

Biomonitoring is the use of biological variables to survey the environment (Gerhardt, 2000). The first step in this type of monitoring is to find the ideal bioindicator whose presence, abundance and behavior reflects the effect of a stressor on biota (Bonada et al., 2006b).

Benthic macroinvertebrates are considered as good indicators of local scale conditions (Metcalfe, 1989; Freund & Petty, 2007). These invertebrates live on the bottom of aquatic ecosystems at least part of their life cycle and can be collected using aquatic nets of 500 μm or less (Hauer & Resh, 1996) or ISO 7828 (EN 27828, 1994). They include molluscs, crustaceans, leeches, worms, flatworms and insects (especially larval stages of some orders). Aquatic macroinvertebrates are used to bioassess aquatic ecosystem quality due to their great diversity of form and habits (Rosenberg & Resh, 1993). According to Johnson et al. (1993) a biological indicator has to fulfil different characteristics:

- to be taxonomically easy and well-known,
- to be widely distributed,
- to be abundant and easy to capture,
- to present low genetic and ecological variability,
- to be preferably big,
- to have low mobility and a long life cycle,
- to present well-known ecological characteristics,
- to have the possibility of being used in laboratory studies

Different sampling protocols and metrics are used to evaluate the water quality of rivers and streams. Among them, biotic indices are the most used because they are highly robust, sensitive, cost-effective, easy to apply, and easy to interpret (Bonada et al., 2006a; Chessman et al., 1997; Dallas, 1995, 1997). Biotic indices are tools for assessing quality based on the different response of organisms to environmental changes (Ministry of Environment, 2005). There are many biotic indices developed for different regions, for instance the TBI (Trent Biotic Index, Woodiwis, 1964), the BMWP (Biological Monitoring Working Party) and the ASPT (Average Score per Taxon) (Armitage et al., 1983; National Water Council, 1981) for the UK, the BBI - Belgian Biotic Index (De Pauw & Vanhooren, 1983; Gabriels et al., 2005) for Belgian rivers, the FBILL (Foix, Besòs i Llobregat, Prat et al., 1999) and the IBMWP (Alba-Tercedor et al., 2002) for Spain, or the HES (Hellenic Evaluation System) (Artemiadou & Lazaridou, 2005) for Greece. Many of these indices have to be adapted when they are used in different regions from where they were developed.

One of the most used biotic indices in the Iberian Peninsula is the IBMWP (Iberian Biological Monitoring Working Party) formerly BMWP' (Alba-Tercedor & Sánchez-Ortega, 1988), which is an adaptation of the British BMWP (Armitage et al., 1983) for the Iberian Peninsula. The taxonomic resolution of this index is mostly at family level, and in some cases is even considered at a higher level. Each benthic macroinvertebrate family (or higher taxa) has a score in relation to their tolerance to pollution, so the sum of the scores of the different taxa found in one site gives a total score allowing this sampling site to be classified in one of the five water quality classes (Alba-Tercedor et al., 2002):

- Class I: very good (≥ 101)
- Class II: good (61-100)
- Class III: acceptable (36-60)
- Class IV: poor (15-35)
- Class V: bad (< 15)

Fig. 2. Central power station of the DevaPO River (Pontevedra, NW Spain).

4. Case study

4.1 Introduction

The negative influence of human impacts, especially pollution, on macroinvertebrate fauna has been described in different studies (Blasco et al., 2000; Dahl et al., 2004; Elbaz-Poulichet et al., 1999; Nummelin et al., 2007; Smolders et al., 2003; Yoshimura, 2008; Artemiadou et al., 2008). In Spain, in recent years such studies have increased considerably and several papers studying these types of impacts in different regions of the country have been published. Amongst them we highlight Alonso (2006) in Madrid, Bonada et al. (2000) and Ortiz et al. (2005) in Catalonia, García Criado & Fernández Aláez (2001) in León, or Marqués et al. (2003) in the Basque Country.

Despite its importance, few studies describe the effects of the impact of hydroelectric power stations, for instance Bredenhand & Samways (2009) in South Africa, Kubecka et al. (1997) in the Czech Republic, Jesus et al. (2004) in Portugal, Lessard & Hayes (2003) in Michigan (USA), Stanley et al. (2002) in Wisconsin (USA), Thomson et al. (2005) in Pennsylvania (E.E.U.U.) or Tonkin et al. (2009) in New Zealand. In Spain we must highlight the study by Oscoz et al. (2006) in Navarre, north of the country, which explored both the impacts of pollution and the impacts of hydroelectric power stations.

So far, in Galicia there are few studies to assess human impact on invertebrate fauna. Some of them have analyzed the effects of anthropogenic impacts on water beetle fauna, one of the most important groups of invertebrates. These studies focused mainly on the effects of water pollution (Benetti & Garrido, 2010; Benetti et al., 2007; Pérez-Bilbao & Garrido, 2009), while Sarr (2011) explores the impact of small hydroelectric stations. The results of these studies provide a basis for conducting this study, which focuses on the impact on the entire fauna of macroinvertebrates. Additionally, this study is partly based on technical reports of different environmental monitoring programs developed by the research group at the University of Vigo and the Ingenieria y Ciencia Ambiental S.L. Company (Garrido et al., 1999, 2000a, 2000b, 2000c, 2003, 2005).

Concerning the macroinvertebrates this study (a) assesses the importance of invertebrate fauna as an indicator of water quality in these rivers; (b) identifies the response of macroinvertebrates to human activities; (c) and denotes the responsible factors for the differentiation of the studied rivers.

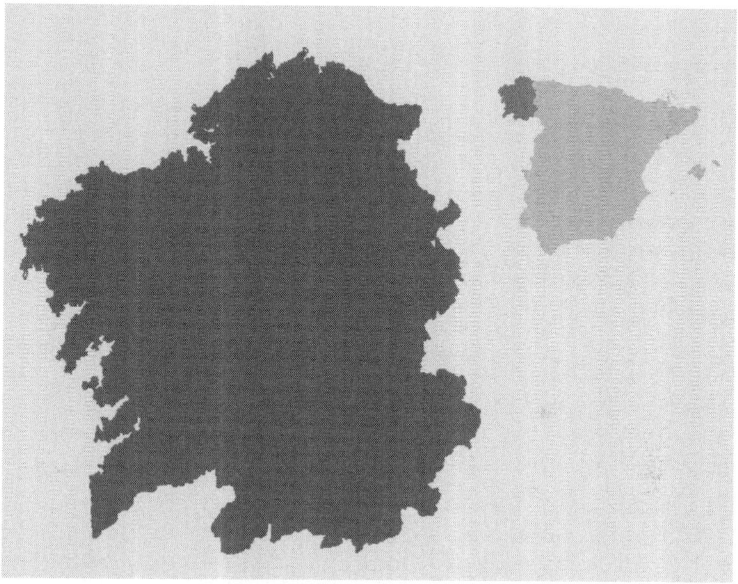

Fig. 3. Map of the study area showing the location of the sampling sites.

4.2 Material and methods

4.2.1 Study area

The study area comprised 10 rivers located in the Autonomous Region of Galicia (Northwestern Spain) (Figure 3). According to the Köppen-Geiger Climate Classification, the climate of the study area is warm temperate, with dry summers and mild temperature (Kottek et al., 2006). This territory belongs to the Atlantic and the Mediterranean biogeographical regions. Altitude ranges from the sea level to around 1,000 m in the highlands. Due to its geographical location, orography and climate, this area has a large number of running waters, from large rivers to small streams (Figure 1). The landscape consists of woodlands (temperate broadleaf forest, pine and eucalyptus reforestations), farmlands, urban and industrial areas. According to the WFD (Annex II, System A), the studied rivers were classified as: Iberian-macaronesian ecoregion, siliceous/calcareous, lowland/mid-altitude and small/medium catchment area.

These rivers are located in areas with different degrees of anthropogenic impacts. The Lagares River (Figure 4) runs almost entirely through the urban area of Vigo, a city with approximately 300,000 inhabitants. This river has undergone a profound change in its structure, especially channelling, as a result of the growth of the city and rapid industrial development with the consequent establishment of industries on its banks. The rest of the rivers run mostly through rural and semi-urban areas. The course of the rivers Furnia and Miñor has not been altered very much, but the other rivers (Avia, DevaOU, DevaPO, Limia, Tambre, Tea and Tuño) have at least one small hydroelectric power station (Figure 2).

4.2.2 Sampling methods and variables measured

In a period of 10 years, between June 1998 and April 2008, 10 rivers were sampled. Each river was sampled four times, in spring, summer, autumn and winter. In each river 3 sites were selected, located in the upper, middle and low stretches (Table 1). The selection of sampling sites was based on land uses (woodlands, agriculture, transport system, urban areas, and industrial activities) and the position of the hydraulic infrastructures, for the regulated rivers. In these cases the first site was located upstream the dam, the second between the dam and the power station, and the third downstream the station.

Sampling was carried out in all types of substrate for a standardized time (one minute). Fauna was collected with an entomological water net of 30 cm diameter, 60 cm depth and 0.5 mm mesh. The specimens were stored in a 4% formaldehyde solution and taken to the laboratory, where they were sorted and identified. After being studied, they were conserved in 70° alcohol and deposited in the scientific collection of the Aquatic Entomology Laboratory at Vigo University.

The following water quality parameters were measured at each site: temperature, dissolved oxygen, pH, electrical conductivity and total dissolved solids (TDS). Additionally, we measured the altitude (meters above sea level) for each site. These above parameters are considered fundamental in the typology of rivers by the WFD.

4.2.3 Data analysis

The structure of the assemblages was assessed using different diversity indices: richness (S), rarefied richness (ES), abundance (N) and the Shannon-Wiener diversity index (H'). The IBMWP biological index was also calculated. The values of rarefied richness were calculated for 100 individuals ES (100). ES and H' were calculated using PRIMER version 6. Analysis of variance (two-way ANOVA) was used to test for significant differences between seasons in both diversity indices and environmental variables. ANOVA was run using SPSS version 19.

		UTM Coordinate			
River	Site	X	Y	Altitude	Sampling Date
Ávia	AV1	567297	4709038	569	1998-1999
Ávia	AV2	571104	4693658	113	1998-1999
Ávia	AV3	572157	4690485	108	1998-1999
Deva OU	DEOU1	575784	4662895	552	1998-1999
Deva OU	DEOU2	574167	4665277	329	1998-1999
Deva OU	DEOU3	572991	4666511	192	1998-1999
Deva PO	DEPO1	558785	4674976	508	2001-2002
Deva PO	DEPO2	558972	4671670	350	2001-2002
Deva PO	DEPO3	558650	4670559	220	2001-2002
Furnia	FU1	524677	4655852	77	2007-2008
Furnia	FU2	525442	4652027	36	2007-2008
Furnia	FU3	525603	4649199	6	2007-2008
Lagares	LA1	529266	4675118	240	2001-2002
Lagares	LA2	524323	4673627	34	2001-2002
Lagares	LA3	520918	4673068	10	2001-2002
Limia	LI1	593494	4654101	610	2003
Limia	LI2	591737	4652248	600	2003
Limia	LI3	588536	4651147	550	2003
Miñor	MI1	525456	4667235	330	2001-2002
Miñor	MI2	519199	4662061	9	2001-2002
Miñor	MI3	517156	4662213	2	2001-2002
Tambre	TA1	536133	4756871	210	1998-1999
Tambre	TA2	534947	4758037	190	1998-1999
Tambre	TA3	534574	4759534	179	1998-1999
Tea	TE1	550473	4678468	192	1999
Tea	TE2	550950	4677584	114	1999
Tea	TE3	550656	4677503	109	1999
Tuño	TU1	580854	4665249	730	1998-1999
Tuño	TU2	580506	4667324	484	1998-1999
Tuño	TU3	579643	4671803	372	1998-1999

Table 1. Sampling sites, with their code, location in UTM coordinates, altitude and sampling date.

Relationships between environmental variables with diversity indices and fauna were determined by a Pearson correlation test. Prior to this, the Kolmogorov-Smirnov test was used to verify the normal distribution of the data. Variables not following the normal distribution were logarithmically transformed (log_{10}). These analyses were performed using SPSS version 19.

Canonical correspondence analysis (CCA) was used to analyze fauna-environment relationships in order to identify environmental factors potentially influencing macroinvertebrate assemblages. A Monte Carlo permutation test was performed to assess the statistical significance of the environmental parameters and the full model to arrive at the significance of the first two canonical axes (Heino, 2000). The environmental factors used were pH, water temperature, dissolved oxygen and conductivity. TDS was not considered because it was redundant with conductivity. CCA was carried out on global abundances, that is, total number of individuals collected at a site over the sampling period. Taxa with less than 10 individuals were removed from the analysis, which was carried out using the CANOCO 4.5 program (Ter Braak & Šmilauer, 2002).

Complete linkage cluster analysis with Bray-Curtis coefficient was used to cluster the rivers into groups and thus be able to verify differences in assemblage composition. The results were represented graphically by Multidimensional scaling (MDS). SIMPER analysis was used to identify which species generate the most similarity within each MDS group. For the SIMPER routine, the raw data were square root transformed and reporting was limited to species with more than 2.5% contribution to dissimilarity. This analysis was carried out with PRIMER version 6.

Fig. 4. Lagares River (Pontevedra, NW Spain) impacted by pollution.

4.3 Results

4.3.1 Diversity and biological indices

In total, 115 taxa (mostly family) of 7 phyla were collected (Table 3). The most representative groups were insects (86 families), especially the orders Diptera (20 families), Trichoptera (19 families) and Coleoptera (12 families).

Only the family Elmidae was recorded in all sites. Other very common families were Baetidae and Chironomidae (recorded in 29 sites), and Nemouridae and Simuliidae (recorded in 28 sites). Besides, 10 families (Bithyniidae, Capniidae, Hebridae, Heteroceridae, Libellulidae, Noteridae, Pyralidae, Scatophagidae, Sciomyzidae and Sisyridae) were only recorded in one site.

If we analyse these results by samples, we can see that site 3 in the Limia River (LI3) presented the highest value of invertebrate richness in autumn with 68 taxa recorded, followed by FU2 (Furnia River) in spring and DEOU1 (DevaOU River) in winter with 55 taxa each. On the other hand, the lowest richness was found in LA3 (Lagares River) in spring, where only 7 taxa were collected, followed by TA2 (Tambre River) in winter and in autumn with 8 and 11 taxa recorded respectively.

Total abundance was 217,577 individuals (185,287 insects, 14,931 Annelida, 9,430 Mollusca and 7,929 other groups). The greatest abundance was observed in the third site in the Limia River in spring, with 29,931 individuals collected. On the other hand, the lowest value was obtained in the third site of the Lagares River in spring with 59 individuals. The most abundant family of macroinvertebrates was Chironomidae with 40,584 individuals recorded. Other abundant families were Ephemerellidae (25,788 individuals), Baetidae (22,860), Elmidae (15,171) and Simuliidae (13,611).

Rarefied species richness is the expected number of species for a given number of randomly sampled individuals and facilitates comparison of areas in which densities may differ (McCabe & Gotelli, 2000). The highest values correspond to the Limia River, sites LI3 (24.18) and LI2 (24.13) in autumn, and to the Furnia River, site FU2 in autumn (23.51) and spring (23.27). On the contrary, the Lagares River had the lowest values in site LA2 (6.62) and site LA3 (6.92).

The Shannon-Wiener index (H'(log2)) revealed that most of the studied rivers presented high diversity values. The lowest diversity was recorded in DEPO3 (DevaPO River) in winter (0.86) and the highest in LI3 (Limia River) in autumn (4.40). In general, the diversity values were high, greater than 3 in 60% of the samples.

According to the IBMWP index, most samples (95%) presented good water quality (> 60, class II), even 87% of the samples presented very good quality, because the index values were above 100 (class I). The highest value (338) was obtained at site FU2 in the Furnia River and the lowest (26) at site LA3 in the Lagares River. Only 2 samples, belonging to the Lagares and Tambre rivers, obtained low values, below 35 and therefore classified as poor quality (class IV). No samples presented bad water quality (< 15, class V).

Table 2 shows the mean minimum and maximum values of the diversity indices for the studied rivers. There were no significant differences ($p < 0.05$) among seasons in any of the diversity indices, as evidenced by the ANOVA.

Richness measures	Mean ± SD	Minimum	Maximum	ANOVA F	p
Richness S	30.98 ± 11.84	7	68	0.128	0.943
Rarefied Richness ES (100)	15.70 ± 4.62	6.62	24.18	0.502	0.682
Abundance N	2175.77 ± 3445.50	59	23931	1.579	0.202
Diversity H'(log2)	3.04 ± 0.75	0.86	4.40	0.535	0.660
IBMWP	174.51 ± 70.47	26	338	0.181	0.909

Table 2. Mean, SD and ranges of biological and diversity indices of the samples and ANOVA with season as factor.

4.3.2 Environmental variables

Table 4 shows mean minimum and maximum values of the environmental variables measured in the 10 studied rivers. The main result to highlight is the high value of conductivity measured in some sites, especially in the Lagares River, higher than in the other surveys. ANOVA showed no significant differences ($p<0.05$) among seasons in almost all environmental variables. This analysis only showed significant differences among seasons in temperature, as expected.

4.3.3 Influence of environmental variables on macroinvertebrate assemblages

The Pearson correlation test was performed to assess the relation between the environmental variables and the taxa and diversity indices. We found several significant correlations ($p<0.05$), but most of them were low ($r<0.5$). We only highlight those that were higher ($r>0.5$). Regarding the diversity indices, there were significant negative correlations between conductivity and rarefied richness ($r=-0.52$) and diversity ($r=-0.50$). For the taxa we found a significant negative correlation between oxygen concentration and Naididae ($r=-0.81$), and a significant positive correlation between conductivity and Hydrobiidae ($r=0.51$).

Figure 5 shows the results of the CCA. The eigenvalues for axes 1-4 were 0.400, 0.198, 0.092 and 0.074 respectively. Correlations for axes III and IV with environmental variables were low ($r<0.5$), and only axes I and II were used for data interpretation. The cumulative percentage of variance for the species-environmental relation for these two axes was 76.1%. The first two canonical axes were significant, as shown by the Monte Carlo permutation test ($p=0.004$).

The first principal axis is positively correlated with conductivity ($r=0.892$) and temperature ($r=0.788$), and negatively with altitude ($r=-0.532$). This component describes water quality and can be an indicator of contamination, since all Lagares river sites are located at the positive end of the axis. The second axis is positively correlated with temperature ($r=0.396$) and negatively with oxygen ($r=-0.786$). Rivers Limia and Miñor are located at the negative end of this axis, indicating low oxygen values. According to the CCA analysis, the family Naididae is related to sites with low oxygen values and the family Hydrobiidae to sites with high conductivity values, while Enchytraeidae prefer sites with high pH values.

Higher taxa	Taxa	Ávia	Deva OU	Deva PO	Furnia	Lagares	Limia	Miñor	Tambre	Tea	Tuño
Hydrozoa	Hydridae		5			1					
Turbellaria	Dugesiidae	16				2	159				
Turbellaria	Planariidae	37	366	256	72	43	14	27			14
Nematoda	Nematoda		11		2	82	53	3			5
Nematomorpha	Gordiidae		2								1
Hirudinea	Erpobdellidae	55	1	16	2	20	84	2	30		
Hirudinea	Glossiphoniidae	10	32			1	15		8	17	753
Hirudinea	Hirudinidae		6		1						
Oligochaeta	Enchytraeidae	1621	735	595		1957	1386	45	726	127	413
Oligochaeta	Lumbricidae	65		162			53		25		
Oligochaeta	Lumbriculidae		25			27	320			3	40
Oligochaeta	Naididae	1	7	492		646	2609	1261	25	220	
Oligochaeta	Oligochaeta unidentified				61						
Oligochaeta	Proppapidae						62				
Oligochaeta	Tubificidae	84		13			72				
Bivalvia	Sphaeriidae	10	23	1	1	11	193	9	104		11
Gastropoda	Ancylidae	20	620	11	10	15	282		33	8	111
Gastropoda	Bithyniidae		1								
Gastropoda	Hydrobiidae	6	84			6960		15	218	2	2
Gastropoda	Lymnaeidae	79	8		13		119		148		5
Gastropoda	Physidae	6				39	213		32		
Gastropoda	Planorbidae	1	1		1		1				
Gastropoda	Valvatidae		2						1		
Crustacea	Cladocera					1	45				
Crustacea	Copepoda		12			10	184				
Crustacea	Ostracoda				1	31		1			
Crustacea	Asellidae		17				233				5
Crustacea	Gammaridae	2	8	11	534	37		70	805		421
Arachnida	Hydracarina	66	1244	188	34	83	2311	65	130	187	22
Collembola	Collembola				8	20	163	3			
Odonata	Aeshnidae	24	64	17	14	2	17		44	22	28
Odonata	Calopterygidae	18	47	26	55	2	21		60	7	17
Odonata	Coenagrionidae	2					14				
Odonata	Cordulegasteridae	12	31	5	5	4	92	3	1	11	47
Odonata	Gomphidae	102	6	3	57	1	709	13	40	54	1
Odonata	Lestidae				2		20				
Odonata	Libellulidae						2				
Plecoptera	Capniidae						1				
Plecoptera	Chloroperlidae	133	7	131	41		6	32	1	6	1
Plecoptera	Leuctridae	430	623	161	272		1789	46	419	73	1458
Plecoptera	Nemouridae	670	871	301	264	11	58	368	30	102	403
Plecoptera	Perlidae	11	48	6	8		2		26	37	50
Plecoptera	Perlodidae	1	2	35	84		2	2		13	33

Ephemeroptera	Baetidae	551	3765	2638	371	2411	6078	3117	1146	850	1933
Ephemeroptera	Caenidae	401				1	1094		256	4	
Ephemeroptera	Ephemerellidae	213	864	591	278	28	22547	42	578	257	390
Ephemeroptera	Ephemeridae		328	4	6		20			3	87
Ephemeroptera	Heptageniidae	180	933	57	45	35	2909	37	82	27	184
Ephemeroptera	Leptophlebiidae	1004	505	2342	258		119	132	13	112	428
Ephemeroptera	Oligoneuriidae		1				253		9		3
Ephemeroptera	Potamanthidae			532							
Hemiptera	Aphelocheiridae				1		215		112	54	
Hemiptera	Corixidae	6					33		1	4	
Hemiptera	Gerridae	13	16	54	17	25	84	11	76	40	20
Hemiptera	Hebridae						3				
Hemiptera	Hydrometridae		9	5			16		2		
Hemiptera	Mesoveliidae	2	1				1				
Hemiptera	Naucoridae				1		47				
Hemiptera	Nepidae				1		4				
Hemiptera	Notonectidae						4				
Hemiptera	Veliidae				1		2		2		
Coleoptera	Dryopidae	3	13	9	1		48	1	36	11	19
Coleoptera	Dytiscidae	9	14	79			799	1	8		4
Coleoptera	Elmidae	663	1899	3615	448	22	4474	200	1702	298	1850
Coleoptera	Gyrinidae		1	51	73		112		11	9	11
Coleoptera	Haliplidae					2	165				
Coleoptera	Helophoridae			1			6		1		
Coleoptera	Heteroceridae						40				
Coleoptera	Hydraenidae	395	500	396	126	1	414	24	15	33	239
Coleoptera	Hydrochidae		1				44				
Coleoptera	Hydrophilidae	4	1	12	1	1	139	2	23	3	
Coleoptera	Noteridae						1				
Coleoptera	Scirtidae		20	25	62		2	1		48	2
Diptera	Anthomyiidae					8	268	1	1	9	
Diptera	Athericidae	132	298	31	35	23		30	200	60	24
Diptera	Blephariceridae		3		3		17				
Diptera	Ceratopogonidae	28	106	189	21	41	2	9	10	8	27
Diptera	Chironomidae	1597	9418	13729	1326	3575	376	5287	1750	620	2906
Diptera	Culicidae						7471	1			
Diptera	Dixidae	6	30	2	5		35			3	4
Diptera	Dolichopodidae	8	123				2		5	6	5
Diptera	Empididae	22	348	221	208	94	1	74	13	81	133
Diptera	Limoniidae	62	57	18	19		1005	11		10	66
Diptera	Psychodidae	11	31	402	13	39	4	13		29	56
Diptera	Ptychopteridae						21				
Diptera	Rhagionidae	3	7	3	3	2	26				1

Diptera	Scatophagidae				2						
Diptera	Sciomyzidae	11									
Diptera	Simuliidae	787	3832	2829	1020	550	2	2740	559	168	1124
Diptera	Stratiomyidae						51				
Diptera	Tabanidae	2	7	2	1		762	3		9	2
Diptera	Thaumaleidae						3				
Diptera	Tipulidae	20	25	9	1	4	2		11	6	12
Lepidoptera	Pyralidae						9				
Megaloptera	Sialidae	29	22	2			2			1	
Neuroptera	Sisyridae						37				
Trichoptera	Beraeidae	5	4					3	19	13	1
Trichoptera	Brachycentridae	18		1496	34		651	6	21		10
Trichoptera	Calamoceratidae	5	24		5		64		175		25
Trichoptera	Ecnomidae		2				4				6
Trichoptera	Glossosomatidae	1	63	8	34	1	1	3			3
Trichoptera	Goeridae	5	148	7	1					1	1
Trichoptera	Helicopsychidae	43	107	15	24	1	1	3			17
Trichoptera	Hydropsychidae	291	837	567	466	82	4367	43	482	118	199
Trichoptera	Hydroptilidae	2		9	14		38	1			
Trichoptera	Lepidostomatidae	25	43	40	111	3		41	1	22	2
Trichoptera	Leptoceridae	49	69	71	12	5	20	3		9	5
Trichoptera	Limnephilidae	39	76	172	7		4416	1	3	5	12
Trichoptera	Philopotamidae	61	98	179	569	4	5421	63		22	9
Trichoptera	Phryganeidae		4				32				
Trichoptera	Polycentropodidae	27	45	94	11		663	4	19	5	53
Trichoptera	Psychomyiidae	5	5	73	5	1	10				
Trichoptera	Rhyacophilidae	59	235	147	56	20	200	33	58	14	124
Trichoptera	Sericostomatidae	59	463	222	124	1	8	3	2	67	173
Trichoptera	Uenoidae	20	16	2	19			20			1

Table 3. Abundance of macroinvertebrates recorded in 10 rivers between 1998 and 2008.

				ANOVA	
Variable	Mean ± SD	Minimum	Maximum	F	p
Water Temperature °C	13.90 ± 3.38	7.90	22.00	70.536	0.000
pH	6.43 ± 0.63	5.20	8.17	0.297	0.828
Conductivity μS/cm	60.95 ± 38.12	20.00	186.00	0.975	0.410
Oxygen %	97.83 ± 5.76	80.40	107.00	1.436	0.239

Table 4. Mean, SD, minimum and maximum values of the environmental variables measured in the studied rivers and ANOVA with season as factor.

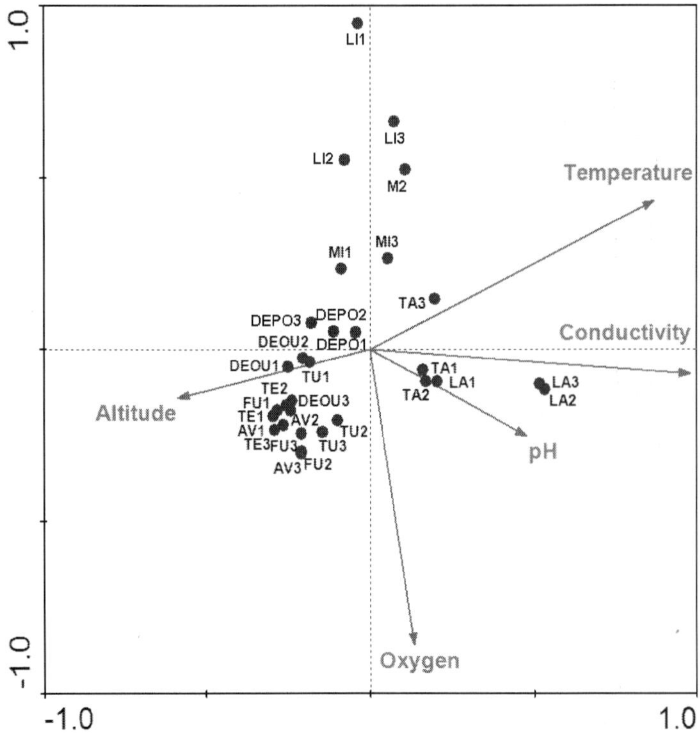

Fig. 5. Results of the canonical correspondence analysis (CCA) based on the invertebrates assemblages with respect to environmental variables. Arrows represent the environmental variables and circles the sites.

4.3.4 Assemblage composition

The Bray-Curtis coefficient was used to calculate the affinity between rivers and the results ranged from 0 to 100. If the value obtained is close to 100, populations will be more similar. According to this coefficient the studied rivers had a 36.32 average faunal affinity. The greatest degree of affinity was observed between the rivers DevaOU and DevaPO (68.20), and between the Tambre and Tuño (61.95). On the other hand, the rivers faunistically farthest were the Limia and Tea, with 7.66 of affinity. MDS analysis provided alternative insights into the similarity of sites with regard to macroinvertebrate assemblage composition. Figure 6 shows the formation of four clearly separated groups with low faunal affinity between them, less than 40%.

The Limia River has an affinity of less than 25% with all the other rivers, forming a group apart and completely separated from the rest. The Lagares River also forms a separate group, with a greater affinity with the Miñor River. Finally, we identified two other groups, one formed by the rivers Miñor, DevaPO and DevaOU, and another formed by the rivers Avia, Furnia, Tambre, Tea and Tuño.

Table 5 lists the species that contributed the greatest amount of similarity in groups A and B. We found a relatively high level of overall similarity in group A (59.26%) and B (50.81%) with the SIMPER analysis. The taxa that contributed most to similarity were the same in both groups. The most important taxon was Chironomidae, contributing with 42.64% similarity in the group A and 23.73% similarity in the group B. Other important taxa were Baetidae, Elmidae and Simuliidae.

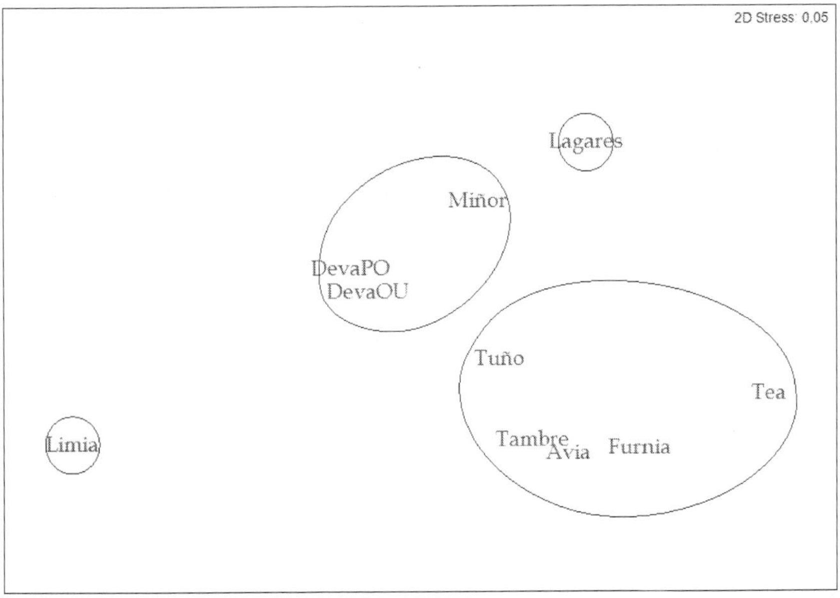

Fig. 6. Multidimensional scaling (MDS) from complete linkage clustering based on the Bray-Curtis coefficient.

4.4 Discussion

The importance of using macroinvertebrates as bioindicators of the water quality of rivers has already been highlighted by several authors (Alba-Tercedor et al., 2002; Alonso, 2006; Bonada et al., 2000; Ortiz et al., 2005; Oscoz et al., 2006). The importance of this group is also reflected in this study because we were able to evaluate the conservation status of these rivers and assess the degree of impact that they are subjected to, whether by pollution or the construction of hydroelectric power stations.

In general, the biological and diversity values observed in the studied sites were high. However, in some places, especially in the Lagares River, these indices values are considerably low in comparison with other rivers in northern Spain (Álvarez-Troncoso, 2004; Fernández-Díaz, 2003; García-Criado, 1999; Paz, 1993; Pérez-Bilbao & Garrido, 2009). We also found a negative correlation between conductivity and richness parameters with a significant decrease in rarefied richness and diversity in sites with high values of this variable, especially in the Lagares River. The reduction in macroinvertebrate richness and abundance in stretches with high values of variables indicative of pollution reflect anthropogenic impacts and corroborates the data obtained by the biological index, which showed poor quality. In relation to this, the results obtained in this study are in concordance with others reported in other papers, which have often documented a decrease in richness and diversity in water bodies with high values of chemical variables (Heino, 2000, Prenda & Gallardo-Mayenco, 1996; Thorne & Williams, 1997).

Taxa	Mean Abundance	Contribution to similarity (%)
Group A (Average similarity: 59.26)		
Chironomidae	9478	42.64
Baetidae	3173.33	18.86
Simuliidae	3133.67	18.48
Elmidae	1904.67	4.34
Group B (Average similarity: 50.81)		
Chironomidae	1639.8	23.73
Baetidae	970.2	13.15
Elmidae	992.2	11.24
Simuliidae	731.6	9.99
Ephemerellidae	343.2	5.74
Leuctridae	530.4	4.7
Hydropsychidae	311.2	4.57
Enchytraeidae	577.4	3.7
Nemouridae	293.8	2.84
Leptophlebiidae	363	2.79
Gammaridae	352.4	2.66

Table 5. SIMPER analysis of macroinvertebrates assemblages of groups A (Miñor, DevaOU and DevaPo rivers) and B (Avia, Furnia, Tambre, Tea and Tuño rivers).

IBMWP index values obtained in the samples of this study, mainly above 100, highlighted the good preservation state of the studied rivers. Besides, it is important to note that the highest value was obtained in the Furnia River, a very little impacted river with excellent water quality. On the contrary, the lowest value was obtained in the Lagares River, which is highly polluted and affected by anthropogenic pressure. The importance of using this index was also pointed out by other studies conducted in the Iberian Peninsula (Alba-Tercedor et al., 2002; Bonada et al., 2006a; Poquet et al., 2009).

In addition to the reduction in richness, abundance and biological indices, pollution also causes a change in faunal composition, as reflected by the separation of the Lagares River from the others in the faunal affinity analysis. In this sense, it is important to note that the dominance of certain taxa, e.g. Naididae and Chironomidae, and absence of others, e.g. Plecoptera, at some sites may indicate the existence of alterations in them (Oscoz et al., 2006).

According to Beasley & Kneale (2003), increasing urbanization and industrialization generates different non-point sources of contamination, causing impairment of water quality of rivers. This environmental impact can be seen in the city of Vigo and its surroundings. High anthropogenic pressure on aquatic ecosystems in this region is a consequence of the ever-increasing population and establishment of industries, especially on the banks of rivers (Benetti & Garrido, 2010).

The response of macroinvertebrates to water pollution seems to define, at least, the species typical of non-contaminated sites. In this sense, ordination analysis identified a group of sensitive taxa especially evident for those most abundant, whose numbers fall considerably in impacted sites. According to CCA, there are also tolerant taxa, for example the Hydrobiidae family, correlated with high values of conductivity, and Naididae, correlated with low values of dissolved oxygen. These results are in agreement with those found in other studies. According to Brinkhurst & Gelder (2001), several aquatic oligochaetes, including Naididae species, have red blood pigments which aid oxygen uptake and transport, thus they can live in environments with low oxygen. Pérez-Quintero (2007) documented that some Hydrobiidae species were salinity-tolerant, so possibly also conductivity-tolerant, as these variables are closely correlated.

Most studies about impacts in rivers (Benetti & Garrido, 2010; Dahl et al., 2004; Nummelin et al., 2007) are mainly focused on the impact assessment of different sources of water pollution, without considering the impacts of infrastructures that change riverbeds, such as the case of small hydro. One of the best ways to evaluate the impact produced by hydroelectric power stations on wildlife is to check changes in the faunal composition and mainly their feeding habits (Argyroudi et al., 2010). In the studied rivers, we observed that the impact produced by hydroelectrics brings about a change in the hydrological regime, mainly the damming, and consequently a change in the community of invertebrates, especially altering its faunal composition, something already noted by other authors for different groups of invertebrates (Bredenhand & Samways, 2009; Jesus et al., 2004; Sarr, 2011; Stanley et al., 2002; Yoshimura, 2008). The structure of macroinvertebrate assemblages is influenced by factors such as the hydrological regime, substrate stability, type and abundance of trophic resources, or land use in the river basin (Dessaix et al., 1995; Quinn et al., 1997; Zamora-Muñoz & Alba-Tercedor, 1996). In our study, both natural characteristics (geology, substrate, water flow) and those artificially created by the impact of hydroelectric infrastructures, determined the structure of the invertebrate assemblages.

The regulation of rivers and hydropower development changes the habitat structure (Dessaix et al., 1995; Dolédec et al., 1996; Fjellheim & Raddum, 1996; Oscoz et al., 2006) and causes the loss of more sensitive taxa and thus imbalance in community structure (Fjellheim et al., 1993). Also, this impact causes the disappearance of many species and otherwise artificially created microhabitats are colonized by other species, perhaps more tolerant to changes or perhaps better adapted to new habitats formed in the river bed and that are not characteristic of their original bed. This may be the case of the Limia River, isolated from the rest and with low faunal affinity, especially in site LI1 (Figure 6), situated upstream the power station, as damming causes the slowdown of the water flow and low levels of oxygen, similar to that observed in stagnant water environments.

In conclusion, in the studied rivers the main factors that determined the macroinvertebrate fauna were (1) water pollution, which directly affected water quality and (2) damming, a disturbance that has caused a change in water flow. In this sense, the negative effect of the anthropogenic impact, especially water contamination, on macroinvertebrates in some of the studied rivers is evident, as shown by the decrease in richness attributes and the IBMWP biological index in impacted sites. As anticipated, there is a loss of species as land use changes from rural to urban. Besides, the water quality in rivers located in natural areas and without strong impacts is better than in rivers located in areas with urban land uses. We also have demonstrated that macroinvertebrates can be used as indicators of environmental impacts in rivers. Their responses to impacts in rivers differ; the majority of taxa are not tolerant to increasing contamination and changes in river structure, but some taxa seem to have adapted to these changes and become dominant in highly disturbed sites. As expected, rare taxa appear to be unmistakeably associated with good water quality, which highlights the importance of conserving freshwater habitats.

Fig. 7. Site LI1 of the Limia River (Ourense, NW Spain), with slow flow caused by damming.

5. References

Abellán, P.; Sánchez-Fernández, D.; Millán, A.; Botella, F.; Sánchez-Zapata, J. A. & Giménez, A. (2006). Irrigation pools as macroinvertebrate habitat in a semi-arid agricultural landscape (SE Spain). *Journal of Arid Environmenats*, 67, 255-269.

Alba-Tercedor, J. & Sánchez-Ortega, A. (1988). Un método rápido y simple para evaluar la calidad biológica de las aguas corrientes basado en el de Hellawell (1978). *Limnetica*, 4, 51–56.

Alba-Tercedor, J. (1996). Macroinvertebrados acuáticos y calidad de las aguas de los ríos. *IV Simposio del Agua de Andalucía (SIAGA)*, 2, 203-213.

Alba-Tercedor, J.; Jáimez-Cuéllar, P.; Álvarez, M.; Avilés, J.; Bonada, N.; Casas, J.; Mellado, A.; Ortega, M.; Pardo, I.; Prat, N.; Rieradevall, M.; Robles, S.; Sáinz-Cantero, C.E.; Sánchez-Ortega, A.; Suárez, M.L.; Toro, M.; Vidalabarca, M.R.; Vivas, S. & Zamora-Muñoz, C. (2002). Caracterización del estado ecológico de ríos mediterráneos ibéricos mediante el índice IBMWP (antes BMWP'). *Limnetica*, 21(3-4), 175-185.

Alonso, A. (2006). Valoración del efecto de la degradación ambiental sobre los macroinvertebrados bentónicos en la cabecera del río Henares. *Ecosistemas*, 15(2), 1-5.

Álvarez-Troncoso, R., (2004). *Estudio faunístico y ecológico de los tricópteros de la cuenca del río Avia (Ourense, España)*, Degree Thesis, Universidad de Vigo, Vigo, Spain.

Argyroudi, A.; Poirazidis, K. & Lazaridou, M. (2010). Macroinvertebrate communities of intermittent and ephemeral streams of Dadia - Lefkimi - Soufli Forest National Park and the impact of small release dams. In: *The Dadia - Lefkimi - Soufli Forest National Park, Greece: Biodiversity, Management and Conservation*, Catsadorakis, G. & Källander, H. (eds). 2010. WWF Greece, Athens.

Armitage, P.D.; Moss, D.; Wright, J.F. & Furse, M.T. (1983). The performance of a new biological water quality score system based on macroinvertebrates over a wide range of unpolluted running water sites. *Water Research*, 17, 333–347.

Artemiadou, V. & Lazaridou, M. (2005). Evaluation Score and Interpretation Index for the ecological quality of running waters in Central and Northern Hellas. *Environmental Monitoring and Assessment*, 10, 1–40.

Artemiadou, V.; Statiri, X.; Brouziotis, T.H. & Lazaridou, M. (2008). Ecological quality of small mountainous Mediterranean streams (river type R-M4) and performance of the European intercalibration metrics. *Hydrobiologia*, 605(1), 75-88.

Baeza, D.; Martínez-Capel, F. & García de Jalón, D. (2003). Variabilidad temporal de caudales: aplicación a la gestión de ríos regulados. *Ingeniería del Agua*, 10(4), 469-478.

Beasley, G. & Kneale, P. (2003). Investigating the influence of heavy metals on macroinvertebrate assemblages using Partial Canonical Correspondence Analysis (pCCA). *Hydrology and Earth Systems Sciences*, 7(2), 221-233.

Beasley, G. & Kneale, P. (2004). Assessment of heavy metal and PAH contamination of urban streambed sediments on macroinvertebrates. *Water, Air and Soil Pollution*, 4, 563-578.

Belmar, O.; Velasco, J.; Martínez-Capel, F. & Marín, A.A. (2010). Natural flow regime, degree of alteration and environmental flows in the Mula stream (Segura River basin, SE Spain). *Limnetica*, 29(2), 353-368.

Benejam, L.; Angermeier, P.L.; Munné, A. & García-Berthou, E. (2010). Assessing effects of water abstraction on fish assemblages in Mediterranean streams. *Freshwater Biology*, 55: 628-642.

Benetti, C.J. & Garrido, J. (2010). The influence of stream habitat and water quality on water beetles assemblages in two rivers in northwest Spain. *Vie et milieu*, 60(1), 53-63.

Blasco, J.; Sáenz, V. & Gómez-Parra, A. (2000). Heavy metal fluxes at the sediment-water interface of three coastal ecosystems from south-west of the Iberian Peninsula. *Science of the Total Environment*, 247, 189-199.

Bonada, N.; Dallas, H.; Rieradevall, M.; Prat, N. & Day, J. (2006a). A comparison of rapid bioassessment protocols used in 2 regions with Mediterranean climates, the Iberian Peninsula and South Africa. *Journal of the North American Benthological Society*, 25(2), 487–500.

Bonada, N.; Prat, N.; Resh, V.H. & Statzner, B. (2006b). Devolopments in aquatic insect biomonitoring: A Comparative Analysis of Recent Approaches. *Annual Reviews of Entomology*, 51, 495 - 523.

Bonada, N.; Rieradevall, M. & Prat, N. (2000). Temporalidad y contaminación como claves para interpretar la biodiversidad de macroinvertebrados en un arroyo mediterráneo (Riera de Sant Cugat, Barcelona). *Limnetica*, 18, 81-90.

Bredenhand, E. & Samways, M.J. (2009). Impact of a dam on benthic macroinvertebrates in a small river in a biodiversity hotspot: Cape Floritic Region, South Africa. *Journal of Insect Conservation*, 13(3), 297-307.

Brinkhurst, R.O. & Gelder, S.R. (2001). Annelida: Oligochaeta inlcuding Branchiobdellidae, In: *Ecology and classification of North American freshwater invertebrates*, Thorp, J. P. & Covich, M. P., 431-464, Academic Press, Orlando, USA.

Buffagni, A. & Furse, M. (2006). Intercalibration and comparison - major results and conclusions from the STAR project. *Hydrobiologia*, 566, 357–364.

Chatzinikolaou, Y. & Lazaridou, M. (2007). Identification of the self-purification stretches of the Pinios River, Central Greece. *Mediterranean Marine Science*, 8 (2), 19-32.

Cheimonopoulou M.; Bobori D.; Theocharopoulos I. & Lazaridou M. (2011). Assessing Ecological Water Quality with Macroinvertebrates and Fish: A CaseStudy from a Small Mediterranean River. *Environmental Management*, 47, 279–290.

Chessman, B.C.; Growns, J.E. & Kotlash, A.R. (1997). Objective derivation of macroinvertebrate family sensitivity grade numbers for the SIGNAL biotic index: application to the Hunter River system, New South Wales. *Australian Journal of Marine and Freshwater Research*, 48, 159–172.

Council of the European Communities (2000). Directive 2000/60/EC of the European Parliament and of the Council of 23 October 2000 establishing a framework for Community action in the field of water policy. *Official Journal of the European Communities*, L327, 1-72.

Dahl, J.; Johnson, R.K. & Sandin, L. (2004). Detection of organic pollution of streams in southern Sweden using benthic macroinvertebrates. *Hydrobiologia*, 516, 161-172.

Dallas, H.F. (1995). *An evaluation of SASS (South African Scoring System) as a tool for the rapid bioassessment of water quality*, MSc Thesis, University of Cape Town, Cape Town, South Africa.

Dallas, H.F. (1997). A preliminary evaluation of aspects of SASS (South African Scoring System) for the rapid bioassessment of water quality in rivers, with particular reference to the incorporation of SASS in a national biomonitoring programme. *South African Journal of Aquatic Sciences*, 23, 79–94.

De Pauw, N. & Vanhooren, G. (1983). Method for biological quality assessment of watercourses in Belgium. *Hydrobiologia*, 100, 153-168.

Dessaix, J.; Fruget, J.F.; Olivier, J.M. & Beffy, J.L. (1995). Changes of the macroinvertebrate communities in the dammed and by-passed sections of the French upper Rhône after regulation. *Regulated Rivers: Research & Management*, 10, 265-279.

Dolédec, S.; Dessaix, J. & Tachet, H. (1996). Changes within the Upper Rhône River macrobenthic communities after the completion of three hydroelectric schemes: anthropogenic effects or natural change? *Archiev für Hydrobiologie*, 136, 19-40.

Dudgeon, D.; Arthington, A.H.; Gessner, M.O.; Kawabata, Z.I.; Knowler, D.J.; Lévêque, C.; Naiman, R.J.; Prieur-Richard, A.H.; Soto, D.; Stiassny, M.L.J. & Sullivan, C.A. (2006). Freshwater biodiversity: importance, threats, status and conservation challenges. *Biological Reviews of the Cambridge Philosophical Society*, 81, 163-182.

Elbaz-Poulichet, F.; Morley, N.H.; Cruzado, A.; Velasquez, Z.; Achterberg, E.P. & Braungardt, C.B. (1999). Trace metal and nutrient distribution in an extremely low pH (2.5) river-estuarine system, the Ria of Huelva (South-West Spain). *Science of the Total Environment*, 227, 73-83.

EN 27828 (1994). Water quality - Methods of biological sampling - *Guidance on handnet sampling of aquatic benthic macro-invertebrates* (ISO 7828:1985).

European Commission, (2008). Commission decision of 30 October 2008 establishing, pursuant to Directive 2000/60/EC of the European Parliament and of the Council, the values of the Member State monitoring system classifications as a result of the intercalibration exercise (notified under document number C (2008) 6016) (Text with EEA relevance) (2008/915/EC). *Official Journal of the European Communities* L332/20, Luxemburg.

European Commission, (2010). Common Implementation Strategy for the Water Framework Directive (2000/60/EC). Guidance document no. 14. Guidance on the intercalibration process 2008 – 2011.
http://circa.europa.eu/Public/irc/env/wfd/library?l=/framework_directive/guidance_documents/intercalibration_1/_EN_1.0_&a=d

Fernández Díaz, M. (2003). *Estudio faunístico y ecológico de los coleópteros acuáticos (Adephaga y Polyphaga) en la cuenca del río Ávia (Ourense, NO España): Distribución espacial y temporal*, Degree Thesis, Universidad de Vigo, Vigo, Spain.

Fernández-Díaz, M.; Benetti, C.J. & Garrido, J. (2008). Influence of iron and nitrate concentration in water on aquatic Coleoptera community structure: application to the Avia river (Ourense, NW, Spain). *Limnetica*, 27 (2), 285-298.

Fjellheim, A. & Raddum, G.G. (1996). Weir building in a regulated west Norwegian river: longterm dynamics of invertebrates and fish. *Regulated Rivers: Research & Management*, 12, 501-508.

Fjellheim, A.; Havardstun, J.; Radum, G.G. & Schenell, O.A. (1993). Effects of increased discharge on benthic macroinvertebrates in a regulated river. *Regulated Rivers: Research & Management*, 8, 179-187.

Freund, G.J. & Petty, J.T. (2007). Response of fish and macroinvertebrate bioassessment indices to water chemistry in a mined Appalachian Watershed. *Environmental Management*, 39, 707-720.

Fulan, J.A.; Raimundo, R.; Figueiredo, D. & Correia, M. (2010). Abundance and diversity of dragonflies four years after the construction of a reservoir. *Limnetica*, 29(2), 279-286.

Gabriels, W.; Goethals, P.L.M. & De Pauw, N. (2005). Implications of taxonomic modifications and alien species on biological water quality assessment as exemplified by the Belgian Biotic Index method, *Hydrobiologia*, 542, 137–150.

García-Criado, F. Fernández Aláez, M. (2001). Hydraenidae and Elmidae assemblages (Coleoptera) from Spanish river basin: good indicators of coal mining pollution? *Archiv für Hydrobiologie* 150, 641-660.

García-Criado, F.; Fernández Aláez, C. & Fernández Aláez, M. (1999). Environmental variables influencing the distribution of Hydraenidae and Elmidae assemblages (Coleoptera) in a moderately-polluted river basin in north-western Spain. *European Journal of Entomology*, 96, 37-44.

Garrido, J.; Alonso, A.I.; Álvarez-Troncoso, R.; Benetti, C.; Fernández-Díaz, M. & Gayoso, A. (2003). *Programa de vigilancia ambiental de la minicentral del Río Deva (Pontevedra),* Technical Report, Universidad de Vigo, Vigo, Spain.

Garrido, J.; Alonso, A.I.; Álvarez-Troncoso, R.; Benetti, C.; Fernández-Díaz, M. & Gayoso, A.; González, A.; Penín, L.; Pérez-Bilbao, A.; Rodríguez, I. & Vidal, M. (2005). *Programa de vigilancia ambiental de la minicentral de Ponteliñares (Ourense),* Technical Report, Universidad de Vigo, Vigo, Spain.

Garrido, J.; Alonso, A.I.; Álvarez-Troncoso, R.; Fernández-Díaz, M. & Vidal, M. (2000a). *Evaluación de la calidad del agua mediante parámetros biológicos y fisicoquímicos en los ríos de la cuenca del Avia (Ourense),* Technical Report, Universidad de Vigo, Vigo, Spain.

Garrido, J.; Alonso, A.I.; Álvarez-Troncoso, R.; Fernández-Díaz, M.; Vidal, M. & Benetti, C. (2002). *Estudio y seguimiento de la recuperación biológica de la cuenca del río Lagares (Vigo, Pontevedra),* Technical Report, Universidad de Vigo, Vigo, Spain.

Garrido, J.; Benetti, C.J. & Pérez-Bilbao, A. (2008). *Evaluación de la calidad del agua del Río Furnia (Tuy, pontevedra) mediante indicadores biológicos, antes de la puesta en marcha de la estación depuradora de aguas residuales,* Technical Report, Universidad de Vigo, Vigo, Spain.

Garrido, J.; Membiela, P. & Vidal, M. (1998). Calidad biológica de las aguas del río Barbaña. *Tecnología del agua,* 175, 50-54.

Garrido, J.; Vidal, M.; Alonso, A.I.; Cuadrado, D.; Outerelo, J.A. & Fernández-Díaz, M. (1999). *Programa de vigilancia ambiental del aprovechamiento hidroeléctrico de fecha en el Río Tambre (A Coruña). Fase de funcionamiento,* Technical Report, Universidad de Vigo, Vigo, Spain.

Garrido, J.; Vidal, M.; Alonso, A.I.; Cuadrado, D.; Outerelo, J.A. & Fernández-Díaz, M. (2000b). *Programa de vigilancia ambiental del aprovechamiento integral del Río Tuño (Ourense),* Technical Report, Universidad de Vigo, Vigo, Spain.

Garrido, J.; Vidal, M.; Alonso, A.I.; Cuadrado, D.; Outerelo, J.A. & Fernández-Díaz, M. (2000c). *Programa de vigilancia ambiental de la minicentral hidroeléctrica del Río Deva (Ourense),* Technical Report, Universidad de Vigo, Vigo, Spain.

Garrido, J.; Vidal, M.; Alonso, A.I.; Gayoso, A.; Cuadrado, D.; Outerelo, J.A. & Fernández-Díaz, M. (2000d). *Programa de vigilancia ambiental de la minicentral del Río Tea, propiedad de Engasa,* Technical Report, Universidad de Vigo, Vigo, Spain.

Gerhardt, A. (2000). *Biomonitoring of Polluted Water: Reviews on Actual Topics.* Trans Tech Publications Inc., Zürich, Switzerland.

Harper, M.P. & Peckarsky, B.L. (2005). Effects of pulsed and pressed disturbances on the benthic invertebrate community following a coal spill in a small stream in northeastern USA. *Hydrobiologia,* 544, 241-247.

Hauer, F.R. & Resh., V.H. (1996). Benthic Macroinvertebrates, In: *Methods in Stream Ecology,* F.R. Hauer & G.A. Lamberti (eds), pp. 339-369, Academy Press, New York, USA.

Heino, J. (2000). Lentic macroinvertebrate assemblage structure along gradients in spatial heterogeneity, habitat size and water chemistry. *Hydrobiologia,* 418, 229-242.

Hirst, H.; Jüttner, I. & Ormerod, S.J. (2002). Comparing the responses of diatoms and macroinvertebrates to metals in upland streams of Wales and Cornwall. *Freshwater Biology,* 47, 1752-1765.

Jenkins, A.; McCartney, M. & Sefton, C. (1993). Impacts of climate change on river quality in the United Kingdom. *Report to Department of Environment. Institute of Hydrology.* Wallingford.

Jesus, T.; Formigo, N.; Santos, P. & Tavares, G.R. (2004). Impact evaluation of the Vila Viçosa small hydroelectric Power plant (Portugal) on the water quality and on the dynamics of the benthic macroinvertebrate communities of the Ardena river. *Limnetica,* 23 (3-4), 241-256.

Johnson, R.K.; Wiederholm, T. & Rosenberg, D.M. (1993). Freshwater biomonitoring using individual organisms, populations, and species assemblages of benthic macroinvertebrates, In: *Freshwater Biomonitoring and Benthic Macroinvertebrates,* D.M. Rosenberg & V.H. Resh (eds), pp. 40-158, Chapman & Hall, New York, USA.

Kottek, M.; Grieser, J.; Beck, C.; Rudolf, B. & Rubel, F. (2006). World Map of the Köppen-Geiger climate classification updated. *Meteorologiche Zeitung,* 15, 259-263.

Kubecka, J.; Matena, J. & Hartvich, P. (1997). Adverse ecological effects of small hydropower stations in the Czech Republic: 1. Bypass plants. *Regulated rivers: research & management,* 13, 1001-113.

Lessard, J.L. & Hayes, D.B. (2003). Effects of elevated water temperature on fish and macroinvertebrate communities below small dams. *River Research Applications,* 19, 721-732.

Lytle, D.A. & Peckarsky, B.L. (2001). Spatial and temporal impacts of a diesel fuel spill on stream invertebrates. *Freshwater Biology,* 46, 693-704.

Marqués, M.J.; Martínez-Conde, E. & Rovira, J.V. (2003). Effects of zinc and lead mining on the benthic macroinvertebrates of a fluvial ecosystem. *Water, Air, and Soil Pollution,* 148, 363 - 388.

McCabe, D.J. & Gotelli, N.J. (2000). Effects of disturbance frequency, intensity, and area on assemblages of stream macroinvertebrates. *Oecologia,* 124, 270-279.

Metcalfe, J.L. (1989). Biological quality assessment of running waters based on macroinvertebrate communities: History and Present Status in Europe. *Environmental Pollution,* 60, 101-139.

Ministry of Environment (2005). *Metodología para el establecimiento del estado ecológico según la Directiva Marco del Agua. Protocolos de muestreo y análisis para invertebrados bentónicos*, Madrid, España.

National Water Council (1981). *River quality: the 1980 survey and future outlook*, London, UK.

Nummelin, M.; Lodenius, M.; Tulisalo, E.; Hirvonen, H. & Alanki, T. (2007). Predatory insects as bioindicators of heavy metal pollution. *Environmental Pollution*, 145, 339-347.

Oller, C. & Goitia, E. (2005). Benthic macroinvertebrate and heavy metals in the Pilcomayo River (Tarija, Bolivia). *Revista Boliviana de Ecología*, 18, 17-32.

Ortiz, J.D.; Martí, E. & Puig, M.A. (2005). Recovery of the macroinvertebrate community below a wastewater treatment plant input in a Mediterranean stream. *Hydrobiologia*, 545, 289–302.

Oscoz, J.; Campos, F. & Escala, M.C. (2006). Variación de la comunidad de macroinvertebrados bentónicos en relación con la calidad de las aguas. *Limnetica*, 25 (3), 683-692.

Paz, C. (1993). *Hydradephaga (Coleoptera) en la cuenca del río Landro (NW Península Ibérica). Estudio faunístico y ecológico*, Ph. D. Thesis, Universidad de Santiago de Compostela, Santiago de Compostela, Spain.

Pérez-Bilbao, A. & Garrido, J. (2009). Evaluación del estado de conservación de una zona LIC (Gándaras de Budiño, Red Natura, 2000) usando los coleópteros acuáticos como indicadores. *Limnetica*, 28 (1), 11-22.

Pérez-Quintero, J.C. (2007). Diversity, habitat use and conservation of freshwater molluscs in the lower Guadiana River basin (SW Iberian Peninsula). *Aquatic Conservation: Marine and Freshwater Ecosystems*, 17, 485–501.

Poquet, J.M.; Alba-Tercedor, J.; Puntí, T.; Sánchez-Montoya, M.M.; Robles, S.; Álvarez, M.; Zamora-Muñoz, C.; Sáinz-Cantero, C.E.; Vidal-Abarca, M.R.; Suárez, M.L.; Toro, M.; Pujante, A.M.; Rieradevall, M. & Prat, N. (2009). The MEDiterranean Prediction And Classification System (MEDPACS): an implementation of the RIVPACS/ AUSRIVAS predictive approach for assessing Mediterranean aquatic macroinvertebrate communities. *Hydrobiologia*, 623, 153–171.

Prat, N.; Munné, A.; Solá, C.; Rieradevall, M.; Bonada, N. & Chacón, G. (1999). *La qualitat ecològica del Llobregat el Besòs i el Foix*, Àrea de Medi Ambient, Diputació de Barcelona, Barcelona, España.

Prenda, J. & Gallardo-Mayenco, A. (1996). Self-purification, temporal variability and the macroinvertebrates community in small lowland Mediterranean streams receiving crude domestic sewage effluents. *Archiv für Hydrobiologie*, 136, 159-170.

Quinn, J.M.; Cooper, A.B.; Davies-Colley, R.J.; Rutherdford, J.C. & Williamson, R.B. (1997). Land use effects on habitat, water quality, periphyton, and benthic invertebrates in Waikato, New Zealand, hill-country streams. *New Zealand Journal of Marine and Freshwater Research*, 31, 579-597.

Rahel, F.J. (2002). Homogenization of freshwater faunas. *Annual Review of Ecology and Systematics*, 33, 291–315.

Richter, B.D.; Braun, D.P.; Mendelson, M.A. & Master, L.L. (1997). Threats to imperilled freshwater fauna. *Conservation Biology*, 11(5), 1081-1093.

Rosenberg, D.M. & Resh, V.H. (1993). Introduction to freshwater biomonitoring and benthic macroinvertebrates, In: *Freshwater Biomonitoring and Benthic Macroinvertebrates*, D.M. Rosenberg & V.H. Resh (eds), pp. 1-9, Chapmann & Hall, New York, USA.

Sarr, A.B. (2011). *Evaluación del estado de conservación de ríos afectados por minicentrales hidroeléctricas mediante el estudio de coleópteros acuáticos*. PhD Thesis, University of Vigo, Vigo, Spain.

Smolders, A.J.P.; Lock, R.A.C.; Van der Velde, G.; Medina Hoyos, R.I. & Roelofs, J.G.M. (2003). Effects of mining activities on heavy metal concentrations in water, sediment, and macroinvertebrates in different reaches of the Pilcomayo River, South America. *Archives of Environmental Contamination and Toxicology*, 44, 314-323.

Smolders, A.J.P.; Van Hengstum, G.; Loermans, J.; Montes, A. Rizo, H. &. Castillo, I. (1999). Efectos de la Contaminación Minera sobre la Composición de la Macrofauna Bentónica en el Río Pilcomayo. *Revista Boliviana de Ecología*, 6,229-237.

Song, M.Y.; Leprieur, F.; Thomas, A.; Lek-Ang, S.; Chon, T.S. & Lek, S. (2009). Impact of agricultural land use on aquatic insect assemblages in the Garonne river catchment (SW France). *Aquatic Ecology*, 43, 999-1009.

Stanford, J.A. & Ward, J.V. (1979). Stream regulation in North America, In: *The ecology of regulated streams*, J.V. Ward & J.A. Stanford (eds.), pp. 215-236, Plenum Press, New York, USA.

Stanley, E.H.; Luebke, M.A.; Doyle, M.W. & Marshll, D.W. (2002). Short-term changes in channel form macroinvertebrate communities following low-head dam removal. *Journal of the North American Benthological Society*, 21, 172-187.

Strayer, D.L. (2006). Challenges for freshwater invertebrate conservation. *Journal of the North American Benthological Society*, 25, 271-287.

Ter Braak, C.J.F. & Šmilauer, P. (2002). *CANOCO: Referente manual and CanoDraw for Windows User's guide: Software for Canonical Community Ordination (version 4.5)*. Microcomputer Power, Ithaca.

Thomson, J.R.; Hart, D.D.; Charles, D.F.; Nightengale, T.L. & Winter, D.M. (2005). Effects of removal of a small dam on downstream macroinvertebrate and algal assemblages in a Pennsylvania stream. *Journal of the North American Benthological Society*, 24(1), 192-207.

Thorne, R.J. & Williams, W.P. (1997). The response of benthic macroinvertebrates to pollution in developing countries: a multimetric system of bioassessment. *Freshwater Biology*, 37, 671-686.

Tonkin, J.D.; Death, R.G. & Joy, M.K. (2009). Invertebrate drift patterns in a regulated river: dams, periphyton biomass or longitudinal patterns? *River Research and Applications*, 25(10), 1205-1338.

Velasco, J.; Millán, A.; Hernández, J.; Gutiérrez, C.; Abellán, P.; Sánchez, D. & Ruiz, M. (2006). Response of biotic communities to salinity changes in a Mediterranean hypersaline stream. *Saline Systems*, 2 (12), 1-15.

Woodiwis, F.S. (1964). The biological system of stream classification used by Trent River Board. *Chemistry & Industry*, 11, 443-447.

Yoshimura, M. (2008). Longitudinal patterns of benthic invertebrates along a stream in the temperate forest in Japan: in relation to humans and tributaries. *Insect Conservation and Diversity*, 1, 95–107.

Zamora-Muñoz, C. & Alba-Tercedor, J. (1996). Bioassessment of organically polluted Spanish river, using a biotic index and multivariate methods. *Journal of the North American Benthological Society,* 15 (3), 332- 352.

6

Posidonia oceanica and *Zostera marina* as Potential Biomarkers of Heavy Metal Contamination in Coastal Systems

Lila Ferrat[1] et al.*

[1]*University of Corsica, Sciences for Environment, France*

1. Introduction

In the early 1960s recognition of the adverse effects of environmental contamination due to industrial, pesticide, and agricultural pollution led to the emergence of the field of ecotoxicology (Ramade, 1992). Today, marine estuary and inshore ecosystems continue to be negatively impacted by environmental contamination (Short & Wyllie-Echeverria 1996; Orth et al., 2006; Osborn & Datta, 2006). In order to reduce these negative impacts, bio-surveillance programs are needed to monitor environmental conditions so that changes in ecosystem processes, structure, and the physiological condition of species can be assessed (Blandin, 1986; Tett et al., 2007). An important characteristic of these programs is that indicator species must be capable of rapidly detecting significant changes in the ecosystem so that the cause of deterioration can be addressed early (e.g. Hemminga & Duarte, 2000).

Mussels (Goldberg et al., 1983) and fish (Reichert et al., 1998; Stephensen et al., 2000) are frequently used as indicators of chemical contamination in long-term environmental monitoring programs. However, these programs can be deficient because they only provide information about water column contamination, and these organisms can have limited ranges and often must be introduced to a site as part of the monitoring program. To offset these deficiencies widely distributed indicator organisms in coastal systems that have the capacity to provide contamination information from both water column and sediment environments are needed. Consequently, there is increasing interest in the use of marine macrophytes because they grow in most coastal and estuarine systems (see Green & Short, 2003). These rooted vascular plants interact with the chemical properties of the water column and surface sediment environments within site-specific and basin-wide locations (Brix, 1997; Brix et al., 1983; den Hartog, 1970; Lange and Lambert, 1994; Rainbow and

*Sandy Wyllie-Echeverria[2], G. Cates Rex[3], Christine Pergent-Martini[1], Gérard Pergent[1], Jiping Zou[3], Michèle Romeo[4], Vanina Pasqualini[1] and Catherine Fernandez[5]

[1]*University of Corsica, Sciences for Environment, France*

[2]*Friday Harbor Laboratories, University of Washington USA*

[3]*Brigham Young University, Department of Biology, USA*

[4]*University of Nice Sophia-Antipolis, ROSE, France*

[5]*University of Aix-Marseille, IMEP, France*

Phillips, 1993). For this study the focus was on the seagrasses *Posidonia oceanica* (L.) Delile (Posidoniaceae) and *Zostera marina* L. (Zosteraceae). These were chosen because they are the dominant species in the regions of our inquiry which were, respectively, the Mediterranean Sea (Lipkin et al., 2003; Procaccini et al., 2003) and the Pacific Northwest (Wyllie-Echeverria & Ackerman, 2003). Both species can form vast meadows across the intertidal-subtidal gradient in their respective ecosystems (Molinier & Picard, 1952; Phillips, 1984).

1.1 *P. oceanica* and *Z. marina* as indicators of environmental quality

The potential for these species to provide an early warning of deteriorating environmental quality has been noted for *P. oceanica* and *Z. marina* where both species were found useful at detecting environmental deterioration within local and basin-wide locations (Augier, 1985; Dennison et al., 1993; Pergent, 1991; Pergent-Martini et al., 1999). For example, *P. oceanica* accumulates certain metal pollutants, notably mercury (Pergent-Martini, 1998), which is one of the most abundant marine pollutants. Within the Mediterranean Sea elevated mercury levels have been reported in certain regions (Maserti et al., 1991), and correlations have been drawn between mercury levels in plant tissue and the concentrations of mercury in the water column (Pergent-Martini, 1998). In laboratory studies Lyngby & Brix (1984) and Brix & Lyngby (1984) demonstrated that *Z. marina* can accumulate heavy metals in concentrations above natural levels, and that these concentrations inhibited growth. In addition, based on extensive sampling along the coastline of Limfjord, Denmark, these authors noted that *Z. marina* could be used to monitor heavy metal contamination. Also, a related species *Z. capricorni* has provided valuable information in monitoring iron, aluminium, zinc, chromium, and copper contamination (Prange & Dennison, 2000).

Indicator species that provide an early warning of ecosystem change will likely be those that reveal first order changes in organism function. Molecular, biochemical, and/or cellular changes triggered by pollutants are measurable in biological mediums such as cells, tissues, and/or cellular fluids (McCarthy & Shugart, 1990). For example, oxidation is known to be a significant factor in stress-related organismal weakening, and antioxidant molecules have been used to evaluate organism health (Chen et al., 2007). One group of antioxidant molecules are the widely studied phenolic compounds (Ferrat et al., 2003a) which are known to be induced by reactive oxygen species (Rice-Evans et al., 1995; Vangronsveld et al., 1997).

1.2 Physiological and ecological roles of phenolics and volatile compounds

Phenolic compounds produced via the Shikimic Acid Pathway, and volatiles produced via the Mevalonate Pathway, are known to be important to plant health and survival (Cates, 1996; Fierer et al., 2001; Hartman, 2007; Phillips, 1992; Schimel et al., 1996). They are found in terrestrial higher plants, most notably angiosperms (Goodwin & Mercer, 1983; Hadacek, 2002), some seagrasses (Verges et al., 2007; Zapata and McMillan, 1979), and have a wide range of chemical structures and activities (Hadacek, 2002; Hartman, 2007). Phenolic and volatile compounds contribute significantly to the antioxidant activity of plants, have the capacity to bind heavy metals (Emmons et al., 1999), and are an important mechanism in protecting plants against stress (Swain, 1977). Volatile compounds (e.g. monoterpenes, sesquiterpenes) have been found to serve as energy sources in plants (Croteau & Sood,

1985), are important in the defensive system of higher plants (Cates, 1996; Langenheim, 1994), and influence ecosystem processes such as nutrient cycling (Horner et al., 1988; White, 1986). The production of phenolics and volatiles is under genetic control (Croteau & Gershenzon, 1994; Hartman, 2007), but their qualitative and quantitative production is affected by various environmental factors (Bryant et al., 1983; Gershenzon, 1984; Hartman, 2007; Macheix, 1996; Quackenbush et al., 1986; Ragan & Glombitza, 1986). However, as with other seagrasses, only a very limited number of studies deal with the role of phenolic and volatile compounds from *Posidonia oceanica* (Heglmeier & Zidorn, 2010) and *Zostera marina* (Short & Willie Echeverria, 1996). Only were investigated the impacts of interspecific competition (Dumay et al., 2004), nutrient variation, diseases (Vergeer & Develi, 1997) and grazing (Cannac et al., 2006), or general anthropization of water masses (Short & Wyllie-Echeverria, 1996, Agostini et al., 1998).

1.3 Objective

The objective of this study was to determine if *P. oceanica* and *Z. marina* might be reliable candidates as bio-surveillance organisms with regard to heavy metal pollution. We choose to consider different environmental conditions and to monitor physiological changes through two different seasons. Our assumption was that heavy metal contamination would adversely impact adult *P. oceanica* and *Z. marina* plants, and that plant response to these impacts could be assessed by differences in phenolic and volatile compound content of tissue from impacted and non-impacted sites.

We assessed differences in heavy metal content of plant tissues from sites with documented heavy metal pollution versus controls with no sources of heavy metal pollution. Then, we tested the hypothesis that the presence of identified contaminants could induce a bio-indicator response in these seagrass species. To do this we measured changes in total phenolic content in the leaf and sheath tissue of *P. oceanica,* and total phenolic and volatile compound content in above-and below-ground tissue of *Z. marina.*

2. Materials and methods

2.1 Site location and sample collection

In June 2000 and January 2001, 30 adult shoots of *P. oceanica* were collected by SCUBA at ~10 m in the sub-tidal region at two sites located in the northwestern Mediterranean Sea. The Bay of Bonifacio, a control site, is a pristine area relatively free of industrial pollution located in the south of Corsica (Tonnara - France; 41.4000 N; 9.0830 E; Capiomont et al., 2000). The Bay of Rosignano site south of Livorno (Italy; 43.4000 N; 10.4166 E) is a polluted site. At this site, a chlor-alkali plant has discharged industrial wastes rich in mercury since 1920 (130 kg per year; Ferrara et al., 1989). Water temperature ranged from 18°C in June 2000 to 14°C in January 2001 at all sites but salinity was relatively constant at 38.5 PSU within the study zone (i.e., 10 m depth contour; Villefranche sur Mer Observatory and Di Martino, personal communication).

For *P. oceanica,* foliage leaf and sheath tissue was analyzed for mercury and phenolic content. Tissue was obtained by separating the foliage leaf and sheath tissue from the roots and rhizomes following the procedure of Giraud (1977); root and rhizome tissue was discarded. The chlorophyllous foliage leaves were then separated from non-chlorophyllous sheaths that are located at the leaf base. Foliage leaves from three adult shoots were dissected according to Giraud (1977) and combined to form one sample. Sheaths from the same three shoots were combined to form each sheath sample. After epiphytes were removed from leaf and sheath samples using a glass slide, each sample was rinsed with ultra-pure water and frozen (-20°C) until analysis. To determine mercury and phenolic content, we extracted 0.5 g dry wt. of each tissue sample (n=10).

During maximum low tide, *Z. marina* adult shoots were hand-collected from the lower intertidal region of two sites in Northern Puget Sound, Washington, USA in April and June 2000. The site located near Anacortes, WA (48.4263 N; 122.5897 W) was documented as having heavy metal pollution (http://www.ecy.wa.gov/programs/wq/permits/permit_pdfs/dakota/factsheet.pdf), and the other was a pristine location with no industrial activity on the southeast side of Shaw Island (48.33942 N; 122.55448 W) that served as the control site (Wyllie-Echeverria & Ackerman, 2003). Water temperature ranged between 9°C in April to12°C in June 2000 at all sites, and salinity was relatively constant at 30 PSU during this time period (Wyllie-Echeverria, unpublished data).

Three samples were collected from each site, and each sample consisted of at least 0.5 g dry wt (Cuny et al., 1995) of eight to ten sterile (non-reproductive) shoots which were separated into above- and below-ground parts. Above-ground tissue consisted of the foliage leaf (i.e. basal leaf sheath and distal leaf blade; Kuo & den Hartog, 2006) excised from the rhizome at the node primordia (Tomlinson, 1974). The remaining rhizome and associated nodes and roots formed the below-ground sample. Epiphytes were scraped from the above-ground tissue and sediment was rinsed from the roots and rhizomes (Brackup & Capone, 1985). Each above- and below-ground sample was placed in labelled bags, kept moist and cool in a refrigerator, and shipped overnight to the Chemical Ecology Laboratory at Brigham Young University. Three replicate samples of above-ground tissue from each site, and three replicates of below-ground tissue from each site, were frozen at -80°C until extracted for heavy metals, phenolics or volatiles. Samples were stored at -80°C to preserve the volatile compounds in the tissues.

2.2 Qualitative and quantitative analysis of plant tissues for heavy metal content

Foliage leaves and sheaths of *P. oceanica* and above- and below-ground tissues of *Z. marina* were analyzed qualitatively and quantitatively for heavy metals. For *P. oceanica*, only mercury content, which is the predominant heavy metal pollutant at the Rosignano site (Lafabrie et al., 2007), was analyzed. Three individual shoots (three foliage leaves and three sheaths) that had been separately freeze dried were ground to a powder, and an aliquot of 0.05 g dry wt was digested. Digestion was performed in a 100-ml Teflon® advanced composite vessel reactor with 5 ml HNO_3 and 1 ml H_2O_2 (30%). Microwave digestion (Mars 5, CEM Chemistry, Engineering and Microwave, Matthews, NC, USA) was carried out using a temperature ramp of 8 min up to 200°C followed by a heating plateau of 20 min at 200°C. After digestion, the samples were increased to 25 ml with ultra-pure water and then filtered. Total mercury was determined using a flameless atomic absorption spectrophotometer flow injection (Perkin-Elmer System 100; Norwalk, CT, USA). The procedure consisted of reduction with 1.1% tin chloride ($SnCl_2$, $2H_2O$) in 3% HCl and 0.5% hydroxylammonium chloride (NH_2OH, HCl). A standard addition method for total mercury was used to calibrate the protocol. The analytic procedure was verified using a moss as the certified reference material (*Lagarosiphon major*, Certified Reference Material 60, Community Bureau of Reference, Commission of the European Community, Brussels, Belgium). Data are expressed as ng per g dry wt.

For Z. *marina*, heavy metal content was analyzed using the EPA Method 3052 Procedure. All elements were wet-ashed to prevent loss of elements and reduce the potential of confounding data due to silica content. Above- and below-ground tissue (0.5 g dry wt) was placed in a 50 ml folin tube, and 5 ml concentrated nitric acid was added. Samples sat overnight, and then were placed on a block digester at 200°C for 5-10 minutes. Tubes were removed, cooled, and then digested with 1 ml hydrofluoric acid. Samples were placed back on the block digester for 45-60 minutes. Tubes were removed and brought to a 50 ml volume with distilled water. Stoppered tubes were shaken and then analyzed by inductively coupled plasma atomic emission spectrometry (Iris Intrepid II XSP, model 14463001; Thermo Electron Corporation, Franklin, MA) equipped with an ASX-520 autosampler. Data are expressed as ppm (Table 2).

2.3 Extraction and determination of phenolic content in the tissues of *P. oceanica*, and phenolic and volatile content of above-ground tissues of *Z. marina*

Total phenolic content for both species, and total volatile content for Z. *marina*, were determined to ascertain whether tissue collected from impacted (heavy metal pollution for both species) and control sites differed. A different method is used for the definition of the phenolic and volatile compounds, because the measurements were realized in different labs. For P. *oceanica*, extraction of total phenolic compounds was carried out on 0.5 g dry wt freeze-dried foliage leaf or sheath tissue. Extraction followed Cuny et al. (1995) and consisted of infusing each sample at 40°C in 50 % (v/v) aqueous ethanol in darkness for 3 h. The extract was acidified with a few drops of 2N HCl, the ethanol was evaporated under vacuum, and the aqueous residue extracted with ethanol/acetic acid. The organic phase was dried using anhydrous Na_2SO_4. Concentration of total phenolic compounds was measured by colorimetry (Swain & Hillis, 1959) using Folin-Denis reagent (Folin and Dennis, 1915). Phloroglucinol (Frantzis, 1992) was used for elaboration of standard curves. For Z. *marina*, phenolics were extracted using $MeOH/CH_2CH_2$ (50/50) from 200 mg dry wt of freeze dried above-ground tissue, filtered using VWR grade 415 filter paper, and blown dry using nitrogen gas to prevent oxidation. After resdisolving in $MeOH/CH_2CH_2$ (50/50), the extract was again filtered, placed in an auto-sampler vial (Chromatography Research Supplies, Addison, IL) and injected into a high pressure liquid chromatograph (HPLC) (HP Model 1100; Agilent 1100 Series, Model G1313A; Santa Clara, CA) equipped with a diode-array detector (Model G1316A) and a C_{18} reverse phase 5μm column (Phenomenex, Torrance, CA). The HPLC solvents were A = water/acetic acid (98:2); B = acetonitrile/acetic acid (98:2). Temperature was 50°C, flow rate 1ml/min, and wavelength of the detector set at 280 nm (optimized for Z. *marina* phenolic compounds). Phenolic content is expressed as total peak height /200 mg dry wt. To obtain volatile compound content in Z. *marina* samples, 3 g fresh wt of above-ground tissue was ground to a fine powder in liquid nitrogen and hexane. The extract was then filtered, and the filtrate injected into a capillary gas chromatograph (HP Model 6890) equipped with a head-space sampler (Perkin-Elmer HS 40 XL; Waltham, MA) and a HP-1 column. Oven temperature was 80°C, needle temperature 85°C, transfer temperature 120°C, thermostat time 10 min, pressurizing time 0.6 min, injection time 0.2

min, and withdrawal time 0.5 min. The ramp GC program was 40-210°C at intervals of 3°C ramp/min. Total volatile compound content is expressed as total peak height per 3 g fresh wt tissue.

2.4 Statistical analysis

Data from *P. oceanica* samples were analyzed using a three-way ANOVA to allow comparisons between the phenolic compounds and mercury levels according to tissue, site and sampling period. Since the interaction among these factors was significant, one-way analyses followed by a Tukey test (for analyses over the annual cycle) or Student-t test (for analyses of tissue and site factors at given months) were performed (Zar, 1999). Normality and homoscedasticity were verified by Shapiro Wilks and Bartlett tests, respectively (Zar, 1999). The relationships between phenolic compounds and mercury level were assessed using correlation and regression analyses in Statgraphics plus (ver 3.1) for Windows. Data from *Z. marina* are expressed as ppm for heavy metals, total peak area per 200 mg freeze dried tissue for phenolics, and total peak area per 3 g fresh wt for volatiles. Since all samples were randomly collected along a transect, each sample is treated as an independent experimental unit. Comparison of heavy metal content between impacted and control sites in *Z. marina* above- and below-ground tissues, and for phenolic and volatile content in above-ground tissues, was conducted using a one-way ANOVA, SAS GLM program (SAS, 1996).

3. Results

3.1 Site and tissue differences in heavy metal contamination

Foliage leaf and sheath tissue of *P. oceanica* from the industrially impacted Rosignano site showed large and significant ($p<0.05$) differences in mercury content when compared to the control Tonnara site (Table 1).

Tissue Type		Mercury impacted site (Rosignano)	Control site (Tonnara)
Foliage Leaves	June 2000	233 ± 23	77 ± 11
	January 2001	317 ± 41	79 ± 15
Sheaths	June 2000	368 ± 26	64 ± 8
	January 2001	215 ± 16	80 ± 19

Table 1. Mercury levels (ng/g dw) in foliage leaf and sheath tissues of *P. oceanica* collected at different sites and different sampling periods.

Samples of above-ground tissue collected in April 2000 from *Z. marina* plants growing in the impacted site were higher in iron, aluminium, and copper when compared to tissue from the control site (Table 2). However, above-ground tissue from the control site was significantly higher in zinc, nickel, molybdenum, and mercury when compared to the impacted site (Table 2). For the July 2000 samples, the only significant differences were that nickel and copper were in highest concentration in plants from the impacted site when compared to plants from control site (Table 2). For below-ground tissue of *Z. marina* in April, samples from the industrially impacted site were significantly higher ($p<0.05$) for iron, aluminium, nickel, manganese, copper, cadmium, chromium, and lead when compared to the control site (Table 2). None of the heavy metals was higher in concentration in the control site for samples taken in April 2000. For the July 2000 samples, barium, iron, aluminium, zinc, manganese, copper, cadmium, arsenic, and chromium were higher in the plants from the impacted site when compared to the control site, and cobalt and strontium were higher in plants from the control site (Table 2).

Heavy Metals	Site (ppm)*							
	Above-ground Tissue				Below-ground Tissue			
	April		July		April		July	
	Industrially impacted site	Control site	Industrially impacted site	Control site	Industrially impacted site	Control site	Industrially impacted site	Control site
Barium	323(41)^{a}	364(32)^{a}	279(107)^{a}	312(55)^{a}	466(136)^{a}	420(100)^{a}	570(148)^{a}	315(84)^{b}
Iron	320(127)^{a}	180(58)^{b}	204(87)^{a}	142(94)^{a}	5801(2846)^{a}	1068(540)^{b}	5591(1503)^{a}	576(263)^{b}
Aluminum	183(75)^{a}	119(43)^{b}	88(42)^{a}	100(81)^{a}	1626(1341)^{a}	665(435)^{b}	1737(494)^{a}	503(336)^{b}
Zinc	100(11)^{a}	119(13)^{b}	102(22)^{a}	110(15)^{a}	134(45)^{a}	133(44)^{a}	169(46)^{a}	96(16)^{b}
Nickel	55(21)^{a}	104(25)^{b}	45(16)^{a}	23(15)^{b}	127(78)^{a}	34(13)^{b}	63(20)^{a}	64(31)^{a}
Manganese	37(6)^{a}	42(6)^{a}	48(16)^{a}	51(7)^{a}	38(33)^{a}	11(6)^{b}	26(7)^{a}	10(4)^{b}
Copper	14(2)^{a}	12(2)^{b}	16(4)^{a}	10(1)^{b}	40(21)^{a}	19(29)^{b}	43(12)^{a}	10(3)^{b}
Molybdenum	5(2)^{a}	6(1)^{b}	7(2)^{a}	8(1)^{a}	0(0)	----**	0(0)^{a}	0(1)^{a}
Cadmium	2(1)^{a}	1(1)^{a}	2(1)^{a}	2(1)^{a}	13(8)^{a}	4(2)^{b}	11(3)^{a}	3(1)^{b}
Arsenic	4(2)^{a}	3(2)^{a}	3(2)^{a}	3(2)^{a}	8(7)^{a}	8(1)^{a}	10(6)^{a}	1(2)^{b}
Cobalt	2(1)^{a}	4(1)^{a}	2(1)^{a}	2(1)^{a}	2(1)^{a}	1(1)^{a}	1(1)^{a}	2(1)^{b}
Mercury	1(1)^{a}	2(1)^{b}	0(1)^{a}	1(1)^{a}	3(3)^{a}	2(1)^{a}	4(1)^{a}	4(2)^{a}
Strontium	1(2)^{a}	2(3)^{a}	4(3)^{a}	4(2)^{a}	3(4)^{a}	6(4)^{a}	0(0)^{a}	1(2)^{b}
Chromium	1(1)^{a}	2(1)^{a}	1(0)^{a}	1(0)^{a}	7(4)^{a}	2(1)^{b}	6(2)^{a}	1(1)^{b}
Lead	0(0)^{a}	1(2)^{a}	0(0)^{a}	0(0)^{a}	13(23)^{a}	6(3)^{b}	0(1)^{a}	0(1)^{a}

Table 2. Differences in accumulation of heavy metals in above- and below- ground tissues of *Z. marina* between impacted and control sites [April, July 2000; x, -]. *Means followed by different letters are significantly different at $p < 0.05$; Means followed by the same letter (i.e. "a") are not significantly different at $p < 0.05$. **Insufficient sample for analysis.

3.2 Production of phenolic and volatile compound content in plant tissues between impacted and control sites

Foliage leaves from Tonnara (20.5 $mg.g^{-1}$) were significantly higher (Tukey test, $p< 0.05$) in phenolic content in January 2001 compared to plants from the mercury impacted Rosignano site (13.2 $mg.g^{-1}$), but were not significantly different in the June 2000 samples (Fig. 1). For sheaths, the levels of total phenolic compounds from Tonnara plants in June and January (9.2 and 15.2 $mg.g^{-1}$, respectively) were significantly higher than those measured in plants at the Rosignano site (5.0 and 6.4 mg.g^{-1}, respectively) (Tukey test, $p<0.05$). Phenolic content was higher across sites and sampling times in *P. oceanica* foliage leaves compared to sheaths in all comparisons (Mann and Whitney test, $p> 0.05$).

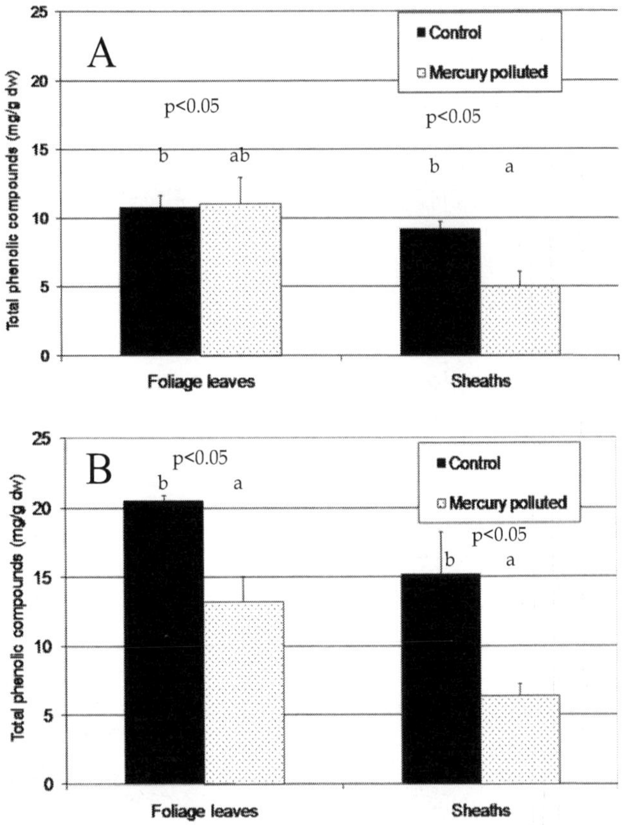

Fig. 1. Total phenolic concentration (mg.g^{-1} dw) in foliage leaf and sheath tissues of *P. oceanica* in Tonnara (control) and Rosignano (mercury polluted) in June 2000 (A) and January 2001 (B).

For *Z. marina* total phenolic content in above-ground tissues collected from plants at the control site always was higher when compared to above-ground tissues collected from the impacted site for both April and July 2000 (Fig. 2). However, the only significant difference was in July where the control site produced a higher amount of total phenolic (65.8 vs 50.8 peak area / 200 mg dry wt, respectively; $p<0.05$). Total volatile compound production also was higher at both sampling periods, but the only significant difference occurred in the April 2000 sampling where above-ground tissues from the control site showed an average peak area of 551 per 200 mg dry wt tissue compared to 352 at the impacted site ($p<0.05$).

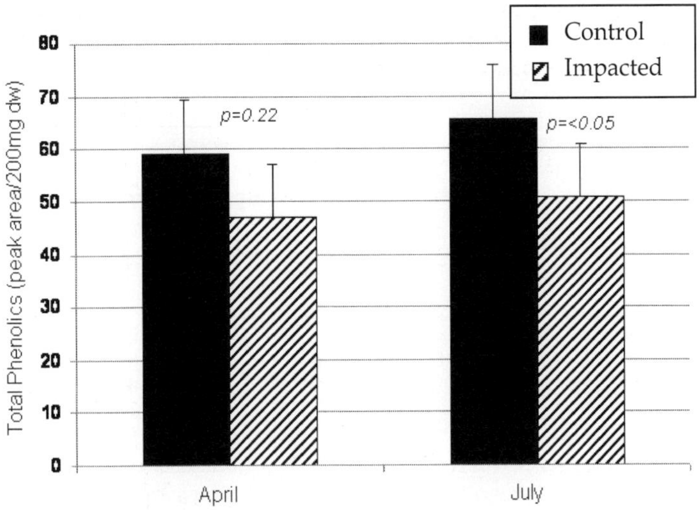

Fig. 2. Total phenolic content in above-ground tissue from *Z. marina* plants growing in heavy metal impacted and control sites (April and July, 2000).

4. Discussion

4.1 Tissue and site differences in heavy metal content

Results presented indicate that plant tissues of *P. oceanica* and *Z. marina* significantly accumulated high levels of heavy metals when growing on heavy metal-impacted sites (Tables 1 & 2). At the Rosignano site, when compared to the control Tonnara site, foliage leaves and sheaths contained two to over six times the amount of mercury. These patterns of accumulation are consistent with findings by other authors who have studied the same sites (Capiomont et al., 2000; Ferrat, 2001; Ferrat et al., 2003b; Maserti & Ferrara, 1991).

Z. marina plants from the heavy-metal impacted site accumulated significantly higher concentrations of iron, aluminum, nickel, and copper in their above-ground tissues when compared to the control site (Table 2). In addition, below-ground tissue of *Z. marina* plants from the industrially-impacted site accumulated over three, and up to five, times the levels of heavy metals compared to plants from the control site. A striking difference between above- and below-ground tissue, is that below-ground tissue from the impacted site accumulated 12 heavy metals (barium, iron, aluminum, zinc, nickel, manganese, copper, cadmium, arsenic, cobalt, chromium, lead; Table 2) while above-ground tissue only accumulated four heavy metals (iron, aluminum, nickel, copper) (Table 2). Another major difference is that the quantity of heavy metals accumulated in the below-ground tissue was higher for most of the heavy metals compared to that in the above-ground tissue.

Variation in metallic accumulation between above- and below-ground seagrass tissue has been discussed by various authors (see synthesis in Pergent Martini & Pergent, 2000), and could be a function of differences in binding sites or seasonal translocation between above- and below-ground structures (Libes & Boudouresque, 1987; Ward, 1987). The level of environmental contamination within a particular site also may be an important factor. For example Capiomont et al. (2000) found that mercury content was higher in the interstitial water than in the water column at our Rosignano sampling location.

Heavy metals are known to have adverse affects on the physiology of *P. oceanica* and *Z. marina* as well as other seagrasses (Ward, 1987). Lyngby & Brix (1984) have shown that the order of heavy metal inhibition of growth of *Z. marina* from greatest to least is mercury, copper, cadmium, zinc, chromium, and lead. Interestingly mercury was not significantly accumulated by *Z. marina* at our impacted site but the other five generally followed the pattern described by Lynby & Brix (1984) (Table 2).

4.2 Phenolic and volatile compound production in plant tissues between impacted and control sites

Our results suggest that total phenolic compound levels within seagrass tissue could be an indicator of site quality. Differences in production of phenolics in tissues from both species were noted between impacted and control sites. For foliage leaves and sheaths of *P. oceanica* collected in January, and above-ground tissue of *Z. marina* collected in July, total phenolic content was significantly lower in plants collected from industrial sites (Fig. 1 & 2). This is supported by Vergeer et al. (1995) who concluded that a decrease of total phenolic compounds in the tissue of *Z. marina* indicated plants may be growing in unsuitable environmental conditions. Noteworthy is that correlation analysis indicated a significant ($p< 0.05$) inverse relationship between heavy metal content and the health of plants as measured by phenolic content for *P. oceanica* ($r^2 = 69.8$ %, linear model of regression: mercury = 0.22 - 0.0055 * phenol for sheaths).

Additionally, gas chromatography analysis of volatile compounds from *Z. marina* indicated that above-ground tissue from plants growing in the impacted site was significantly lower in volatiles from the April collection, when growth begins in Northern Puget Sound (Phillips, 1984) compared to tissue from the control site (Fig. 3). However, no significant differences occurred in volatile compound production between impacted and control sites in the July collection.

4.3 Phenolic compound production with regard to tissue and time collection

For *P. oceanica,* the concentration of phenolic compounds differed between foliage leaves and sheaths being higher in leaf tissue regardless of site. Similarly, Agostini et al. (1998) found higher concentrations (6 $mg.g^{-1}$) in the apical parts and youngest leaves and lower concentrations in sheaths (0.1 $mg.g^{-1}$). Also, in our study significant variation was observed between seasons; for example, phenolic levels were found to be higher in the January 2001 samples compared to the June 2000 samples.

Differences occur in the natural products analyzed depending on month of collection for both *P. oceanica* and *Z. marina (Fig. 1-3)*. For example, *P. oceanica* foliage leaves and sheaths in January 2001 were higher in phenolic content than those collected in June 2000 (Fig. 1). While phenolic content in above-ground *Z. marina* tissue was similar in concentration between April and July (Fig. 2), but volatile compounds in above-ground tissue collected in April were significantly higher than those collected in July (Fig. 3). April and July were selected as sampling times for *Z. marina* because they represent early and mature tissue growth in the Northern Puget Sound (Phillips, 1984). However, in a preliminary study in which *Z. marina* shoots were collected in February 2000, plants from the heavy metal-impacted site produced only 19% of the total phenolic content when compared to plants from the control site (Zou et al., unpublished data). In order to establish when phenolics and volatiles may best indicate plant health, experimental designs need to involve sampling plants every two months throughout the year.

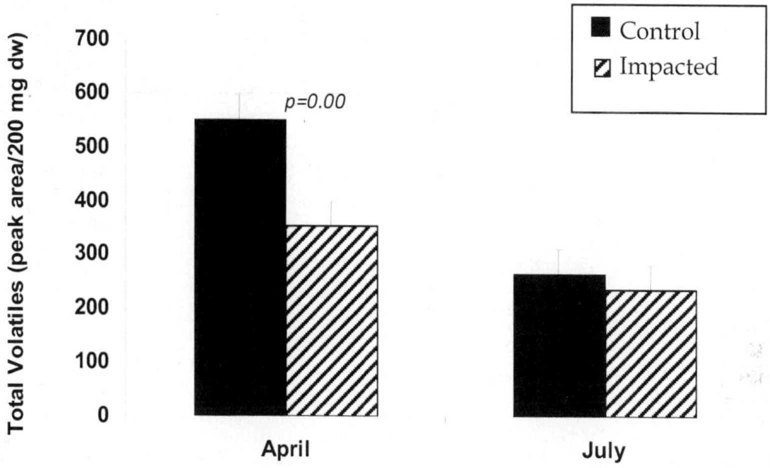

Fig. 3. Total volatile content of above-ground tissue from *Z. marina* plants growing in heavy metal impacted and control sites (April and July, 2000).

Finally, based on the response of different seagrass genotypes to disturbance (e.g. Ehlers et al., 2008; Hughes & Stachowicz, 2009; Wyllie-Echeverria et al., 2010), we suspect that variation in the type and concentration of heavy metal uptake may exist within different genotypes. However, this aspect of heavy metal accumulation in needs investigation in controlled conditions with seagrass species from different locations.

5. Conclusions

Significant differences were found in the accumulation of mercury in leaf and sheath tissues of *P. oceanica* when plants were growing on impacted sites as compared to sites not impacted heavily by mercury (Table 1). *Z. marina* plants growing in a site impacted by heavy metals associated with industrial pollution accumulated significantly higher amounts of iron, aluminium, nickel, and copper in above-ground tissues as compared to a non-impacted site, and higher amounts of barium, iron, aluminium, zinc, nickel, manganese, copper, cadmium, arsenic, chromium, and lead in below-ground tissues at the impacted site (Table 2). For *P. oceanica,* total phenolics were significantly higher in leaves at the control site when compared to the mercury impacted site for the January sampling period (Fig. 1). For sheath tissue total phenolics from the control site were significantly higher when compared to the mercury impacted site for both sampling periods (Fig. 1). For *Z. marina,* total phenolic content was higher in both sampling periods at the non-impacted site compared to the control site, but only significantly so for the July 2000 sampling period (Fig. 2). Total volatile content also was higher at the control site for both sampling periods, but only significantly higher for the April sampling period (Fig. 3). These results support the hypotheses that *P. oceanica* and *Z. marina* accumulate significant amounts of heavy metals from impacted sites, and that these accumulations are associated with reduced total phenolic and volatile compound content. Based on these supportive data, we conclude that *P. oceanica* and *Z. marina* are potential candidates as bio-surveillance organisms especially with regard to heavy metal pollution of coastal and estuarine ecosystems.

Since we observed variation in the production of phenolics and volatiles with regard to sampling time and season, a priority is the identification of individual phenolic and volatile compounds in the tissue of these two species. In our labs we have identified in one or both species using gas chromatography/mass spectroscopy and high pressure liquid chromatography several cinnamic acid and benzoic acid derivatives; these results are comparable to those found by Quackenbush et al. (1986). Additionally, these analyses indicate not only a quantitative decrease in total phenolic and volatile compounds, but also qualitative differences between plants growing on impacted and non-impacted sites (Ferrat et al., unpublished data for *P. oceanica*; Zou et al., unpublished data for *Z. marina*). Finally, since various environmental perturbations may adversely affect seagrass health (impact of human activity reviewed in Short & Wyllie-Echeverria, 1996), and thereby phenolic and volatile compound production, collaboration among scientists working at a diversity of sites would greatly facilitate progress toward this bio-surveillance effort.

6. Acknowledgments

We thank Tina, Victoria, Rebecca, and Tessa Wyllie-Echeverria and Carl Young for assistance with sample collection in Washington State. RGC and JZ thank the Department of Botany for the professional development funds supporting this research, and RGC is grateful for additional funding from the Karl G. Maeser Research Award. SWE gratefully acknowledges support from the Russell Family Foundation during the writing portion of this investigation.

7. References

Agostini, S.; Desjobert, J.M. & Pergent, G. (1998). Distribution of phenolic compounds in the seagrass *Posidonia oceanica*. *Phytochemistry*, Vol.48, No.4, (June 1998), pp. 611-617, ISSN 0031-9422

Augier, H. (1985). L'herbier à *Posidonia oceanica*, son importance pour le littoral Méditerranéen, sa valeur comme indicateur biologique de l'état de santé de la mer, son utilisation dans la surveillance du milieu, les bilans écologiques et les études d'impact. *Vie Marine*, Vol.7, (1985), pp. 85-113

Blandin, P. (1986). Bioindicateurs et diagnostic des systèmes écologiques. *Bulletin d'Ecologie*, Vol.17, No.4, (April 1986), pp. 211-307, ISSN 0395-7217

Brackup, I. & Capone, D.G. (1985). The effect of several metal and organic pollutants on nitrogen-fixation (acetylene reduction) by the roots and rhizomes of *Zostera marina* L. *Environmental and Experimental Botany*, Vol.25, No.2, (May 1985), pp. 145-151, ISSN 0098-8472

Brix, H.; Lyngby, J.E. & Schierup, H.H. (1983). Eelgrass (*Zostera marina* L.) as an indicator organism of trace metals in the Limfjord, Denmark. *Marine Environmental Research*, Vol.8, No.3, (March 1983), pp. 165-181, ISSN 0141-1136

Brix, H. & Lyngby, J.E. (1984). A survey of the metallic composition of *Zostera marina* L. in the Limfjord, Denmark. *Archiv für Hydrobiologie*, Vol.99, No.3, (March 1984), pp. 347-359, ISSN 0003-9136

Brix, H. (1997). Do macrophytes play a role in constructed treatment wetlands? *Water Science and Technology*, Vol.35, No.5, (May 1997), pp. 11-17, ISSN 0273-1223

Bryant, J.P.; Chapin F.S. & Klein, D.R. (1983). Carbon/nutrient balance of boreal plants in relation to vertebrate herbivory. *Oikos*, Vol.40, No.3, (March 1983), pp. 357-368, ISSN 0030-1299

Capiomont, A.; Piazzi, L. & Pergent, G. (2000). Seasonal variations of total mercury in foliar tissues of *Posidonia oceanica. Journal of the Marine Biological Association of the United Kingdom*, Vol.80, No.6, (June 2000), pp. 1119-1123, ISSN 0025-3154

Cates, R.G. (1996). The role of mixtures and variation in the production of terpenoids in conifer-insect-pathogen interactions, In: *Phytochemical Diversity and Redundancy in Ecological Interactions, Serie Recent Advances in Phytochemistry*, Romeo, J.T.; Saunders, J.A.; Barbosa, P., Vol.30, pp. 179-216, Plenum Press, ISBN 978-0-306-45500-1, New York

Chen, C.; Arjomandi, M.; Balmes, J.; Tager, I. & Holland, N. (2007). Effects of chronic and acute ozone exposure on lipid peroxidation and antioxidant capacity in healthy young adults. *Environmental Health Perspectives*, Vol.115, No.12, (December 2007), pp. 1732-1737, ISSN 0091-6765

Croteau, R. & Gershenzon, J. (1994). Genetic control of monoterpene biosynthesis in mints (*Mentha*: Lamiaceae), In: *Genetic Engineering of Plant Secondary Metabolism, Serie Recent Advances in Phytochemistry*, Ellis, B.E.; Kuroki, G.W.; Stafford, H.A., Vol.28, pp. 193-229, Plenum Press, ISBN 978-0-306-44804-1, New York

Croteau, R. & Sood, V. (1985). Metabolism of monoterpenes. Evidence for the function of monoterpenes and catabolism in peppermint (*Mentha piperita*). *Plant Physiology*, Vol.77, No.4, (April 1985), pp. 801-806, ISSN 0032-0889

Cuny, P.; Serve, L.; Jupin, H. & Boudouresque, C.F. (1995). Water soluble phenolic compounds of the marine phanerogam *Posidonia oceanica* in a Mediterranean area colonized by the introduced chlorophyte *Caulerpa taxifolia. Aquatic Botany*, Vol.52, No.3, (December 1995), pp. 237-242, ISSN 0304-3770

den Hartog, C.D. (1970). *The Seagrasses of the World*, Verhand Koninklijke Nederl Akad. Wetenschap Afd. Nat. Tweede Reeks, North-Holland Publication, Amsterdam

Dennison, W.C.; Orth, R.J.; Moore, K.A.; Stevenson, J.C.; Carter, V.; Kollar, S.; Bergstrom, P.W. & Batiuk, R.A. (1993). Assessing water quality with submerged aquatic vegetation: Habitat requirements as barometers of Chesapeake Bay health. *BioScience*, Vol.43, No.2, (February 1993) pp. 86-94, ISSN 0006-3568

Ehlers, A.; Worm, B. & Reusch, T.B.H. (2008). Importance of genetic diversity in eelgrass *Zostera marina* for its resilience to global warming. *Marine Ecology Progress Series*, Vol.355, (February 2008), pp. 1-7, ISSN 0171-8630

Emmons, C.L.; Peterson, D.M. & Paul, G.L. (1999). Antioxidant capacity of oat (Avena sativa L.) extracts. 2. In vitro antioxidant activity and contents of phenolic and tocol antioxidants. Journal of Agricultural and Food Chemistry, *Vol.47, (December 1999), pp. 4894-4898, ISSN 0021-8561*

Fierer, N.; Schimel, J.P.; Cates, R.G. & Zou, J. (2001). Influence of balsam poplar tannin fractions on carbon and nitrogen dynamics in Alaskan taiga floodplain soils. *Soil Biology and Biochemistry*, Vol.33, No.12-13, (October 2001), pp. 1827-1839, ISSN 0038-0717

Ferrat, L. (2001). *Réactions de la phanérogame marine Posidonia oceanica en réponse à des stress environnementaux*, Thèse de doctorat, Université de Corse, France

Ferrat, L.; Pergent-Martini C. & Roméo M. (2003a). Assessment of the use of biomarkers in aquatic plants for the evaluation of environmental quality: application to seagrasses. *Aquatic Toxicology*, Vol.65, No.2, (October 2003), pp. 187-204, ISSN 0166-445X

Ferrat, L.; Gnassia-Barelli, M.; Pergent-Martini, C. & Roméo, M. (2003b). Mercury and non-protein thiol compounds in the seagrass *Posidonia oceanica*. *Comparative Biochemistry and Physiology c*, Vol.134, No.1, (January 2003), pp. 147-155, ISSN 0742-8413

Ferrara, R.; Maserti, B.E. & Paterno, P. (1989). Mercury distribution in marine sediment and its correlation with the *Posidonia oceanica* prairie in a coastal area affected by a chlor-alkali complex. *Toxicological and Environmental Chemistry*, Vol.22, No.1-4, (April 1989), pp. 131-134, ISSN 0277-2248

Folin, O. & Denis, W. (1915). A colorimetric method for the determination of phenols (and phenol derivatives) in urine. *Journal of Biological Chemistry*, Vol.22, (1915), pp. 305-308, ISSN 0021-9258

Frantzis, A. (1992). *Etude expérimentale des niveaux de consommation et d'utilisation des macrophytes et des détritus dérivés par deux invertébrés benthiques Paracentrotus lividus et Abra ovata*, Thèse de doctorat, Université Aix Marseille II, France

Gershenzon, J. (1984). Changes in the level of plant secondary metabolite production under water and nutrient stress. In: *Phytochemical adaptation to stress, Serie Recent Advances in Phytochemistry*, Loewus F.A.; Timmermann B.N.; Steelink C., Vol.24, pp. 273-320, Plenum Press, ISBN 0306417200, New York

Giraud, G. (1977). *Contribution à la description et à la phénologie quantitative des herbiers à Posidonia oceanica (L.) Delile*, Thèse de Doctorat 3ème cycle, Université Aix-Marseille II, France

Goldberg, E.D.; Koide, M. & Hodeg, V. (1983). U.S. Mussel watch: 1977-1978 results on trace metals and radionuclides. *Estuarine Coastal and Shelf Science*, Vol.16, No.1, (January 1983), pp. 69-93, ISSN 0272-7714

Goodwin, T.W. & Mercer, E.I. (1983). *Introduction to Plant Biochemistry*, Pergamon Press, Oxford, ISBN 0080249221 9780080249223 0080249213 9780080249216, England

Green, E.P. & Short, F.T. (2003). *World Atlas of Seagrasses*, UNEP World Conservation Monitoring Centre, University of California Press, Berkeley, ISBN 0-520-24047-2, USA

Hadacek, F. (2002). Secondary Metabolites as Plant Traits: Current Assessment and Future Perspectives. *Critical Reviews in Plant Sciences*, Vol.21, No.4, (April 2002), pp. 273-322, ISSN 0735-2689

Hartman, T. (2007). *From waste products to ecochemicals: fifty years research of plant secondary metabolism.* Phytochemistry, Vol.68, No.22-24, (November-December 2007), pp. 2831-2846, ISSN 0031-9422

Heglmeier, A. & Zidorn, C. (2010). Secondary metabolites of *Posidonia oceanica* (Posidoniaceae). *Biochemical Systematics and Ecology*, Vol.38, No.5, (May 2010), pp. 1-7, ISSN 03051978

Hemminga, M. & Duarte, C. (2000). *Seagrass Ecology*, Cambridge University Press, ISBN 0-521-66184-6, USA

Horner, J.D.; Gosz, J.R. & Cates, R.G. (1988). *The role of carbon-based plant secondary metabolites in decomposition in terrestrial ecosystems.* American Naturalist, Vol.132, No.6, (June 1988), pp. 869-883, ISSN 0003-0147

Hughes, A.R. & Stachowicz, J.J. (2009). Ecological impacts of genetic diversity in the clonal seagrass *Zostera marina*. *Ecology*, Vol.90, No.5, (May 2009), pp. 1412-1419, ISSN 0012-9658

Kuo, J. & den Hartog, C. (2006). Seagrass morphology, anatomy and ultrastructure, In: *Seagrasses: Biology, Ecology, and Conservation,* Larkum, A.W.D.; Orth, R.J.; Duarte, C.M., pp. 51-87, Springer-Verlag, ISBN 978-1402029424, The Netherlands

Lafabrie, C.; Pergent, G.; Kantin, R.; Pergent-Martini, C. & Gonzalez, J-L. (2007). Trace metals assessment in water, sediment, mussel and seagrass species - Validation of the use of *Posidonia oceanica* as a metal biomonitor. *Chemosphere*, Vol.68, No.11, (August 2007), pp. 2033-2039, ISSN 0045-6535

Lange, C.R. & Lambert, K.E. (1994). Biomonitoring. *Water Environmental Research*, Vol.66, No.4, (June 1994), pp. 642-651, ISSN 1061-4303

Langenheim, J. (1994). Higher plant terpenoids: A phytocentric overview of their ecological roles. *Journal of Chemical Ecology*, Vol.20, No.6, (June 1994), pp. 1223-1280, ISSN 0098-0331

Libes, M. & Boudouresque C.F. (1987). Uptake and long-distance transport of carbon in the marine phanerogram *Posidonia oceanica. Marine Ecology Progress Series*, Vol.38, (June 1987), pp. 177–186, ISSN 0171-8630

Lipkin, Y.; Beer S. & Zakai D. (2003). The seagrases of the Eastern Mediterranean and the Red Sea, In: *World Atlas of Seagrasses,* Green, E.P.; Short F.T., pp. 65-73, UNEP World Conservation Monitoring Centre, University of California Press, Berkeley, ISBN 0-520-24047-2, USA

Lyngby, J.E. & Brix, H. (1984). The uptake of heavy metals in eelgrass *Zostera marina* and their effect on growth. *Ecological Bulletins*, Vol.36, (1984), pp. 81-84

Macheix, J.J. (1996). Les composés phénoliques des végétaux : quelles perspectives à la fin du XXème siècle? *Acta Botanica Gallica*, Vol.143, No.6, (June 1996), pp. 473-479, ISSN 1253-8078

Maserti, B.E. & Ferrara, R. (1991). Mercury in plants, soil and atmosphere near a chlor-alkali complex. *Water, Air, Soil Pollution*, Vol.56, No.1, (April 1991), pp. 15-20, ISSN 1567-7230

Maserti, B.E.; Ferrara, R. & Morelli, M. *(*1991*)*. *Posidonia oceanica*: uptake and mobilization of mercury in the Mediterranean basin. In: *Proceedings of the FAO/UNEP/IAEA Worshop on the biological effects of pollutants on marine organisms*, Gabrielides, G.P., Vol. 59, pp. 243–249, MAP Technical Reports Series, Athens, Greece

McCarthy, J.F. & Shugart, L. (1990). *Biomarkers of Environmental Contamination*, Lewis Publishers: Boca Raton, ISBN 0873712846, Florida, USA

Molinier, R. & Picard, J. (1952). Recherches sur les herbiers de phanérogames marines du littoral méditerranéen français. *Annales de l'Institut Océanographique*, Vol.27, No.3, (1952), pp. 157-234, ISSN 0078-9682

Orth, R.J.; Carruthers, T.J.B.; Dennision, W.C.; Duarte, C.M.; Fourqurean, J.W.; Heck, Jr.K.L.; Hughes, A.R.; Kendrick, G.A.; Kenworthy, W.J.; Olyarnik, S.; Short, F.T.; Waycott, M. & Williams, S.L. (2006). A global crisis for seagrass ecosystems. *Bioscience*, Vol.56, No.12, (December 2006), pp. 987-996, ISSN 0006-3568

Osborn, D. & Datta A. (2006). Institutional and policy cocktails for protecting coastal and marine environments from land-based sources of pollution. *Ocean and Coastal Management*, Vol.49, No.9-10, (Januray 2006), pp. 576-596, ISSN 0964-5691

Pergent, G. (1991). Les indicateurs écologiques de la qualité du milieu marin en Méditerranée. *Oceanis*, Vol.17, No.4, (April 1991), pp. 341-350, ISSN 0182-0745

Pergent-Martini, C. (1998). *Posidonia oceanica*: a biological indicator of past and present mercury contamination in the Mediterranean Sea. *Marine Environmental Research*, Vol.45, No.2, (March 1998), pp. 101-111, ISSN 0141-1136

Pergent-Martini, C. & Pergent G. (2000). Are marine phanerogams a valuable tool in the evaluation of marine trace-metal contamination: example of the Mediterranean sea? *International Journal of Environment and Pollution*, Vol.13, No.1-6, (January 2000), pp. 126-147, ISSN 0957-4352

Pergent-Martini, C.; Pergent, G.; Fernandez, C. & Ferrat, L. (1999). Value and use of *Posidonia oceanica* as a biological indicator, In: *Proceedings MEDCOAST-EMECS 99 Joint Conference Land-ocean interactions: managing coastal ecosystems*, Vol.1, pp. 73-90, MEDCOAST, Middle East Technical University Publication, ISBN 975-429-142-X, Ankara, Greece

Phillips, D. (1992). Flavonoids: plant signals to soil microbes. In: *Phenolic Metabolism in Plants, Serie Recent Advances in Phytochemistry*, Stafford, H.; Ibrahim, R., Vol.26, pp. 201-231, Plenum Press, ISBN 9780306442315, New York, USA

Phillips, R.C. (1984). *The ecology of eelgrass meadows in the Pacific Northwest: A community profile*, FWS/OBS-84/24, U.S. Dept. of the Interior, Washington, USA

Prange, J.A. & Dennison, W.C. (2000). Physiological responses of five seagrass species to trace metals. *Marine Pollution Bulletin*, Vol.41, No.7-12, (July 2000), pp. 327-336, ISSN 0025-326X

Procaccini, G.; Buia, M.C.; Gambi, M.C.; Perez, M.; Pergent, G.; Pergent-Martini, C. & Romero, J. (2003). The seagrasses of the Western Mediterranean, In: *World Atlas of Seagrasses*, Green, E.P.; Short F.T., pp. 48-58, UNEP World Conservation Monitoring Centre, University of California Press, Berkeley, ISBN 0-520-24047-2, USA

Quackenbush, R.C.; Bunn, D. & Lingren, W. (1986). HPLC determination of phenolic acids in the water-soluble extract of *Zostera marina* L. (eelgrass). *Aquatic Botany*, Vol. 24, No.1, (January 1986), pp. 83-89, ISSN 0304-3770

Ragan, M.A. & Glombitza, K.W. (1986). Phlorotannins, brown algal polyphenols. In: *Progress in Phycological Research*, Round, F.E.; Chapman D.J., Vol.4, pp. 130-241, Elsevier Biomedical Press, ISBN 094873700X, 9780948737008, UK

Rainbow, P.S. & Phillips, D.J.H. (1993). Cosmopolitan biomonitors of trace metals. *Marine Pollution Bulletin*, Vol.26, No.11, (November 1993), pp. 593-601, ISSN 0025-326X

Ramade, F. (1992). *Précis d'écotoxicologie*, Masson, Collection d'écologie, ISBN 2-225-82578-5, France

Reichert, W.L.; Myers, M.S.; Peck-Miller, K.; French, B.; Anulacion, B. F.; Collier, T. K.; Stein, J.E. & Varanasi, U. (1998). Molecular epizootiology of genotiocic events in marine fish: Linking contaminant exposure, DNA damage, and tissue-level alterations. *Mutation Research*, Vol.411, No.3, (November 1998) pp. 215-225, ISSN 0027-5107

Rice–Evans, C.A.; Miller, N.J.; Bolwell, P.G.; Bramley, P.M. & Pridham, J.B. (1995). The relative antioxidant activities of plant-derived polyphenolic flavonoïds. Free Radical Research, Vol.22, No.4, (April 1995), pp. 375-383, ISSN 1071-5762

SAS Institute, Inc. (1996). *SAS User's Guide: Statistics*, Carey, ISBN 0-917382-01-3, USA

Schimel, J.P.; Cates, R.G.; Clausen, T.P. & Reichardt, P.B. (1996). Effects of balsam poplar (*Populus balsamifera*) tannins and low molecular weight phenolics on microbial activity in taiga floodplain soil: Implications for changes in N cycling during succession. *Canadian Journal of Botany*, Vol.74, No.1, (January 1996), pp. 84-90, ISSN 0008-4026

Short, F.T. & Wyllie-Echeverria, S. (1996). Human-induced and disturbance in seagrasses. *Environmental Conservation*, Vol.23, No.1, (January 1996), pp. 17-27, ISSN 0376-8929

Stephensen, E.; Savarsson, J.; Sturve, J.; Ericson, G.; Adolfsson-Erici, M. & Forlin, L. (2000). Biochemical indicators of pollution exposure in shorthorn sculpin (*Myoxocephalus scorpius*), caught in four harbours on the southwest coast of Iceland. *Aquatic Toxicology*, Vol.48, No.4, (April 2000), pp. 431-442, ISSN 0166-445X

Swain, T. & Hillis, W.E. (1959). *The phenolic constituents of* Prunus domestica. *I. The quantitative analysis of phenolic constituents.* Journal of the Science of Food and Agriculture, Vol.10, No.1, (1959), pp. 63-68, ISSN 0022-5142

Swain, T. (1977). Secondary compounds as protective agents. *Annals Review of Plant Physiology*, Vol.28, (1977), pp. 479-501, ISSN 0066-4294

Tett, P.; Gowen, R.; Mills, D.; Fernandes, T.; Gilpin, L.; Huxham, M.; Kennington, K.; Read, P.; Service, M.; Wilkinson, M. & Malcolm, S. (2007). Defining and detecting undesirable disturbance in the context of marine eutrophication. *Marine Pollution Bulletin*, Vol.55, No.1-6, (2007), pp. 282-297, ISSN 0025-326X

Tomlinson, P.B. (1974). Vegetative morphology and meristem dependence: The formation of productivity in seagrasses. *Aquaculture*, Vol.4, (1974), pp. 107-130, ISSN 0044-8486

Vergeer, L.H.T.; Aarts, T.L. & Groot, J.D. (1995). The wasting disease and the effect of abiotic factors (light intensity, temperature, salinity) and infection with *Labyrinthula zosterae* on the phenolic content of *Zostera marina* shoots. *Aquatic Botany*, Vol.52, No.1, (September 1995), pp. 35-44, ISSN 0304-3770

Vangronsveld, J.; Mocquot, B.; Mench, M. & Clijsters, H. (1997). Biomarqueurs du stress oxydant chez les végétaux, In: *Biomarqueurs en écotoxicologie - Aspects fondamentaux*, Lagadic L.; Caquet T.; Amiard J.C.; Ramade F., pp. 165-181, Dunod, ISBN 2225830533, Paris, France

Verges, A.; Becerro, M.A.; Alcoverro, T. & Romero, J. (2007). Experimental evidence of chemical deterrence against multiple herbivores in the seagrass *Posidonia oceanica*. *Marine Ecology Progress Series*, Vol.343, (August 2007), pp. 107-114, ISSN 0171-8630

Ward, T.J. (1987). Temporal variation of metals in the seagrass *Posidonia australis* and its potential as a sentinel accumulator near a lead smelter. *Marine Biology*, Vol.95, No.2, (February 1987), pp. 315-321, ISSN 0025-3162

White, C.S. (1986). Volatile and water-soluble inhibitors of nitrogen mineralization and nitrification in a ponderosa pine ecosystem. *Biology and Fertility of Soils,* Vol.2, (1986), pp. 97-104, ISSN 0178-2762

Wyllie-Echeverria, S. & Ackerman J.D. (2003). The seagrasses of the Pacific Coast of North America, In: *World Atlas of Seagrasses,* Green, E.P.; Short F.T., pp. 199-206, UNEP World Conservation Monitoring Centre, University of California Press, Berkeley, ISBN 0-520-24047-2, USA

Wyllie-Echeverria, S.; Talbot S.L. & Rearick J.R. (2010). Genetic structure and diversity of *Zostera marina* (eelgrass) in the San Juan Archipelago, Washington, USA. *Estuaries and Coasts,* Vol.33, No.4, (July 2010), pp. 811-827, ISSN 1559-2723

Zapata, O. & Mc Millan, C. (1979). Phenolic acids in seagrasses. *Aquatic Botany,* Vol.7, (1979), pp. 307-317, ISSN 0304-3770

Zar, J.H. (1999). *Biostatistical analysis,* Prentice-Hall International, ISBN 9780130815422, U.K

7

Biofilms Impact on Drinking Water Quality

Anca Farkas[1], Dorin Ciatarâş[1] and Brânduşa Bocoş[2]
[1]*Someş Water Company,*
[2]*National Public Health Institute, Regional Public Health Center of Cluj,*
Cluj-Napoca,
Romania

1. Introduction

A paradigm shift currently occurs in microbiology, with significant impacts in a variety of environmental, medical and industrial applications. The old misconception of free floating microbes is invalidated by a different knowledge pattern: the great majority of terrestrial microorganisms live in communities associated to surfaces, called biofilms (Costerton et al, 1987; Flemming, 2008; Muntean, 2009). This organisation mode is associated to all surfaces in contact with water in drinking water processing, storage and distribution. Such biofilms are represented by structured consortia of sessile microorganisms characterized by surface attachment, self-produced exopolymeric matrix, structural, functional and metabolic heterogeneity, capable of intercellular communication by quorum-sensing and plurispecific composition.

Biofouling in drinking and industrial water systems has detrimental effects such as microbiological and chemical deterioration in water quality, corrosion inducing, drinking water treatment yield loss, efficiency reducing in cooling and heating exchange and transport, as well as in membrane processes (White et al, 1999; Flemming et al, 2002; LeChevallier and Au, 2004; Coetser and Cloete, 2005).

Biofilms are playing a major role in drinking and waste water treatment processes due to their enhanced properties of mineralization, bioaccumulation and bioadsorbtion. Despite the beneficial effects of the biological filter known as *schmutzdecke* in slow sand filtration or of the bio-sand filters, biofilms occurrence in other treatment stages, in drinking water networks and reservoirs represents a continuous challenge to water professionals. Drinking water associated biofilms induce residual disinfectants depletion and may cause aesthetic problems consisting in colour, odour and taste degradation due to chemical compounds released and more important, they pose a threat to human and animal health by hosting pathogenic or toxins producing bacteria, viruses, protozoa, algae, fungi and invertebrates. The great majority of water related health problems are the result of microbial contamination (Riley et al, 2011). Considering these aspects, naturally occurring biofilms in contact with drinking water were identified and described as microbial reservoirs for further contamination (Szewzyk et al, 2000; Wingender and Flemming, 2011).

The complex structure of drinking water associated biofilms is influenced by the microbial composition of source water and sediments (LeChevallier et al, 1987; Szewzyk et al, 2000;

Emtiazi et al, 2004). They may enter the distribution network, escaping the treatment and disinfection processes (known as breakthrough) and multiply in bulk water or biofilms. The two modes of multiplication are defined as regrowth (recovery of disinfectant injured cells) and aftergrowth (microbial growth in a distribution system) processes (Characklis, 1988; van der Kooij, 2003). Pathogenic microorganisms of concern may also emerge in drinking water systems by intrusion, due to external contamination events in different steps of water treatment, storage and transportation: cross connections, backflow events, pipe breaks, negative pressure and because of improper flushing and disinfection procedures.

2. Drinking water biofilms, emerging pathogens and opportunistic pathogens

The most alarming consequences as a result of biofouling in drinking water distribution systems consist in the presence, multiplication and dispersion into water of bacterial pathogens, opportunistic pathogens, parasitic protozoa, viruses and toxins releasing fungi and algae. They may appear as primary colonizers promoting the adhesion at the interface and subsequent biofilm formation (Costerton, 1994), but more often as secondary colonizers in ecological microniches offered by the existent attached community.

Emerging pathogens are those that have appeared in a human population for the first time, or have occurred previously but are increasing in incidence or expanding to areas where they have not previously been reported, usually over the last 20 years. They include: bacteria (pathogenic *E. coli, Helicobacter pylori, Campylobacter jejuni, Mycobacterium* avium complex), parasitic protozoa (*Cryptosporidium* spp., *Cyclospora cayetanensis, Toxoplasma gonidii*), viruses (noroviruses, hepatitis E) and toxic cyanobacteria (Hunter et al, 2003). Opportunistic pathogens are commonly members of water microbiota that would be normally harmless to a healthy individual but can infect a compromised host (US EPA, 2002).

Such microorganisms were detected worldwide in drinking water and associated biofilms and in raw water and sediments. In this context, establishing health-based targets, drinking water quality assurance implies effective preventive measures, such as sources water protection in order to reduce contamination risk in these strategic environments, as well as corrective actions stipulated in water safety plans prepared by water suppliers.

When cases of illness are registered, epidemiological studies are conducted in order to demonstrate similarities in genetic profiles of strains isolated from clinical and the environmental specimens, to track the source of infection. Drinking water and associated biofilms are often among the prime candidates tested when gastrointestinal diseases and different types of infections are recorded.

Developing countries are facing a major lack of safe drinking water, the transmission via faecal-oral route causing enormous numbers of severe water related illness and cases of deaths, especially in infants (Riley et al, 2011). Microbial source tracking approach is particularly important for drinking water sources, in order to identify the origin of contamination. Increasing population's access to clean drinking water and sanitation facilities is one of the first priorities of local and global authorities.

Even if the access to high quality drinking water is provided, neither humans in rich countries have definitely won the fight against microbes, yet. Tap water quality assurance is facing new challenges, consisting mainly in biofouling issues, emergent waterborne pathogens, toxins releasing and opportunistic pathogens occurrence. Besides households, different branches of industry, especially food and pharmaceutical, health care facilities, schools, nursing homes and other critical areas are carefully included in monitoring assays. Recently is seriously considered the ability of some species, referred as opportunistic pathogens, to induce disease under certain circumstances in immunocompromised individuals: immunosuppressed, malnourished, diabetic, burn, cancer, AIDS, on haemodialysis, respiratory, with organ transplants patients (Rusin et al, 1997; Payment and Robertson, 2004). Sensitive subpopulations such as young children, elderly persons or pregnant women are also vulnerable to infections caused by opportunistic pathogens (Reynolds et al, 2008). Other special categories of exposed subjects consist in patients with indwelling cannulae and catheters, implant devices and contact lenses wearers. Opportunistic pathogens are becoming a major issue, causing from allergy or superficial infections to life-threatening systemic infections, since the ascending trend in congenital and acquired immunodeficiency affecting global population. Such species presence is often investigated, in addition to routine monitoring, within drinking water and associated biofilms, as wide-occurring bacteria of concern in the continuous increasing category of hospitalised and ambulatory immunocompromised persons (Glasmacher et al, 2003).

Even in drinking water carefully treated and distributed at high standards, pathogenic contamination and disease outbreaks may occur (Szewzyk et al, 2000; Wingender & Flemming, 2011) demonstrating the imperative requirement for comprehensive water safety plans implementation.

2.1 Drinking water quality assessment - Microbiological aspects and biofilms

Tap water supposes not to be and is not sterile, microbial load in bulk water consisting mainly in inoffensive heterotrophs, presumably coming from associated biofilms by detachment during dispersion. Routine monitoring of raw water sources, finishing water at the exit from treatment plant, drinking water in pipe networks, service reservoirs and finally at the consumers implies periodically investigations of a number of water samples collected with a frequency depending on the population deserved. According to European regulations, microbial indicators assessed by standardised conventional culturing techniques are: colony count at 37°C, colony count at 22°C, total coliforms, *Escherichia coli*, intestinal enterococci and *Clostridium perfringens*. The greatest microbial risk being associated with ingestion of water contaminated with human or animal faeces, thus the potential presence of pathogenic bacteria, viruses and cysts of protozoan parasites; faecal indices (*E. coli*, intestinal enterococci and *C. perfringens*) presence is routinely investigated.

The shortcomings of water quality monitoring based on faecal indicators and heterotrophic plate count, resulting in underestimation of drinking water microbial populations in numbers and composition are discussed worldwide considering the following:

- Only a small volume approximated to represent from 2×10^{-7} to 5×10^{-7}% of delivered drinking water is examined in routine monitoring (Allen, 2011);
- In drinking water systems, the high majority of bacteria, estimated at 95%, are located attached at the surfaces, while only 5% are found in water phase and detected by sampling as commonly used for quality control (Flemming et al, 2002). Other studies

are indicating bacterial numbers characterizing the biomass in pipe biofilms being 25 times more abundant than the suspended cells (Servais et al, 2004). Although, the common notion that biofilms dominate the distribution systems has been proven to be not true under all conditions by Srinivasan (2008), whose findings suggest that bulk bacteria may dominate in network sections containing chlorine residuals lower than 0.1mg/L and having residence time longer than 12 hours.

- A significant percent of water and biofilm bioburden may be in a viable but non cultivable state, unable to grow on artificial growth media but alive and capable of renewed activity and so hygienically relevant (Moritz et al, 2010). A small fraction of waterborne microorganisms (0.01%) are estimated to be culturable heterotrophic bacteria (Watkins and Jian, 1997; Exner et al, 2003);
- Limitations of detection methods (Lehtola et al, 2006; 2007; September et al, 2007). The investigation of drinking water associated biofilms from four European countries (France, Great Britain, Portugal and Latvia) confirmed *E. coli* presence by culturing techniques in one out of five pipes whereas all networks except one were positive for *E. coli* using the PNA FISH methods; their viability was also demonstrated. *E.coli* contributed with percents from 0.001% to 0.1% in the total bacterial numbers (Juhna et al., 2007);
- Faecal indicators are the best predictors of potential risks, but their concentrations rarely correlate perfectly with those of pathogens (Payment and Locas, 2011). Although in freshwater significant correlations have been established between faecal indices and pathogenic species, their presence in drinking water showed limited or no correlation with different species of pathogenic or opportunistic pathogenic bacteria, viruses, protozoa and fungi. Water quality assessment based only on the investigation of faecal indicators' presence proved to be insufficient when many waterborne outbreaks emerged. Still, until more reliable indices and methods of detection will be wider implemented, the well- known standardised procedures are applied in routine monitoring across the globe.

Many studies targeting attached microbial communities have been performed for quantification of the total number of germs, by different methods. They offer an unspecific overview upon microbial load, bringing certain information about drinking water treatment process efficiency and distribution system integrity. Still, the real composition and dynamics of microbial populations within drinking water associated biofilms represents a continuous challenge. Experimental biofilm succession monitored for a long term development indicated a stable population state after 500 days in a model drinking water distribution system. A homogenous composition of the population in the mature biofilm could mask a dynamic situation at a smaller scale (Martiny et al., 2003). Quantitative and prescriptive evaluation is the next target of scientific community. Prediction of microorganisms' behaviour in the distribution system water and biofilms requires greater understanding of the effects in microbial attachment, detachment, survival, multiplication and viability of three groups of abiotic and biotic factors: substratum physicochemical properties (type of materials), biofilm composition (microbial intra- and interspecific interactions) and bulk water characteristics (disinfectants residuals, oxygen and nutrients concentrations, system hydraulics, temperature).

2.1.1 Occurrence of bacteria in drinking water and associated biofilms

Among the nuisance bacteria regularly found in drinking water and biofilms, species that are not characteristic to the water environment may appear due to contamination events, with major impacts upon human health. Enteric bacteria such as *Escherichia coli, Klebsiella pneumoniae, K. oxytoca, Enterobacter cloacae, E. aglomerans, Helicobacter pylori, Campylobacter* spp., *Shigella* spp., *Salmonella* spp., *Clostridium perfringens, Enterococcus faecalis, E. faecium*, as well as environmental bacteria becoming opportunistic pathogens *Legionella pneumophila, Pseudomonas aeruginosa, P. fluorescens, Aeromonas hydrophila, A. caviae, Mycobacterium avium, M. xenopi*, together with other waterborne agents have been indicated to live in ecological microniches offered by drinking water associated biofilms (table 1).

When compared with planktonic counterparts, biofilm bacteria and other inhabitants display superior characteristics due to specialization within this emergent structure and to complex relationships established (Costerton, 1994). Community belonging, from a microbial perspective represents a benefit materialized in increasing the chances of survival in this oligotrophic environment by offering ecological microniches, establishing intra- and interspecific cooperation relationships by communication via quorum sensing and perpetuating individuals' resistance to disinfection agents. Even species not able to survive and most of them incapable of growth and multiplication in water were identified in associated biofilms; recent studies have demonstrated their ability to grow in those microniches. For example, *Legionella pneumophila* survives but does not multiply in sterile drinking water, its proliferation being dependent on parasitic relationship with other microorganisms: 14 species of amoebae, two species of ciliated protozoa, and one slime mould - *L. pneumophila* being described as protozoonotic bacteria (Murga et al, 2001; Fields et al, 2002; Declerck, 2010). In many outbreaks, the presence of pathogenic bacteria was not detected by routine monitoring, the correlation with faecal indicators found in tap water samples being defective. *E. coli* bacillus, the most popular faecal indicator, was chosen inter alia based on its incapacity of growth in water. Recent studies have shown its ability to multiplicate in drinking water associated biofilms under strictly anaerobic conditions (Latimer et al., 2010), so the indicative value of the faecal index of choice becomes questionable.

One of the advantages offered by drinking water biofilm organization to its members is represented by the enhanced resistance to disinfection residuals. The four hypothetical mechanisms of biofilm resistance involve slow antimicrobial penetration, deployment of adaptative stress responses, physiological heterogeneity in biofilm population and the presence of phenotypic variants or persister cells (Chambless et al, 2005). Another benefit from the microbial perspective consists in the emergence of genetically encoded resistance to biocides and antibiotics, and the spread of antimicrobial resistance genes in bacterial populations via mobile genetic elements, by lateral gene transfer. Integrons are genetic elements possessing a site-specific recombination system for assembling of resistance genes in gene cassettes. They play a major role in the rapid spread of antibiotic resistance in clinical environments. Gene cassettes encoding resistance to quaternary ammonium compounds (*qac*) and integron-integrase (*intlI*) genes characteristics for class 1 integron were recently recovered from environmental samples, including biofilms from a groundwater treatment plant (Gillings et al, 2009). The proximity of individuals in biofilm consortia and the extremely short generation times in bacteria multiplication are prerequisites for intensive rates of lateral gene transfer and thus resistance spreading and perpetuation in diverse natural or artificial ecosystems.

Bacteria	Samples type/Origin	Country	References
Escherichia coli	Biofilm - WDS Biofilm - WDS Biofilm – WDS Biofilm, Water - WDS Biofilm - DWTP	USA Germany France, England, Portugal, Latvia South Africa Romania	LeChevallier et al, 1987 Schmeisser et al, 2003 Juhna et al, 2007 September et al, 2007 Farkas et al, 2011
Faecal enterococci	Water - WDS Water - WDS Biofilm - WDS Biofilm - DWTP	Korea South Africa Portugal Romania	Lee et al, 2006 September et al, 2007 Menaia et al, 2008 Farkas et al, 2011
Clostridium spp.	Biofilm - WDS Water - WDS Biofilm - DTP	Portugal Greece Romania	Menaia et al, 2008 Kormas et al, 2010 Farkas et al, 2011
Klebsiella spp.	Biofilm - WDS Water - WDS Biofilm, Water - WDS	USA Korea South Africa	LeChevallier et al, 1987 Lee et al, 2006 September et al, 2007
Pseudomonas spp.	Biofilm - WDS Biofilm - WDS Water - WDS Biofilm, Water - WDS Biofilm - WDS Biofilm - DWTP	Germany Germany Korea South Africa Portugal Romania	Schmeisser et al, 2003 Emtiazi et al, 2004 Lee et al, 2006 September et al, 2007 Menaia et al, 2008 Farkas et al, 2011
Aeromonas spp.	Water - WDS Biofilm, Water - WDS Biofilm - WDS Water - WDS Biofilm, Water - WDS Water - WDS Biofilm - DWTP	Scotland USA Australia Korea South Africa Brasil Romania	Gavriel et al, 1998 Chauret et al, 2001 Bomo et al, 2004 Lee et al, 2006 September et al, 2007 Razzolini et al, 2008 Farkas et al, 2011
Vibrio spp. *V. cholerae*	Water - WDS Biofilm, Water - WDS Biofilm - pond Water - reservoirs	Korea South Africa Bangladesh Sudan	Lee et al, 2006 September et al, 2004; 2007 Alam et al, 2007 Shanan et al, 2011
Mycobacterium spp.	Biofilm - WDS Biofilm - WDS Water - WDS Water - WDS	Germany South Africa Greece USA	Schmeisser et al, 2003 September et al, 2004 Kormas et al, 2010 Marciano-Cabral et al, 2010
Shigella spp., *Salmonella* spp.	Biofilm - WDS Water - WDS	Germany Korea	Schmeisser et al, 2003 Lee et al, 2006
Campylobacter spp.	Water - WDS Raw water	Finland France	Hänninen et al, 2002 Gallay et al, 2006
Helicobacter pylori	Biofilm - WDS Biofilm - WDS	England Portugal	Watson et al, 2004 Bragança et al, 2005
Legionella pneumophila	Biofilm - WDS Water - WDS Biofilm - WDS Water - WDS	Germany The Netherlands Portugal USA	Emtiazi et al, 2004 Diederen et al, 2007 Menaia et al, 2008 Marciano-Cabral et al, 2010

Table 1. Pathogenic and opportunistic pathogenic bacteria detected in association with drinking water; WDS – water distribution systems, DWTP - drinking water treatment plant.

Experimental studies emphasized on bacteria ability of colonization, survival and multiplication in water associated biofilms, followed by dispersion in water phase in a planktonic state. The findings of Banning et al. (2003) suggested that the ability of *P. aeruginosa* to survive longer than *E. coli* in water associated biofilm could not be attributed to the association with the biofilm, rather than to the ability to utilize a wider range of organic molecules as carbon and energy sources compared to other *Enterobacteriaceae*. An increment in available nutrients may reduce *E. coli* survival in enhanced competition for nutrients and increased antagonism by the indigenous microbial population.

Lehtola et al. (2007) investigated the survival of faecal indices versus pathogenic bacteria and viruses, in drinking water biofilms experimentally infested with *E. coli, L. pneumophila, Mycobacterium avium* and canine calcivirus (as a surrogate for human norovirus). The results proved that pathogenic bacteria and virus particles entering water distribution systems can survive in biofilms for weeks, even in conditions of high-shear turbulent flow and may pose a risk to the consumers. Meanwhile, *E. coli* registered a limited survival to a few days in water and in biofilms, being a poor indicator of certain pathogens in biofilms. The study also showed that standard culture methods may seriously underestimate the real numbers of bacteria in water and biofilms.

Comparative evaluation of classical techniques involving bacterial growth on specific selective media and molecular methods based on 16s rDNA sequence identity reveals a high discrepancy between what was expected to grow and the species isolated from specific selective growth media. Bacterial analyses of water based on selective isolation and culturing approach is recommended to be interpreted with caution (September et al, 2007).

Experimental studies revealed also low detectable numbers by culture-based technique in case of potable water biofilms infected with *Campylobacter jejuni* (Lehtola et al, 2006). *C. jejuni* and *C. coli* waterborne epidemics registered in Finland (Hänninen et al, 2000) and France (Gallay et al, 2000) were associated to consumption of contaminated tap water with origin in polluted sources.

Severe outbreaks such as cholera caused by the ingestion of water contaminated with *Vibrio cholerae*, typhoid and paratyphoid enteric fevers caused by *Salmonella enterica* subsp. *enterica* serovar *Typhi*, respective serovar *Paratyphi*, shigellosis due to infections with *Shigella* species still occur in countries with insufficient access to safe water. But even in developed countries, outbreak events involving emerging pathogenic bacteria like *Legionella pneumophila*, waterborne E.coli O157:H7 and foodborne *E.coli* O104:H4 demonstrate the microbes' versatility and the fragility of humanity's victory over the nature.

2.1.2 Occurrence of protozoa in drinking water and associated biofilms

The food web in drinking water microbial consortia is based on heterotrophic bacteria, the next trophic level being represented by protozoa. Species of parasitic protozoa, including free living amoebae associated with infections in humans have been isolated from source waters and drinking water (table 2). Their presence represents a double threat to human health, being also related to amoeba-resisting bacteria, such as *Legionella* spp. and *Mycobacterium* spp., which proliferate in protozoa thus increasing the probability of causing diseases in humans (Marciano-Cabral et al, 2010).

Some protozoa, for example *Giardia* spp. and *Cryptosporidium* spp. may persist to hostile environment in drinking water, resist to different disinfection procedures and accumulate in biofilms under the form of cysts, respective oocysts. Experimental introducing *Cryptosporidium* oocysts for the prediction of behaviour in drinking water distribution system showed surface attachment and subsequent intermittent detachment, with exposure to high doses of chlorine (20mg/L) needed for the removal of substantial numbers of oocysts attached to pipe walls (Warneke et al, 2006). The study of Helmi et al. (2008) investigating the interaction of *Giardia lamblia* and *Cryptosporidium parvum* (oo)cysts in drinking water biofilms, revealed that protozoa are able to attach in biofilm matrix from the first day and survive extended periods of time, longer for *Cryptosporidium*. Viable (oo)cysts were recovered from biofilm and water phase for the whole period of investigation, of 34 days, turbulent shear stress influencing the detachment.

Protozoa	Samples type/Origin	Country	References
Flagellates: *Giardia lamblia*	Filtered water - DWTP Water - WDS Water - WDS Water - WDS	USA Canada Australia Spain	LeChevallier et al, 1991 Chung et al, 1998 Hellard et al, 2001 Carmena et al, 2007
Apicomplexa (Sporozoans): *Cryptosporidium parvum*	Filtered water - DWTP Water - WDS Water - WDS	USA Canada Spain	LeChevallier et al, 1991 Chung et al, 1998 Carmena et al, 2007
Amoebae: *Naegleria fowleri* *Acanthamoeba* spp. *Hartmannella* spp. *Vahlkampfia* spp.	 Well water Biofilm, Water - WDS Biofilm, Water - WDS Water reservoir	 USA USA USA Sudan	 Blair et al, 2008 Marciano-Cabral et al, 2010 Shoff et al, 2010 Shanan et al, 2011

Table 2. Protozoa detected in raw water sources, water treatment plants (DWTP) and drinking water networks (WDS) and associated biofilms.

2.1.3 Occurrence of viruses in drinking water and associated biofilms

Sources of drinking water were investigated for the presence of enteric viruses, especially when gastrointestinal outbreaks occurred, and the results revealed episodes of faecal contamination in raw water. Epidemiological studies conducted supported the association between drinking water consumption and illness (table 3).

Viruses	Samples type/Origin	Country	References
Hepatitis A virus Hepatitis E virus	Well water Water - WDS	USA India	Bloch et al, 1990 Hazam et al, 2010
Noroviruses	Well water Spring water Groundwater Groundwater	USA Finland New Zeeland Korea	Parshionikar et al, 2003 Maunula et al, 2005 Hewitt et all, 2007 Koh et al, 2011
Coxsakie A viruses	Raw water	Taiwan	Hsu et al, 2009
Adenoviruses Rotaviruses	Drinking water sources	West Africa	Verheyen et al, 2009

Table 3. Viruses identified in raw water sources and water distribution networks (WDS).

The presence of enteric viruses associated with inadequate water supplies, poor sanitation and hygiene is mostly affecting developing countries (Ashbolt et al, 2004). Episodes of gastroenteritis caused especially by noroviruses attributed to contaminated drinking water have been reported also in developed countries. Inefficient raw water treatment and secondary contamination of distribution systems with sewage are of high concern, enteric viruses being generally more resistant than enteric bacteria to widely used free chlorine, chlorine dioxide and monochloramine disinfectants (LeChevallier & Au, 2004). Although no complete investigations regarding faecal indicators presence were performed in the considered studies (especially for intestinal enterococci as index of viruses), coliforms and *E. coli* have been detected in water samples in many cases of gastroenteritis outbreaks investigated (Parshionikar et al, 2003; Hewitt et al, 2007; Koh et al, 2011).

There is no evidence of viruses ability of multiplication in environmental biofilms, but they may survive for extended periods of time trapped in the matrix, similarly to protozoan (oo)cysts, and be detached in water column, where remain inactive until they find a host. Experimental studies using pilot scale systems demonstrated the ability of viruses to attach and accumulate into drinking water biofilms within one hour after inoculation, while their detachment in water phase is influenced by flow velocity (Lehtola et al, 2007; Helmi et al, 2008). The viral genomes were detected in biofilms over the whole period of both experiments (for 21, respectively 34 days). Helmi and co-workers, investigating the poliovirus infectivity, recovered the infectious viruses only for 6 days, when flow velocity increment from laminar to turbulent regimen was applied, concluding that detection of viral genome in biofilms is not sufficient to assess a risk associated with the presence of infectious particles.

2.1.4 Occurrence of fungi in drinking water and associated biofilms

Initially considered to be airborne, fungal infections in immunocompromised patients hospitalized in controlled atmospheric conditions raised the hypothesis of waterborne origin of aspergillosis (Anaissie & Costa, 2001). Opportunistic pathogens, potentially causing superficial or systemic infections, allergenic or toxigenic species of fungi (yeasts and moulds) have been isolated from drinking water worldwide, their presence being primary attributed to the ability of surfaces colonization as biofilms (table 4).

Fungi	Samples type/Origin	Country	References
Paenicillium spp.	Biofilm - WDS	USA	Doggett, 2000
Aspergillus spp.	Water - WDS	Germany	Göttlich et al, 2002
Cladosporium spp.	Biofilm, water - WDS	USA	Kelley et al, 2003
Epicoccum spp.	Water - WDS	Norway	Hageskal et al, 2006
Alternaria spp.	Water - WDS	Portugal	Gonçalves et al, 2006
Trichoderma spp.	Water WDS	Brazil	Pires-Gonçalves et al, 2008
Acremonium spp.	Water - WDS	Australia	Sammon et al, 2010
Exophiala spp.			
Phialophora spp.			
Fusarium spp.			
Mucor spp.			
Candida spp.			

Table 4. Fungi identified in drinking water distribution systems (WDS) and associated biofilms.

In some studies, the correlation with standard hygiene indicators was not found (Göttlich, 2002), other authors described negative correlations between bacteria and filamentous fungi, which may be explained either by competition for nutrients either by inhibiting toxins produced (Gonçalves et al, 2006) while in other investigations positive significant correlations were found between the presence of filamentous fungi, yeasts and bacteria in drinking water (Sammon et al, 2010). Regarding filamentous fungi behaviour in water distribution systems, deposition is attributed to highly resistant spores, while mycotoxins, taste and odour changing compounds producing implies germination and hyphal growth in biofilms. The occurrence of fungi in drinking water systems may have significant impact due to health effects of mycotoxins (such as aflatoxins): mutagenic, teratogenic, oestrogenic, carcinogenic and allergenic, although no reports of disease attributed to mycotoxins produced in the water distribution systems have been reported (Sonigo et al, 2011).

2.1.5 Occurrence of algae in drinking water and associated biofilms

Algae are assumed not to be characteristic to water distribution system biofilms due to the absence of light (Wingender and Flemming, 2011), but algal biomass is a major component of biofilms in surface source waters, water treatment and storage, in areas exposed to air and light. Experimental research designed by Chrisostomou et al (2009) emphasized on air-dispersed phytoplankton diversity and colonization potential of algal taxa in drinking water reservoir systems. Algal communities are associated to biofilms and may support bacterial growth, for example *Legionella* species (Declerck, 2010). Few recent studies investigating the presence of algae in drinking water are available (table 5).

Algae	Samples type/Origin	Country	References
Oocystis spp. *Xenococcus* spp.	Water - WDS	Spain	Codony et al, 2003
Anabaena spp. *Microcystis* spp. *Oscillatoria* spp. and many more	Water - WDS	Argentina	Ricardo et al, 2006
Microcystis aeruginosa *Chroococcus dispersus*	Water reservoirs	Greece	Lymperopoulou et al, 2011

Table 5. Algae identified in drinking water distribution systems (WDS).

Algal toxins, of which the most dangerous for humans is cyanobacterial microcystin, are considered chemical hazards in drinking water, especially when open-air reservoirs are used in water storage (Lymperopoulou et al, 2011). Algal growth and eutrophication in surface waters are widely investigated, with respect to ecological effects. In drinking water sources and throughout the water treatment process, distribution and storage, algal blooms raise issues about toxins releasing and aesthetic problems inducing, such as colour and smell. Algal removal in drinking water treatment is recommended to be carefully performed, in order not to disrupt the cells and release toxins in drinking water (LeChevallier and Au, 2004). Epidemiological studies are conducted worldwide in order to demonstrate the evidence of algal toxins in the environment and to evaluate their relatedness to illness in humans. Possible linkages between algae toxins in drinking water and health effects, including liver problems and diarrhoea in children were indicated by a survey in Namibia, although microcystin never exceeded the tolerable daily intake (Gunnarsson and Sanseovic, 2001).

3. Drinking water and associated biofilms – Chemical aspects

Detrimental effects of biofouling in drinking water distribution systems include chemical aspects, involving organic and inorganic compounds produced by the microorganisms inhabiting water phase, biofilms and sediments. Different volatile compounds, organic and inorganic acids, metal oxides and enzymes resulted in microbial metabolism or decay may cause aesthetic problems in water: colour, taste and odours and may also have an impact on the substratum, leading to microbially influenced corrosion.

3.1 Drinking water aesthetic problems

Aesthetic and organoleptic characteristics of water may be affected by a series of chemical substances, resulting in colour, odour and taste degradation. Such substances originate in microbial activity and decomposition in source waters and in distribution systems, disinfectants used in water treatment, materials used in pipes and joints in water networks. A list of these substances, related to microbial activity and decay that may be produced in the journey of drinking water from drinking water sources to the tap, that may influence consumers perception, is presented in table 6 (after the UK Environment Agency, 2004). These chemical compounds are usually attributed to microbial biofilms associated to drinking water processing and distribution.

Investigating the sources of taste and odour in drinking water in order to find their sources and mitigation strategies, Peter (2008) concluded that low concentrations in chlorine residuals, stagnant water, plastic pipes and particles accumulation in distribution systems may increase the generation of taste and odour compounds by favouring biofilm formation and microbial activity. Other sources of aesthetic problems in water may reside in the activity of bacteria involved in sulphur cycle, producing sulphur odours and yellow discoloration (US EPA, 2002). Oxidation and reduction of soluble metals may produce metal oxides, leading to consumer complaints about the metallic taste and yellow, black or brown staining water (Cerrato et al, 2006).

3.2 Microorganisms – Surface interactions and microbially influenced corrosion

Biofouling proved to be interdependent on surface characteristics. Investigations of microbial reversible and irreversible attachment in primary or secondary colonization and in drinking water biofilms composition concluded as following:

- The hydrophobic/hydrophilic properties of the substrate are influencing biofilm formation. Exopolysacharides produced by some bacteria facilitate cell adhesion to hydrophilic surfaces, while exopolymers of other bacteria may show a preference for hydrophobic substrata (Beech et al, 2005).

Regarding the influence of the substratum on biofilm composition, copper materials appear to be colonized just by *L. pneumophila* in low numbers, inhibiting *P. aeruginosa* integration, while drinking water biofilms on elastomeric and polyethylene materials proved to be a better support for pseudomonads (Moritz et al, 2010).

- Pipe materials may be corroded, influencing disinfection effectiveness: corrosion products in iron pipes react with free chlorine and lead to residual disinfectants depletion. LeChevallier et al. (1987) detected high concentrations of coliforms only in tubercles formed on iron pipes and suggested few possible explanations: coliform growth stimulated by iron oxides; nutrient syntropy; sourface roughness; protection from disinfection. Iron pipes may be a better support for fungi also, when compared to PVC pipes (Doggett, 2000).

Drinking water flowing through PVC pipes contains three times the aqueous concentration of soluble manganese and 35 times the concentration of total manganese than present in the drinking water transported by iron pipes (Cerrato et al, 2006).

Microorganisms	Chemical substances produced	Aesthetic effects
Microbial decomposition	Indole, skatole, putrescine, cadaverine, β-phenylethylamine, butyric, propionic and stearic acids	Fishy, grassy, woody tastes Faecal, rotten, cheese, pungent odours
Algae decomposition	Mercaptan, dimethyl sulphide, polysulphides	Fishy, swampy, septic odours
Algae decomposition/activity	n-hexanal, n-heptanal, isomers of decadienal sulphur compounds, terpenes, aromatic compounds, esters	Fishy odours Rotten eggs odours Aromatic odours
Pseudomonas spp. *Flavobacterium* spp. *Aeromonas* spp. *Paenicillium caeseicolum*	Dimethyl polysulphides	Swampy odours
Fungi *Chaetomium globosum* *Basidiobolus ranarumi* *Actinomycetes*	Geosmin Cadin-4-ene-1-ol 2-isopropyl-3-methoxypyrazine	Earthy, musty taste and odour Woody, earthy odour Musty, mouldy potato odour
Actinomycetes: *Streptomyces* spp. *Nocardia* spp. *Microbiospora* spp Cyanobacteria: *Anabaena* spp. *Microcystis* spp. *Oscillatoria* spp. *Aphanizomenon* spp. Algae: Chlorophyceae Bacillariophyceae	Geosmin 2-methylsorboneol	Earthy, musty taste and odours
Sulphur oxidizing/reducing bacteria Sulphate reducing bacteria	Sulphuric acid, sulphates, sulphur, methyl mercaptan, hydrogen sulphide Metal sulphides (ferrous sulphide)	Rotten eggs, rotten cabbage odours Yellow, brown, black staining
Metals oxidizing/ reducing bacteria	Metal oxides	Rusty or metallic taste Brown, black staining

Table 6. Chemical compounds produced by microbial decomposition and metabolism, affecting taste and odour of drinking water.

Microbially influenced corrosion represents another undesirable impact of biofilms associated to drinking water treatment and distribution, involving metallic or non-metallic materials deterioration as a result of pipes inner surface biofouling.

Physiological groups of bacteria classified on account of the ability to use different substrates in their nutrition or in respiration are summarized in table 7 (Drăgan-Bularda & Kiss, 1986; Drăgan-Bularda & Samuel, 2006; Muntean, 2009). Their representative species may belong to microbial communities of source waters and sediments, enter drinking water treatment plants and distribution systems in a planktonic state and adhere to surfaces or become members of established biofilms (Costerton, 1994). Their metabolites have significant impacts on drinking water quality, either being released in bulk water where may react with other compounds, for example with disinfectants, leading to toxic disinfection-by products (as trihalomethanes), or by remaining in biofilm matrix where acting upon pipes surfaces and inducing corrosion.

Physiologycal groups of bacteria	**Representatives**	**Metabolites produced**
Ammonifying bacteria	*Bacillus* spp. *Clostridium* spp. *Pseudomonas* spp. *Burkholderia* spp.	Ammonium Ammonia
Nitrosifiers (Ammonia oxidizing bacteria)	*Nitrosomonas* spp. *Nitrocystis* spp. *Nitrospira* spp. *Nitrosolobus* spp. *Nitrosovibrio* spp.	Nitrite ions
Nitrifying bacteria (Nitrite oxidizing bacteria)	*Nitrobacter* spp *Nitrococcus* spp. *Nitrospira* spp. *Nitrospina* spp.	Nitrate ions
Denitrifying bacteria	*Paracoccus denitrificans* *Pseudomonas stutzeri* *Thiobacillus denitrificans* *Alcaligenes* spp. *Bacillus* spp.	Nitrous oxide Nitrogen
Sulphur reducing bacteria	*Desulfuromonas* spp. *Proteus* spp.	Hydrogen sulphide
Sulphate reducing bacteria	*Desulfovibrio desulfuricans* *Desulfovibrio sulfodismutans* *Desulfotomaculum* spp. *Desulfonema* spp. *Desulfosarcina* spp. *Desulfobacter* spp. *Desulfococcus* spp. *Desulfomicrobium* spp.	Hydrogen sulphide
Sulphur oxidizing bacteria:	*Thiobacillus* spp. *Sulfolobus* spp. *Beggiatoa* spp. *Thiothrix* spp.	Sulphuric acid Sulphates Sulphur
Iron reducing bacteria	*Sphaerotilus natans* *Leptothrix ochracea* *Crenothrix polyspora*	Iron (Fe^{2+}) oxides
Iron oxidizing bacteria	*Galionella feruginea* *Ferrobacillus ferooxidans* *Thiobacillus ferooxidans*	Iron (Fe^{3+}) oxides
Manganese oxidizing/reducing bacteria	*Sphaerotilus discophorus* *Pseudomonas* spp. *Metallogenium* spp. *Pedomicrobium* spp. *Bacillus* spp. *Micrococcus* spp. *Vibrio* spp.	Manganese oxides

Table 7. Physiological groups of bacteria, their representatives and metabolites.

Chemical and enzymatic microbial products resulted in biofilms activity may induce corrosion and related effects by different mechanisms:

- oxygen concentration cells and anaerobic sites generation (promoting growth of anaerobic bacteria);
- formation of iron concentration cells by the activity of iron and manganese oxidizing bacteria;
- metabolites such as acids produced by bacteria have corrosive action upon the surface;
- production of depolarizing enzymes within the biofilm matrix, which may persist longer than viable cells;
- exopolymers produced by slime forming bacteria stimulate biofilm formation and biomass accumulation;
- the binding capacity of biofilm matrix which may lead to deposits accumulation with clogging effects (Beech et al, 2005; Coetser and Cloete, 2005).

Some of the recommended strategies in drinking water associated biofilm control are: source waters protection, appropriate treatment, infrastructure contamination prevention, pipes and reservoirs maintenance, corrosion control, appropriate disinfection practices, nutrient levels reducing, water quality monitoring, personnel training, water safety plans implementation.

We are still living in an age of surfaces, even the remark was first said by Oscar Wild's character in 1895. Having in mind the virtual idea of self-cleaning surfaces, researchers in nanotechnology field are targeting innovative repellent materials with a wide range of applications, for the biofouling control. The superhydrophobicity models such as "the lotus effect" characterizing the lotus (*Nelumbo nucifera*) leaf, offered by natural patterns are investigated at nanoscale. The interdependence between surface roughness, reduced particle adhesion and water repellence proved to be the keystone in the self-cleaning mechanism of many biological surfaces (Barthlott & Neinhuis, 1997).

4. Conclusions

The present review emphasize on the following recent and relevant findings:

- Biofilms associated with drinking water are ubiquitous, harbouring bacterial pathogens, opportunistic pathogens, parasitic protozoa, viruses, toxins releasing fungi and algae;
- Microbial consortia in contact with drinking water have significant impacts upon water quality and may threat human health when contamination events occur;
- Access to safe water continues to be a target for developing countries, unfulfilled at the moment;
- Even in developed countries, where substantial efforts are submitted in order to ensure population's access to a high quality drinking water, microbial versatility represents an endless source of problems, with respect to opportunistic pathogens emergence;
- Microbial communities in water networks and biofilms represent complex ecosystems; their ecology is influenced by a series of abiotic and biotic factors: raw water sources quality, temperature, flow rate and system hydraulics, nutrient concentration, pipe material, particles accumulation, ingress and intrusion, water treatment, water disinfection and microbial interactions;
- Further research is needed in order to understand attached microbial consortia for biofouling prevention and control in drinking water industry, as a matter of public security.

5. References

Alam, M.; Sultana, M.; Nair, G.B.; Siddique, A.K.; Hasan, N.A.; Sack, R.B.; Sack, D.A.; Ahmed, K.U.; Sadique, A.; Watanabe, H.; Grim, C.J.; Huq, A. & Colwell, R.R. (2007). Viable but nonculturable *Vibrio cholerae* O1 in biofilms in the aquatic environment and their role in cholera transmission. *Proceeding of the National Academy of Sciences of the USA*, Vol.4, No.5, pp. 17801-17806, www.pnas.org/cgi/content/full/0705599104/DC1

Allen, M.J. (2011). *Escherichia coli* - the most relevant microbial public health indicator for drinking water. *Faecal indicators: problem or solution?* Conference presentation, Edinburgh, UK

Anaissie, E.J. & Costa, S.F. (2001). Nosocomial aspergillosis is waterborne. *Clin Infect Dis*, Vol.33, pp. 1546-1548

Ashbolt, N.J. (2004). Microbial contamination of drinking water and disease outcomes in developing countries. *Toxicology*, Vol.198, pp. 229-238

Barthlott, W. & Neinhius, C. (1997). Purity of the sacred lotus, or escape from contamination in biological surfaces. *Planta*, Vol.202, pp. 1-8

Banning, N.; Toze, S. & Mee, B.J. Persistence of biofilm-associated *Escherichia coli* and *Pseudomonas aeruginosa* in groundwater and treated effluent in a laboratory model system. *Microbiology*, Vol.149, pp. 47-55

Beech, I.B.; Sunner, J.A. & Hiraoka, K. (2005). Microbe-surface interactions in biofouling and biocorrosion processes. *Int Microbiol*, Vol.8, pp. 157-168

Blair, B; Sarkar, P.; Bright, K.R.; Marciano-Cabral, F. & Gerba,C.P. (2008). *Naegleria fowleri* in well water. *Emerg Infect Dis*, Vol.14, No.9, pp. 1499-1500

Bloch, A.B.; Stramer, S.L.; Smith, J.D.; Margolis, H.S.; Fields, H.A.; McKinley, T.W.; Gerba, C.P.; Maynard, J.E. & Sikes, R.K. (1990). Recovery of hepatitis A virus from a water supply responsible for a common source outbreak of hepatitis A. *Am J Public Health*, Vol.80, No.4, pp. 428-430

Bomo, A.M.; Storey, M.V. & Ashbolt, N.J. (2004). Detection, integration and persistence of aeromonads in water distribution pipe biofilm. *J Water Health*, Vol.2, No.2, pp. 83-96

Bragança, S.M.; Azevedo, N.F.; Simões, L.C.; Vieira, M.J. & Keevil, C.W. (2005). Detection of *H. pylori* in biofilms formed in a real drinking water distribution system using peptide nucleic acid fluorescence in *situ* hybridization. *Biofilm Club*, pp. 231-239

Carmena, D.; Aguinagalde, X.; Ziggoraga, C.; Fernández-Crespo, J.C. & Ocio, J.A. (2007). Presence of *Giardia* cysts and *Cryptosporidium* oocysts in drinking water supplies in northern Spain. *J Appl Microbiol*, Vol.102, pp. 619-629

Cerrato, J.M.; Reyes, L.P.; Alvarado, C.N. & Dietrich, A.M. (2006). Effect of PVC and iron materials on Mn(II) deposition in drinking water distribution systems. *Water Res*, Vol.40, pp. 2720-2726

Chambless, J.D.; Hunt, S.M. & Stewart, P.S. (2005). A three-dimensional computer model of four hypothetical mechanisms protecting biofilms from antimicrobials. *Appl Environ Microbiol*, Vol.72, No.3, pp. 2005-2013

Characklis, W.G. (1988). *Bacterial regrowth in distribution systems*, Project Report, American Water Works Association, USA, http://waterrf.org/ProjectsReports/Executive SummaryLibrary/90532_102_profiles.pdf

Chauret, C.; Volk, C.; Creason, R.; Jarosh, J.; Robinson, J. & Wrnes, C. (2001). Detection of *Aeromonas hydrophila* in a drinking-water distribution system: a field and pilot study. *Can J Microbiol,* Vol.47, No.8, pp. 782-786

Chrisostomou, A.; Moustaka-Gouni, M.; Sgardelis, S. & Lanaras, T. (2009). Air-dispersed phytoplankton in a Mediterranean river-reservoir system (Aliakmon-Polyphytos, Greece). *J Plankton Res,* Vol.31, pp. 877-884

Chung, E.; Aldom, J.E.; Chagla, A.H.; Kostrzynska, M.; Palmateer, G.; Trevors, J.T.; Unger, S. & De Grandis, S. (1998). Detection of *Cryptosporidium parvum* oocysts in municipal water samples by the polymerase chain reaction. *J Microbiol Meth,* Vol.33, No.2, pp. 171-180

Codony, F.; Miranda, A.M. & Mas, J. (2003). Persistence and proliferation of some unicellular algae in drinking water systems as a result of their heterotrophic metabolism. *Water SA,* Vol.29, No.1, pp. 113-116, http://www.wrc.org.za

Coetser, S.E. & Cloete, T.E. (2005). Biofouling and biocorrosion in industrial water systems. *Crit Rev Microbiol,* Vol.31, No.4, pp. 213-232

Costerton, J.W.; Cheng, K.J.; Gessey, G.G.; Ladd, T.I.; Nickel, J.C.; Dasgupta, M. & Marrie, T.J. (1987). Bacterial biofilms in nature and disease. *Annu Rev Microbiol* , Vol.41, pp. 435-464

Costerton, J.W. (1994). Structure of biofilms, In: *Biofouling and biocorrosion in industrial water systems,* Geesey, G.G.; Lewandowski, Z. & Flemming, H.C. (Ed.) CRC Press, ISBN 0 87371 928 X, USA

Declerck, P. (2010). Biofilms: the environmental playground of *Legionella pneumophila. Environ Microbiol,* Vol.12, No.3, pp. 557-566

Diederen, B.M.W.; de Jong, C.M.A. & Aarts, I. (2007). Molecular evidence for the ubiquitous presence of *Legionella* species in Dutch tap water installations. *J Water Health,* Vol.5, No.3, pp. 375-383

Doggett, M.S. (2000). Characterization of fungal biofilms within a municipal water distribution system. *Appl Environ Microb,* Vol. 66, No. 3, pp. 1249-1251

Drăgan-Bularda, M. & Kiss, S. (1986). *Soil microbiology* (in Romanian), Babeş-Bolyai University of Cluj-Napoca Press, Romania

Drăgan-Bularda, M. & Samuel, A.D. (2006). *General microbiology* (in Romanian), University of Oradea Press, ISBN 10 973 759 056 2

Emtiazi, F.; Schwartz, T.; Marten, S.M.; Krolla-Sidenstein, P. & Obst, U. (2004). Investigation of natural biofilms formed during the production of drinking water from surface water embankment filtration. *Water Res,* Vol.38, pp. 1197-1206

Exner, M.; Vacata, V. & Gebel, J. (2003). Public health aspects of the role of HPC - an introduction, In: *Heterotrophic plate counts and drinking water safety. The significance of HPCs for water quality and human health,* Bartram, J.; Cotruvo, J.; Exner, M.; Fricker, C. & Glasmacher, A., (Ed.), IWA Publishing, ISBN 1 84339 025 6, London, UK, pp. 12-19

Farkas, A.; Drăgan-Bularda, M.; Ciatarâş, D.; Bocoş, B. & Ţigan, Ş. (2011): Opportunistic pathogens and faecal indicators assessment in drinking water associated biofilms in Cluj, Romania (in preparation)

Fields, B.S.; Benson, R.F. & Besser, R.E. (2002). *Legionella* and Legionnaire's disease: 25 years of investigation. *Clin Microbiol Rev,* Vol.15, No.3, pp. 506-526

Flemming, H.C.; Percival, S.L. & Walker, J.T. (2002). Contamination potential of biofilms in water distribution systems. *Water Supp,* Vol.2, No.1, pp. 271-280

Flemming, H.C. (2008). Why microorganisms live in biofilms and the problem of biofouling. *Biofouling,* pp. 3-12, doi: 10.1007/7142

Gallay, A.; De Valk, H.; Cournot, M.; Ladeuil, B.; Hemery, C.; Castor, C.; Bon, F.; Mégraud, F.; Le Cann, P. & Desenclos, J.C. (2006). A large multi-pathogen waterborne community outbreak linked to faecal contamination of a groundwater system, France, 2000. *Clin Microbiol Infec*, Vol.12, No.6, pp 561-570

Gavriel, A.A.; Landre, J.P.B. & Lamb, A.J. (1998). Incidence of mesophilic *Aeromonas* within a public drinking water supply in north-east Scotland. *J Appl Microbiol*, Vol.84, pp. 383-392

Gillings, M.R.; Xuejun, D.; Hardwick, S.A. & Holley, M.P. (2009). Gene cassettes encoding resistance to quaternary ammonium compounds: a role in the origin of clinical class 1 integrons? *The ISME Journal*, Vol. 3, pp. 209-215

Glasmacher, A.; Engelhart, S. & Exner, M. (2003). Infections from HPC organisms in drinking-water amongst the immunocompromised, In: *Heterotrophic Plate Counts and Drinking-water Safety*, Bartram, J.; Cotruvo, J.; Exner, M.; Fricker, C. & Glasmacher, A., (Ed.), IWA Publishing, ISBN 1 84339 025 6, London, UK, pp. 137-145

Gonçalves, A.B.; Paterson, R.M. & Lima, N. (2006). Survey and significance of filamentous fungi from tap water. *Int J Hyg Environ Health*, Vol.209, pp. 257-264

Göttlich, E.; van der Lubbe, W.; Lange, B.; Fiedler, S.; Melchert, I.; Reifenrath, M.; Flemming, H.C. & Hoog, S. (2002). Fungal flora in groundwater-derived public drinking water. *Int J Hyg Environ Health*, Vol.205, No.4, pp. 269-279

Gunnarsson, H. & Sanseovic, A.M. (2001). *Possible linkages between algae toxins in drinking water and related illnesses in Windhoek, Namibia*, Thesis, Kristianstad University, Sweden, hkr.divaportal.org/smash/get/diva2:231226/FULLTEXT01

Hageskal, G.; Knutsen, A.K.; Gaustad, P.; de Hoog, G.S. & Skaar, I. (2006). Diversity and significance of mold species in Norvegian drinking water. *Appl Environ Microbiol*, Vol.72, No.12, pp. 7586-7593

Hazam, R.K.; Singla, R.; Kishore, J.; Singh, S.; Gupta, R.K. & Kar, P. (2010). Surveillance of hepatitis E virus in sewage and drinking water in a resettlement colony of Delhi: what has been the experience? *Arch Virol*, Vol.155, No.8, pp. 1227-1233

Hänninen, M.L.; Haajanen, H.; Pummi, T.; Wermundsen, K.; Katila, M.L.; Sarkkinen, H.; Miettinen, I. & Rautelin, H (2003). Detection and typing of *Campylobacter jejuni* and *Campylobacter coli* and analysis of indicator organisms in three waterborne outbreaks in Finland. *Appl Environ Microb*, Vol.69, No.3, pp. 1391-1396

Hellard, M.E.; Sinclair, M.I.; Forbes, A.B. & Fairley, C.K. (2001). A randomized, blinded, controlled trial investigating the gastrointestinal health effects of drinking water quality. *Environ Health Persp*, Vol.109, No.8, pp. 773-778

Helmi, K.; Skraber, S.; Gantzer, C.; Williame, R.; Hoffmnn, L. & Cauchie, H.M. (2008). Interactions of *Cryptosporidium parvum*, *Giardia lamblia*, vaccinal poliovirus type 1 and bacteriophages φX174 and MS2 with a drinking water biofilm and a wastewater biofilm. *Appl Environ Microb*, Vol.74, No.7, pp. 2079-2088

Hewitt, J.; Bell, D.; Simmons, G.C.; Rivera-Aban, M.; Wolf, S. & Greening, G.E. (2007). Gastroenteritis outbreak caused by waterborne Norovirus at a New Zeeland ski resort. *Appl Environ Microb*, Vol.73, No.24, pp. 7853-7857

Hsu, B.M.; Chen, C.H.; Wan, M.T.; Chang, P.J & Fan, C.W. (2009). Detection and identification of enteroviruses from various drinking water sources in Taiwan. *J Hydrol*, Vol.365, No.1-2, pp. 134-139

Hunter, P.R.; Payment, P.; Ashbolt, N & Bartram, J. (2003). Assessment of risk, In: *Assessing microbial safety of drinking water*, IWA Publishing, ISBN 1 84339 036 1, London, UK

Juhna, T.; Birzniece, D.; Larsson, S.; Zulenkovs, D.; Sharipo, A.; Azevedo, N.F.; Menard-Szczebara, F.; Castagnet, S.; Feliers, C. & Keevil, C.W. (2007). Detection of *Escherichia coli* in biofilms from pipe samples and coupons in drinking water distribution networks. *Appl Environ Microb* Vol.73, No.22, pp. 7456–7464

Kelley, J.; Kinsey, G.; Peterson, R. & Brayford, D. (2003). Identification and control of fungi in distribution systems. *Water Quality Technology Conference Proceedings*, Awwa Research Foundation Denver, USA, www.biosan.com/pubs/Identification Significance.pdf

Koh, S.J.; Cho, H.G.; Kim, B.H. & Choi, B.Y. (2011). An outbreak of gastroenteritis caused by Norovirus-contaminated groundwater at a waterpark in Korea. *Infectious Diseases, Microbiology & Parasitology*, Nr.26, pp. 28-32

Kormas, K.A.; Neofitou, C.; Pachiadaki, M. & Koufostathi, E. (2010). Changes of the bacterial assemblages throughout an urban drinking water distribution system. *Environ Monit Assess*, No.165, pp. 27-38

Latimer, J.; McLeod, C.; Jackson, R.; Bunch, J.; Graham, A.; Stokes S. & Poole, R. (2010). *Escherichia coli* biofilms: gene expression and elemental heterogeneity. *Biofilms 4 Conference Handbook*, Winchester, UK, pp. 86

LeChevallier, M.W.; Babcock, T.M. & Lee, R.G. (1987). Examination and characterization of distribution system biofilms. *Appl Environ Microb*, Vol.53, No.12, pp. 2714-2724

LeChevallier, M.W.; Norton, W.D. & Lee, R.G. (1991). *Giardia* and *Cryptosporidium* in filtered drinking water supplies. *Appl Environ Microb*, Vol.57, No.9, pp. 2617-2621

LeChevallier, M.W. & Au, K.K. (2004). *Water treatment and pathogen control*, IWA Publishing, ISBN 1 84339069 8, London, UK

Lee, D.G.; Kim, S.J. & Park, S.J. (2006). Effect of reservoirs on microbiological water qualities in a drinking water distribution system. *J Microbiol Biotechn*, Vol.16, pp. 1060-1067

Lehtola, M.J.; Pitkänen, T.; Miebach, L. & Miettinen, I.T. (2006). Survival of *Campylobacter jejuni* in potable water biofilms: a comparative study with different detection methods. *Water Sci Technol*, Vol.54, No.3, pp. 57-61

Lehtola, M.J.; Torvinen,E.; Kusnetsov, J.; Pitkänen, T.; Maunula,L.; von Bonsdorff, C.H.; Martikainen, P.J.; Wilks, S.A.; Keevil, C.W & Miettinen, I.T. (2007). Survival of *Mycobacterium avium, Legionella pneumophila, Escherichia coli*, and calciviruses in drinking water-associated biofilms grown under high-shear turbulent flow. *Appl Environ Microb*, Vol.73, No.9, pp. 2854-2859

Lymperopoulou, D.S.; Kormas, K.A.; Moustaka-Gouni, M. & Karagouni, A.D. (2011). Diversity of cyanobacterial phylotypes in a Mediterranean drinking water reservoir (Marathonas, Greece). *Environ Monit Assess*, Vol.173, pp.155-165

Marciano-Cabral, F.; Jamerson, M. & Kaneshiro, E.S. (2010). Free-living amoebae, *Legionella* and *Mycobacterium* in tap water supplied by a municipal drinking water utility in the USA. *J Water Health*, Vol.8, No.1, pp. 71-82

Martiny, A.C.; Jørgensen, T.M.; Albrechtsen, H.J.; Arvin, E. & Molin, S. (2003). Long term succession of structure and diversity of a biofilm formed in a model drinking water distribution system. *Appl Environ Microb*, Vol.69, No.11, pp. 6899-6907

Maunula, L.; Miettinen, I.T. & von Bonsdorff, C.H. (2005). Norovirus outbreaks from drinking water. *Emerg Infect Dis*, Vol.11, No.11, pp. 1716-1721

Menaia, J.; Benoliel, M.; Lopes, A.; Neto, C.; Ferreira, E.; Mesquita, E.& Paiva, J. (2008). Assessment of Lisbon drinking water distribution network biofilm colonization and associated hazards. *Wa Sci Technol*, Vol.8, No.4, pp. 421-426

Moritz, M.M.; Flemming, H.C. & Wingender, J. (2010). Integration of *Pseudomonas aeruginosa* and *Legionella pneumophila* in drinking water biofilms grown on domestic plumbing materials. *Int J Hyg Environ Health*, Vol.213, No.3, pp. 190-197

Muntean, V. (2009). *General microbiology* (in Romanian), Babeş-Bolyai University of Cluj-Napoca Press, Romania

Murga, R.; Forster, T.S.; Brown, E.; Pruckler, J.M.; Fields, B.S. & Donlan, R.M. (2001). Role of biofilms in the survival of *Legionella pneumophila* in a model potable-water system. *Microbiology*, Vol.147, pp. 3121-3126

Parshionikar, S.U.; William-True, S.; Fout, G.S.; Robbins, D.E.; Seys, S.A.; Cassady, J.D. & Harris, R. (2003). Waterborne outbreak of gastroenteritis associated with a Norovirus. *Appl Environ Microb*. Vol. 69, No.9, pp. 5263-5268

Payment, P. & Robertson, W. (2004). The microbiology of piped distribution systems and public health, In: *Safe piped water: Managing Microbial Water Quality in Piped Distribution Systems*, Ainsworth, R. (Ed.), IWA Publishing, ISBN 1 84339 039 6, London, UK

Payment, P. & Locas, A. (2011). Pathogens in water: value and limits of correlation with microbial indicators. *Ground Water*, Vol.49, No.1, pp 4-11

Peter, A. (2008). *Taste and odor in drinking water: Sources and mitigation strategies*, PhD Thesis, Swiss Federal Institute of Technology Zurich, Switzerland, http://e-collection.ethbib.ethz.ch/eserv/eth:30628/eth-30628-02.pdf

Pires-Gonçalves, R.H.; Sartori, F.G.; Montanari, L.B.; Zaia, J.E.; Melhem, M.S.C.; Mendes-Giannini, M.J.S. & Martins C.H.G. (2008). Occurrence of fungi in water used at a haemodialysis centre. *Lett Appl Microbiol*, Vol.46, pp. 542-547

Razzolini, M.T.P.; Bari, M.; Sanchez, P.S. & Sato, M.I.Z. (2008). *Aeromonas* detection and their toxins in drinking water from reservoirs and drinking fountains. *J Water Health*, Vol.6, No.1, pp. 117-123

Reynolds, K.A.; Mena, K.D. & Gerba, C.P. (2008). Risk of waterborne illness via drinking water in the United States. *Rev Environ Contamin T*, Vol.192, pp. 117-158

Ricardo, E.; Leda, G. & Luis, F. (2006). Drinking water: problems related to water supply in Bahia Blanca, Argentina. *Acta Toxicol Argent*, Vol.14, No.2, pp. 23-30.

Riley, M.R.; Gerba, C.P. & Elimelech, M. (2011). Biological approaches for addressing the grand challenge of providing access to clean drinking water. *J Biol Eng*, Vol.5, No.2

Rusin, P.A.; Rose, J.B.; Haas, C.N. & Gerba, C.P. (1997). Risk assessment of opportunistic bacterial pathogens in drinking water. *Rev Environ Contam T*, Vol.152, pp. 57-83

Sammon, N.B.; Harrower, K.M.; Fabbro, L.D. & Reed, R.H. (2010). Incidence and distribution of microfungi in a treated municipal water supply system in sub-tropical Australia. *Int J Environ Res Public Health*, Vol.7, pp. 1597-1611

Schmeisser, C.; Stockigt, C.; Raasch, C.; Wingender, J.; Timmis, K.N.; Wenderoth, D.F.; Flemming, H.C.; Liesegang, H.; Schmitz, A.; Jaeger, K.E. & Streit, W.R. (2003). Metagenome survey of biofilms in drinking water networks. *Appl Environ Microb*, Vol.69, No.12, pp. 7298-7309

September, S.M.; Els, F.A.; Venter, S.N. & Brozel, V.S. (2007). Prevalence of bacterial pathogens in biofilms of drinking water distribution systems. *J Water Health*, Vol.5, No.2, pp. 219-227

Servais, P.; Anzil, A.; Gatel, D. & Cavard, J. (2004). Biofilm in the Parisian suburbs drinking water distribution system. *J Water Supply Res T*, Vol.53, pp. 313-324

Shanan, S.; Abd, H.; Hedenström, I.; Saeed, A. & Sandström, G. (2011). Detection of *Vibrio cholerae* and *Acanthamoeba* species from same natural water samples collected from different cholera endemic areas in Sudan. *BioMed Central Research Notes*, Vol.4, No.109, http://www.biomedcentral.com/1756-0500/4/109

Shoff, M.E.; Rogerson, A.; Kessler, K.; Schatz, S. & Seal, D.V. (2008). Prevalence of *Acanthamoeba* and other naked amoebae in South Florida domestic water. *J Water Health*, Vol.6, No.1, pp. 99-104

Sonigo, P.; De Toni, A. & Reilly, K. (2011). *A review of fungi in drinking water and the implications for human health*, Report, Bio Intelligent Service, France

Srinivasan, S. (2008). *Managing bacterial regrowth and presence in drinking water distribution systems*, Ph.D. thesis, University of Wisconsin-Madison, USA

Szewzik, U.; Szewzyk, R.; Manz, W. & Schleifer, K.H. (2000). Microbiological safety of drinking water. *Annu Rev Microbiol*, Vol.54, pp. 81-127

UK Environment Agency (2004). *The microbiology of drinking water (2004) - Part 11 – Taste, odour and related aesthetic problems*, London, UK, http://www.environment-agency.gov.uk/static/documents/Research/mdwpart112004_859972.pdf

US Environmental Protection Agency (1992). *Control of biofilm growth in drinking water distribution systems*, Seminar publication, Washington, USA, www.epa.gov/nrmrl/pubs/625r92001/625r92001.html

van der Kooij, D. (2003). Managing regrowth in drinking water distribution systems, In: *Heterotrophic plate counts and drinking water safety. The significance of HPCs for water quality and human health*, Bartram, J.; Cotruvo, J.; Exner, M.; Fricker, C. & Glasmacher, A., (Ed.), IWA Publishing, ISBN 1 84339 025 6, London, UK, pp. 199-232

Verheyen, J.; Timmen-Wego, M.; Laudien, R.; Boussaad, I.; Sen, S.; Koc, A.; Uesbeck, A.; Mazou, F. & Pfister, H. (2009). Detection of adenoviruses and rotaviruses in drinking water sources used in rural areas of Benin, West Africa. *Appl Environ Microb*, Vol.75, No.9, pp. 2798-2810

Warneke, M. (2006). *Cryptosporidium oocyst interactions with drinking water pipe biofilm*, Research report, Cooperative Research Center for Water Quality and Treatment, Adelaide, Australia,
www.wqra.com.au/publications/document-search/ ?download=30

Watkins, J. & Jian, X. (1997). Cultural methods of detection for microorganisms: recent advances and successes. In: *The microbiological quality of water*, Sutcliffe, D.W (Ed.), Freshwater Biological Association, ISBN 0 900386 57 6, Ambleside, UK, pp 19-27

Watson, C.L.; Owen, R.J.; Said, B.; Lai, S.; Lee, J.V.; Surman-Lee, S. & Nichols, G. (2004). Detection of *Helicobacter pylori* but not culture in water and biofilm samples from drinking water distribution systems in England. *J Appl Microb*, No.97, pp. 690-698

White, D.C.; Kirkegaard, R.D.; Palmer, R.J.; Flemming, C.A.; Chen, G.; Leung, K.T.; Phiefer, C.B. & Arrage, A.A., (1999). The biofilm ecology of microbial biofouling, biocide resistance and corrosion, In: *Biofilms in the aquatic environment*, Keevil, C.W.; Godfree, A.; Holt, D. & Dow, C., (Ed.), The Royal Society of Chemistry, ISBN 0 85404 758 1, Cambridge, UK, pp. 120-130

Wingender, J. & Flemming, H.C. (2011). Biofilms in drinking water and their role as reservoir for pathogens. *Int J Hyg Environ Heal*, Vol.213, pp. 190-197

8

Water Quality After Application of Pig Slurry

Radovan Kopp
Mendel University in Brno, Department of Fisheries and Hydrobiology
Czech Republic

1. Introduction

Pig slurry is a complex organic-mineral fertilizer with high fertilising efficiency and it is comparable with farmyard manure. The pig house with excremental ending and subsequent use of slurry is very popular in agriculturally advanced countries because of economic efficiency, work culture and hygiene. Pig house is operationally by 30 - 40 % cheaper than using bedding. Properly made and treated slurry means a significant source of organic matter, nutrients, bacteria and stimulators which if correctly applied, increase fertility and provide a significant cost saving. These slurry nutrients are easily usable for phytoplankton. Nitrogen is found mainly in inorganic form (50 – 60%), just 10% is nitrate and rest forms organic nitrogen. Phosphorus is bound to organic matter and potassium is included in the urine. Chemically the high fertilising efficiency of slurry depends on C:N ratio, which is usually 4-8:1. This ratio affects the mineralization of organic matter, release of N from organic fixation and energy utilization for multiplication of microorganisms. Composition of the slurry is in the Table No.1.

Dry matter	pH	Conductivity	ANC	COD_{Cr}	Ca^{2+}
%		$mS.m^{-1}$	$mmol.l^{-1}$	$g.l^{-1}$	$mg.l^{-1}$
1.4 - 7.5	6.21 – 7.90	1918 - 2715	179 - 227	16.0 – 18.5	300 - 1300
TP	**TN**	**$N\text{-}NH_4$**	**$N\text{-}NO_3$**	**$N\text{-}NO_2$**	**K^+**
$mg.l^{-1}$	$mg.l^{-1}$	$mg.l^{-1}$	$mg.l^{-1}$	$mg.l^{-1}$	$mg.l^{-1}$
33 - 2200	1440 - 6000	1000 - 4348	16.9 – 36.2	1.28 – 3.34	315 - 7686

Table 1. Values of physical and chemical parameters in pig slurry applied into ponds. (min-max interval, amended by more authors), ANC - acid neutralization capacity, COD - chemical oxygen demand, TP - total phosphorus, TN – total nitrogen

An additional fertilizer in intensively managed fish-ponds usually does not mean that significant increase of fish production as observed in nutrient-poor waters. Moreover, excessive application of fertilizer may lead to environmental pollution and sanitary problems. Fairly high fish production can be achieved by organic fertilizers. The application of pig slurry to ponds as a fertilizer is widely used in many countries in order to increase plankton production and fish growth. That's why manuring is considered to be a cheap way to increase carp production in the pond. Some fish species also feed directly upon these waste, which is enriched by microorganisms with high protein value. Further, different types of manure can be also added into the low-level protein feeds. The type of pig slurry, quality, quantity and the season affects water quality and production of the pond and finally on well-being and growth of fish. Additionally, recycling of pig slurry could be interesting for fish farming in order to reduce the impact of intensive pig farming on environment.

Pig slurry applied to ponds as organic fertilizer necessary needs a decomposition before their nutritional contents are released, assimilated and utilized by plankton. The nutrient content of pig slurry may vary with time, and nutrient availability to phytoplankton growth remains unclear. The rate of nutrients released from animal manure is an important factor to regulate the frequency and amount of manure required to fertilize fish ponds. Pig slurry is destined mainly for refilling of carbon into water and modification of proportion basic biogenic elements (C, N, P). During the water treatment by slurry the amount of carbon dioxide in pond was relatively higher (Fig. 1). After the last application of pig slurry in early may in 2001 alkalinity increased and value of pH decreased which caused a significant increased in values of TIC due to the weather conditions and lower intensity of photosynthesis. On the other hand another TIC value decrease occurred in 2002 (at the end of March) due to higher phytoplankton development. This expansion decreased carbonate content, alkalinity and increased pH.

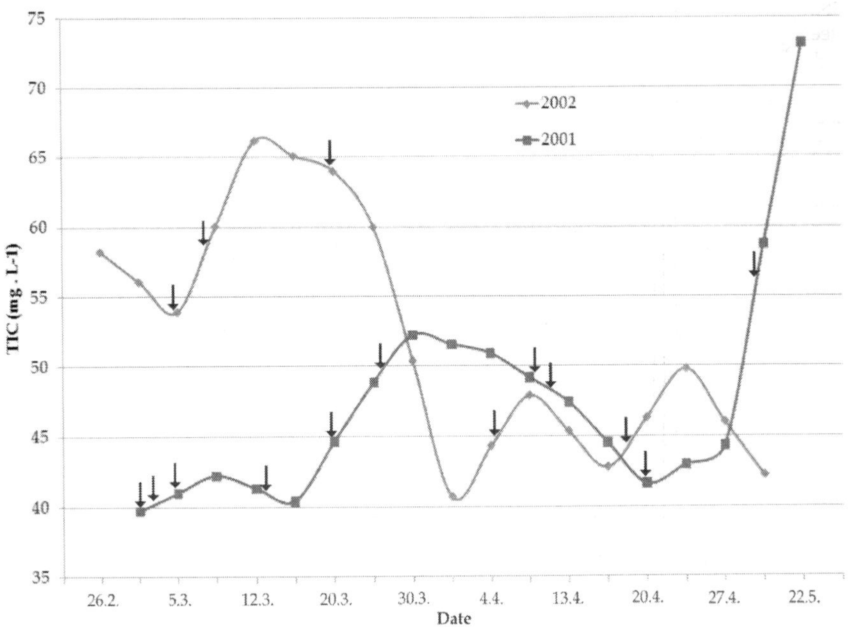

Fig. 1. Variation of total inorganic carbon (TIC) during application of pig slurry into the pond. Level of total doses 16 kg . m^{-2} in the year 2001 and 2002. (Arrows indicate term of pig slurry application)

Negative balance of carbon dioxide in ponds, evoked by plants assimilation at high content of nutrients in water, leads to high values of pH that could be the cause of fish gill necrosis. When using the pig slurry for ponds treatment, some bacteria and other microorganisms are also required to commence decomposition of the slurry and they have positive influence on zooplankton development, especially cladocerans. Bacteria that come into the water environment together with pig slurry are also a direct food source of zooplankton.

2. Fertilizer management

The amount of applied pig slurry depends on natural production of the pond, altitude, depth, control of the water inflow, fish stock etc. Due to the fish stock safety it is better to apply a slurry before vegetation season when a lower water temperature reduces the risk of oxygen deficit (Olah et al. 1986). Abnormal phytoplankton development is reduced in cold weather us well, which is associated with low pH and limited effect of toxic ammonia nitrogen on aquatic organisms. To maintain the stability of the pond ecosystem is not suitable to use a slurry in case of abnormal growth of submerged vegetation or water bloom, when water transparency is less than 40 cm (measured by Secchi disc), in case of abundance of basic nutrients or zooplankton overgrowth. For stability of the physical - chemical parameters it is necessary to properly organize an areal distribution, dose and interval of slurry application (Wohlfarth and Schroeder, 1979).

Knud-Hansen and Batterson (1994) tested four fertilization frequencies (daily, twice per week, weekly, and once every two weeks). The study showed that fertilization frequency had a strong linear relationship to net primary production. Zhu et al. (1990) showed that applying pig slurry daily produced higher yields of fish than five or seven day application. Garg and Bhatnagar (2000) also stated that fertilization frequency influenced fish yield. Variation of manuring frequency often depended on convenience, labour saving, the need to dispose the waste and availability of fertilizer, rather than increasing of the productivity of pond ecosystem. Under the experimental conditions it appears that the frequency of fertilizers application to maintain optimum primary production, should be every eight to ten days (Kestemont, 1995; Kumar et al. 2004). Planning of the fertilization frequencies should predict the time frame between each treatment which has got the potential of production enhancing and also reducing costs. There is a factor to be considered when fertilizing with manure: possibility of de-oxygenation of the water, which is quite likely when large quantities of manure are added at a time.

Fertilizers batching at doses 0.5-2.0 $kg.m^{-2}$ of ponds water increased live weight gain of fish about 30-450 $kg.ha^{-1}$ without supplementary feeding of fish (Hartman et al., 1973; Dhawan and Kaur, 2002). The effect of organic fertilizers from animal breeding farm is higher in ponds with polycultural fish stock where is the highest weight-gain per fish (Buck et al., 1978; Dhawan and Kaur, 2002; Zoccarato et al., 1995). Woynarovich (1976) reported pig slurry conversion of 3–5 % into fish body mass. In the case of polycultural fish stock is ratio higher than in the case of carp monoculture (Prinsloo and Schoonbee, 1984). Zhu et al. (1990) mentioned demand of 8.3 kg dry weight of pig slurry to weight-gain of 1 kg fish flesh. The maximal increases in fish production from fertilizers is especially in the tropics, on average, 4 kg fresh pig slurry can produce 1 kg of fish (Biro, 1995). Variously high doses of pig slurry were tried in ponds experiments. Pig slurry doses around 15-16 $kg.m^{-2}$ were acceptable but they should be applied mainly before vegetation season, when a water temperature is lower (Kopp et al. 2008). Ponds have had concentrate polycultural fish stock during the experiment and fish were intensively fed by cereals. The increased production of the pond supported by pig slurry was 270–630 kg.ha^{-1}. In these experiments the fish production in the high fertilized ponds appears the same values that obtained other researchers.

3. Autotrophs and heterotrophs

A successful fertilization programme in ponds should develop adequate amounts of food organisms for fish. Organic fertilizer (e.g. pig slurry) has been used to stimulate the development of heterotrophs (bacteria), autotrophs (algae) and other food organisms (zooplankton) to increase fish production in ponds (Schroeder, 1978). Bacteria and algae are important food organism for the herbivorous zooplankton that is consumed by many fish species (Wylie and Currie, 1991). In additional to the bulk of photosynthetic production passing directly to higher trophic levels, in some cases more than 50% of primary production is derived from a microbial loop (Hepher, 1992; Qin et al. 1995; Biro, 1995). Bacterivorous flagellates are a key link in the microbial loop to transfer organic carbon from bacteria to zooplankton and fish (Sanders and Porter, 1990). An increasing amount of evidence shows that bacteria together with algae provide an efficient energy pathway from low trophic levels to zooplankton. In natural waters, bacterial production can be enhanced by increasing organic matter loading, but the excessive application of organic matter into fish ponds can reduce dissolved oxygen and can cause fish kills (Qin and Culver, 1992). It is crucial to keep the amounts of organic fertilizer within limits such that adequate amount of heterotrophic organisms are produced yet without oxygen depletion occurring.

The reduction of algal biomass through grazing of zooplankton may indirectly regulate the amount of substrate available for bacterial growth. Bacterial and algal abundances were negatively correlated in fertilized ponds. This may have resulted from the die-off of algae that released dissolved organic compounds into the water, which in turn stimulated bacterial growth (Qin et al. 1995). In natural lakes, zooplankton populations crash after the decline of algae. In ponds, the zooplankton population might be sustained after algal decline by adding organic fertilizer to promote bacterial growth. Thus a direct input of pig slurry could increase bacterial productivity, which in turn could serve as a valuable food source for zooplankton.

When organic fertilizers are applied to ponds the first link of the food chain is heavily influenced, principally phytoplankton (Dhawan and Toor, 1989). High doses of pig slurry supported typically groups of phytoplankton favour higher amount of organic matter. Dominant algal genera often are *Cryptomonas, Chroomonas, Monodus* and *Euglena*. Generally also are occurred small genera of centric diatoms (*Stephanodiscus, Cyclotella*) and chlorococcal green algae (*Desmodesmus, Scenedesmus, Chlorella, Monoraphidium*). Cyanobacteria usually are not dominant groups in fertilizing ponds. High nitrogen to phosphorus ratios non favour dominance by cyanobacteria in pond phytoplankton (Smith, 1983). Typically genus *Microcystis, Aphanizomenon* and *Dolichospermum* in many fish-ponds and eutrophic lakes, cyanobacteria constitute the greater part of the summer phytoplankton biomass, causing regular water blooms and massive fish mortalities by depletion of oxygen after the bloom collapsed are not common in fertilizing ponds. Most often are occurred genus of cyanobacteria *Pseudanabaena, Planktothrix* and *Aphanocapsa* (Kopp and Sukop, 2003; Qui et al. 1995; Terziyski et al. 2007). Higher fish stock, sufficiency of carbon dioxide, suitable ratio of nutrients (C, N, P) in ponds after applications of pig slurry handicapped cyanobacteria forming water blooms. Cyanobacteria are of poor food value to zooplankton, their large size making them inaccessible to the filter-feeding entomostraca. Even the substances produced by many species of cyanobacteria are toxic to aquatic plants and animals (Sevrin-Reyssac and Pletikosic, 1990).

In ponds treated with slurry they have positive influence on zooplankton development. Though, the results of the experiments indicate that the fish stock with its predaceous pressure exerts a more significant influence on zooplankton than application of the slurry. Nevertheless, the initial development of large species of cladocerans was in ponds fertilized by slurry the more intensive to compare with control ponds (Sukop, 1980). When comparing different treatments, zooplankton was significantly higher in ponds manured with pig slurry. Higher zooplankton density as a result of manuring has also been reported in ponds receiving pig dung (Govind et al. 1978; Dhawan and Kaur, 2002). Dominant of zooplankton genera often are Copepoda (*Cyclops, Thermocyclops* and *Acanthocyclops*), nauplius and copepodits stages, small and middle genus of cladocerans (*Bosmina, Chydorus, Moina* and *Daphnia*) and Rotatoria (*Brachionus, Keratella* and *Asplanchna*) (Kopp and Sukop, 2003; Terziyski et al. 2007). Development biomass of planktonic communities in pond after pig slurry application is demonstrated in Figures 2-4.

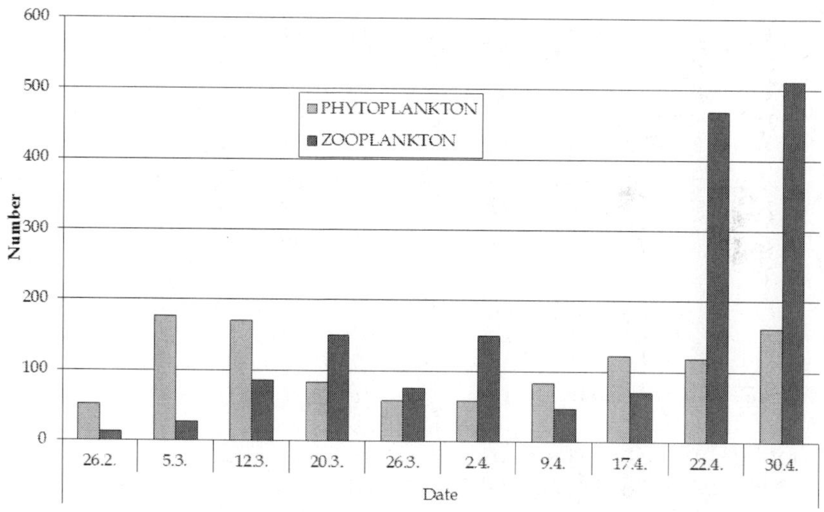

Fig. 2. Quantity of phytoplankton (individuals $10^3 \cdot mL^{-1}$) and zooplankton (individuals . L^{-1}) in pond during applications of pig slurry in total doses 16 kg . m^{-2}. (Application began at 5.3.)

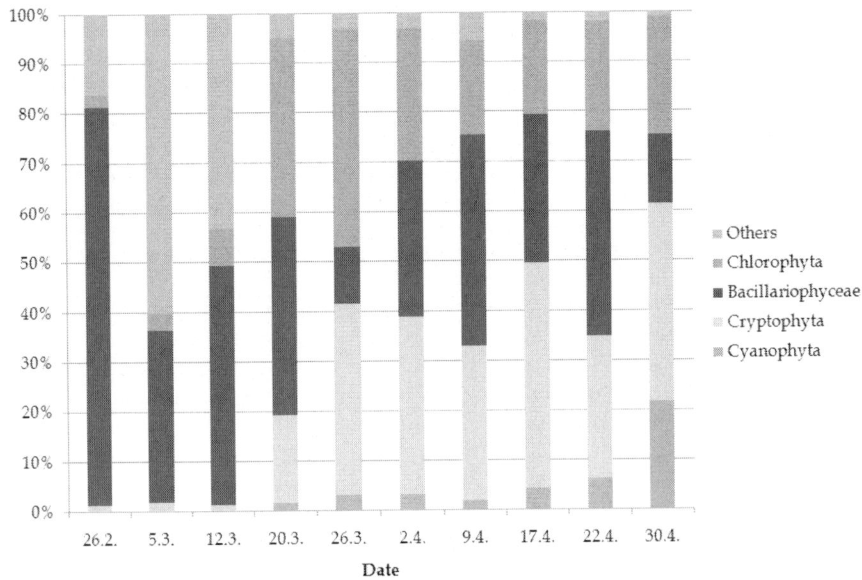

Fig. 3. Quantity of main groups of phytoplankton in pond during applications of pig slurry in total doses 16 kg . m-2. (Application began at 5.3.)

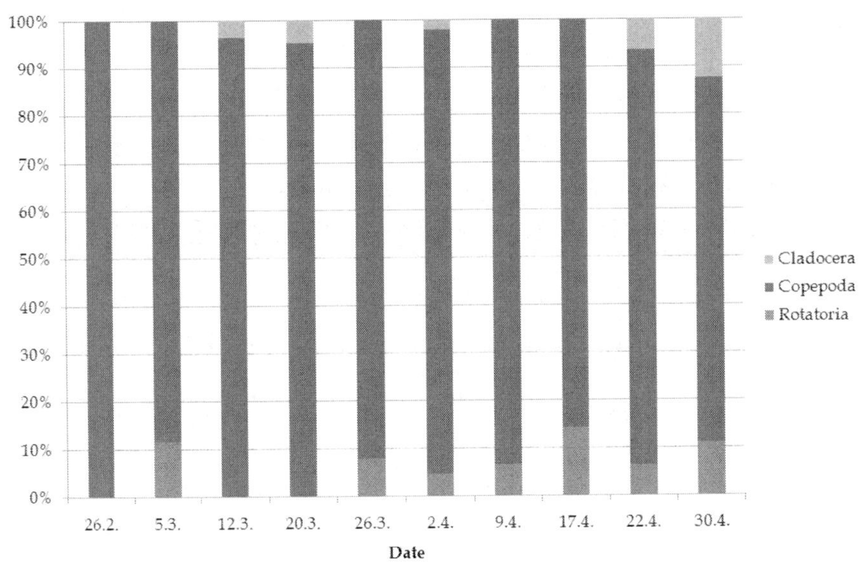

Fig. 4. Quantity of main groups of zooplankton in pond during applications of pig slurry in total doses 16 kg . m-2. (Application began at 5.3.)

4. Organic matter

Application of pig slurry to the water environment is bringing a huge amount of organic matter. Compare to other organic fertiliser the pig slurry has got lower content of organic substances, usually not exceed 10% in total (Hennig and Poppe, 1975). Majority of organic substances in pig slurry is biodegradable and relatively quickly mineralized by microorganisms. During the higher doses of manure to the fish ponds is important to control oxygen content which is intensively consumed due the decomposition of organic matter (Kopp et al. 2008). Application of organic fertiliser into the pond is immediately reflected in the increase of biological oxygen demand (BOD) and chemical oxygen demand (COD). As shows Figures 5 and 6, values change in wide range depending on dosage and elapsed time since last application. Important role for degradation of organic matter plays mainly water temperature and amount of microorganisms that decompose present organic matter. Significant reduction of COD and BOD observed during pig slurry application (Fig. 5 and 6) were mainly caused by changes of water temperature and the rate of degradation of organic substances. In water unloaded organic substances is value of BOD under 8 mg . L^{-1} and COD under 35 mg . L^{-1}, due to manuring could be these values doubled (Zalud, 2008). Just one month after last application of pig slurry into the pond are decreasing values of organic load to a level comparable with ponds without fertilization (Kopp et al. 2008).

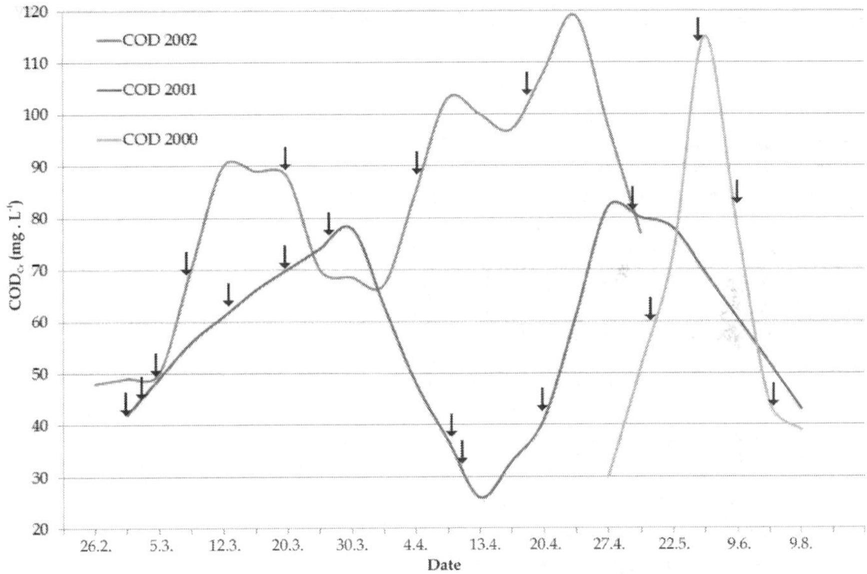

Fig. 5. Variation values of chemical oxygen demand (COD_{Cr}) during application of pig slurry into the ponds. Level of total doses 16 kg . m^{-2} (2001, 2002) and 0.10 kg . m^{-2} (2000). (Arrows indicate term of application of pig slurry)

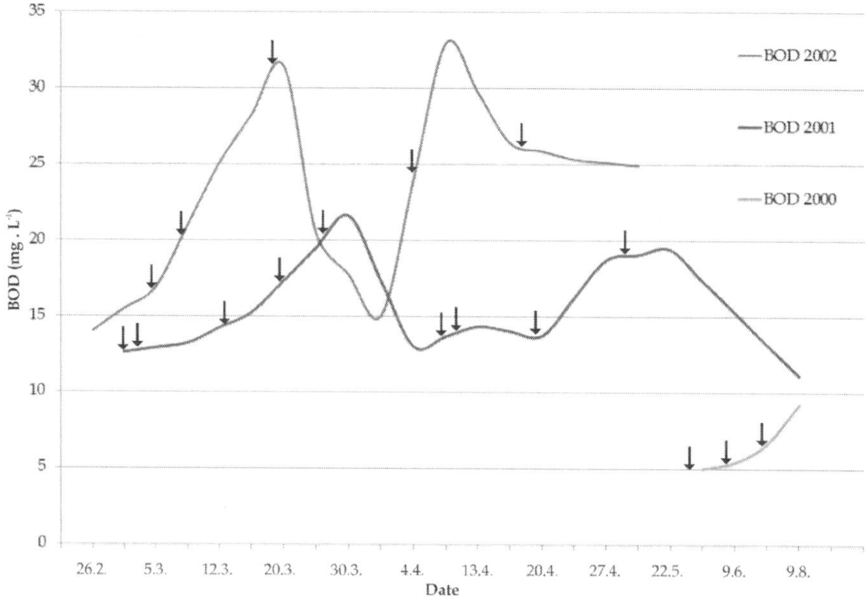

Fig. 6. Variation values of biological oxygen demand (BOD_5) during application of pig slurry into the ponds. Level of total doses 16 kg . m^{-2} (2001, 2002) and 0.10 kg . m^{-2} (2000). (Arrows indicate term of application of pig slurry)

5. Chlorophyll-a

Chlorophyll content in surface water is an important indicator of trophic level and one of the most often used indicators of biomass of primary producers. Concentration of chlorophyll in clear water usually does not exceed 10 μg . L^{-1}. In fish ponds and water reservoirs are values during the vegetation season in the tens to hundreds μg . L^{-1}. In case of hypertrophic water area with massive cyanobacteria water bloom could be chlorophyll content over thousands μg . L^{-1} (Zalud, 2008).

The average amount of phytoplankton in ponds, expressed as chlorophyll-a concentration, has increased since 1960s from 30-35 μg . L^{-1} to 140-150 μg . L^{-1} till then 1990s. The impact of organic fertilization on increasing values of chlorophyll-a is well documented (Potužák et al., 2007). The application of organic fertilizer at regular intervals caused regular fluctuations of chlorophyll-a in the ponds.

After the application of pig slurry is increasing phytoplankton biomass as well as content of chlorophyll-a. The subsequent increase of zooplankton is lowering phytoplankton biomass by predation and also cause decreasing of chlorophyll–a. Fluctuations in value of chlorophyll–a during the pig slurry application is shows in Figure 7. Despite bigger doses of pig slurry are values of chlorophyll–a not so high. High stock of carp fish is increasing inorganic water turbidity by the feed pressure on the ponds bottom, decreasing transparency and limiting abnormal phytoplankton development due to inadequate lighting conditions. Lower water temperature and predation of zooplankton is limiting phytoplankton as well. (Kopp et al. 2008).

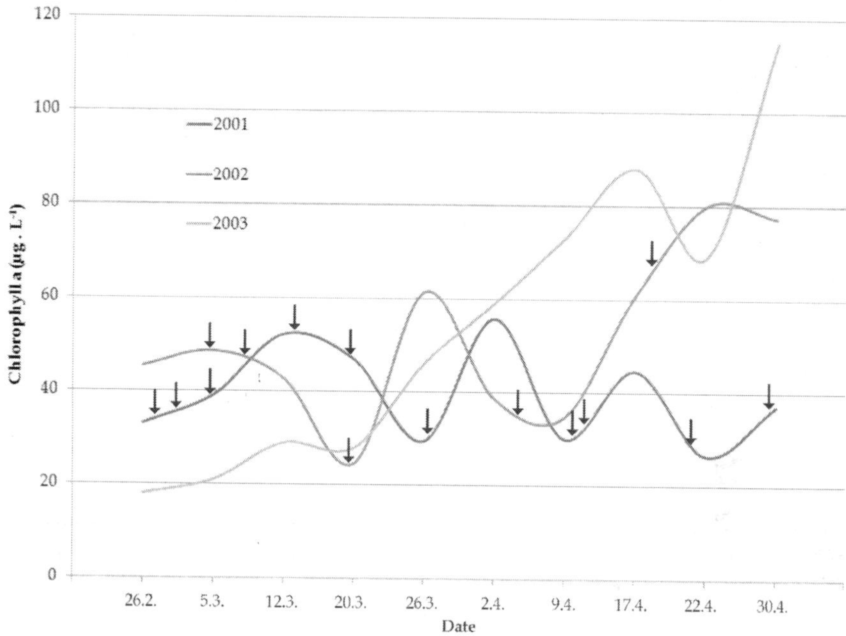

Fig. 7. Variation values of chlorophyll-a during the application of pig slurry into the pond. Level of total doses 16 kg . m^{-2} (2001, 2002) and without application of pig slurry (2003). (Application began at 28.2. in the year 2001 and 5.3. in the year 2002 respectively)

6. Dissolved oxygen

Dissolved oxygen concentrations in the ponds are most affected by phytoplankton biomass. Greater oxygen production and consumption occurs in ponds with higher phytoplankton biomass. The use of pig slurry increased the biological oxygen demand in ponds and could result in periods of low dissolved oxygen levels. The aerobic decomposition of organic matter by bacteria is an important drain of oxygen supplies in ponds. When added to ponds organic fertilizers exert an oxygen demand and an excessive application may result in depletion of dissolved oxygen (Schroeder, 1974, 1975). Qin et al., (1995) observed that dissolved oxygen in fish ponds with organic fertilizer was lower than in those without organic fertilizer, and dissolved oxygen in enclosures decreased with the increase of organic fertilizer loading.

Oxygen consumption in pond is regulated by biological and sediment oxygen demand and fish. In intensive fish aquaculture, BOD reached 0.29 mg O_2 . L^{-1} . h^{-1} in Israel (Schroeder, 1975), and 0.14 mg O_2 . L^{-1} . h^{-1} in fertilized ponds in USA (Boyd 1973). Qin et al., (1995) described that the BOD value varied with organic matter loadings and temperature. The BOD of fertilized ponds water varied from 0.19 to 0.34 mg O_2 . L^{-1} . h^{-1}. Similar results described Kopp et al. (2008) from Czech Republic, when the BOD value varied from 0.12 to 0.34 mg O_2 . L^{-1} . h^{-1} in pond during application high doses of pig slurry (Figure 8). It is certain that excessive organic fertilizer inputs can be a major factor depleting oxygen in ponds, especially during the clear-water phase when algae are less abundant (Qin and Culver, 1992). Analogous situation can be turn up during application of organic fertilizers, when higher development of zooplankton reduced the quantity of phytoplankton and dissolved oxygen concentrations declined. At high temperatures a heavy organic fertilizer application should also be avoided, otherwise a fish kill is likely to occur due to oxygen depletion.

On the second hand Dhawan and Kaur, (2002) reported that pig slurry even at higher dose (36 t . ha^{-1} . yr^{-1}) had no adverse effect on the dissolved oxygen content, which is an important parameter for the survival and growth of fish.

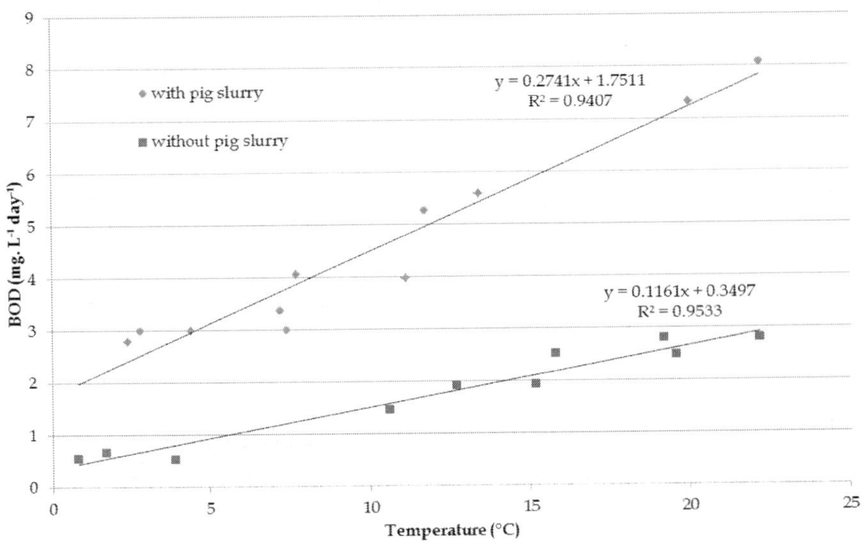

Fig. 8. Regression of temperature with biological oxygen demand (BOD). Pond water without pig slurry was used from pond a year before application and pond water with pig slurry was used during application of 16 kg . m^{-2} of pig slurry (February-May)

Sharma and Das, (1988) also reported that in a fish-pig pond, even higher organic loading by pig slurry did not deteriorate the oxygen content of water. This may be attributed to the higher plankton abundance in manure-loaded ponds, leading to higher dissolved oxygen level by photosynthetic activity especially during day. The values of dissolved oxygen varied during day and night in large interval and after application of pig slurry in during higher development of phytoplankton biomass can be exceed to 250 % saturation within day

(Behrends et al. 1980; Kopp et al., 2008). Figure 9 shows fluctuations in oxygen content during the application of pig slurry at low and high values. In both cases oxygen varied in wide range but has not fall below 50% of saturation, which is limit value ensure survival of carp fish. A wide range of values of oxygen during the application of pig slurry is determined by the intensity of photosynthesis, which is mainly influenced by the amount of available nutrients, water temperature and feed pressure of the zooplankton.

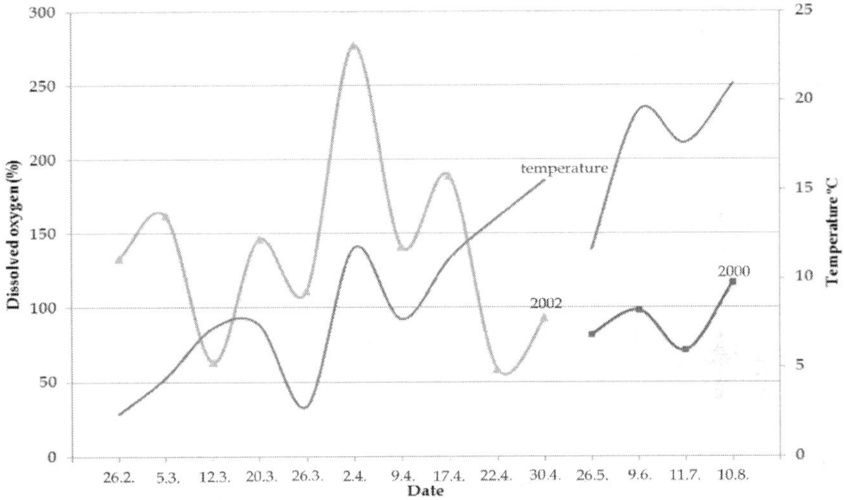

Fig. 9. Variation values of dissolved oxygen during application of pig slurry into the ponds. Level of total doses 16 kg . m^{-2} (2002) and 0.10 kg . m^{-2} (2000) (application began at 9.5. in the year 2000 and 5.3. in the year 2002 respectively)

7. pH

The pH value of water depends on the alkalinity and hardness of the water. Algal photosynthesis causes an increase in pH due to removal of H^+ ions. Algal assimilation of nitrate and its subsequent reduction within the algal cell to ammonium also increases the pH. On the second hand pH decreases with organic fertilization because bacterially generated carbon dioxide from manure decomposition. It was clear that the fertilization regimes were able to maintain adequate buffering capacity so that the pH fluctuations were within the desired limits for fish culture.

Properly chosen dosage and interval of slurry application should keep pH in optimum range for fish farming. To avoid pH increasing (over 8.5) due to intensive photosynthesis of phytoplankton what can happen especially in pond with a low alkalinity it is recommended to add calcium. Dose of calcium fertilizer is set up along of water alkalinity and depends on amount and timing of slurry application. In intensively fertilized ponds is pH usually under 9.0 (Kopp et al. 2008). Figure 10 shows pH fluctuations during the high dose application of pig slurry into intensively managed pond for two years. Significant changes in pH depending on phytoplankton development were apparent in 2001. First two years after slurry application is decreasing pH due to reduced nutrients supply and lower development of primary producers.

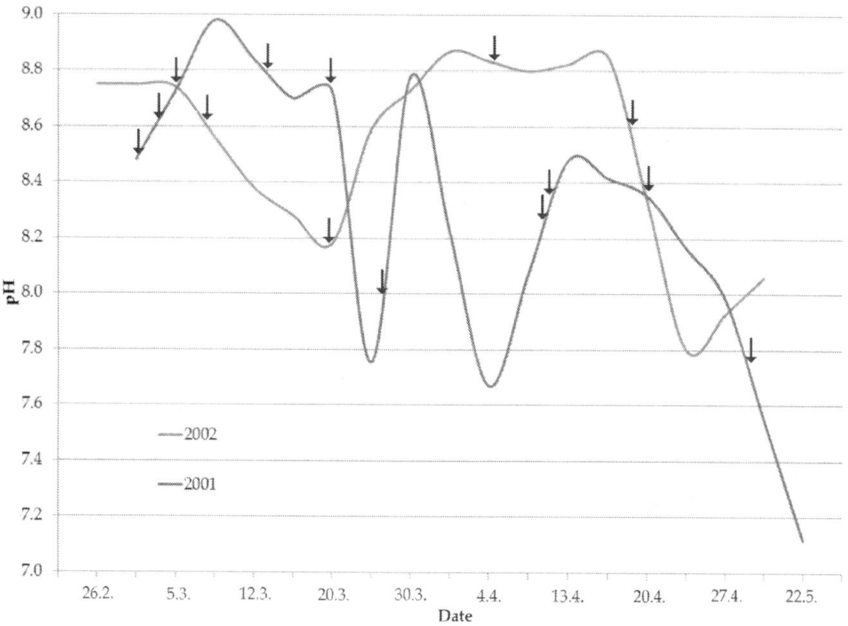

Fig. 10. Variation values of pH during application of pig slurry into the pond. Level of total doses 16 kg . m^{-2} (2001, 2002). (Arrows indicate term of pig slurry application)

8. Transparency

Application of fertilizer at regular intervals affects the transparency of the water. The transparency was significantly worsened within several weeks in the ponds after pig slurry application. Organic particles from manure had a significant influence in water turbidity. Minimum transparency was measured in ponds with abnormal phytoplankton biomass and higher fish stock of cyprinids, which reducing transparency by feeding by pressure on the bottom. High abundance of autotrophs may cause higher development of zooplankton and better transparency. During applications of pig slurry was transparency usually in interval 20-40 cm (Kumar et al. 2004; Kopp et al. 2008). Figure 11 shows different transparency during the high dose application in 2001 and 2002 compare to water transparency in 2003 without pig slurry application. Higher transparency in 2003 was due to lower level of nutrients and lower phytoplankton development in cooler period of the year compare to years with pig slurry application.

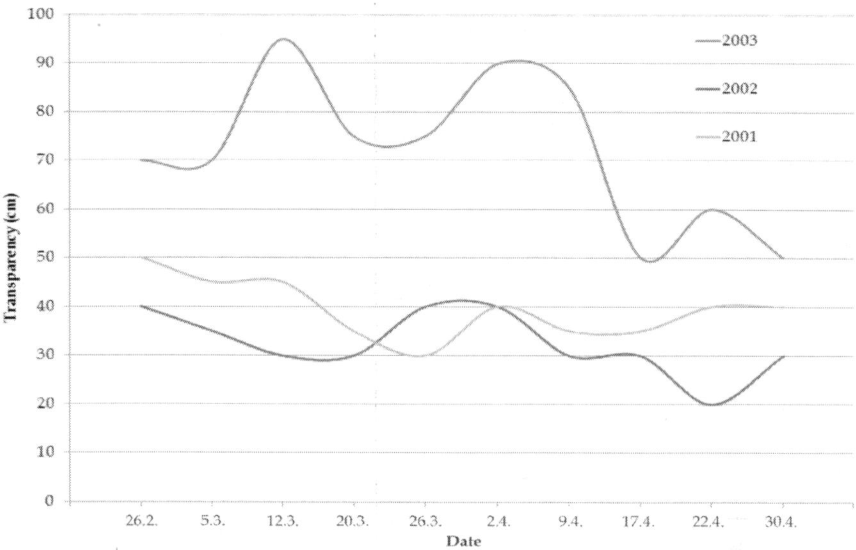

Fig. 11. Variation values of transparency during application of pig slurry into the pond. Level of total doses 16 kg . m^{-2} (2001, 2002) and without application of pig slurry (2003). (Application began at 28.2. in the year 2001 and 5.3. in the year 2002 respectively)

9. Alkalinity

The alkalinity of the water is associated with the level of carbonates and bicarbonates content. The bicarbonate and carbon dioxide in the water has a buffering effect, preventing sudden changes in pH. Low alkalinity (under 1 mmol . L^{-1}) leads to poor buffering capacity and generally, a high fluctuation of pH. A high buffering capacity prevents fluctuation in the acid-base equilibrium caused by strong photosynthetic activity. Alkalinity increases with pig slurry application because bacterially generated carbon dioxide from manure decomposition dissolves calcium carbonate present in the pond sediments and slurry (Kumar et al. 2005). The gradual reduction in alkalinity occurred during the prime period of primary production. According to Boston et al. (1989), a decrease in pond water alkalinity can occur when algae remove bicarbonates. This may occur during periods of high photosynthetic activity which is promoted by application of pig slurry. Figure 12 shows fluctuations in alkalinity during the application of pig slurry into the fishpond. Alkalinity rises at the beginning of application (in accordance with literature) and after that change due to phytoplankton biomass development. Significant decrease of alkalinity during the pig slurry application in 2002 was caused by abnormal plankton development, high intensity of photosynthesis and depletion of carbonates. The alkalinity was quite stable or fluctuates during the year in small interval only in the fishponds that was not fertilizer.

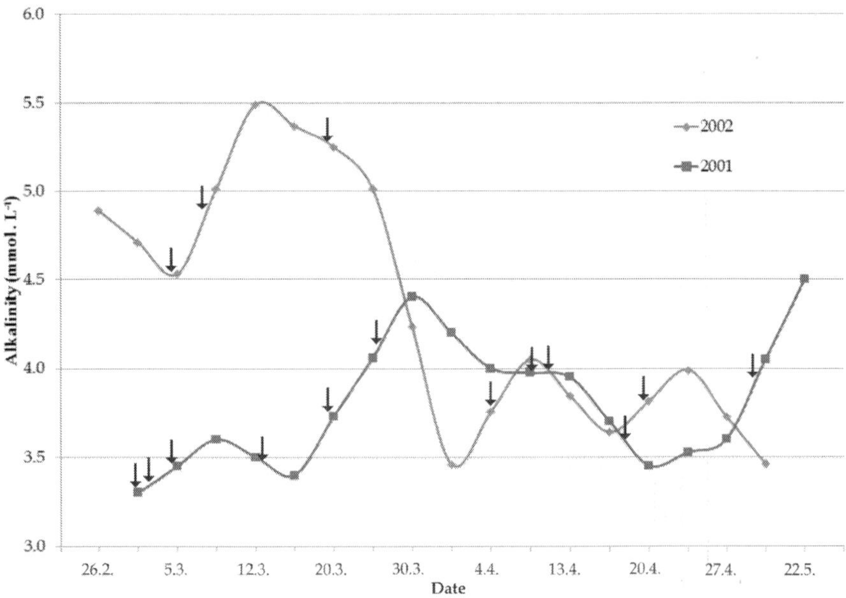

Fig. 12. Variation values of alkalinity during application of pig slurry into the ponds. Level of total doses 16 kg . m^{-2} (2001, 2002). (Arrows indicate term of application pig slurry)

10. Nutrients

The nutrient content of animal manure may vary with time an d the quantity of nutrient released from the manures over time is a key factor in determining the fertilization regime. According to Muck and Steenhuis (1982), manure nutrient concentrations, and the percentage of P, N and C which become available for algal uptake, depend primarily on the animal´s diet, whether the manure is liquid or solid, and the age and storage conditions of the manure. The primary productivity pattern in the pond indicated a gradual built-up phase, a peak phase followed by a decline phase, this can be related to the nutrient release pattern. In pig slurry application, the primary productivity attained its peak 5-7 days after the application of slurry (Kumar et al. 2004).

Culver (1991) and Culver et al. (1993) introduced a fertilization regimen which attempts to optimize the algal composition by maintaining nitrogen and phosphorus at a specific ratio in ponds. They suggested that green algae could be maintained whereas cyanobacterial blooms could be suppressed in ponds by manipulating the N:P ratio (20:1) with fertilizer. Oin et al. (1995) showed that it is difficult to reach the N:P ratio recommended by Culver et al. (1993) by using only organic fertilizer, because most organic fertilizers have a low N and P content and inappropriately low N:P ratios.

Sediment is usually considered the major sink of orthophosphates in fishponds. The sediments absorb soluble forms of P from pond water until fully saturated. The rate of absorption decreases with an increase in partial saturation of sediments (Boyd and Musig, 1981). The P probably accumulates in the newly created sludge layer. Thus, the retained P does not appear to seep into the underground water, but remains in the pond. The P-containing sediment is the basis of the fertility of the pond. The higher P- input into the pond, the more P was retained (Knosche et al. 2000). Pig slurry has a low P content relatively, and application of pig slurry to pond has not negatively impact on higher value of P in outlet water (Kopp et al., 2008; Blažková et al., 1987). Balances of total P and N during the application of pig slurry into the ponds are demonstrated in the Figure 13 and 14.

Nitrogen is one of the most important agents of eutrophication, but it is not the most decisive factor in the eutrophication process of surface waters. Its balance is also influenced by the fixation of N from the air caused by cyanobacteria and by bacterial denitrification. The N balances in carp pond shows the similar picture as the P-retention, i.e. a clear increase in retention with N load (Olah et al. 1994). On the second hand Pechar (2000) reported that the rise in P and N in carp ponds during the 1990s is a result of the higher loading of organic fertilizers, particularly pig and cattle manure. Higher doses of pig slurry increase amount of nitrogen in outlet water from manure pond, especially ammonium for define time (Kopp et al. 2008).

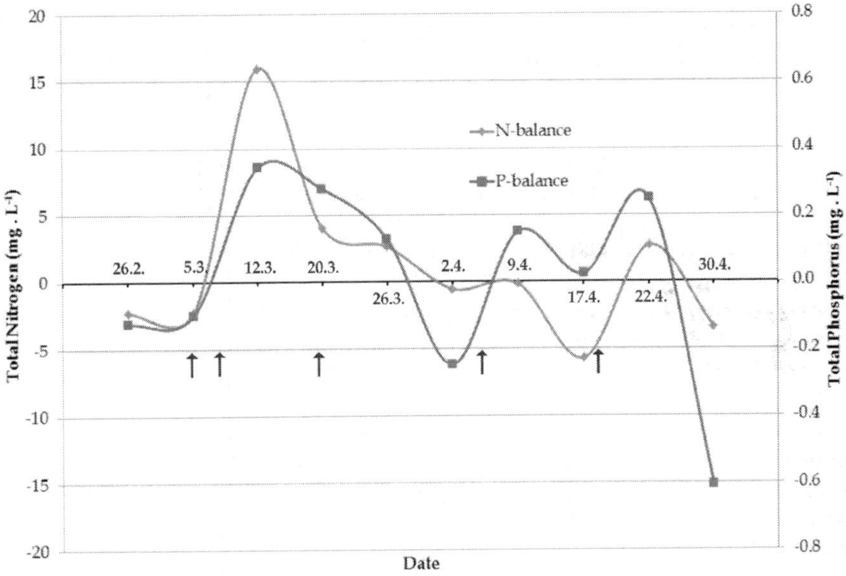

Fig. 13. Balance of total P and N (output by outflow minus input by inflow) during the application of pig slurry (total doses 16 kg . m-2) into the pond (February-April), (Arrows indicate term of application pig slurry)

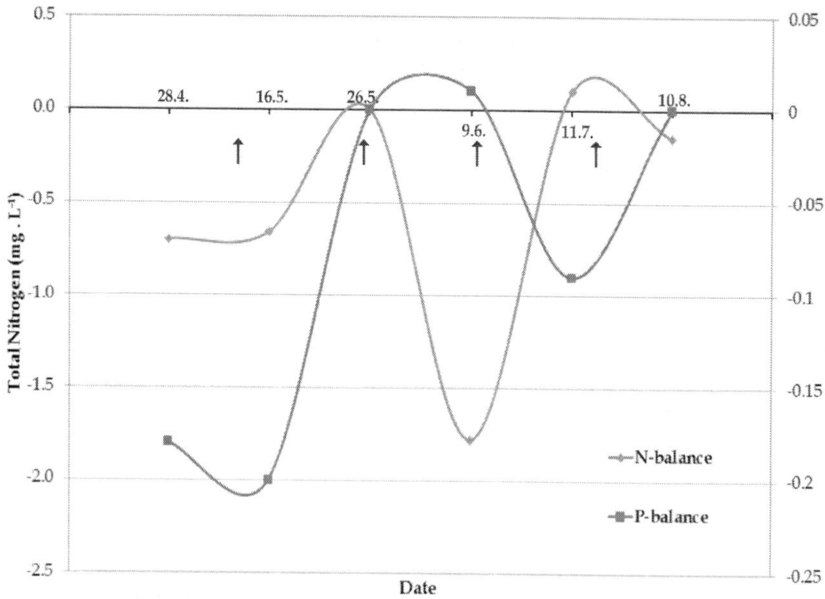

Fig. 14. Balance of total P and N (output by outflow minus input by inflow) during the application of pig slurry (total doses 0.10 kg . m^{-2}) into the pond (May-July), (Arrows indicate term of application pig slurry)

11. Ammonia nitrogen

The critical water quality parameter during application of pig slurry is ammonia nitrogen, a principal excretory product of fish metabolism, which in the un-ionized form is highly toxic to fish. The oxidation of ammonia requires oxygen in the water, for this reason during the initial period of pig slurry application, low levels of nitrate were recorded.

The amount of toxic ammonia depends mainly on pH and water temperature. The maximum allowable (safe) concentration for carp is very low (0.05 mg . L^{-1} NH_3), LC_{50} (lethal concentration) for carp is 1.0 -1.5 mg . L^{-1} NH_3 (Svobodová et al. 1971; Russo and Thurson, 1991). During the application of higher dose of pig slurry has been level of ammonia nitrate increased for a short time which could endanger fish stock by not just acute toxic effects of ammonia but by autointoxication of fish body as well (Kopp et al. 2008). The dosage of pig slurry must follow hydrochemical conditions in pond and application is not possible in high pH or higher water temperature.

Figure 15 shows fluctuations of ammonia nitrogen during the high dose application of pig slurry. The graph clearly demonstrates the significant increase of ammonia in 2002, when were applied higher doses of slurry then a year ago. Despite high level of toxic ammonia in the water there was no fish kills or noticeable reduction of food intake but we can expect some negative impact of sublethal level of ammonia on fish organism. Arrilo et al. (1981)

observed biochemical changes in fish after 48 hour exposure to a low concentration of toxic ammonia (0.02-0.04 mg . L-1). Higher value of toxic ammonia in combination with high pH is show by reduced weight, lower ability to survive and worse feed conversion (Máchová et al. 1983).

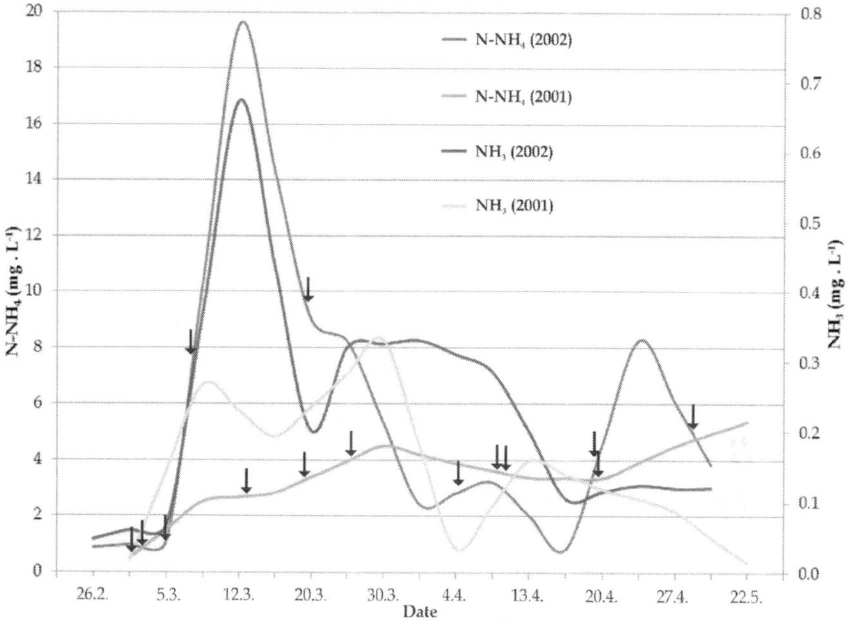

Fig. 15. Variation values of ammonia nitrogen during application of pig slurry into the ponds. Level of total doses 16 kg . m-2 (2001, 2002). (Arrows indicate term of application pig slurry)

12. Nitrate and nitrite

The amount of nitrates which gets into the aquatic environment during the slurry application is relatively small compare to ammonia nitrate. Primary producers are using nitrates as sources of nitrogen therefore is negative balance between nitrates intake (water supply, organic fertilizer) and drain from the fishpond. High concentration of nitrates has not negative effect on fish stock but significantly contributes on water eutrophication. In Figure 16 is well show an increase of nitrogen after the start of slurry application and decrease after fertilization. Level of nitrate nitrogen in eutrophic pond often decreases under measurable value due to high assimilation of phytoplankton during the vegetation season (Kestemont, 1995).

Nitrite is an intermediate stage in the oxidation of ammonium to nitrate. Elevated nitrite concentration in water is a potential threat to freshwater fish since nitrite is actively taken up via the gills in competition with chloride and causing elevation of methaemoglobin levels (Russo and Thurson, 1991). Application of pig slurry to the water is bringing a lot of nitrate nitrogen. As show Figure 16, even during the high dose of slurry the value of nitrates is not dangerous for fish organism. Due to their chemical and biochemical instability nitrites in toxic environment are quickly transformed into nitrates and use by phytoplankton. Even in case of short term high value of nitrites due to single application of pig slurry fish are protected against the toxic effect of chloride ions which are a normal part of surface water (Jensen, 2003).

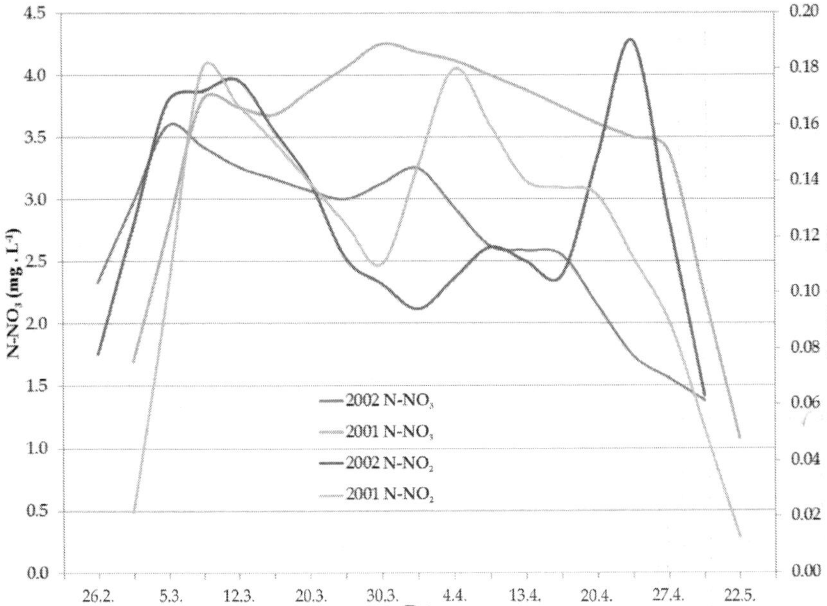

Fig. 16. Variation values of nitrate and nitrite during application of pig slurry into the pond. Level of total doses 16 kg . m^{-2} (2001, 2002). (beginning application at 28.2. in the year 2001 and 5.3. in the year 2002 respectively)

13. Conclusion

Dhawan and Kaur, (2002) presented that pig dung, as pond manure which even at higher dose did not adversely affect water quality. Likewise Kajak and Rybak, (1980) advised, than even the heaviest load of nutrients from pig slurry which was added to water did not destroyed the functioning of the ecosystem until an "ecological catastrophe" resulted. A significantly higher plankton production was also recorded in this treatment. However, indiscriminate use of pig dung may deteriorate the water quality and hence decrease plankton production (Boyd and Doyle 1984). Preliminary determination of doses and composition of organic fertilizer to be introduced seems advisable. Another solution will might be able to allow enough time to recover the desired water quality after the introduction of organic fertilizer. The enormous quantities of organic fertilizers and the high fish stock densities in ponds resulted the presence of very small-sized group of zooplankton.

The phytoplankton blooms elevated pH levels and decreased the N:P ratio. Cyanobacteria, *Planktothrix agardhii* and *Limnothrix redekei,* typical of hypertrophic waters have become common (Pechar, 1995).

Pig slurry is used especially to supplement carbon into water. Due to change of environmental condition by human, eutrophication of fish ponds is increasing. Uncontrolled and unbalanced input of nutrients causes a disparity of basic biogenic elements N, P, which are in excess to carbon. High assimilation of plants makes carbon a limiting element for another production. This situation is common especially in ponds with intensive fish management (Sukop, 1980). It is necessary to implement the ameliorative intervention in colder period of the year considering higher hazard of variations of decisive hydrochemical parameters at higher water temperature. Unsuitable influence of high single doses of pig slurry on hydrochemical parameters is evident especially by higher values of toxic ammonia (Kopp et al. 2008). Pig slurry application has only short-term influence on water quality. Values describing that high organic pollution are noted immediately after the application. Adequate doses of pig slurry and acceptable form of its distribution to ponds does not cause permanent decline of water quality. Influence on values of physical and chemical parameters in water is not permanent (Sharma and Olah, 1986; Blažková et al., 1987).

From the water management point of view, ponds are not a burden to the environment, but generally improve the water quality downstream of the ponds. A claim for the reduction of production intensity in pond aquaculture cannot be justified from the water quality concerns. Carp ponds generally release better-quality water than these water bodies receive as inflow. Additionally, ponds act as water storage basin and improve the microclimate.

Public attention in recent years is focused on negative impact of agriculture on nature. High eutrophication of the water environment is decreasing biodiversity, causes abnormal water bloom development and fluctuations in physico-chemical parameters. So the application of organic fertiliser could be a problematic issue thought positive impact on fish production. Considering to the pressure of the human society on better environment the application of this kind of matter (i.e. pig slurry) into the water is strictly restricted in many countries.

14. Acknowledgment

This chapter was supported by the Research plan No. MSM6215648905 "Biological and technological aspects of sustainability of controlled ecosystems and their adaptability to climate change", which is financed by the Ministry of Education, Youth and Sports of the Czech Republic.

15. References

Arillo, A., Margiocco, C., Melodia, F., Mensi, P., Schenone, G. (1981). Biochemical aspects of water quality criteria: the case of ammonia pollution. Environment Technology Letters 2:285-292

Behrends, L.L., Maddox, J.J., Madewell, E.C., Pile, S.R. (1980). Comparison of two methods of using liquid swine manure as an organic fertilizer in the production of filter-feeding fish. *Aquaculture* 20:147-153

Biro, P. (1995). Management of pond ecosystems and trophic webs. *Aquaculture* 129:373-386

Blažková, D., Kočková, E., Žáková, E. (1987). Effect of slurry application and changes of water quality in ponds. In: *Intensification of fish production and water quality*. Velké Meziříčí 8.-9. December 1987, s. 56–61 (in Czech)

Boston, H.L., Adams, M.S., Madsen, J.D. (1989). Photosynthetic strategies and productivity in aquatic system. *Aquatic Botany* 34:27-57

Boyd, C.E. (1973). The chemical oxygen demand of waters and biological material from ponds. *Transactions of the American Fisheries Society* 102:606-611

Boyd, C.E., Musig Y. (1981). Orthophosphate uptake by phytoplankton and sediment. *Aquaculture* 22:165-173

Boyd, C.A., Doyle, K.M. (1984). The timing of inorganic fertilization of sunfish ponds. *Aquaculture* 37:169-177

Buck, D.H., Baur, R. J., Rose, C.R. (1978). Utilization of swine manure in a polyculture of Asian and North American fishes. *Transactions of the American Fisheries Society* 107, 1: 216–222

Culver, D.A. (1991). Effects of the N:P ratio in fertilizer for fish hatchery ponds. *Verhandlungen Internationale Vereinigung fur Theoretische und Angewandte Limnologie* 24: 1503-1507

Culver, D.A., Madon, S.P., Qin, J. (1993). Percid pond production techniques: timing, enrichment, and stocking density manipulation. *Journal of Applied Aquaculture* 2:9-31

Dhawan, A., Toor, H.S. (1989). Impact of organic manure and supplementary diet on plankton production and fish growth and fecundity of an Indian major carp, *Cirrhina mrigala* (Ham.) in fishponds. *Biological Wastes* 29: 289-298

Dhawan, A., Kaur, S. (2002). Effect of pig dung on water quality and polyculture of carp species during winter and summer. *Aquaculture International* 10 (4): 297–307

Garg, S.A., Bhatnagar, A. (2000). Effect of fertilization frequency on pond productivity and fish biomass in still eater ponds stocked with *Cirrhinus mrigala* (Ham.). *Aquaculture Research* 31:409-414

Govind, B.V., Raja, G., Singh, G.S. (1978). Studies on the efficacy of organic manures as fish feed producers. *Journal of Inland Fisheries Society* 10:101-106

Hartman, P., Lavický, K., Červinka, S., Pokorný, J., Komárková, J., Reichard, S. (1973). The use of pig slurry for ponds fertilization. – *Report of State fishing* Č. Budějovice, 24 p. (in Czech)

Hepher, B. (1962). Primary production in fishponds and its application to fertilization experiments. *Limnology and Oceanography* 7: 131-137

Hennig, A., Poppe, S. (1975). Abprodukte tierischer Herkunft als Futtermittel. VEB Deutcher Landwirtschaftsverlag, DDR 104, Berlin, 302 pp. (in German)

Jensen, F.B. (2003). Nitrite disrupts multiple physiological functions in aquatic animals. *Comparative Biochemistry and Physiology* – Part A, 135: 9-24

Kajak, Z., Rybak, J. (1980). The effect of pig manure and mineral fertilization on a eutrophic lake ecosystem. *Development of hydrobiology* 2: 337-346

Kestemont, P. (1995). Different systems of carp production and their impacts on the environment. *Aquaculture* 129:347-372

Knösche, R., Schreckenbach, K., Pfeifer, M., Weissenbach, H. (2000). Balances of phosphorus and nitrogen in carp ponds. *Fisheries Management and Ecology* 7:15-22

Knud-Hansen, C.F., Battevrson, T.R. (1994). Effect of fertilization frequency on the production of Nile tilapia (*Oreochromis niloticus*). *Aquaculture* 123: 271-280

Kopp, R., Sukop, I. (2003). The development of plankton communities at applications of pig liquid manure on Jarohněvický pond. *Acta Facultatis Ecologiae*, 10, Suppl. 1: 271-273 (in Czech)

Kopp, R., Ziková, A., Mareš, J., Vítek, T. (2008). Variations of chemical parameters in hypertrophic pond within pig slurry application. *Acta Universitatis agriculturae et silviculturae Mendelianae Brunensis* 2: 95-99

Kumar, S.M., Luu, T.L., Ha, V.M., Dieu, Q.N. (2004). The Nutrient Profile in Organic Fertilizers: Biological Response to Nitrogen and Phosphorus Management in Tanks. *Journal of Applied Aquaculture* 16:45-60

Kumar, S.M., Binh, T.T., Luu, T.L., Clarke, M.S. (2005). Evaluation of Fosh Production Using Organic and Inorganic Fertilizer: Application to Grass Carp Polyculture. *Journal of Applied Aquaculture* 17:19-34

Máchová, J., Peňáz, M., Kouřil, J., Hamáčková, J., Macháček, J., Groch, L. (1983). The Effects of Different pH Values and Increased Ammonia Concentrations on the Growth and Ontogenetic Development of Carp Fry. *Bulletin VÚRH Vodňany* 3: 3-14 (in Czech)

Muck, R.E., Steenhuis, S.T. (1982). Nitrogen losses from manure storage. *Agricultural Wastes* 4:41-54

Olah, J., Sinha, V.R.P., Ayyappan, S., Purushothaman, C.S., Radheyshyam, S. (1986). Primary production and fish yields in fish-ponds under different management practices. *Aquaculture* 58: 111-122

Olah, J., Pekar, F., Szabo, P. (1994). Nitrogen cycling and retention in fish-cum-livestock ponds. *Journal of Applied Aquaculture* 10: 342-348

Pechar, L. (1995). Long-term changes in fish pond management as an uplanned ecosystem experiment: importance of zooplankton structure, nutrients and light for species composition of cyanobacterial blooms. *Water Science and Technology* 32:187-196

Pechar, L. (2000). Impacts of long-term changes in fishery management on the trophic level water quality in Czech fish ponds. *Fisheries Management and Ecology* 7:23-31

Potužák, J., Hůda, J., Pechar, L. (2007). Changes in fish production effectivity in eutrophic fishponds-impact of zooplankton structure. *Aquaculture International* 15:201-210

Prinsloo, J.F., Schoonbee, H.J. (1984). Observation on fish growth in polyculture during late summer and autumn in fish ponds at the Umtata Dam Fish Research Centre, Transkei. Part I: The use of pig manure with and without pelleted fish. *Journal Water SA* 10: 15-23

Qin, J., Culver, D.A., Yu, N. (1995). Effect of organic fertilizer on heterotrophs and autotrophs: implications for water quality management. *Aquaculture Research* 26:911-920

Qin, J., Culver, D.A. (1992). The survival and growth of larval walleye, *Stizostedion vitreum*, and trophic dynamics in fertilized ponds. *Aquaculture* 108:257-276

Russo, R.C., Thurson, R.V. (1991). Toxicity of ammonia , nitrite and nitrate to fishes. Pages 58-59 in D.E. Burne and J.R. Tomasso, editors. Aquaculture and water quality. Word Aquaculture society, Baton Rouge, Lousiana

Sanders, R.W., Porter, K.G. (1990). Bacterivorous flagellates as food resources for the freshwater crustacean zooplankter *Daphnia ambigua*. *Limnology and Oceanography* 35:188-191

Schroeder, G.L. (1974). Use of fluid cowshed manure in fish ponds. *Bamidgeh* 26: 84-96

Schroeder, G.L. (1975). Night-time material balance for oxygen in fish ponds receiving organic wastes. *Bamidgeh* 27: 65-74

Schroeder, G.L. (1978). Autotrophic and heterotrophic production of microorganisms in intensively-manured fish ponds, and related fish yields. *Aquaculture* 14: 303-325

Sevrin-Reyssac, J., Pletikosic, M. (1990). Cyanobacteria in fish ponds. *Aquaculture* 86:1-20

Sharma, B.K., Olah, J. (1986). Integrated fish-pig farming in India and Hungary. *Aquaculture* 54:135-139

Sharma, B.K., Das, M.K. (1988). Studies on integrated fish-livestock carp farming system. *Fishing Chimes* 7:15-27

Smith, V.H. (1983). Low nitrogen to phosphorus ratios favour dominance by blue-green algae in lake phytoplankton. *Science* 221: 669-671

Sukop, I. (1980). Effect of Application of Poultry Slurry and Cererite on the Zooplankton Development in Fingerling Ponds. *Czech Journal of Animal Science* 25, 11: 847–855

Svobodová, Z., Groch, L. (1971). Possibilities of the Diagnosis of Ammonia-Intoxication of Fish. *Bulletin VÚRH Vodňany* 1: 9-18 (in Czech)

Terziyski, D., Grozev, G., Kalchev, R., Stoeva, A. (2007). Effect of organic fertilizer on plankton primary productivity in fish ponds. *Aquaculture International* 15:181-190

Wohlfarth, G.W., Schroeder, G.L. (1979). Use of manure in fish farming - a review. *Agricultural Wastes* 1: 279-299

Woynarovich, E. (1976). The possibility of combining animal husbandry with fish farming, with special reference to duck and pig production. - FAO *Techn. Conf. Aquacult.*, Kyoto, Japan 1976, R6: 11 s.

Wyllie, J.L., Currie, D.J. (1991). The relative importance of bacteria and algae as food sources for crustacean zooplankton. *Limnology and Oceanography* 36: 708-728

Zalud, Z. (Eds.) (2008). Biological and technological aspects of sustainability of controlled ecosystems and their adaptability to climate change- indicators of ecosystem services. *Folia Universitatis Agriculturae et Silviculturae Mendelianae Brunensis* 4: 176 pp. (in Czech)

Zhu, Y., Yang, Y., Wan, J., Hua, D., Mathias, J. A. (1990). The effect of manure application rate and frequency upon fish yield in integrated fish farm ponds. *Aquaculture* 91 (3-4): 233–251

Zoccarato, I., Benatti, G., Calvi, S.L., Bianchini, M.L. (1995). Use of pig manure as fertilizer with and without supplement feed in pond carp production in Northern Italy. - *Aquaculture* 129, 1–4: 387–390

9

Diatoms as Indicators of Water Quality and Ecological Status: Sampling, Analysis and Some Ecological Remarks

Gonzalo Martín and María de los Reyes Fernández
University of Seville
Spain

1. Introduction

Diatoms are amazing microscopical algae whose typical feature is a siliceous coverage, called frustule, extremely diverse in shape. Diatoms live in almost all types of superficial waters. Depending on their habitats, diatoms are either planktonic (living suspended on the water), benthic (growing associated to a substrate), or both planktonic and benthic.

On one hand, planktonic diatoms usually have fine frustules and/or long appendices to facilitate floatability. Many marine diatoms, like the genus *Bacteriastrum* Shadbolt, are good examples of this characteristic. The chain formation facilitates the planktonic life too, such as in *Skeletonema costatum* (Greville) Cleve or *Fragilaria crotonensis* Kitton.

On the other hand, benthic diatoms do not need delicate structures because they live on a substrate. Therefore they do not have to worry about sinking. They can be motile when living on sediment like many species of *Navicula* Bory, grow closely attached to a substrate like *Cocconeis* Ehrenberg (Fig. 1), or live on top of auto-produced mucilaginous stalks as many species of *Gomphonema* Ehrenberg do.

The algal species that develop in an area depend on different environmental factors: salinity, temperature, pH, water velocity, shading, depth, availability of substrata to grow on, water chemistry, etc. Thus, the species that can be found in a water body will inform about some characteristics of the water. Because algae are good indicators of the features of the water, they are used to monitor water quality. Among algae, diatoms have the advantage of being easily identifiable at the species level without the need of cultures and they are very easy to collect and store due to their hard frustule. The ecological requirements of many diatom species are known and therefore many diatom-based indexes of water quality have been developed.

Such indexes have been created for benthic diatoms of continental running waters. There is a reason why benthic diatoms are more reliable for this purpose—they remain in their location unless disturbed by some catastrophic event, like a big flood. This means that they reflect the characteristics of the water from the area in which they live.

On the contrary, planktonic diatoms "get old with the water"—as the water runs, they move with it. Therefore, when a researcher collects planktonic diatoms that grew upstream, such algae may reflect other water characteristics than the ones in the sampling site.

Fig. 1. Four valves of *Cocconeis* Ehrenberg.

In this chapter we will explore the use of diatoms to monitor water quality. To achieve this, this chapter has two main objectives:

1. To be a guide for those who work with this type of algae: The chapter shows how sampling must be carried out and how samples have to be treated and analysed.
2. To warn about the risk of using the indexes alone, without any further information. There are some situations that can lead to incorrect conclusions about the health of an ecosystem if only the values of the indexes are observed. For this reason, we show two cases in which diatom indexes reflect something different from the actual state of the ecosystem.

2. Water quality and ecological status of a water body

In recent decades, a significant effort has been put forth all over the world to assess water quality attending to not only to chemical parameters (nutrients, metals, pesticides, etc.), which are obviously important, but also to biological indicators. In fact, one of the undesirable consequences of pollutants is their effect on biota. In this case, the direct study of the effects of pollution on biota is of great interest.

However, it is not so easy to define the extent to which an ecosystem is damaged. There are obvious situations in which heavily polluted waters give the ecosystems such a structure and characteristics that there is no doubt about the heaviness of the human pressures and no effort is necessary to get conclusions about the health of the system. However, there are intermediate situations where it is difficult to draw a conclusion about the water quality without further evaluation.

In this context, the European Water Framework Directive (2000) has a simple philosophy, easy to understand but difficult to put in practice: An ecosystem should be evaluated according to how it was expected to be under no human pressure. Such an ecosystem would be considered to have a good ecological status and could be used as a pattern - the so-called reference conditions - with which other similar ecosystems can be compared. There are five levels of quality, which are identified by colours, according to the closeness of the ecosystem to its reference condition: red, very poor; orange, poor; yellow, moderate; green, good; blue, excellent. Despite these five standards of water quality levels, it is hard to know how each ecosystem would be under pristine conditions, which leaves much work to be done.

There are different biological indicators that can be used to determine the ecological status of a water body: macroinvertebrates, aquatic plants, and algae, especially diatoms, are all used in water quality research (Triest et al., 2001; Hering et al., 2006; Torrisi et al. 2010). The ecological information given by diatoms is usually summed up through one or more diatom-based indexes, which indicate the trophic level with just one number. However, the indexes alone do not determine the water quality. Moreover, the water quality is not the only factor that determines the ecological status. Other factors such as the structure of the vegetation or the hydrodynamics of the ecosystem must also be taken into account.

In summary, the use of the values of the diatom indexes alone may not be enough to determine if an ecosystem has got a good status or not and sometimes can be insufficient even to determine the water quality, as will be shown at the end of this chapter.

3. Diatoms sampling

Currently, the diatom sampling process is mostly normalised, since there are norms to ensure the accuracy of the procedure. The European Standard EN 13946 (2003), as well as the recomendations provided by Kelly et al. (1998) are well known and widely followed.

In most cases, the diatom sampling is carried out in rivers in order to monitor water quality through the calculation of one or more indexes based on these algae. Firstly, we will focus on rivers, and secondly, we will show some other variations to sample diatoms in wetlands. The following sections summarize the sampling procedure in order to serve as a quick guide.

3.1 Sampling rivers

Sampling sites must be located up and downstream of any impact in the basin of which the researchers are aware but the final selection of the sampling sites depends on the objectives of the work.

The sampling must be carried out in a lightly shadowed by riparian vegetation segment of the river of 10 m length or more. This segment has to include rapids, since stretches of river with very slow currents (<20 cm/s) allow the buildup of loosely attached diatoms, silt, and other debris. Samples have to be collected in rapids at a distance from the riverbanks to avoid waters somehow isolated from the main current that would not reflect the characteristics of the site along with any substrata that may had recently remained out of the water and could contain aerophilous diatoms.

The next step is to choose the type of substrate to sample. There are many works explaining how the sampled substrata can influence the species that can be found on the rivers

(Danilov & Ekelund, 2001; Albay & Akcaalan, 2003; Potapova & Charles, 2005; Towsend & Gell, 2005). For example, some diatoms, such as *Epithemia adnata* (Kützing) Brebisson or *Cocconeis pediculus* Ehrenberg are more likely to be found growing on plants or macroalgae rather than on rocks.

Natural rocks are the best substrate for sampling springs (Round, 1991). However, researchers should avoid any large rock difficult to remove from the river because it would make the sampling process too difficult. Pebbles are not appropriate either, because they might have come recently from upstream and they can mislead the sample. The best choice is an average-size rock of about 15 to 30 cm. However, when this type of substrata is not available, the researcher should take samples in the following order of preference: natural rocks, artificial surfaces (such as bridge pillars), artificial substrates, and plants.

Artificial substrates must be placed in the riverbed at least one month before the sampling is executed to ensure the proper development of a mature algal community. Some researches use slides, which have the advantage of giving a known surface and let the direct observation of the algal growing and the biofilm architecture under a microscope. However, it is preferable to use tiles or bricks in routine biomonitoring work because they provide rough surfaces resembling natural rocks that facilitate the settlement of algae. Unfortunately, all these substrates are often lost in the riverbed due to strong currents and human interaction with the river. Moreover, these artificial substrates can be found completely covered of fine sediments instead of algae, which is quite common since these substrata are likely to be left in rivers whose riverbed has only fine sediment and no natural rocks. A stake can be driven into the sediment through the holes of several bricks, which should be piled up at the bottom of the river. It ensures a vertical surface colonizable by algae on which little mud would sediment. Moreover, it will be easy to localize when returning to the river.

When the researcher decides to sample natural rocks, he or she should randomly collect between five and ten of them. Then, the surface of the rocks should be scrapped off and removed with a toothbrush. The brushed area of the stones, as well as the thoothbrush covered with algae, can be cleaned on a plastic tray with water from the river if it is clear or with tap water if the river runs muddy. The obtained brownish suspension constitutes the sample. Tiles and bricks must be treated as natural rocks.

When sampling diatoms growing on plants, the researcher can use either emergent plants– also called helophytes, such as *Thypha* L. or *Phragmites* Adans. - or submerged plants - the so-called hygrophyes, like *Ceratophyllum* L. or *Myriophyllum* L. Not all plant species facilitate the growing of the same diatoms, in quantity and diversity. Some plants are harder to be colonised by algae for many reasons, such as being too smooth for the algae to grow, etc. For example, *Typha* L. typically exhibits a thicker biofilm than *Scirpus* L. In general, it is recommended to always sample on the same plant species in order to make the data comparison from different sites easier. It is often impossible to do this, since the elected species may not grow in all sampling sites, but this is not a matter of extreme importance, according to Cejudo-Figueiras et al. (2010)

The algae growing on plants have to be collected by gathering leaves that are alive and stem sections, discarding either those parts that had been recently out of the water or those that are near the bottom and covered by sediment. Cuttings of about 5-cm length must be carefully placed together in a bottle filled with tap water. The stems and leaves have to be gently scraped with a cover slip in the laboratory, while softly washed with water. The obtained brown or greenish suspension contains the diatoms that will be used in the research. This procedure is suitable with helophytes - *Typha, Scirpus, Phragmites,* etc. - but not with fine and delicate hygrophytes such as *Myriophyllum*, which would be completely smashed with the coverslip. In that case, some fragments of plants must be introduced into a plastic recipient with some water (from the river if it is clear or tap water otherwise). The recipient must be closed and shaked vigorously. The water will become muddy, brownish or greenish. The plants can be removed out of the jar and replaced with other fragments. After repeating this operation three or four times, the water from the recipient will constitute the sample and it will be blurred because of all the epiphitic algae.

3.2 Sampling wetlands

Small springs are quite easy to sample, since their riverbed is not muddy and usually have many rocks of different sizes that are easily removable. However, wetlands, as well as wide rivers, only have plants and sediments as the only available substrata, and in some cases even plants are absent. If there are plants, they should be treated as described above for rivers. As a general rule, plants have demonstrated to be better than stones in wetlands because plants provide a vertical surface mostly free of mud (Blanco et al., 2004). Stones, if present, typically have a thick layer of sediment, making them useless for sampling. However, the allocation of artificial substrata is a good choice too, if they are disposed vertically (Sekar et al., 2004).

Sediment is not a suitable substrate because diatoms growing among its particles can get nutrients from the intersticial water, having an extra source of nutrients that other algae living out of the sediment lack. This means that trophic levels would be overestimated when analysing diatoms from sediments. That is why rocks are preferred, since they cannot provide extra nutrients to the algae, forcing them to obtain the nutrients needed from the open water.

Nevertheless, some studies may recommend sampling sediment if the goal is not the application of water quality indexes but the knowledge of the algal flora. For example, there are wide marshlands in which there are large extensions of sediment (under the water or just in wet areas) free of plants. Sometimes the primary production of this sediment is very important for the whole system and, very often, a greenish highly productive biofilm can grow on it. If the researcher wants to study the algal composition of this marshland, attention must be focused on sediments, which is what characterizes these types of ecosystems, rather than any other type of substrata.

What is the best way to sample algae from sediments? If researchers want to perform quantitative studies they must sample a known surface. Half of a Petri dish must be driven into sediment to take the superficial layer, using the aid of a scrapper to facilitate the process. After homogenising the sample at the laboratory, small subsamples can be taken from it to count algae under the microscope or to measure chlorophyll values. Any subsample must be extrapolated to the volume or weight of the whole sample and to the area of the Petri dish.

Analysing sediment under the microscope is terribly weary and exasperating. If the objective of the study is to identify the species with a rough estimate of their proportions, there is a strategy that not only facilitates this task but also ensures that only living diatoms are observed. Diatoms living in sediment, like many species of *Nitzschia* Hassall, *Gyrosigma* Hassall or *Navicula* Bory (Fig. 2) are motile, which implies that the sample - or a subsample after homogenising the whole sample - can be left in the Petri dish with a thin layer of water for about 12 hours under good illumination, which acts like a lure attracting diatoms towards the light. A thin green biofilm will appear on the mud that can be carefully picked up and disposed on a slide with a drop of water to be observed under the microscope. Moving algae may adhere to some cover slips if they are left on the wet sediment, although they tend to accumulate around the slips better than under them. Therefore, the analysis of these cover slips is easier because they will have less mud.

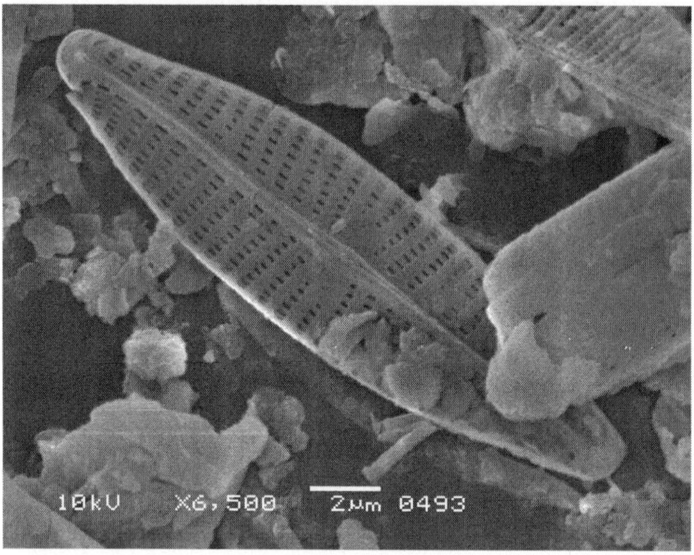

Fig. 2. A single valve of *Navicula* among particles of sediment.

In this case previously mentioned, diatoms have to be alive in order to move among the particles of the sediment. In any other case, if the collected material is not going to be treated in the lab within one day- no matter the substrate or the type of sampled ecosystem— it must be homogenized and fixed with 4% (v/v) formaldehyde.

4. Diatoms treatment

Diatoms can be identified to species level though morphological features of their frustules but these will hardly be seen without a previous treatment consisting of making all the organic matter disappear, leaving their siliceous frustule empty. There are different techniques for diatom cleaning but the hydrogen peroxide method (30% H_2O_2 solution) is the most widely used.

Firstly, if formaldehyde has been added, it should be removed. The sample has to be homogenized by shaking and a tube for centrifugation has to be filled with it. Centrifuge the sample at 2,500 rpm for 15 minutes, remove the supernatant, refill the tube with distilled water and homogenize it again. This process should be repeated three times but after the first one the supernatant must be revised to prevent any loss of diatoms.

Once the supernatant is removed for the third time, add a few drops of hydrogen peroxide to the pellet. Be careful because a lot of foam may appear! Some more peroxide has to be added carefully until half of the tube is filled with it. Then, the tube can be either left covered for several days until bubbles stop flowing or heated in a sand bath or a hotplate at about 90 °C during 1 to 4 hours. In both cases, the brownish suspension has to become whitish.

After that, add a few drops of HCl and repeat the centrifuge process previously described to remove all the hydrogen peroxide.

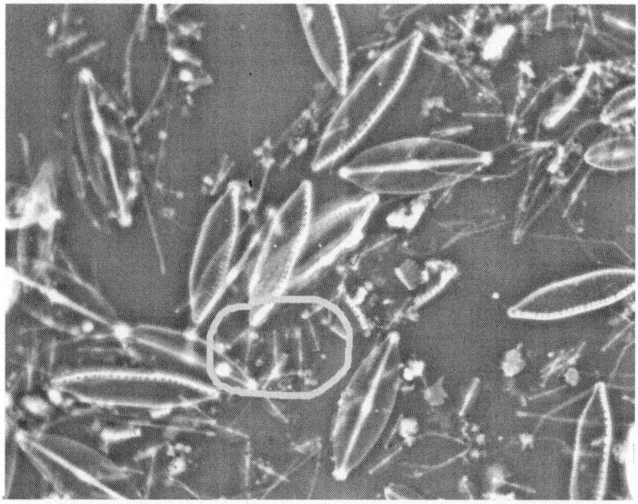

Fig. 3. An excessive density of valves makes the counting difficult. The yellow shape surrounds seven small valves of *Fistulifera saprophila* (Lange-Bertalot & Bonik) Lange-Bertalot. This species is almost impossible to detect without phase contrast. Notice that these diatoms are abundant in the sample and are widespread all over the image.

When diatoms are cleaned and suspended in water, dispense a few drops of this water on a cover slip and let it dry. Once dried, examine the cover slip under the microscope to ensure that diatoms can be easy to count. A good number of valves are eight per field but it depends on their size and shape. If there are too many, the sample must be diluted (Fig. 3); if there are only a few diatoms, add more drops and let the cover slip dry again. If there are not enough diatoms on the cover slip but there is too much inorganic matter on it, the researcher has to make a decision whether or not to add more sample to the cover slip: the addition of more sample would reduce the time of analysis but diatoms would be covered by more inorganic matter, which makes the identification of species harder.

After getting an appropriated number of diatoms on the cover slip, dispense a small drop of high refractive medium (such as Naphrax™) on a slide and place the cover slip on it, making sure that diatoms are in contact with the drop. Heat the preparation for about one minute; some bubbles will appear under the cover slip when the drop of high refractive medium boils but they will disappear as the slide cools again. Once cold, the cover slip should be closely adhered to the slide; if not, the preparation has to be heated again.

5. Diatom analysis

The next step is to identify diatoms from the sample. Some investigators obtained good results identifying to the genus level (Wu & Kow, 2002) either when samples had only a few species of each genus (Growns, 1999) or when a quick assessment of ecological quality is required (Rimet & Bouchez, 2011). However, we recommend reaching to species level when the goal of the study is one of the following: To obtain an accurate bio-assessment, to apply diatom-based water quality indexes, or to define the ecological status of a body of water. Reaching the species level the first time that samples are analysed is time-consuming but provides significantly more information than a genus-level analysis. It also permits data aggregation in higher taxonomical groups if necessary. If other than species level is reached, the information obtained is likely to be too poor, thus forcing researchers to analyse the samples again to identify and count species. In the end, this task would be much more time-consuming.

There are genera with a large number of species and these species are water quality indicators that should be taken into consideration when analysing the samples.

For example, the big genus *Nitzschia* Hassall comprises species such as *Nitzschia fonticola* Grunow (Fig. 4), which is typically present in clean water, as well as *Nitzschia capitellata* Hustedt (one of the largest species from Fig. 3), frequently found in polluted water. The observation of *Nitzschia* in one sample does not provide information neither about water quality nor the features of the ecosystems, since *Nitzschia* can be easily found in almost any water body in the world.

Therefore, it is necessary to reach to species level when calculating diatom indexes. There are some other indexes that even require variety levels. There are many previous studies that can be referenced when determining diatoms. The most widely followed are Krammer and Lange-Bertalot (1986, 1988, 1991a, b). It is useful to have more guides and bibliography when identifying diatoms but we consider these ones to be sufficient, at least as an initial approach to diatom identification.

The scientific names from these guides are not up to date but the old names of the diatoms are listed with the founding author's name and are updated in some websites such as www.algaebase.org. In relation to synonymy, there are differences in taxonomical criteria among authors; therefore, it is important referring the authors following the scientific names of diatoms. From a taxonomical point of view, the simple determination of the genus sometimes can be insufficient to ensure its own genus identity, due to the evolution of the taxonomy. For example, *Nitzschia* Hassall is split into different genera (*Nitzschia sensu stricto*, *Grunowia* Rabenhorst, *Tryblionella* W. Smith, *Psammodictyon* Mann, etc.). Knowing that a *Nitzschia* has been found is not enough to determine if that diatom is actually *Nitzschia sensu stricto* or a member of any other genus coming from *Nitzschia*, such as *Grunowia* (Fig. 5).

Fig. 4. External view of *Nitzschia fonticola*'s valve Grunow.

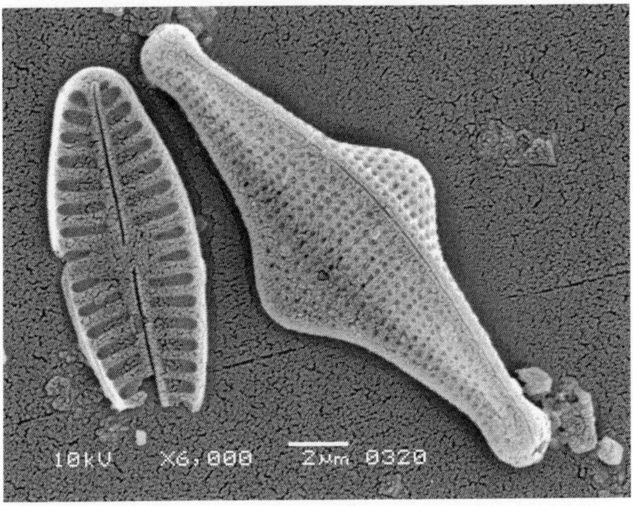

Fig. 5. The right valve is an external view of *Nitzschia sinuata* var. *tabellaria* (Grunow) Grunow, whose updated name is *Grunowia tabellaria* (Grunow) Rabenhorst.

Moreover, if species level is reached, the authors following the diatom names should be observed to update the identifications. For example, if one researcher indentifies *Nitzschia constricta,* without providing any reference regarding the authors, it would be impossible to know whether he or she has actually found *Nitzschia constricta* (Gregory) Grunow (=*Psammodictyon constrictum* (Gregory) Mann) or *Nitzschia constricta* (Kützing) Ralfs (=*Tryblionella apiculata* Gregory). A diatom analysis that determines the presence of *Nitzschia constricta* (Kützing) Ralfs gives enough evidence to indicate that we are actually referring to *Tryblionella apiculata* Gregory, in spite of not having used updated taxonomical criteria.

The diatoms must be identified at a high magnification (100X) while using immersion oil. Non-experienced researchers tend to maximize the contrast of the image observed through the microscope to detect the limits of the valves clearly but some structures, like striae, can sometimes go unnoticed because of this excessive contrast. In these cases, it would be better to increase the lighting over the sample and to reduce the contrast in order to easily observe the diatoms in detail.

It is highly recommended to combine phase contrast with bright field techniques in order to see fine structures (striae, processes, etc.). When a microscope is not equipped with a phase contrast, put an opaque object (a pencil would be enough) between the lamp and the sample, as shown in the figure 6. Light will be deflected and it would work similarly to a phase contrast. The quality of the image can be improved if the object is slightly moved, changing its orientation in relation to the observed valve. The results of this technique are not as good as they would be when using phase contrast but it will provide similar images, plus it is cheaper than purchasing a phase contrast.

Fig. 6. An opaque object - a pencil- is placed between the lamp and the sample to simulate the phase contrast technique.

Figure 7 shows four photographs of the same valve with different types of illumination. Notice that the shape of the valve is perfectly visible in the first image, but the striae are not.

At least 400 valves have to be counted in each sample and the relative proportions of the species have to be calculated. This number provides a good estimate of the composition of the diatom flora and this is a requirement for the application of the biotic indexes.

5.1 Indexes calculation

Once the analysis is finished, the next step is to calculate biotic indexes in order to determine the water quality. Most of the widely used indexes come from the Zelinka and Marvan equation (Zelinka and Marvan, 1961), which is:

$$\text{Index} = \Sigma\,(A_i \times v_i \times j_i)\ /\ A_i v_i \quad (1)$$

Where A is the relative abundance of species i, v is the indicator value of the species and j its sensitivity.

The indexes differ in both diatom species and the values v and j assigned for each index.

Existing indexes must be tested when applied to a basin different from the ones (Prygiel et al., 1999). This testing is usually done by comparing the values given by the indexes with the physicochemical data from the same sites. The Spearman correlation between an index and chemical variables (phosphate, DIN, COD) is enough to determine whether that index can be applied to the basin or not.

There are many studies regarding this issue and it has been proved that these indexes are applicable and work in different parts of the world (Torrisi & Dell'Uomo, 2006; Atazadeh et al., 2007; Taylor et al, 2007). However, most researches tend to state that it is important to test the indexes before using them to verify their accuracy (Potapova & Charles, 2007). As a matter of fact, new indexes have been developed for different basins after having found a weak correlation between chemical data and diatom indexes.

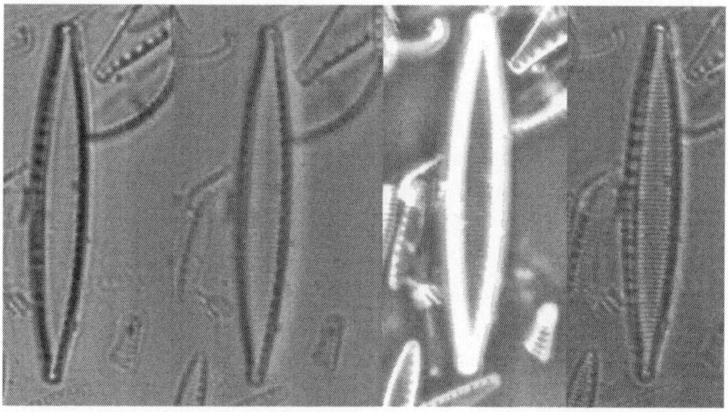

Fig. 7. Four photographs of the same valve of *Nitzschia fonticola* Grunow under different types of illumination. From left to right, bright field with a high contrast, bright field with low contrast, phase contrast, and simulation of phase contrast with an opaque object.

Index calculation is a quick and easy task if the software Omnidia, which is widely used among researchers, is available. It is an expensive computer program but it is extremely useful and worthy. It performs fast index calculations and transforms them into a scale that ranges values from 0 to 20, which facilitates comparisons. There is also a free Excel file, available at http://omnidia.free.fr/update/IBD_NEW.XLS, that permits the calculation of the IBD index. However, we recommend the acquisition of the Omnidia software.

Can diatom-based indexes be used carelessly if they correlate with the chemical data? It is not possible to do that, even though its applicability has been demonstrated all over the world. There are some situations in which indexes can present spurious values that have to be analyzed in order to avoid false conclusions. The rest of this chapter will show some examples where diatom-based indexes failed to work.

6. High index values in bad water quality and the relationship with the ecological status: The case of the Guadiamar River

This case shows a real situation in which some diatom indexes, applied in a river with poor water quality and bad ecological status, return the high values that can be found in a pristine river. When facing a similar situation, researchers must consider other ecological elements in order to determine what is actually happening in the ecosystem.

The Guadiamar River is located in the South of the Iberian Peninsula. The river has been suffering human impact for decades by receiving non-treated wastewaters, which have drastically reduced the quality of its waters. The diatom-based indexes showed low values indicating bad water quality due to the extended presence of organic matter and eutrophy in the water, as they were expected to do.

However, the Guadiamar River has also received acidic waste with a high content of heavy metals—mainly Fe, Zn, Cd, and Cu— from a nearby mine. This impact is located upstream the organic inputs. The pollution from this mine reaches the Guadiamar River through one of its tributaries, the Agrio River. This chemical pollution became more serious after a catastrofic accident in 1998, when a dam spilt tons of contaminated mud and water with high levels of heavy metals into the river.

Fortunately, the situation has improved since the accident. After hard and expensive work, today the river is a green corridor, a path communicating two different protected natural areas. Wastewater was treated in sewage treatment plants and most heavy metals were removed. Nevertheless, the water still has lower quality than desired because: 1. the eutrophy level is still too high and 2. part of the heavy metals remained within the sediment, which risk of mobilization after a big flood.

In relation to diatoms, it is interesting to analyze the composition of the flora on the areas not affected by the organic matter and eutrophy but only by the mine. We will briefly present the characteristic appearance of the river after the spilling of the mine and the subsequent cleaning, between the years 2001 to 2006. After the spill, the pH values of the water decreased and the amount of dissolved heavy metals increased drastically. This situation worsened during summer because in this season the water levels diminish, increasing the concentration of pollutants and changing the water pH to below 5.

Benthic algae were sampled in the Agrio and Guadiamar Rivers following the previously explained procedure. In the Guadiamar River, sampling was carried out upstream, in relation to the mine, in the confluence of the two rivers, and downstream.

The samples obtained upstream from the mine spilling showed small cyanobacteria, different green algae and many diatoms from different genera (*Melosira* Agardh, *Ulnaria* (Kützing) Compère, *Achnanthidium* Kützing, *Planothidium* Round et Bukhtiyarova, *Cocconeis*

Ehrenberg, *Cymbella* Agardh, *Encyonema* Kützing, *Navicula* Bory, *Gomphonema* Ehrenberg, *Nitzschia* Hassall, *Tryblionella* W. Smith and others). After the confluence of the Agrio River and the Guadiamar, the algal flora was completely different. Researchers observed a strong development of filamentous green algae *Klebsormidium* Silva, Mattox & Blackwell, *Ulothrix* Kützing and *Mougeotia* Agardh in the acidic waters along with large quantities of diatoms belonging to only a few species. The most characteristic species from the flora in this part of the river were *Brachysira vitrea* (Grunow) Ross and different taxa belonging to the complex of *Achnanthidium minutissimum* (Kützing) Czarnecki

Some diatom indexes where calculated in all these sites using the software Omnidia.

Table 1 shows the results of one sampling performed in April 2006, where the indexes suggest that both Guadiamar and Agrio rivers had good water quality. In terms of organic matter or eutrophy, the information given by the indexes was accurate. But we cannot consider the general quality of the water to be good, since it was highly polluted by acidity and metals.

Some features of the flora were more indicative of the actual state of the river. The observed green masses of *Klebsormidium* or *Ulothrix* are typical of acidic water with high levels of metals and are often found in these types of ecosystems (Stevens et al., 2001; Niyogi et al., 2002; Martín et al. 2004). Moreover, the richness of diatoms (see number of species in Table 1) was much lower in the Agrio River and downstream of the confluence of both rivers than in the upper part, upstream of the mine.

The genus *Achnanthidium* Kützing, very abundant in these rivers, is also noteworthy. This genus comprises small pennate diatoms. The most relevant ecological feature of these diatoms is an extraordinary ability to colonize, as a pioneering species, any system submitted to some kind of perturbation that leads to the desaparition of the flora of a site. This perturbation can be natural, like a flood after a heavy rain. Likewise, this genus can dominate in some samples just because the biofilm is not mature (Ács et al., 2004).

Moreover, these diatoms can live under a wide range of environmental conditions and resist different types of pollutants. On one hand, *Achnanthidium saprophilum* (Kobayasi & Mayama) Round & Bukhtiyarova was first described in waters polluted with organic matter (Kobayasi & Mayama, 1982), whereas *Achnanthidium minutissimum* (Kützing) Czarnecki (Fig. 8) and *A. biassolettianum* (Grunow) Round & Bukhtiyarova are considered indicative of clean water by the diatom indexes. The exact differentiation among these species and some others is often difficult (Potapova & Hamilton, 2007; Fig. 9) and their utilization in the indexes can sometimes be misleading (Martín et al., 2010).

On the other hand, different authors have found these diatoms growing in waters polluted by metals, as in the Guadiamar River (Sabater, 2000; Seguin et al., 2001; Szabó et al., 2005; Luís et al., 2011). Actually, this genus is frequently seen in this basin, independent of the presence of the mine, since it was abundant upstream. However, at the same time, this genus is the one that better resists the inputs from the mine. In fact, not only did *Achnanthidium* not disappear from the affected sites – as did *Amphora pediculus* (Kützing) Grunow, for example –, but also grew more than any other diatom. This developing can be linked to the fact that *Achnanthidium* may not be favoured by heavy metals and acidity but less disfavoured than other diatoms. Thus, *Achnanthidium* can grow without almost any competitors and dominate the biofilm. This could be the reason why some authors even consider *Achnanthidium* to be an indicator of heavy metal pollution (Nakanishi et al., 2004), although *Achnanthidium* is not exclusive to these environments.

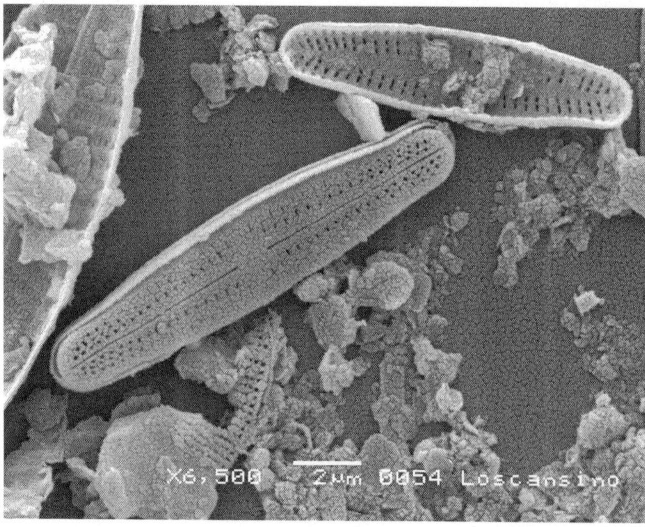

Fig. 8. Internal and external view of two valves of *Achnanthidium minutissimum* (Kützing) Czarnecki.

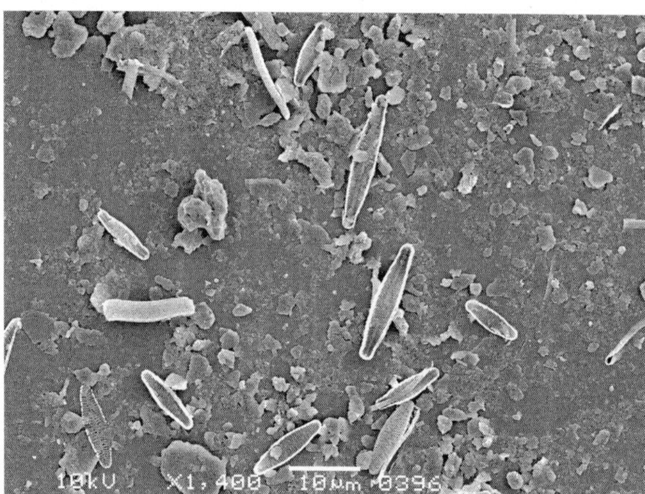

Fig. 9. A group of valves of *Achnanthidium*. Despite the different sizes, they might belong to the same species.

SPECIES	Upstream of the mine	Agrio River	Confluence of both rivers	Downstream of the confluence
Achnanthidium affine (Grunow) Czarnecki	0,0	0,0	4,8	0,0
Achnanthidium jackii Rabenhorst	3,5	0,0	21,6	1,1
Achnanthidium minutissimum (Kützing) Czarneki	30,4	52,7	0,3	0,0
Achnanthidium saprophilum (Kobayasi et Mayama) Round & Bukhtiyarova	0,0	34,4	9,3	73,9
Complex of *Achnanthidium minutissimum*	**33,9**	**87,0**	**35,9**	**75,0**
Amphora pediculus (Kützing) Grunow	22,3	0,0	0,9	0,0
Amphora veneta Kützing	0,0	0,0	2,4	0,0
Brachysira vitrea (Grunow) Ross	0,0	10,9	30,2	0,0
Cocconeis pediculus Ehrenberg	2,0	0,0	0,3	0,0
Cocconeis placentula var. euglypta (Ehrenberg) Grunow	6,3	0,0	0,0	0,0
Cyclotella meneghiniana Kützing	2,2	0,0	0,3	0,2
Cymbella amphicephala Naegeli	0,0	0,2	9,3	0,2
Encyonopsis microcephala Krammer	4,1	0,0	0,0	0,0
Eolimna minima (Grunow) Lange-Bertalot	0,0	0,0	0,3	19,8
Gomphonema angustum Agardh	0,0	0,0	2,1	0,0
Gomphonema olivaceum (Hornemann) Brèbisson	6,5	0,0	0,3	0,0
Nitzschia palea (Küt.) W. Smith	0,0	0,0	2,1	0,0
Staurosirella pinnata (Ehrenberg) Williams & Round	2,6	0,0	0,0	0,0
Ulnaria acus (Kützing) Aboal	0,0	0,0	2,1	0,0
Total number of species	**42**	**10**	**45**	**16**
IPS	**16.4**	**16.6**	**15.9**	**11.0**
IBD	**13.2**	**13.6**	**10.6**	**6.7**
EPI-D	**14.9**	**17.5**	**12.8**	**12.3**

Table 1. Species found in the river Guadiamar and Agrio during April 2006 and their percentages. Only species that contribute to the 80% of the total of diatoms in any site are included in the table and these contributions are shown in the grey cells. The percentages of the species belonging to the complex of *Achnanthidium minutissimum* are added and shown, as well as the total number of species and the values of IPS, IBD and EPI-D indexes. The colors indicate the category of water quality determined by the indexes, according to this key: red, very poor; orange, poor; yellow, moderate; green, good; blue, excellent.

A large abundance of deformed valves also indicates that the river was affected by heavy metals (Fig. 10). Some authors have also found these features in diatoms living under high levels of metals (Morín et al., 2008; Tapia, 2008).

This example shows how indexes can sometimes led us to inaccurate conclusions in situations in which pollutants are not nutrients or biodegradable organic matter. In the case of pollution by heavy metals, other features of the flora (growing of green macroalgae resistant to this type of pollution, low richness and diversity, abundance of tolerant diatoms, high percentage of deformed valves, etc.) are more reliable than biotic indexes to detect the impacts of the pollutants.

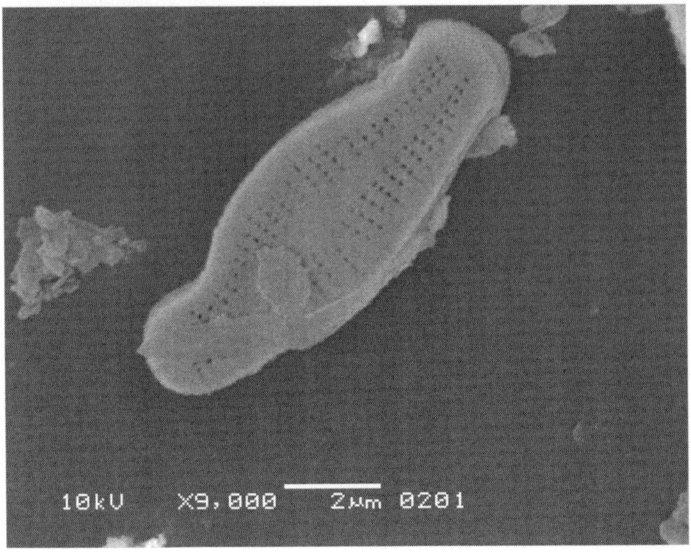

Fig. 10. A deformed valve of *Achnanthidium* from the Guadiamar River, downstream from the mine.

7. Low indexes values but good ecological status: The case of El Picacho pond

This second case study shows the opposite situation with respect to the previously analysed case study. The diatom-based indexes were high in the Guadiamar and Agrio Rivers despite the very poor water quality. In this second case, we find an ecosystem with a very good ecological status that can be underestimated by diatom indexes.

In general, a good ecological status of a water body requires good water quality. Nevertheless, some water bodies pass through different periods during its annual succession. One period can include, without man's influence, high levels of eutrophy, thus giving bad values of water quality by diatom indexes. But if it concerns the nature of the ecosystem, we cannot consider that it has a poor ecological status.

We illustrate this point with the following example. There is a small pond in the mountainous areas in the South of Spain called El Picacho (Fig. 11). It is located in a protected area (Los Alcornocales Natural Park) and has almost no human impact. Its flora and fauna are basically as they are supposed to be under no human pressures. Every spring, its crystalline waters are full of submerged plants, mainly *Myriophyllum alterniflorum* de Candolle and *Ranunculus peltatus* Schrank, the latter covering the whole pond with little beautiful white flowers. The growing of the submerged vegetation is so large that, at the end of the spring, there are not open waters at all.

The pond starts to dry when summer comes until there is no water left in the middle of the season. During the drying, the vegetation starts to die. Decaying plants produce an elevation of the organic matter and eutrophy, and the algal species that develop are indicative of this situation.

Fig. 11. El Picacho pond in August 2011.

We sampled the pond twice during summer 2011, in June and August. The most abundant submerged species was *Myryiophyllum alterniflorum* De Candolle, so it was the chosen substrate for sampling diatoms. The sampling procedure and the work carried out in the laboratory were the same as previously described in this chapter.

Table 2 shows the species found in each sample, the values of the biotic indexes IPS, IBD and EPI-D, and the water quality class determined by the indexes. As in the anterior case, the indexes were calculated using the Omnidia software.

The pond does not suffer any human impact. In this sense, it can be considered as a reference site for other small Mediterranean ponds with a similar annual succession. If the pond can be considered a reference site because of having no impacts, one could expect to have high values of the indexes (green or blue), but the indexes do not actually have a good or excellent rating. In June, when the drying has just started, the values are good (green) for EPI-D and moderate (yellow) for IPS and IBD, although they are in the limit of the poorer category (yellow). However, in August, when there is only a little water left in the pond, the values of the indexes are worse.

According to this, El Picacho pond does not achieve the European Water Framework Directive requirements, at least with respect to this indicator. But the truth is that the pond has a high ecological status because of its natural, mostly untouched condition.

This does not mean that the indexes do not work. In fact, they do work, since the water during this season is actually eutrophic. The values of the indexes, if desired to determine ecological status, have to be analysed in relation to the ecosystem characteristics and nature so as not to reach a misleading conclusion.

SPECIES	**% JUNE**	**% AUGUST**
Gomphonema gracile Ehrenberg	20.3	1.1
Gomphonema parvulum (Kützing) Kützing	29.8	9.4
Nitzschia gracilis Hantzsch	17.9	71.2
Nitzschia palea (Kutzing) W.Smith	2.1	6.8
Ulnaria acus (Kützing) Aboal	8.8	3.2
IPS	**12.3**	**9.9**
IBD	**10.5**	**9.5**
EPI-D	**13.2**	**11.9**

Table 2. Species found in El Picacho during summer 2011 and the respective percentages of each. The species that contribute less to the 5% of the total in both June and August have been deleted in order to reduce the table size. The values of IPS, IBD and EPI-D have been calculated using the Omnidia software. Colours indicate the category of water quality determined by the indexes, according to this key: Red equals very poor quality, orange-poor, yellow-moderate, green-good, and blue-excellent.

8. Conclusion

The diatom-based indexes are widely used and have proved to work in many areas of the world. However, they should be tested before being applied in a basin that was never previously studied. They are mainly used to detect organic pollution and eutrophy.

The sampling procedures to ensure a good calculation, treatment and analysis of the samples are normalised.

However, these indexes are only a working tool, meaning that the information that they provide should be compared to other sources of information from the same sites, such as chemical analysis, diversity of diatoms, and other elements of the flora, among others. Otherwise, indexes could be misinterpreted when estimating water quality and the ecological status of an ecosystem.

9. Acknowledgements

We thank Mr. Andrés Martín and and Ms. Kelly C. Korpiel for their valuable comments on the manuscript as well as their kind review of our English.

10. References

Ács, E.; Szabó, K.; Tóth, B. & Kiss, T. (2004). Investigation of benthic algal communities in connection with reference conditions of the Water framework Directives. *Acta Botanica* Hungarica, Vol. 46, N° 3-4, pp. 255-277.

Atazadeh, I.; Sharifi, M. & Kelly, M.G. (2007) Evaluation of the Trophic Diatom Index for assessing water quality in River Gharasou, western Iran. *Hydrobiologia*, 589: 165-173.

Blanco, S., Ector, L & Becares, E. (2004). Epiphytic diatoms as water quality indicators in Spanish shallow lakes. *Vie et Milieu*, Vol. 54, pp. 71-79.

Cejudo-Figueiras, C.; Álvarez-Blanco, I.; Bécares, E. & Blanco, S. (2010). Epiphytic diatoms and water quality in shallow lakes: the neutral substrate hypothesis revisited. *Marine and Freshwater Research*, Vol. 61, pp. 1457-1467.

European Standard EN 13946. (2003). Water quality - Guidance Standard for the routine sampling and pretreatment of benthic diatoms from rivers. European Committee for Standarization, Brussels

Growns, I. (1999). Is genus or species identification of periphytic diatoms required to determine the impacts of river regulation? *Journal of Applied Phycology*, Vol. 11, pp. 273-283.

Hering, D.; Johnson, R.K.; Kramm, S.; Schmutz, S. ; Szoszkiewicz, K. & Verdonschot, P.F.M. (2006). Assessment of European streams withn diatoms, macrophytes, macroinvertebrates and fish: a comparative metric-based analysis of organism response to stress. *Freshwater Biology*, Vol. 51, pp. 1757-1785.

Kelly, M.G.; Cazaubon, A.; Coring, E.; Dell'Uomo, A.; Ector, L.; Goldsmith, B.; Guasch, H.; Hürlimann, J.; Jarlman, A.; Kawecka, B.; Kwandrans, J.; Laugaste, R.; Lindstrom, E.A.; Leitao, M.; Marvan, P.; Padisák, E.; Pipp, E.; Prygiel, J.; Rott, E.; Sabater, S.; van Dam, H. & Vizinet, J. (1998). Recommendations for the routine sampling of diatoms for water quality assessments in Europe. *Journal of Applied Phycology*, Vol. 10, pp. 215-224.

Kobayasi, H. & Mayama, S. (1982) Most pollution-tolerant diatoms of severly polluted rivers in the vecinity of Tokyo. *Japanese Journal of Phycology*, Vol. 30, pp. 188-196.

Krammer, K. & Lange-Bertalot, H. (1986). *Süsswasserflora von Mitteleuropa. Band 2/1. Bacillariophyceae. Naviculaceae.* Stuttgart: Gustav Fischer.

Krammer, K. & Lange-Bertalot, H. (1988). *Süsswasserflora von Mitteleuropa. Band 2/2. Bacillariophyceae. Epithemiaceae, Bacillariaceae, Surirellaceae.* Stuttgart: Gustav Fischer.

Krammer, K. & Lange-Bertalot, H. (1991a). *Süsswasserflora von Mitteleuropa. Band 2/1. Bacillariophyceae. Centrales, Fragilariaceae, Eunotiaceae.* Stuttgart: Gustav Fischer.

Krammer, K. & Lange-Bertalot, H. (1991b). *Süsswasserflora von Mitteleuropa. Band 2/1. Bacillariophyceae. Achnanthaceae. Kritische Ergänzungen zu Navicula (Lineolatae) und Gomphonema. Gesamtliteraturverzeichnis für Teil 1-4.* Stuttgart: Gustav Fischer.

Luís, A.T.; Teixeira, P.; Almeida, S.F., Matos, J.X. & da Silva, E.F. (2011). Environmental impact of mining activities in the Lousal area (Portugal): chemical and diatom characterization of metal-contamined stream sediment and surface water of Corona stream. *Science of the Total Environment,* Vol. 409, N° 20, pp. 4312-25.

Martín, G.; Alcalá, E.; Solá, C.; Plazuelo, A., Burgos, M.D.; Reyes, E. & Toja, J. (2004). Efecto de la contaminación minera sobre el perifiton del río Guadiamar. *Limnética,* Vol. 23, pp. 315-330.

Martín, G.; Toja, J.; Sala, S.E.; Fernández, R.; Reyes, I. & Casco, M.A. (2010). Application of diatom biotic indexes in the Guadalquivir River Basin, a Mediterranean basin. Which one is the most appropriated? *Environmental Monitoring and Assessment,* Vol. 170, pp. 519-534.

Morín, S.; Duong, T.T.; Dabrin, A.; Coynel, A.; Herlory, O.; Baudrimont, M.; Delmas, F.; Durrieu, G.; Schäfer, J.; Winterton, P.; Blanc, G. & Coste, M. (2008). Long-term survey of heavy-metal pollution, biofilm contamination and diatom community structure in the Riou Mort watershed, South-West France. *Environmental Pollution,* Vol. 151, N° 3, pp. 532-542.

Nakanishi, Y.; Sumita, M.; Yumita, K.; Yamada, T.; Honjo, T. (2004). Heavy-metal pollution and its state in algae in Kakehashi River and Godani River at the foot of Ogoya mine, Ishikawa Prefecture. *Analytical Sciences,* Vol. 20, N° 1, pp. 73-78.

Niyogi, K.D.; Lewis, W.M. & McKnight. (2002). Effects of stress from mine drainage diversity, biomass and function of primary producers in mountain streams. *Ecosystems,* Vol. 5, pp. 564-567.

Potapova, M. & Charles, D.F. (2005). Choice of substrate in algae-based water-quality assessment. *Journal of the North American Benthological Society,* Vol. 24, N° 2, pp. 415-427.

Potapova, M. & Charles, D.F. (2007). Diatom metrics for monitoring eutrophication in rivers of the United States. *Ecological Indicators,* Vol. 7, pp. 48-70.

Potapova, M. & Hamilton, P.B. (2007). Morphological and ecological variation within the *Achnanthidium minutissimum (Bacillariophyceae)* species complex. *Journal of Phycology,* Vol. 43, pp. 561-565.

Prygiel, J.; Coste, M. & Bukowska, J. (1999). Review of the major diatom-based techniques for the quality assessment of rivers – state of the art in Europe, In: *Use of algae for monitoring rivers III,* J. Prygiel, B.A. Whitton & J. Bukowska, (Eds.), 224-238.

Rimet, F. & Bouchez, A. (2011). Biomonitoring river diatoms: implications of taxonomic resolution. *Ecological indicators*, Vol. 15, pp. 92-99.

Round, F.E. (1991). Use of diatoms for monitoring rivers, In: *Use of algae for monitoring rivers*, B.A. , E. Rott & G. Friedrich, (Eds.), 25-32, Düsseldorf.

Sabater, S. (2000). Diatom communities as indicators of environmental stress in the Guadiamar River, S-W Spain, following a major mine tailing spill. *Journal of Applied Phycology*, Vol. 12, pp. 113-124.

Seguin, F.; Druart, J.C.; Le Cohu, R. (2001). Effects of atrazine and nicosulfuron on periphytic diatom communities in freshwater outdoor lentic mesocosm. *Annales de Limnologie*, Vol. 37, pp. 3-8.

Sekar, R.; Venugopalan, V.P.; Nandakumar, K.; Nair, K.V.K & Rao, V.N.R. (2004) Early stages of biofilm sucesión in a lentic freshwater environment. *Hydrobiologia*, Vol. 512, pp. 97-108.

Stevens, A.E.; McCarthy, B.C. & Vis, M.L. (2001). Metal content of *Klebsormidium*-dominated (Chlorophyta) algal mats from acid mine drainage waters in southeastern Ohio. *Jounal of the Torrey Botanical Society*, Vol. 128, N° 3, pp. 226-233.

Szabó, K.; Kiss, K.T.; Taba, G. & Ács, E. (2005). Epiphytic diatoms of the Tisza River, Kisköre Reservoir and some oxbows of the Tisza River after the cyanide and heavy metal pollution in 2000. *Acta Botanica Croatica*, Vol. 64, N° 1, pp. 1-46.

Tapia, P.M. (2008). Diatoms as bioindicators of pollution in the Mantaro River, Central Andes, Peru. *International Journal of Environment and Health*, Vol. 2, N° 1, pp. 82-91.

Taylor, J.C.; Prygiel, J.; Vosloo, A.; de la Rey, P.A. & van Rensburg, L. (2007). Can diatom based pollution indexes be used for biomonitoring in South Africa? A case of study of The Crocodile West and Marico water management area. *Hydrobiologia*, Vol. 592, pp. 455-464.

Torrisi, M. & Dell'Uomo, A. (2006). Biological monitoring of some Apennine rivers (Central Italy) using the Diatom-based Eutrophication/Pollution Index (EPI-D) compared to other European diatom indexes. *Diatom Research*, Vol. 592, pp. 159-174.

Torrisi, M.; Scuri, E.; Dell'uomo, A. & Cocchioni, M. (2010). Comparative monitoring by means of diatoms, macroinvertebrates and chemical parameters of an Apennine watercourse of central Italy: the River Tenna. *Ecological Indicators*, Vol. 10, N° 4, pp. 910-913

Towsend, S.A & Gell, P. (2005). The role of substrate type on benthic diatom assemblages in the Daly and Roper Rivers of the Australian Wet/Dry Tropics. *Hydrobiologia*, Vol. 548, pp. 101-115.

Triest, L.; Kaur, P.; Heylen, S. & de Pauw N. (2001). Comparative monitoring of diatoms, macroinvertebrates and macrophytes in the Woluwe River (Brussels, Belgium). *Aquatic Ecology*, Vol. 35, N° 2, pp. 183-194.

Wu, J.T. & Kow, L.T. (2002). Applicability of a generic index for diatom assemblages to monitor pollution in the tropical river Tsanwun, Taiwan. *Journal of Applied Phycology*, Vol. 14, pp. 63-69.

Zelinka, M. & Marvan, p. (1961). Zur Prazisierung der biologischen Klassification des Reinheit fliessender Gewasser. *Archiv fur Hydrobiologie*, Vol. 57, pp. 389-407.

10

Interplay of Physical, Chemical and Biological Components in Estuarine Ecosystem with Special Reference to Sundarbans, India

Suman Manna[1], Kaberi Chaudhuri[1], Kakoli Sen Sarma[1], Pankaj Naskar[1], Somenath Bhattacharyya[1] and Maitree Bhattacharyya[2*]

[1]Institute of Environmental Studies and Wetland Management,
[2]Department of Biochemistry, University of Calcutta,
Kolkata
India

1. Introduction

An estuary is a partly enclosed coastal body of water with one or more rivers or streams flowing into it, and with a free connection to the open sea. (Jara-Marin et al., 2009; Crossland et al. 2005). They are the transition zones or ecotones between riverine and marine habitats, which differ both in abiotic and biotic factors (McLusky & Elliott, 2004) but many of their important physical and biological attributes are not transitional, but unique. These highly dynamic and rapidly changing systems form a complex mixture of many different habitat types. These habitats do not exist in isolation, but rather have physical, chemical and biological links between them, for example, in their hydrology, in sediment transport, in the transfer of nutrients and in the way mobile species move between them both in seasonally and single tidal cycles. The most distinctive feature that contrasts estuaries from other biomes is the nature and the variability of the physicochemical forces that influence this ecosystem. In contrast the low diversity, the estuarine ecosystems achieve very high productivities through the continuous arrival of new nutrients supply. They are very productive biomes and support many important ecosystem functions like biogeochemical cycling and movement of nutrients, mitigation of floods, maintenance of biodiversity and biological production (Patrick et. al., 2005). The estuarine environment is characterized by a constant mixing of freshwater, saline seawater, and sediment, which is carried into the estuary from the sea and land. The mixture and fluctuation of salt and freshwater impose challenges to the animals and microbes. Along the gradient of conditions from the open sea into the sheltered estuary the salinity ranges from full strength seawater to freshwater, and sedimentary conditions also varies from fine sediment to coarse sediments. Other changes include nutrient input, pollutant and chemical concentration along with estuarine flows (McLusky & Elliott, 2004). The productivity and variety of estuarine habitats support a wonderful abundance and diversity of species.

*Corresponding Author

The ecosystem of any estuary is dependent on both the natural processes (like tide, current, bathymetry, nutrient influx etc) as well as anthropogenic activities (like agriculture, aquaculture etc in the adjoining land part and/or the number and frequency of mechanized boats, trawlers plying within the estuary etc). The entire process is extremely complicated where a balance is achieved through interaction between different components, not clearly well understood so far in many estuaries. The movement of water mass and consequent circulation pattern within an estuary is dependent and thereby should be considered as a response to astronomical tides, inflow of fresh water (i. e. head ward discharge of the fresh water), winds, density of saline water and consequent stratification of different water column etc. At the same time, the basin topography (bathymetry), air-water interaction, water sedimentation interface, mixing characteristics, frictional loss at the bottom, and the rotational effects of the earth, together with the above mentioned driving forces, constitute an extremely complicated balance that conserves mass, momentum, energy, and conservative solutes in the system.

Estuaries are highly dynamic systems with large seasonal and spatial gradients of biogeochemical compounds and processes. Linking land to the ocean, they are often greatly influenced by human activities, including enhanced organic matter and nutrient loadings. Among other parameters, the balance between organic matter and nutrient loading is critical in determining the balance between autotrophy and heterotrophy at the ecosystem level (Kemp et. al., 1997). Estuarine dynamics has been well studied in temperate system such as Chesapeake Bay (Boynton et. al., 1982), San Francisco Bay (Cloern, 1996) and the Baltic sea (Conley et. al., 2000). Tropical and subtropical estuaries received comparatively less study but are experiencing noticeable anthropogenic alterations (Eyre, 1997). Sundarbans is an example of tropical, nutrient rich and turbid estuary.

2. Unique features of Sundarbans estuary

Sundarbans is the largest deltaic tidal halophytic mangrove forest in the world, (Blasco, 1977) with an area of 10, 200 sq km area, spreading over India (4263 sq km of Reserve forest) and Bangladesh (5937 sq km of Reserve forest) (Fig 1). Sundarbans, world's largest delta (80, 000 sq. km.) formed from sediments deposited by three great rivers, the Ganges, Brahmaputra and Meghna, which converges on the Bengal Basin. The area experiences a subtropical monsoon climate with annual rainfall of about 1600-1800 mm and several cyclonic storms (Manna et. al., 2010).

Indian Sundarbans is known as Hoogly-Matla estuary (Hooghly is the Lower part of River Ganges), where apart from Hoogly and Matla, there are innumerable big & small rivers criss-crossing the Sundarbans namely Bidya, Saptamukhani, Raimangal, Muriganga, Thakuran, Gomor etc. Many rivers have become almost completely cut off from the main freshwater sources (Sanyal & Bal, 1986) as for example Bidya, Matla are devoid of fresh water connection due to siltation in the upstream region and are converted into tidal creeks.

Sundarbans is intersected by a complex network of tidal waterways, mudflats and small islands of salt-tolerant mangrove forests. The waterways in this tiger reserve are maintained largely by the diurnal tidal flow (Lahiri, 1973). Sundarbans is known for the eponymous Royal Bengal Tiger (*Panthera tigris tigris*), as well as numerous fauna including species of birds, spotted deer, crocodiles and snakes.

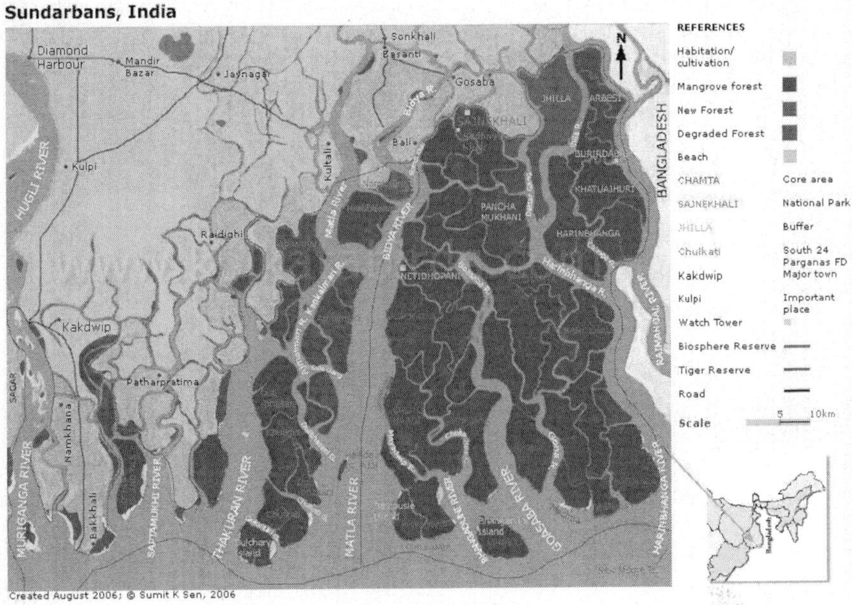

Fig. 1. Geographical location of Sundarban estuary, India

Sundarban delta has the distinction of encompassing the world's largest Mangrove Forest belt and has been identified as World natural heritage site by UNESCO in 1974, National park in 1984 and World heritage site by IUCN in 1989. The mangrove ecosystems in Indian Sundarbans are well known not only for the aerial extent, but also for the species diversity and richness. The biodiversity of Sundarbans includes about 350 species of vascular plants, 250 fishes and 300 birds, along with numerous species of phytoplankton, fungi, bacteria, zooplankton, benthic invertebrates, molluscs, reptiles, amphibians and mammals (Gopal & Chauhan, 2006) It is natural habitat of many rare and endangered species including the Royal Bengal tiger ((*Panthera tigris*) and it is the only mangrove tiger land in the world. Other important species are Estuarine Crocodile (*Crocodilus porosus*), Gangetic Dolphin (*Platinista gangetica*), Snubfin dolphin (*Orcella brevirostris*), River Terrapin (Batagur baska), *Batagur baska, Pelochelys bibroni, Chelonia mydas.*, marine turtles like Olive Ridley (*Lepidochelys olivacea*), Green Sea Turtle (*Chelonia mydas*), Hawksbill Turtle (*Eritmochelys imbricata*), thus making it a natural biodiversity hot spot.

3. Mangrove of Sundarbans

Mangroves are a community of trees and shrubs growing in intertidal forested wetlands restricted to the tropical and subtropical regions (Tomlinson, 1986). Total global area of mangrove forest is estimated to only 18. 1 million ha (Spalding et. al., 1997), against over 570 million ha of freshwater wetlands including peat lands but excluding paddy fields (Spiers et. al., 1999). Mangroves are the only woody halophytes dominated ecosystem situated at the confluence of land and sea, they occupy a harsh environment, being daily subject to tidal changes in temperature, water, salt exposure and varying degree of anoxia (Alongi, 2002). Mangrove forests are recognized as highly productive ecosystems that provide large quantities of organic matter to adjacent coastal waters in the form of detritus and live animals (Holguin et. al., 2001). They provide critical habitat for a diverse marine and terrestrial flora and fauna. Healthy mangrove forests are key to healthy marine ecology. They may be considered as self maintaining coastal, inter-tidal estuarine compartment, which thrives due to constant interaction with terrestrial and marine ecosystem. They are vital to coastal communities as they protect them from damage caused by tsunami waves, erosion, and storms and serve as a nursery for fish and other species that support coastal livelihoods. In addition, they have a staggering ability to sequester carbon from the atmosphere and serve as both a source and repository for nutrients and sediments for other inshore marine habitats.

Sundarbans mangrove estuarine ecosystem is one of the largest detritus-based ecosystems of the world (Pillay, 1958 and Ray, 2008). Litterfall of mangroves supplies the detritus and nutrients regulating the productivity of adjacent Hooghly–Brahmaputra estuarine complex which act as an important nursery ground for many commercially important shell and fin fishes. Due to large scale human intervention from the beginning of last century, several species have become extinct or are in very much threatened or degraded state (Gopal &Chauhan, 2006; Sodhi & Brok, 2007), The loss of the mangroves will have devastating economic and environmental consequence. Royal Bengal tiger, Javan rhino, wild buffalo, hog deer, and barking deer are on the verge of extinction. But any systematic approach towards studying the ecosystem dynamics of Sundarbans has not been attempted so far (Alongi, 2009;Gopal & Chauhan, 2006), where we have attempted to fill in the gaps and create a road map for the sustenance of this World heritage site.

4. Materials and methods

4.1 Physico-chemical analysis

4.1.1 Tidal velocity and current speed

Tide measurement was performed by Valeport MIDAS WTR non directional tide gauge serial no. 34890 (Valeport, U. K). The MIDAS WTR Wave Recorder uses the proven Linear Wave. The MIDAS WTR Wave Recorder uses the proven Linear Wave Theory wave analysis method of measurement. It has high accuracy piezo-resistive pressure sensors and a fast response PRT temperature sensor as standard. Current speed and direction was measured with Aanderra made Doppler Current Sensor 4420 Serial no. 282 Signal type CANbus.

4.1.2 Water quality parameters

Water temperature, pH and conductivity were measured in situ with Hach Portable Meters (HQ40d). Turbidity was measured by using portable turbidity meter (Hach 2100P), salinity was determined in practical salinity units by Knudsen method (Knudson, 1901), dissolved oxygen concentration was studied according to the method of Winkler (JGFOS Protocol, 1994) nutrients like inorganic nitrogen (ammonia, nitrite, nitrate, total nitrogen), soluble phosphate, total phosphate, and reactive silicate were measured according to the same methodology (JGFOS Protocol, 1994).

4.2 Biological analysis

4.2.1 Phytoplankton biomass (Chlorophyll-a)

Chlorophyll samples were filtered through Whatman GF/F (0. 45 μ) filters and extracted in acetone in dark and refrigerated condition. Chlorophyll-a was determined spectrofluorimetrically (Ventrick & Hayward, 1984).

4.2.2 Phytoplankton cell density

Direct estimation of phytoplankton cell abundance and diversity was performed by cell counting method. Surface phytoplankton was collected and the Lugol's preserved subsamples (1-2 liter) were used for quantitative enumeration utilizing a Sedgwick-Rafter counting chamber and Zeiss research microscope according to UNESCO PROTOCOL (1978). Several keys and illustration were consulted to confirm identification (Perry, 2003; Tomas, 1997).

4.2.3 Zooplankton abundance and diversity

Zooplankton samples were collected monthly basis both in night and day time for comparative and quantitative assay. Collection was performed both vertically and horizontally. A long Bongo net, with mesh size 150 μm was used to collect sample. The volume of water flowing through the net was measured by a digital flow meter (Model no. -2030R, General oceanics).

4.2.4 Bacterioplankton abundance

Fluorescence microscope was used to estimate the total number of bacteria. Immediately after sampling, 50 ml of seawater was preserved with 25% gluteraldehyde (0. 2-μm-prefiltered) and stored in cold dark environment to prevent reduction of counts. Cells of bacteria were collected onto a 25-mm black polycarbonate nucleopore membrane with a 0. 2 μm pore size and stained with acridine orange. At least twenty random fields were counted in a Zeiss confocal fluorescence microscope coupled with an image analysis system (Bianchi & Giuliano, 1995; Hobbie et. al., 1976). Viable count of Bacterial colonies was also performed using Luria-Bertani medium by serial dilution method (Cappuccino & Sherman, 2007).

4.2.5 Primary productivity

Primary productivity of water was measured by light and dark bottle method according to the guidance of APHA(1998). Samples collected from each pre selected depth (on the basis of light availability) were taken in triplicate light bottles. Dissolved oxygen of the initial bottles was fixed with NaI-NaOH and $MnCl_2$ in the beginning of incubation period. At the end of the incubation period light and dark bottles were similarly fixed and all the bottles were brought back to laboratory in cold condition for analysis. Then dissolved oxygen concentrations were estimated by Winkler's method (JGFOS Protocol, 1994). NPP (net primary productivity), CR (community respiration) and GPP (gross primary productivity) can be estimated using the following equations

- NPP = Light bottle - Initial bottle
- CR = Initial bottle - Dark bottle with nitrification inhibitor.
- GPP = Net primary productivity + community respiration

4.3 Statistical analysis

The results were expressed as differences between the groups considered significant at $p < 0.05$. Data comparison and influence of the environmental factors on phytoplankton were evaluated by stepwise multiple regression (Manna et. al., 2010) Different statistical analysis and correlation regression analysis were performed using the software STATISTICA.

5. Results and discussions

5.1 Hydrodynamic parameters and their significance

Basic objective of hydrodynamic study is to understand the processes active within the tidal estuaries of Sundarbans; to develop methods so as to quantify the relative importance of river inflow discretely with respect to other forcing parameters like wind, tide, earth's movement etc as a dynamic forcing; to understand the relative importance of flow pattern as well as quantum of water coming as an input/output from different channels (creeks). Water movement and the consequent distributions of nutrients and other chemicals in an estuary are dependent on the hydrodynamic condition of the estuary. A hydrodynamic model is therefore a tool to understand the distribution pattern and availability of different nutrients within the estuary. This model is dependent on many physical parameters like tidal discharge, wave and meteorological forcing. In an estuarine condition, at the seaward boundary, tidal forcing drives the model. Tidal regime in an estuary ultimately determines the amount of sea water to be pushed into the estuary from the open sea carrying sediments as well as nutrients. Thus the boundary tide is most important and is usually specified by a water level time series, a velocity time series or a set of tidal harmonics. All these data need to be acquired from a set of water level measurements using different kinds of tide gauge and current meters. Unfortunately, for Sundarbans, such work has yet to be carried out. The freshwater discharges from rivers at the uplands also play an important role.

Sundarbans can be characterized with six North - South bound estuaries namely, Saptamukhi, Thankuran, Matla, Bidya, Goasaba and Raimongal. Each of these estuaries is different from each other in hydrodynamic set up. Except Raimongal, none of these estuaries are having freshwater discharge at head ward portion. These estuaries are more than 70 kilometers in length starting from its head ward point to the meeting point at the Bay of Bengal. A considerable portion of these estuaries pass through the inhabited islands, from where the supply of nutrients is in the form of agricultural field wash and aquaculture pond effluents and hardly any mangrove detritus is available. Thus, the availability of nutrients in these estuaries depends on the mixing within the estuaries, which in turn depends on tides and current.

Perhaps the most important data required is for tidal and surge information, however such data are almost entirely absent at the moment (Bhattacharyya et al, 2011). Delta Project report of 1968 (Delft, 1968) may be considered as the only published account of tidal range in Sundarbans. One permanent tide gauge station is located at Sagar Island in the Hooghly

Estuary and, although records have been kept here since 1937 these are a made using visual staff record operating only during daylight hours and are therefore incomplete. In order to overcome this problem, automatic tide gauge (Valeport make) instruments were placed within several estuaries of Sundarbans which indicated certain interesting hydrodynamic condition.

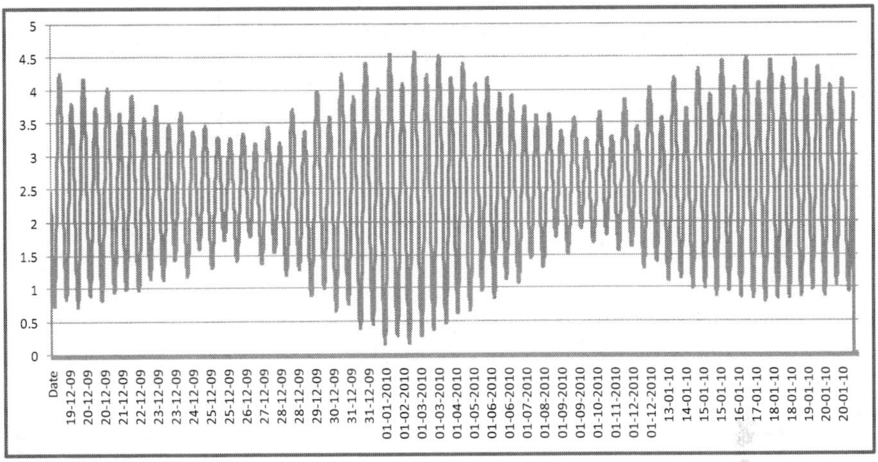

Fig. 2. Tidal fluctuation in Saptamuki mouth near Sitarampur

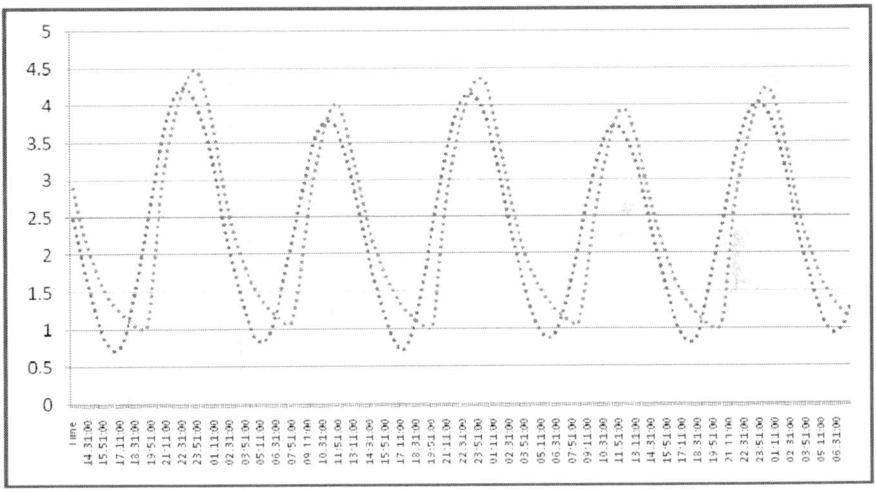

Fig. 3. Saptamukhi tides (in meters) December 2009 Red graph indicates tide at the mouth of Saptamukhi while blue colour indicates tidal fluctuation at Milon More about 50 kilometers away from the Saptamukhi mouth

It was observed that tidal range along the Sundarbans estuaries vary from place to place within the estuary. On an average, tidal fluctuation in Sundarbans estuary is around 5 meters depending on the lunar cycle, as is evident from the Figure. 2. But, tidal fluctuation near the mouth is comparatively smaller in range. As the tidal current pushes within the estuary, tidal range increases due to funneling effect of estuary. In case of Sundarbans, after travelling a certain distance tidal fluctuation starts receding due to bed resistance and ultimately dies down. It is quite interesting to observe the tidal regime along the estuary since ultimately this determines the availability of nutrients within the estuary.

During the period of the present study two automatic tide gauges were simultaneously deployed in the Saptamukhi estuary - one near its mouth (where the width of the estuary is about 1. 3 kilmeters) with Bay of Bengal and the other at the head ward position at a place near Milan More (where the width of estuary is only 60 meters), at a distance of about 50 kilometer northwards from the first tide gauge. Incidentally, Saptamukhi is the westernmost estuary in Sundarbans and a number of mangrove islands including Lothian Island (which is a wildlife sanctuary) are located within it. The variance in tidal behavior pattern is quite obvious (Fig. 3).

It was interesting to note, that near the mouth tide is essentially symmetric in nature indicating that the time taken by high tide to reach the climax is exactly the same as that of low tide. Hence there is no additional residence time of tidal water within the estuary. However, at the Milon More, the tide is asymmetric in nature indicating that time taken by high tide to reach climax is much less compared to that of draining out of tidal water during ebb tide. Current speed within Sundarbans was found to be varying between 140 to 180 cm/sec. The current direction also is also controlled by the geomorphology of the creeks.

The Sundarbans estuary is thus a flood dominated estuary. The nutrient rich tidal water has a more residence time within estuary which is being used by the phytoplankton, the primary producer of this ecosystem. Thus the hydrodynamic set up helps in making this estuary so productive. It should be mentioned that the mangrove islands are all sea facing and water present within the creeks and estuary around the mangrove forest are always rich in nutrient, which is pushed well inside the estuary through tidal water.

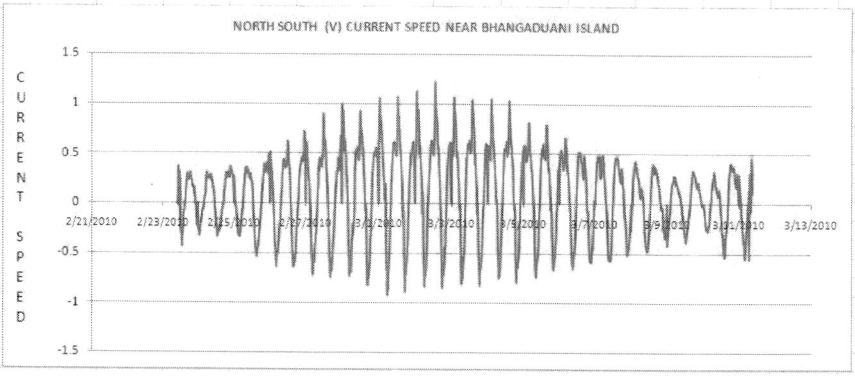

Fig. 4. 'V' Component i. e. North-South bound current at Bhangadunia Island, Sundarbans

During the period of study, several current meters were deployed in different estuaries of Sunderbans to measure the current speed and velocity of tidal current for a period of one month. This is for the first time such an exercise was carried out for Sundarbans. It was observed that the tidal current velocity always depends on the lunar cycle and the phase of tide, which is identical in case of all estuaries.

The U-V component of current the tidal current was then calculated at each spot to assess the North-South current and East-West current. Interesting in all places it was found that the V-component of current is the predominant current over the east-west current. Figure. 4 and Figure. 5 indicate the comparative nature of these two components. The major estuaries of Sundarbans are having a north-south trend and flanked by embankments along east-west sides. Thus, observed current patterns indicate that whatever nutrients are discharged by the mangrove islands along the southern boundary of Sundarbans are carried deep inside the estuaries due to current pattern.

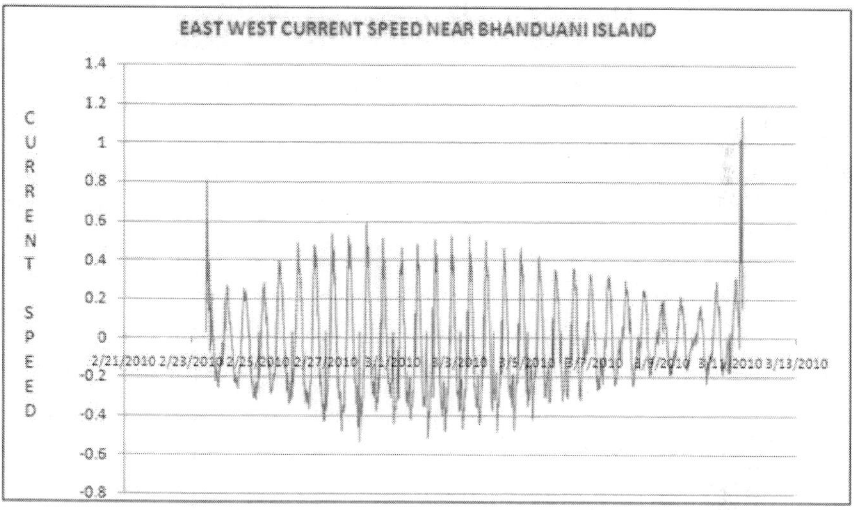

Fig. 5. 'U' component of current velocity i. e. East-West component of currentnear Bhangadunia Island, Sundarbans.

Thus the hydrodynamic studies along estuaries of Sundarbans indicate that the hydrodynamics play the major role for distribution of nutrients along the estuaries of Sunderbans from the southern tip. While current is the driving force for transport of nutrients to the distant parts of the estuaries, tidal regime ensures the availability of the nutrients within the estuary for a longer duration to make it more productive.

5.2 Physicochemical characteristics

Physicochemical characters of estuarine ecosystem mainly depends on three factors, namely dissolved oxygen concentration, salinity and sedimentation load within the water body. Physical factors like temperature, pH, turbidity, salinity and dissolved oxygen in a water body vary quickly relative to biological and chemical transformations. In general, the estuarine environment is oxidizing near the sediment–water interface and more reduced deeper in the sediment. The estuarine circulation movements are the primary mechanism to change the distribution profile of dissolved material in time and space in fresh and ocean water.

5.3 Temperature, pH, turbidity, salinity and dissolved oxygen

Sundarbans are located on shores in the tropics and enjoys tropical monsoon type of climates. The average rainfall of the region amounts to approximately 1750-1800 mm. summer and winter temperature varies between 26^{0}C-40^{0}C . The temperature of the surface water varied between 21^0 C-35^{0}C and significant variation is observed in temperature of different locations in the estuary.

Sundarbans estuarine pH levels generally average from 7. 0 - 7. 5 in the fresher sections, between 8. 0 - 8. 6 in the saline areas. Slightly alkaline pH of seawater is due to the natural buffering from carbonate and bicarbonate dissolved in the water (Volunteer Estuary Monitoring, 2006). The pH values of the Sundarbans estuary were slightly basic and remained almost constant (7. 9-8. 2) except during the monsoon months when a slight but insignificant decrease was noticed.

Turbidity indicates water clarity and it is the measured by light scattering by suspended particles in the water column. Several factors are responsible for water turbidity; suspended soil particles (including clay, silt, and sand); tiny floating organisms (e. g., phytoplankton, zooplankton, and bacterioplankton); and small fragments of dead plants (Voluntary Estuary Monitoring Manual, 2006). In Sundarbans estuary turbidity ranges from 35-150 NTU, and highest index being observed in monsoon.

In estuaries mixing of sea water with fresh water causes brackish water to be more saline than fresh water but less saline than sea water. Salinity of estuaries usually increases away from a freshwater source such as a river, although evaporation sometimes causes the salinity at the head of an estuary to exceed seawater. Vertical salinity structure and nature of salinity variation along an estuary is the unique feature of coastal water ways (Santoroet. al. 1989). Sundarbans estuary situated in the delta of Bay of Bengal showed salinity gradient from the upstream to the downstream part and also margin to central part (Baidya, & Choudhury, 1984). In Sundabans estuary salinity ranges from 11-25 PSU, being highest in dry season and lowest in wet season (Manna et. al., 2010). In Sundarbans estuary tidal action is very strong and practically it is the only regulatory factor, thus water is well mixed from top to bottom and the salinity approaches that of the open sea.

5.4 Dissolved oxygen (DO)

DO is one of the most important controlling factor regulating presence of estuarine species. In addition to support respiration, oxygen is needed for decomposition, an integral part of an estuary's ecological cycle is the breakdown of organic matter (Volunteer Estuary Monitoring Mannual, 2006). DO concentration in the water column is highly dependent on temperature, salinity and biological activity. Tidal estuaries are generally characterized by high DO level. Dissolved oxygen concentration was steady all along the stretch over Sundarban estuary varying between 3. 3 to 9. 5 mg/L with comparatively higher values during November-February (bloom season).

5.5 Nutrients

Autotrophic nutrients are important for estuarine ecosystems and are essential for sustenance of the marine ecosystems because they are the raw materials for primary producers. Main estuarine nutrients include phosphate, nitrate, nitrite, ammonia and silicate. Nitrogen, phosphorus and silica are the key nutrients that generally limit phytoplankton growth in natural waters. Silicate is a primary growth limiting nutrient for diatoms. It is reported that N-limitation is a wide spread phenomenon in tropical lakes rivers and estuaries, e. g., Mandovi-Zuari (Ram et al., 2003), Cochin estuary (Gupta et. al., 2009) and Hoogly estuary (Mukhopadhyay et al., 2006). Sundarbans estuary was phosphorus limited in postmonsoon and nitrogen-limited in premonsoon and monsoon. However, seasonal phosphorous limitation characteristic was found in several estuaries (eg. Gle et al. 2008; Xu et al. 2008). Sundarbans estuary is a nutient rich tropical estuary with high nutrient influx, where a huge quantity of leaf litter is loaded to the estuarine water from adjacent mangrove forests. Besides, land mass wash off during monsoon and drainage waste from shrimp culture farms also contributed to this huge nutrient load. In Sundarban estuary the phosphate concentration varied from 0. 4 to1. 0 μmol/L, the nitrate concentration varied from 2. 4-39. 9 μmol/L, nitrite concentration ranged from 0. 6-1. 6 μmol/L and silicate concentration varied from 4. 8 to 49. 1 μmol/L.

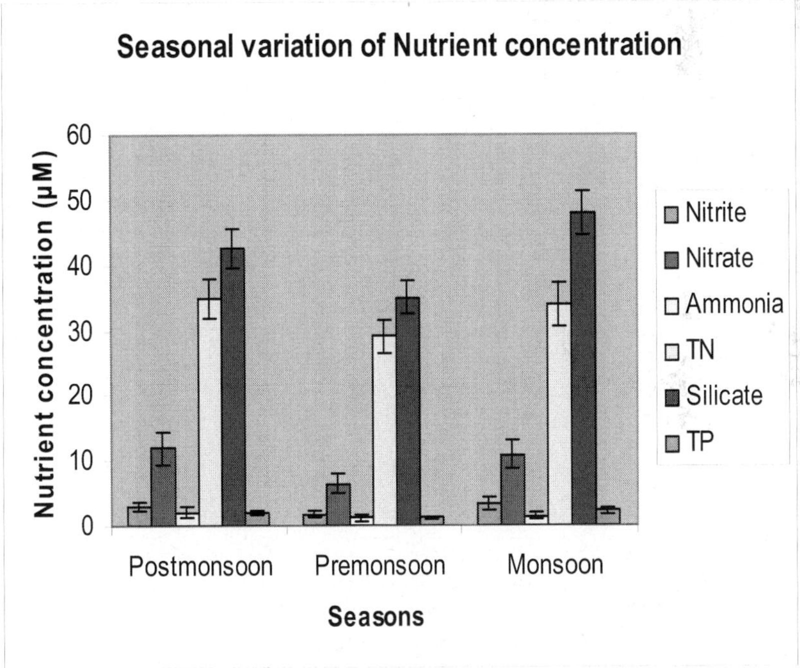

Fig. 6. Seasonal variation of nutrient concentration (μmole L^{-1}) in Sundarban estuary.

6. Biological components

6.1 Autotrophic nutrition and food web

In an estuary, the plants and other primary producers (algae) convert energy into living biological materials (Ryther, 1969). Detritus feeders, plant grazers, and zooplankton are the primary consumers, and the secondary consumers and tertiary consumers include estuarine birds, ducks, invertebrate predators, and fish. Excreta and detritus pass to the decomposer tropic level where microorganisms break down the material. At each stage in this trophic sequence matter and energy are consumed, and some of it is excreted as waste, or converted into body growth or heat after respiration (McLusky and Elliott, 2004).

Plankton is one of the important components of any aquatic ecosystem among which phytoplanktons are the primary source of food in the marine pelagic environment, initiating the food-chain which may culminate even in large mammals (Waniek and Holliday, 2006). Studies of phytoplankton are essential to understand food chain dynamics in aquatic ecosystems (Sieburth and Davis, 1982). In oligotrophic waters, the base of the food chain is composed of very small cells in the size range 0. 2–2 µm ['picoplankton' (Sieburth *et al.*, 1978)] and the microbial loop dominates the pelagic food web (Fenchel, 1988). Conversely, when an import of mineral nutrients takes place, the base of the food chain comprises larger phytoplankton (diatoms and dinoflagellates), which are more readily eaten by zooplanktons; under these conditions, the classic pelagic food chain dominates (Fenchel, 1988). Autotrophic picoplankton ['picophytoplankton' (Fogg, 1986)] is represented by prokaryotic coccoid cyanobacteria, frequently of the genus *Synechococcus*, prochlorophytes (Chisholm *et al.*, 1988) and small eukaryotic cells (Johnson and Sieburth, 1982). Picophytoplankton is important contributors to total phytoplankton biomass and primary production in all aquatic environments (Stockner, 1988). They dominate the total phytoplankton biomass and production in oligotrophic environments (Fogg, 1986; Weisse, 1993).

Phytoplankton biomass and primary production mainly depend on nutrient dynamics of coastal and estuarine ecosystems (Nixon, 1995; Cloern, 1999). Estuarine phytoplankton production is mostly dependent on either nutrient or light availability (Riley, 1967; Williams, 1972; Fisher et al, 1982). In case of nutrient rich turbid estuaries light is the major controlling factor and restricted light availability may alter phytoplankton production (Wofsy, 1983; Pennock, 1985; Harris, 1978;Falkwski, 1980). Estuarine dynamics is well studied in temperate system such as Chesapeake Bay (Bonyton et. al. 1982; Hording, 1994; Kemp et. al., 2005), San Francisco Bay (Review: Cloern, 1996), and the Baltic sea (Graneli et. al., 1990; Conley, 2000).

6.2 Heterotrophic bacterial production

The production and structure of aquatic ecosystems depend on interactions between energy mobilizers, i. e., phytoplankton and bacterioplankton (Jones, 1992) and the abiotic factors that control their activity (Jones, 1998). Bacteria can vary from being consumers of energy released by phytoplankton, to be independent mobilizers of energy to the ecosystem, using carbon compounds imported from the catchment as a source (allochthonous carbon) of energy (Jones, 1992) Consequently, the role of bacteria as an energy mobilizer for the pelagic ecosystem increases with increasing input of allochthonous DOC (Jansson et. al., 2000).

Therefore bacterioplankton may be regarded as a highly important component of the pelagic ecosystem. In case of mangrove dominated Sunderbans estuary where a huge quantity of leaf litters is loaded to the adjacent estuarine water nitrogen may be the main currency in determining overall productivity and heterotrophic bacterial production may exceed phytoplankton primary production.

Many different factors may control the heterotrophic bacterial activity of aquatic ecosystem namely, temperature (Oachs et. al., 1995; Rae & Vincent; 1998), inorganic nutrients (Jansson et. al., 1996; Kroer, 1993), dissolved organic carbon (Tranvik, 1988; Hobbie et. al., 1996) etc. Organic carbon metabolism of clear water bodies like lakes usually dominated by autochthonous sources (Baron et. al., 1991) because of low export of organic matter from surrounding soils whereas in case of nutrient rich estuaries and bays receiving high loading of allochthonous DOC, the bacteria may be relieved of their close dependency on phytoplankton carbon. This may cause bacteria growth to be limited by nutrients (Jansson et. al, 1996; Kroer 1993). Temperature also affects the bacterial growth rate (Oachs et. al., 1995; Rae & Vincent 1998, Tulonen et. al., 1994). In case of nutrient rich Sunderbans estuary bacterial population showed an exponential relationship with temperature (Manna et. al., 2010).

6.3 Community respiration and nitrification

In a water body all the living organisms respire and consume oxygen (O_2). Respiration provides a simple and straightforward measure of heterotrophic activity that can be directly related to the oxidation of organic matter (Willlams, 1981; Hopkinson et al., 1989 & Biddanda et al., 1994) and it is regarded as a key index of the energy used by consumers at a given time and place (Biddanda et. al., 1994 & Pomeroy et. al., 1968).

Episodic oxygen depletion in the water column is a common feature in many coastal and estuarine areas during summer (Turner & Rabalais, 1991; Kemp et al., 1992). In a variety of estuaries nitrification is a major contributor to total pelagic oxygen consumption(Kausch, 1990; Pakulsht, 1995) and records of nitrification rate in addition to primary production and respiration can be an important parameter to understand variations in oxygen concentration in the water column. However, few measurements of the actual rates of pelagic nitrification have been reported from coastal and estuarine waters and studies on factors regulating marine nitrification are still scant in literature (Owens, 1986; Berounsky & Nixon, 1993).

Sunderbans estuary is designated as a moderately productive estuary with an annual integrated phytoplankton production rate of 2. 9 -5. 4 μgC/L/hr and community respiration 1. 75-3. 2 μgC/L/hr (Unpublished data).

Seasons	GPP(μgC/L/hr)	CR (μgC/L/hr)
Postmonsoon	4. 5-5. 4	1. 75-1. 95
Premonsoon	4. 2-4. 4	2. 2-3. 2
Monsoon	2. 9-3. 5	1. 8-2. 07

Table 1. Seasonal variation of primary productivity in Sundarban estuary

6.4 Processes associated with microorganisms

Cycle of energy and matter in estuaries is closely related with microbial activity, half of the aerobic and anaerobic transformations of organic matter in salt marsh are the result of microbial metabolism. Chemical transformation mediated by marine microbes play a critical role in global biogeochemical cycles. Coastal regions show the highest concentration of nutrients and microorganisms and the least light penetration whereas the open ocean is largely olegotrophic (extremely low concentration of nutrients and microorganisms). The concentration of heterotrophic microorganisms determines the Biochemical Oxygen Demand (BOD) i. e. the amount of oxygen removed from water by aerobic respiration. So the coastal region has a high BOD compared to open ocean. A huge energy cycle is carried out by the photosynthetic microbes living on the ocean surface. These microscopic communities are responsible for 98% of primary production (Whitman et. al., 1998, Atlas & Bartha, 1993). This makes ocean microbes one of the major sinks of atmospheric carbon dioxide, a process termed "carbon sequestration". They also mediate all the biogeochemical cycles in the oceans (Atlas & Bartha, 1993).

6.5 Carbon and nitrogen cycles

Bacteria show a variety of metabolic pathways related to carbon flow and cycling. Carbon fixing rate of phytoplankton shows marked seasonal fluctuations in hydrographic and nutrient parameters. As many of the sediment and water-logged soils of estuaries are anoxic, anaerobic decomposition is important. Complex organic matter is used by the fermenters and dissimilatory nitrogenous oxide reducers. The sulfate reducers and methane producers were once thought to have more restricted distributions (John et. al., 1989). Researchers suggested seasonal and inter annual dynamics of free-living bacterioplankton and labile organic carbon available to microbes along the salinity gradient of estuaries. Bacterioplankton abundance may be an important indicator of ecosystem health in eutrophied estuaries, because of the positive relationships between bacterioplankton abundance, microbially labile organic carbon (MLOC), and dissolved oxygen (Leila et. al., 2007).

Nitrogen is a major limiting nutrient for primary production in estuaries. The N-cycling processes that are dominated by microbial activity include nitrification, dissimilatory nitrous oxide reduction, and nitrogen fixation. Nitrogen cycling in estuaries is related to the water mixing and microbial community dynamics.

7. Diversity and distribution of estuarine organisms

7.1 Phytoplankton – The primary producer

Algae belong to a highly diverse group of photoautotrophic organisms with chlorophyll a and unicellular reproductive structures, which are important for aquatic habitats (Ariyadej et. al., 2004) Phytoplankton species composition, richness, population density, and primary productivity vary from coast to coast and sea to sea depending upon the varying hydro biological feature. (Prabhahar et. al., 2011) and seasonal and spatial distribution of plankton in the estuary were discernable). Changes in species composition and dominance of phytoplankton can be mediated by a variety of mechanisms including ambient temperature, light penetration, nutrient supply, and removal by zooplankton (Reynolds 1993).

Sundarbans estuary is formed by a complex network of upstream rivers where the spatial and seasonal variations of some hydrochemical characters are quite prominent and water quality parameters of rivers showed marked variation in different seasons which greatly influence species composition and quantitative abundance of planktons. In Sunderbans phytoplankton abundance ranged from 7. 25 X 10^4 cells / l (June) to 9. 8 x10^6 cells /L (February) (Manna et al., 2010).

Season (Period)			
	Planktonic abundance (Cells/L)	**Biomass (Chlorophyll-a concentration, µg/L)**	***Dominant taxa***
Postmonsoon (Nov - Feb)	**9. 25 x 105 - 9. 8 x 106**	**19. 9 - 36. 5**	*Coscinodiscus sp.,* ***Chaetoceros sp.,*** *Baccteriastrum sp., Thalassiosira sp., Planktoniella sp., Triceratium sp.*
Premonsoon (Mar- Jun)	7. 25 x 104 - 2. 8 x 105	10. 5 - 28. 7	*Navicula sp., Thalassionema sp., Synedra sp. Diatoma sp., Nitzschia sp, Protoperidinium sp., Chlorella sp.* ***Dunaliella sp.***
Monsoon (Jul - Oct)	2. 44 x 105 -5. 2 x 106	6. 9 - 22. 7	*Navicula sp., Thalassionema sp, Cosmarium sp., Closterium sp., Oscillatoria sp. Stigonema sp Pyrocystis sp., Anabena sp* ***Tricodesmium sp.***

Table 2. Distribution and Abundance of Phytoplankton in Sundarban estuary

Sunderbans has a highly diverse algal flora comprised of both benthic and planktonic forms ranging from the freshwater to marine environments (Gopal and Chauhan, 2006) and showed the highest species diversity in all seasons. Noticeable variation of phytoplankton forms was also observed in seasons and sampling locations due to variations of water quality parameters, like pH, salinity, TSS and nutrients and DO. Phytoplankton community was observed to be dominated by diatoms (Biacillariophyceae) followed by Pyrrophyceae (Dinoflagellates) and Chlorophyceae and higest abundance was noticed in postmonsoon (Biswas et al. 2010; Manna et al. 2010) Centric Diatoms predominated in winter months and Pennates in summer whereas Chlorophyceae, Cyanohyceae and Euglenoids dominated the estuary in monsoon. During premonsoon the dominant phytoplankton were species of Ditylum, Ceratium, Biddulphia, Chaetoceros, Coscinodiscus, Thalassiothrix, Rhizosolenia Nitzschia and Thalassionema. However, during postmonsoon phytoplankton species of Bacteriastrum, Biddulphia, Protoperidinium and Ceratium were most dominant (Fig 6a) But in monsoon species of Skeletonema, Fragillaria and some blue green algae, green algae and also euglenoids are quite common. The average phytoplankton load is higher mostly in postmonsoon (Manna et al. 2010). The eastern part of the estuary is dominated by phytoplankton species like Biddulphia diatoms and green and blue green algae, while the central part is dominated by a variety of diatom species viz, Chaetoceros, Coscinodiscus, Bacterioastrum, Cyclotella, Ditylum, Skeletonema, Thallassiothrix, Thalassionema and Triceratium. In contrary, the western region is dominantly represented by species of

Fragillaria, Gyrosigma, Nitzschlia and Bacillaria. The seasonal variation of phytoplankton load indicates that there is a bimodal pattern of distribution with one in premonsoon (May) and other in postmonsoon (November) (Bandyopadhyay, 2003 :Biwas et al. 2010; Manna et al., 2010). Such biomodal seasonal cycle is a typical feature throughout the coastlines in India. (Mani, 1986; Gopinath, 1972; Gouda. 1991) The phytoplankton taxa of Sunderban estuary in general resembles with those of coastal waters, estuarine and near shore region of Goa (Devassy. & Goes, 1988) Cochin back waters, (Devary, & Bhattathiri, 1974) Hooghly river, (Santra et al. 1989; Santra & Pal; 1989), Rushikulya estuary (Gouda, R. & Panigrahy, 1989), Bahuda estuary, (Mishra & Panigrahy, 1995) and Porto Novo (Kannan & Vasantha, 1992). Only difference lies in the occurence of bloom forming diatoms. At high salinity level (25 PSU) halo- tolerant phytoplankton species like Dunaliella salina, Chlorella marina, C. salina, Grammatophora marina, Fragillaria oceanica dominated the estuary which are definitely better adapted to the high saline environment. Bio-indicator species like Polykrikos schwartzil, Dinophysis norvegica and Prorocentrum concavum points to moderately polluted water quality of the Sunderbans estuary (Manna et al., 2010).

Class	Genus	Species
Bacillariophyceae	61	151
Pyrrophyceae	15	53
Cyanophyceae	21	23
Chlorophyceae	34	44
Chrysophyceae	3	8
Euglenophyceae	3	5
Dictyophyceae	3	2
Haptophyceae	1	2
Xanthophyceae	1	1

Table 3. Composition of Phytoplankton Classes.

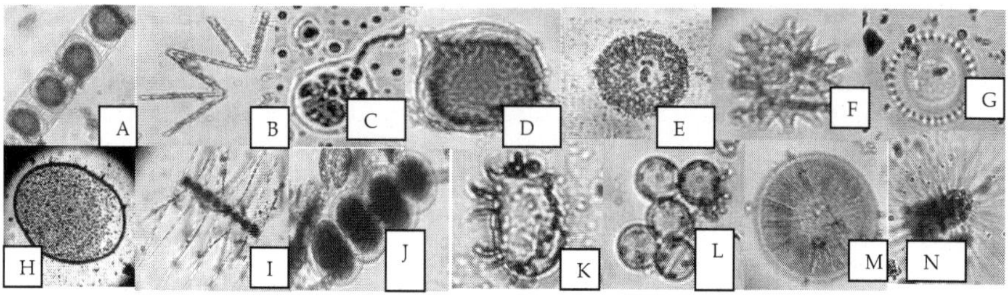

A) *Stephanopyxis palmariana* B) *Thalassionema costatum* C) *Trachelomonas volvocinopsis* D) *Protoperidinium diabolus* E) *Phaeocystis globosa* F) *Pediastrum duplex* G) *Cyclotella stelligera* H) *Aphanothece smithii* I) *Chaetoceros impressus* J) *Peridiniella catenella* H) *Gonyaulax spinifera* L) *Gonium pectorale* M) *Cyclotella striata* N) *Mallomonas tonsulata*

Fig. 7a. Representative phytoplankton taxa of Sundarbans estuary

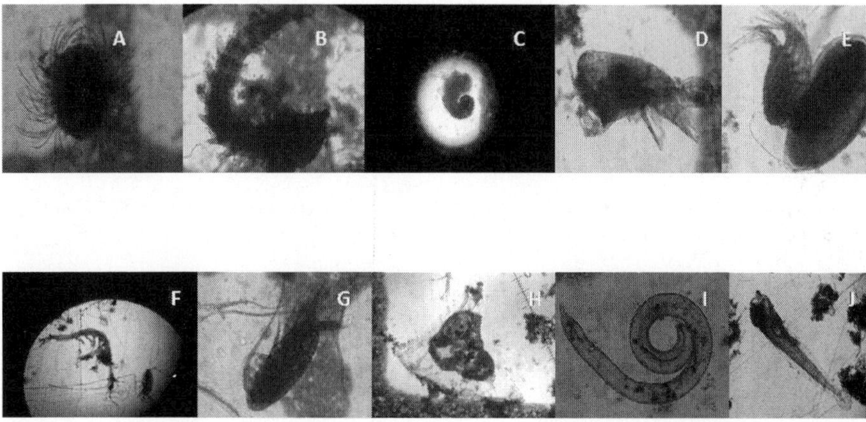

A. *Sabellaria cementarium* B. *Pagurus longicarpus* C. *Gastropod veliger* D. *Corycaeus amizonicus* E. *Cirriped-Cypris* F. *Lucifer sp.* G. *Acrocalanus gracilis* H. *Cyphonautes larva* I. *Nematode larva* J. *Fish larva*

Fig. 7b. Representative zooplankton taxa of Sundarbans estuary

7.2 Zooplankton – Primary consumer

Zooplankton are organism that belonging the secondary level in food web and plays major role from producers level to upper most consumer level. They are more abundant within mangrove water-ways than in adjacent coastal waters (Robertson and Blabber 1992). Zooplankton support many major fisheries and mediate fluxes of nutrients and chemical elements essential to life on Earth.

Copepod dominated Zooplankton abundance in Sunderbans estuary (Fig. 6b) forming 40%-65% of total composition except three months. During February, April and September the dominant species were Tintinnopsis cylindrica, Polychaete larvae(all types)and shrimp larvae respectively. Around 52 taxa of nine major phyla and two minor phyla were identified during the study period. The total counting ranged 1. 92×10^{2}-7. 6×10^{3} cells L^{-1}. Monthly average abundance of Zooplankton ranges indicated the health status of the Sundarbans estuarine aquatic ecosystem. The nutrient upload and water mass upwelling changes the whole pattern of species richness throughout all seasons . Species Richness was highest in Monsoon than the other two seasons where as diversity index was lowest in Post monsoon. Mollusc larvae were dominated in some areas. Mysids and Chaetognath were largely found where as Copepod nauplius, Balanus nauplius, Gastropod veliger and Crab Zoea are more or less same in number in all the stations including Lucifer sp., Protozoans and Bryozoan are distinct Zooplankton which were also found. Nematode larva was a complicated one which only found in estuary but also less in number. In all Estuarine system of World Copepods are the dominating one with respect to other taxa but in case of Indian Sundarbans it fluctuates due to species dominancy in particular time with respect to some cause like nutrient availability or any other environmental factors.

7.3 Fungi – The heterotrophs

The number of fungi living in estuaries is extremely large. Some of fungi are unique in estuaries, while others have a broader range of habitats. Aquatic fungi and yeast dominate species in aquatic environment, few of fungi associate with particles or solid matters in the water. In sediments, the active species of fungi primarily are found in surface aerobic zones. The densities of fungi decrease rapidly with soil depth, but the spores of fungi are found throughout sediments (John et al., 1989). Leaf inhabiting fungi of mangrove plants are known. *Khuskia oryzae* has been reported for the first time from India, among seven species of fungi that exist on mangrove leaf surface of Sunderbanss. There are two new parasitic fungi namely *Pestalotiopsis agallochae* sp. and *Cladosporiummarinum* sp. existing on infected leaves of *Excoecaria agallocha* and *Avicenniamarina* (Pal and Purkayastha, 1992a). Sixteen fungi are isolated from leaves of mangrove plants of Sunderbanss, West Bengal and their growth response to tannin; extracellular pectolytic enzyme (PE) activity and degree of inactivation of PE due to presence of tannin are tested in vitro.

7.4 Bacterioplankton – The detrivore

Microorganisms constitute a huge and almost unexplained reservoir of resources. Microorganisms are the richest repertoire of molecular and chemical diversity in nature. They underlie basic ecosystem processes such as the biogeochemical cycles, food chains and maintain vital relationships between themselves and higher organisms. The enormous number of microbes, their vast metabolic diversity and the accumulation of mutations during the past 3.5 billion years should have led to very high levels of genetic and phenotypic variation (Sogin et. al., 2006).

The world's oceans are teeming with microscopic life forms. Normal microbial cell counts are greater than 10^5 cells per ml in surface sea water (Porter & Feig, 1980). It implies that the oceans harbour 3.6 x 10^{29} microbial cells with a total cellular carbon content of approximately $3x10^{17}$g (Whitman et. al., 1998). In contrast, the total microbial cell number of deep ocean waters (>1000 m) is only $8x10^3$ - $9x10^4$ (Karner et. al., 2001 & Cowen, 2003), half of which are archae (Santelli et. al., 2008). Net primary productivity in the global ocean is estimated to fix 45-50 billion tons of carbon dioxide per year (Falkowski et. al., 1998).

Bacteria are the most abundant organisms in the estuary, averaging between 10^6 to 10^7 per ml organisms in water and 10^8 to 10^{10} per dry weight of sediment. Sediments and salt marsh soil generally harbor more bacteria per unit volume than does the water column. Within the water column, high densities may be found in the surface layer than subsurface layer. Aerobic and facultative anaerobic bacteria are most common, and *Pseudomonads* and *Vibrio* are the most often isolated species. Sediment and waterlogged soils show very high densities of bacteria, which decrease in abundant with depth of soils. Higher bacteria densities have been found in most estuaries than in nearby coastal seawater and river water (John et al., 1989). In Sundarban estuary bacterial population ranged from $1.62X10^3$ - 3. $18X10^6$, lowest in January (3.68×10^6 CFU L^{-1}) and the highest in May (8.9×10^8 CFU L^{-1}) (Manna et. al., 2010).

Marine environment is dominated by microscopic protists and prokaryotes. However, it is widely accepted that current and traditional culture based techniques are inadequate to study microbial diversity from environmental samples. Our understanding of marine microbial communities has increased enormously over the past two decades as result of culture independent phylogenetic studies. Recent advances in molecular techniques are adequate to describe the microbial diversity in a marine sample based on 16S rRNA sequence diversity.

Period	Bacterial count	
	Plate count (CFU ml^{-1})	Fluorescence count (Cells ml^{-1})
Postmonsoon (Nov '08 - Feb '09)	$1.62 \times 10^3 - 4.12 \times 10^4$	$8.15 \times 10^4 - 1.65 \times 10^6$
Premonsoon (Mar '09 - Jun '09)	$3.65 \times 10^4 - 1.85 \times 10^6$	$8.57 \times 10^5 - 6.58 \times 10^7$
Monsoon (Jul '09 - Oct '09)	$6.13 \times 10^4 - 3.18 \times 10^6$	$2.07 \times 10^6 - 9.11 \times 10^7$

Table 4. Seasonal variation of bacterial abundance in Sundarban estuary

It is difficult to grow marine bacteria in culture. Only 0. 1- 10% bacteria are culturable in currently used culture media. Culture-dependent methods do not accurately reflect the actual bacterial community structure. Furthermore, (1) all techniques rely on cultivation are time consuming and expensive as are the physiological and biochemical differentiation tests; (2) after many generations necessary to form plate colonies, the organism may deviate from its physiology, and possibly even from genotypic mix, of the population in nature; (3) though many advances have been made in microbiological culture techniques, it is still not possible to grow a majority of bacterial species using the standard laboratory culturing techniques (Bakonyi et. al., 2003). Only a minor fraction (0. 1-10%) of the bacteria can be cultivated using standard techniques; (4) The biggest drawback in exploring bacterial biodiversity is that phenotypic methods can be applied only on bacteria which can be cultured and (4) it offers a very limited insight into the spatial distribution of the microorganisms (Pace et al., 1986; Holben & Tiedje, 1988; Ward et al., 1992; Amann et al., 1995). Microbial ecologists are turning increasingly to culture-independent methods of community analysis because of the inherent limitations of culture- based methods. Quantitative estimation of community composition can be inferred based on advanced fluorescence microscopic techniques using culture-independent methods. Useful

molecules for such studies include phospholipids, fatty acids and nucleic acids (Morgan & Winstanley, 1997) where as the microscopic techniques involve either the hybridization of fluorescent- labeled nucleic acid probes with total RNA extracted from water or hybridizations with cells in situ. Metagenomics, also known as environmental genomics, is the culture independent study of a community of microorganisms (Steele & Streit, 2005; Fieseler et. al., 2006). This relatively new technology provides a chance to study the other 99% of bacteria (Tringe & Rubin, 2005).

8. Water quality of the estuary

8.1 Nutrient influx and water quality

Estuarine water quality and habitat conditions are directly affected by fluxes of nutrients from the sediments, especially in summer when temperature is high and hypoxic and anoxic events typically occur. The magnitudes of these sediment oxygen and nutrient fluxes also appear to be directly influenced by nutrient and organic matter loading to the estuarine systems (Kemp and Boynton 1992). Both annual and interannual patterns demonstrate that when these external nutrient and organic matter loadings decrease, the cycle of organic matter deposition to the sediments, sediment oxygen demand, and the release of nutrients into the water column also decrease and water quality and habitat conditions improve (Boynton et al., 1995). The chemical parameters of water - primarily nitrates and phosphates - used in combination with productivity determine the trophic condition of estuaries as oligotrophic (low nutrient) through mesotrophic to eutrophic (high nutrient) (Bellinger and Sigee, 2011).

Study in Sundarbans estuary suggests that indicators of inorganic nutrients and plant productivity changed widely during the annual cycle of the estuary as shown below (Manna et. al., 2010; Biswas et al., 2010) (Table 7). The study indicates estuary remained eutrophic during winter and meso-eutrophic during monsoon and premonsoon (Manna et. al., 2010).

Parameter	Postmonsoon	Monsoon	Pre monsoon
Nutrient concentration (μmoll^{-1})			
Total phosphorus	2-2.15	1.1-1.84	.1-1.98
DIN	21-25	6-22	5-15
Chlorophyll a concentration (μg l^{-1})			
Mean concentration	28	19.5	17.5
Maximum concentration	40	28.7	22.7
Secchi depth (cm)			
Mean value	60	25	130
Minimum value	40	20	120

Table 5. Seasonal vaiation of physicochemical parameters in Sunderbans estuary

8.2 Eutrophication of the estuary

Eutrophication seems to be a global problem and nutrient enrichment is one of the most serious threats to near shore coastal ecosystems (Cloern 2001). The balance of water ecosystem is disturbed by eutrophication i. e. excessive fertilization, which, in turn, leads to increases in phytoplankton quantity and primary production. Estuaries receive considerable amounts of freshwater, nutrients, dissolved and particulate organic matter, suspended matter, andcontaminants from land and exchange materials and energy with the open ocean. Estuaries also receive nutrients and organic matter loads from wetlands such as marshes (Meybeck 1982; Kemp 1984, Howarth et al. 1996) and mangroves (Odum and Heald 1972; 1975; Twilley 1988; Robertson et al., 1992) The consequences of nutrient enrichment include algal blooms (Paerl 1997), coral reef degradation (Lapointe 1997& Hughes et al. 2003), loss of diversity and ecosystem resilience (Levine 1998 & Scheffe 2001) and, in extreme cases, the development of "dead zones"(Rabalais et al. 2002). Eutrophication also initiates changes in phytoplankton community structure, decrease in diversity and frequency of harmful algal blooms. It can have significant deleterious effects on the beneficial uses of estuarine and marine waters.

The eutrophication of coastal waters is a problem of epidemic proportion and has disastrous short- and longterm consequences for water quality and resource utilization (Paerl, 1997; Nixon, 1995; Turner and Rabalais, 2003). Metrices based on phytoplankton quantity and productivity is widely used indicators of eutrophication in the status assessment of surface waters (HELCOM:2002; EEA 2007). There are a number of ways in which eutrophication of estuary manifest itself: increase in phytoplankton biomass (Harding & Perry, 1997) and macroalgae (Valiela et al., 1997) anoxia and hypoxia (Rosenberg 1990 & Kiirikki et al. 2006). The most commonly used indicator of eutrophication in waterbody, however, is chlorophyll a (Kauppila 2007). In Sundarbans estuary, chlorophyll-a concentration remained very high (>10 $\mu g\ L^{-1}$) most of the time throughout the year indicating the estuary was in eutrophic condition (Fig 7) (Jones & Fred Lee, 1982).

The Eutrophication may have detrimental effect on the mangrove vegetation. These negative consequences contrast with observations that marine plant growth, including that of tropical mangroves is enhanced with nutrient enrichment (Paerl 1997; Catherine et al. 2009) However, nutrient enrichment favours growth of shoots relative to roots (Grime; 1979, Lambers &, Poorter 1992 & Catherine et al. 2009) thus enhancing growth rates but increasing vulnerability to environmental stresses such as drought, that require large investment in roots for tolerance (Chapin 1991 & Catherine et al. 2009) Thus the benefits of increased mangrove growth in response to coastal eutrophication is offset by the costs of decreased resilience due to mortality during drought, with mortality increasing with soil water salinity along climatic gradient.

8.3 Phytoplanktons as an indicator of water quality

Phytoplankton is the most important producer of organic substances in the aquatic environment and the rate at which energy is stored up by these tiny organisms determine the basic primary productivity of the ecosystem. Phytoplankton satisfy conditions to qualify as suitable pollution indicators in that they are simple, capable of quantifying changes in water quality, applicable over large geographic areas and can also furnish data on background conditions and natural variability (Lee, 1999). Phytoplanktons are characterized for their rapid responses to alterations in environmental conditions (Reynolds 1984; Stolte et. al. 1994) such as anthropogenically introduced eutrophications of coastal waters (Richardson, 1997). The latter characteristic makes phytoplankton sensitive indicators of changes in aquatic ecosystems (Valdes-Weaver et. al., 2006). Their presence or absence from the community indicates changes in physio-chemical environment of the estuary (Rissik, 2009). More so, micro algal components respond rapidly to perturbations and are suitable bio-indicators of water condition which are beyond the tolerance of many other biota used for monitoring (Nwankwo and Akinsoji, 1992). Species diversity indices when correlated with physical and chemical parameters, provide one of the best ways to detect and evaluate the impact of pollution on aquatic communities (Maraglef, 1968) In India, Ganges-Brahmaputra estuary is particularly vulnerable to anthropogenic perturbations due to high nutrient loads from reverine discharge, increasing human population density and rapid economic growth (Seitzinger et. al., 2005; Mukhopadhyay et. al., 2006; Biswas et. al., 2010). Eutrophication as well as presence of toxic Dinoflagellates and Cyanophyceae in the tidal creek of Sunderbans estuary definitely revealed the deteriorated status of the water quality. (Manna et. al. 2010).

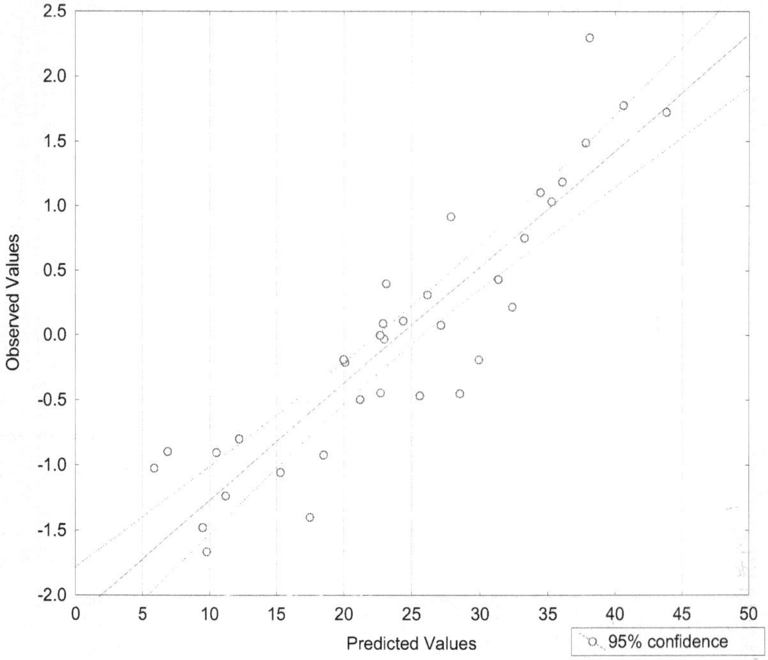

Fig. 8. Plot of observed versus predicted values of chlorophyll-a in Sundarbans estuary (Manna et al.; 2010)

Phytoplankton is one of the most rapid detectors of environmental changes due to their quick response to toxin and other chemicals. Pollution stess reduces the number of algal species but increases the number of individuals. A marked change in algal community severely affects species diversity (Biligrami, 1988). Eutrophication or organic pollution of aquatic ecosystems results in replacement of algal groups. It has been observed that many species are sensitive to nutritional loading, but equally good numbered are pollution tolerant. A numbers of reports are available on pollution -indicating and pollution tolerant algal species (Desei et. al. 2008). Similarly, a good number of indices have also been evolved to determine the trophic level of water ecosystems like Nygaard's algal index, Shannon and Weiner's species diversity indices and Palmer;s pollution index.

In general nutrient-deficient natural water harbouring low populations of algae, on addition of nutrients, increases the growth of the algae. The water appears dark green on excessive algal growth or the algal blooms. These algal blooms occur in highly enriched waters, especially that receiving sewage waste (Trivedi & Goel, 1984). Certain species of phytoplankton grow luxuriantly in eutrophic waters while some species cannot tolerate waters that are contaminated with organic or chemical wastes. Some of the species indicate the clean waters are *Melosira islandica, Cyclotella ocellata* and *Dinobryon*. Pollution indicating plankton includes *Nitzschia palea, aeruginosa* and *Aphanizomenon flosaquae.* The latter two species have been found to produce toxic blooms and anoxic conditions. Some algae were found to cause noxious bloom in polluted water that tastes bad with intolerable odour.

Bioindicators are defined as species or communities which by their presence provide informations on the surrounding physical and chemical environments in a particulary site (Belliger & Sigee 2011) A good indicator species has characteristics like narrow ecological range, wide ranging distribution, easily identifiable characteristics, quick response to environmental stress and well defined taxonomy. The organic pollution influence algal flora more intensely than most other factors including DO, pH, light intensity, hardness of water and other type of pollutants (Palmer, 1969).

Genus	**Family**	**According to Palmer**	**Sundarbans Estuary**
		Pollution index	**Pollution index**
Euglena	Eu	5	5
Oscillatoria	Cy	5	5
Chlamydomonas	Ch	4	4
Scenedesmus	Ch	4	0
Chlorella	Ch	3	3
Nitzschia	Ba	3	3
Navicula	Ba	3	3
Stigeoclonium	Ch	2	2
Synedra	Ba	2	6
Ankistrodesmus	Ch	2	2

Ba-Bacillariophyceae; Cy-Cyanophyceae; Eu-Euglenophyceae: Ch-Chlorophyceae.

Table 6. Palmer's family pollution index (1969) with input from Sundarbans estuary

The algal species are rated on a scale of 1 to 6 (intolerant to tolerant) and index is derived by summing up the scores of all taxa that are present in sample. In computing Palmer's index an alga is considered to be present if there are 50 or more individuals per litre of sample.

Thus, the Palmer's index for Sunderbanss comes to very high (**33**). As any value higher than **20** is condidered to be high in Palmer's index, it tallies quite well with the result of trophic analysis as per OECD norms. Thus, there can be no second opinion about the fact that the estuary is eutrophic for the most time of the year and main reason of eutrophication is high organic load (Manna et. al, 2010).

9. Conclusion

Mangroves are the only woody halophytes dominated ecosystem situated at the confluence of land and sea, they occupy a harsh environment, being daily subject to tidal changes in temperature, water and salt-exposure and varying degree of anoxia (Alongi, 2008). The structure and function of the mangrove food web is unique, driven by both marine and terrestrial components. Little attention has been paid so far to the adaptive responses of mangrove biota to the various disturbances; however they are highly threatened even in this world heritage site, Sundarban. Recently eutrophication seems to be a global problem. The eutrophication of this tidal creek may have detrimental effect on the mangrove vegetation. Nutrient-enrichment i. e. eutrophication of the coastal zone increases the mortality of mangroves by enhancing shoot growth relative to root which makes them vulnerable to environmental stresses like salinity, draught that adversely affect plant water relationships (Catherine et al., 2009). Eutrophication as well as presence of toxic Dinoflagellates and Cyanophyceae in the tidal creek of Sundarban estuary definitely reveals the deteriorated status of the water quality.

Entire mangrove ecosystem of Sundarbans is fragile in nature due to many reasons such as erosion due to sea level rise, increase in salinity, pollution from non-point source like agricultural field wash and point sources like effluent from fishery, Kolkata sewage, loss of biodiversity due to continuous anthropogenic intervention, deforestation due to illegal felling and natural causes, eutrophication etc. Thus, a number of factors are operating in the Sundarbans. Apart from other aspects, different studies on Sundarbans gain its significance for throwing light on one of the most precious natural resource of this biogeographic region. This study indicates that ecosystem dynamics of Sundarbans may facilitate bioinvasion putting a question mark on the sustainability of mangroves. This work will create a roadmap to explore the ecosystem dynamics of Sundarban estuary, which is a great challenge in recent times.

10. Acknowledgement

We are grateful to Ministry of Earth Science, Government of India for the financial support to carry on this work. We express our sincere indebtness to Mr. Arijit Banerjee, Director, IESWM for his continuous support and interest in our work. We acknowledge the instrumental and infrastructural facility provided by UGC and DST, Government of India and ICZM project, West Bengal(World Bank)in the Department of Biochemistry, University of Calcutta. It would not be possible to carry out this immense work without the help and support of the local people of Sundarban. We express our inability to acknowledge them individually.

11. References

[1] Alongi DM. 2009. Life in Tidal Water. The Energetics of Mangrove Forests. Springer Science, pp 65-87.

[2] Alongi DM. 2002. Present state and future of the world's mangrove forests. Environmental Conservation. 29 (3): 331–349.

[3] Amann RI, Ludwig W, Schleifer KH. 1995. Phylogenetic identification and in situ detection of individual microbial cells without cultivation, Microbiol Rev, 59, 143.

[4] Anonymous, 2010 : Annual Report for the year 2009-10 of IESWM on the project entitled 'Oil Spill Trajectory Modeling along Sunderbans and Hooghly estuary' submitted to ICMAM Project Directorate, Ministry of Earth Sciences, Government of India

[5] APHA. 1998. Standard Methods for the Examination of Water and Wastewater. American Public Health Association, Washington, DC.

[6] Ariyadej, C., Tansakul, R., Tansakul, P. and Angsupanich, S. 2004. Phytoplankton diversity and its relationships to the physico-chemical environment in the Banglang Reservoir, Yala Province Songklanakarin J. Sci. Technol, 26(5) : 595-607.

[7] Atlas RM, Bartha R. 1993. Fundamentals and Applications. Benjamin_Cummings, Redwood City, Backwaters. Indian J. Mar. Sci., 3:46-50.

[8] Baidya AU & Chaudhury A. 1984. Ddtibution and abundance of zooplankton ina tidal creek of Sagar Islands, Sunderbanss, West Bengal. Envir. Ecol. 2: No. 4.

[9] Baidya, A. U. & Choudhury, A. (1984) Distribution and abundance of zooplankton in a tidal creek of Sagar Island, Sundarbans, West Bengal. Envir. & Ecol. 2: No. 4.

[10] Bakonyi T, Derakhshifar I, Grabensteiner L, Nowotny N: Development and evaluation of PCR assays for the detection of Paenibacillus larvae in honey samples: comparison with isolation and biochemical characterization. Appl. Environ Microbiol 2003, 69: 1504-1510.

[11] Bandyopadhyay, Ranjini and Sood, AK (2003) Effect of screening of intermicellar interactions on the linear and nonlinear rheology of a viscoelastic gel. In: Langmuir, 19 (8). pp. 3121-312

[12] Baron J, McKnight D, Denning AS. 1991. Sources of dissolved and particulate organic material in Loch Vale watershed, Rocky Mountain national park, Colorado, USA. Biogeochem 15:89–110.

[13] Bartilotti C, Macia A, Queiroga H. 2000. Larval fluxes at saco mangrove creek, Inhaca island (south Mozambique). In:Macrobenthos of Eastern African mangroves:life cycles and reproductive biology of exploited species. Final Report ERBIC 18-CT96-0127, Part B, Florence. pp 97-111.

[14] Bellinger EG. & Sigee DC. Freshwater algae: identification and use as bioindicator. 2010 John Wiley and Sons, -Science . pp 271 .

[15] Berounsky VM, Nixon SW. 1993 Rates of nitrification along an estuarine gradient in Narragansett Bay, Estuaries. 16: 718-730.

[16] Bhattacharyya, S., Pethick, J. & Sen Sarma K., 2011, Managerial response to sea level rise in the tidal estuaries of the Indian Sunderbans : a geomorphological approach, International Journal of Water policy– In press.

[17] Bianchi A, Giuliano L. 1995. Enumeration of Viable Bacteria in the Marine Pelagic Environment. Appl Enviro Microbiol. 62:174-177.

[18] Biddanda B, Opsahl S, Benner R. 1994. Plankton respiration and carbon flux through bacterioplankton on the Lousiana shelf, Limnol Oceanogr. 39: 1259-1275.

[19] Biswas H, Dey M, Ganguly D, De TK, Ghosh S, Jana TK: Comparative Analysis of Phytoplankton Composition and Abundance over a Two- Decade Period at the Land-Ocean Boundary of a Tropical Mangrove Ecosystem. Estuaries and Coasts 2010, 33:384-394.

[20] Blasco, F. 1977. Outlines of ecology, botany and forestry of the mangals of the Indian subcontinent. In:Chapman, V. (ed.) Wet Coastal Ecosystems. Ecosystems of the World. Vol. No. 1. Elsevier Scientific Publishing Co., Amsterdam. pp. 241-257.

[21] Boynton WR, Kemp WM, Keefe CW . 1982. A cornparative analysis of nutrients and other factors influencing estuarine phytoplankton production. Kennedy V(eds.). Estuarine comparisons. Academic Press, New York, pp. 69-90.

[22] Boynton WR, . Garber JH, Summers R and Kemp WM. 1995. Inputs, transformations, and transport of N and P in Chesapeake Bay and selected tributaries. Estuaries 18: 285-314.

[23] Boynton, WR, Kemp WM, Keefe CW. 1982. A cornparative analysis of nutrients and other factors influencing estuarine phytoplankton production. In: Kennedy, V, editor, Estuarine comparisons. Academic Press, New York, pp. 69-90.

[24] Cappuccino JG, Sherman N. 2007. Cultivation of Microorganisms. Microbiology: A Laboratory Manual. Cappuccino JG, Sherman N (eds.). Dorling Kindersley Publishers, India. ISBN 81-317-1437-3 129-134.

[25] Catherine EL, Marilyn CB, Katherine CM, Feller IC. 2009. Nutrient Enrichment Increases Mortality of Mangroves. PLoS ONE 2009, 4(5): e5600.

[26] Chisholm SW, Olson RJ, Zettle JER. Goericke, R., WaterburyJ. B. and Welshmeyer, N. A. (1988) A novel free-living prochlorophyte abundant in the oceanic euphotic zone. Nature. 334:340-343.

[27] Choudhury, A., and Chaudhury, A. B. 1994. Mangroves of the Sunderbanss, Volume one: India. IUCN-The World Conservation Union, Bangkok, Thailand.

[28] Cloern JE . 1996. Phytoplankton bloom dynamics in coastal ecosystems. A review with some general lessons from sustained investigation of San Francisco Bay. California. Rev Geophysics 33: 127-168.

[29] Cloern JE. 2001. Our evolving conceptual model of the coastal eutrophication problem. Mar Ecol Prog Ser. 210: 223–253.

[30] Cloern JE. 1996. Phytoplankton bloom dynamics in coastal ecosystems: A review with some general lessons from sustained investigation of San Francisco Bay. California. Rev Geophysics 33: 127-168.

[31] Cloern JL. 1999. The relative importance of light and nutrient limitation of phytoplankton growth: a simple index of coastal ecosystem sensitivity to nutrient enrichment. Aquatic Ecology 33: 3-16.

[32] Cloern JL. 1999. The relative importance of light and nutrient limitation of phytoplankton growth: a simple index of coastal ecosystem sensitivity to nutrient enrichment. Aquatic Ecology 33: 3-16.

[33] Cole JJ, Caraco NF. 1982. The pelagic microbial food web of oligotrophic lakes. In: Ford T (ed) Aquatic Microbiology. Blackwell Scientific, Cambridge, MA, pp 101–111.

[34] Conley DJ, Kaas F, Mohlenberg F, Rasmussen B, Windolf J (2000) Characteristics of Danish estuaries. Estuaries 23: 820-837.

[35] Conley DJ, Kaas F, Mohlenberg F, Rasmussen B, Windolf J . 2000. Characteristics of Danish estuaries. Estuaries 23: 820-837.

[36] Cowen JP. 2003. Fluids from aging ocean crust that support microbial life. Science. 299: 120-123.

[37] Crossland, C. J. D., Baird, D., Ducrotoy, J. P., Lindeboom, H., 2005. Thecoastalzone—a domain of globalinteractions. In:Crossland, C. J., Kremer, H. H., Lindeboom, H. J., Marshall Crossland, J. I., LeTissier, M. D. A. (Eds.), CoastalFluxesintheAnthropocene. InternationalGeosphere–BiosphereProgrammeSeries. Springer, pp. 1–37.

[38] Delft Hydraulics 1968. Sunderbans Delta Project Phase 1. Report to Rivers Research Institute, Gvt. West Bengal.

[39] Desai SR, Subash Chandran MD and Ramachandra TV. 2008. Phytoplankton Diversity in Sharavati River Basin, Central Western Ghats. The IUP Journal of Soil and Water. Sciences.

[40] Devassy, V. P. & Bhattathiri, P. M. A. 1974. Phytoplankton ecology of the Cochin Backwaters. Indian J. Mar. Sci., 3:46-50.

[41] Dur G, JS Hwang, S Souissi, LC Tseng, CH Wu, SH Hsiao, QC Chen. 2007. An overview of the influence of hydrodynamics on the spatial and temporal patterns of calanoid copepod communities around Taiwan. J. Plankt. Res. 29: 97-116.

[42] EEA: Europe's environment - The fourth assessment. State of the environment report No. 1/2007 2007, 452.

[43] Eyre BD . 1997. Water quality changes in an episodically flushed sub-tropical Australian estuary. A 50 year perspective. Mar chem 59: 177-187

[44] Falkowski PG, Barber RT, Smetacek V. 1998. Biogeochemical controls and feedbacks on ocean primary production. Science. 281: 200-206.

[45] Falkowski PG. 1980. hght-shade adaptation in marine phytoplankton. In: Falkowski, P. G, editor. Primary productivity in the sea. Environmental Science Research. Plenum Press, New York. pp. 99-119.

[46] Fenchel T. 1988. Marine plankton food chains. Ann. Rev. Ecol. Syst. 19: 19–38.

[47] Fieseler L, Quasier A, Schleper C, Hentschel U. 2006. Analysis of the first genome fragment from the marine sponge-associated, novel candidate phylum Poribacteria by environmental genomics. Environ Microbiol 2006, 8: 612-624.

[48] Finnish coastal waters. Monogr Boreal Environ Res. 31:1-57.

[49] Fogg GE. 1986. Picoplankton. Proc. R. Soc. Lond., 228: 1–30.

[50] Gle G, Amo BYD, Laborde SP, and Chardy P. 2008. Variability of nutrients and phytoplankton primary production in a shallow macrotidal coastal ecosystem (Arcachon Bay, France). Estuarine Coastal and Shelf Sciences, **76** : (), 642-656.

[51] Gopal B, Chauhan M. 2006 Biodiversity and its conservation in the Sunderbans Mangrove Ecosystem. Aquatic Sciences. 69:338-354.

[52] Gopinathan, CP. 1972. Seasonal abundance of phytoplankton in the Cochin Backwaters. J. Mar. biol. Ass. India. 14: 568-577.

[53] Gouda, R. & Panigraphy, R. C. 1996. Seasonal distribution of phytoplankton in the surf waters off Gopalpur, east coast of India. Indian J. Mar. Sci., 25:146-150.

[54] Graneli E, Wallstrom K, Larsson U, Graneli W, Elmgren R. 1990. Nutrient limitation of primary production in marine ecosystems in the Baltic Sea. Ambio. 19: 142-151.

[55] Gupta GVM, Thottathil SD, Balachandran, KK, Madhu NV, Madeswaran P, Nair S. 2009. CO2 Supersaturation and net heterotrophy in a tropical estuary (Cochin, India) : Infiuence of Antropqgenic effect. Ecosystems. 12(12): 1145-1157.

[56] Harding LW Jr. 1993. Long-term trends in the distribution of phytoplankton in Chesapeake Bay: rolcs of light, nutrients and streamflow. Mar Ecol Prog Ser. 104: 267-291.

[57] Harding LW, Perry ES Jr. 1997. Longterm increase of phytoplankton biomass in Chesapeake Bay, 1950-1994. Mar Ecol Prog Ser. 157:39-52.

[58] Harris GP. 1978. Photosynthesis, productivity and growth: The physiological ecology of phytoplankton. Arch. Hydrobiol. Beih. Ergeb. limnol. 10:. 10: 1-71.

[59] HELCOM: Environment of the Baltic Sea area 1994-1998. Balt Sea Environ Proc No. 82B 2002, 214.

[60] High Diversity. Trophial conservation Biology, . Blackwell Publishing Ltd, 1-32, ISBN 978-1-4051-5073-6.

[61] Hobbie JE, Daleyr RJ, Jasper S. 1976. Use of Nuclepore Filters for Counting Bacteria by Fluorescence Microscopy. Appl Environ Microbiol., 33:1225-1228.

[62] Hobbie JE, Kling GW, Rublee PA. 1996. Controls of bacterial numbers and productivity in an arctic lake. Bull Ecol Soc Am 77:197.

[63] Holben WE, Tiedje JM. 1988. Applications of nucleic acid hybridization in microbial ecology, Ecology 1988, 69, 561.

[64] Holguin G, Vazquez P, Bashan Y. 2001. The role of sediment microorganisms in the productivity, conservation, and rehabilitation of mangrove ecosystems: an overview. Biol Fertil Soils . 33:265–278.

[65] Hopkmson CS, Sherr B, Wiebe W. 1989. Size fractionated metabolism of coastal microbial plankton, Mar Eco Prog Ser. 51: 155-166.

[66] Hughes TP, Baird AH, Bellwood DR, Card M, Connolly SR. 2003. Climate change, human impacts and the resilience of coral reefs. Science. 301: 929–933.

[67] Hwang JS, HU Dahms, V Aleeksev. 2008. Novel nursery habitat of hydrothermal vent crabs. Crustaceana 81: 375-380.

[68] International Newsletter of Coastal Management-Intercoast Network, Special J. Mar. biol. Ass. India. 14: 568-577.

[69] Jansson M, Bergstrom AK, Blomqvist P, Drakare S. 2000. Allochthonous organic carbon and phytoplankton/bacterioplankton production relationships in lakes. Ecology 81:3250- 3255.

[70] Jansson M, Blomqvist P, Jonsson A, Bergstrom AK. 1996. Nutrient limitation of bacterioplankton, autotrophic and mixotrophic phytoplankton, and heterotrophic nanoflagellates in Lake Ortrasket, a large humic lake in northern Sweden. Limnol Oceanogr 41:1552–1559.

[71] Jara-Marini ME, Soto-Jimenez MF, Paez-Osuna F. 2009. Trophic relationships and transference of cadmium, copper, lead and zinc in a subtropical coastal lagoon food web from SE Gulf of California. Chemosphere. 77(10): 1366-1373.

[72] JGFOS Protocol. 1994. Protocol for the Joint Global Flux Study (JGFOS) Core Management. 178.

[73] John WD, Charles AS, Michael W K, Alejandro YA. 1989. "Estuarine Ecology. " Wiley-Interscience; 1 edition. ISBN 0-10-0471062634.

[74] Johnson PW and Sieburt JM. 1982. In situ morphology and occurrence of eukariotic phototrophs of bacterial size in the picoplankton of estuarine and oceani waters. J. Phycol. 8:318–327.

[75] Jones RA, Fred Lee G. 1982. Recent Advances in Assessing Impact of Phosphorus load on eutrophication-related water quality. Water Research 1982, 16:503-515.

[76] Jones RI. 1992. The influence of humic substances on lacustrine planktonic food chains. Hydrobiol 229:73–91.

[77] Jones RI. 1998. Phytoplankton, primary production and nutrient cycling. In: Hessen DO, Tranvik LJ (eds) Aquatic Humic Substances—Ecology and Biogeochemistry. SpringerVerlag, Berlin, pp 145–175.

[78] Kannan, L. & K. Vasantha, 1992. Microphytoplankton of the Pitchavaram mangals, southeast coast of India: Species composition and population density. In Jaccarini,

V. & E. Martens (eds), The Ecology of Mangrove and Related Ecosystems. Hydrobiologia

[79] Karner MB, Delong EF, Karl DM. 2001. Archaeal dominance in the mesopelagic zone of the Pacific Ocean. Nature. 409: 507-510.

[80] Kauppila P. 2007. Phytoplankton quantity as an indicator of eutrophication in Finnish coastal waters. Monogr Boreal Environ Res. 31:1-57.

[81] Kausch H. 1990. Biological processes in the estuarine environment In Estuarine Water Quality Management edited by Mihaehs W Spnnger-Verlag, Berhn, Coastal and Eatuarine Studies. 36: 353-361.

[82] Kemp WM and. Boynton WR. 1992. Benthic-pelagic interactions: Nutrient and oxygen dynamics, p. 149-209. In D. E. Smith, M. Leffler, and G. Mackiernan (eds.), Oxygen Dynamics in the Chesapeake Bay: A Synthesis of Recent Research. Maryland Sea Grant, College Park, Maryland.

[83] Kemp WM, Bonyton WR, Adolf JE, Boesch DF, Boicourt WC, Brush G, Cornwell JC, Fisher TR, Glibert PA, Hugy JD, Harding LW, Houde Jr. ED, Kimmel DG, Millar WD, Newell RIE, Roman MR, Smith EM. 2005. Stevenson JC: Eutrophication of Chesapeake Bay: Historical trends and ecological interactions. Mar eco Pro Ser. 303: 1-29.

[84] Kemp WM, Sampou PA, Garber J, Tuttle J, Boynton WR. 1992. Seasonal depletion of oxygen from bottom waters of Chesapeake Bay roles of benthic and planktonic respiration and physical exchange processes. Mar Eco Prog Ser. 85: 137-152.

[85] Kiirikki M, Lehtoranta J, Inkala A, Pitkänen H, Hietanen S, Hall POJ, Tengberg A, Koponen J, Sarkkula J. 2006. A simple sediment process description suitable for 3D-ecosystem modelling - Development and testing in the Gulf of Finland. J Mar Syst. 61:55-66.

[86] Knudson M. 1901. Hydrographical tables. G. E. C. Gad Copenhagen 63.

[87] Kroer N. 1993. Bacterial growth efficiency on natural dissolved organic matter. Limnol Oceanogr 38:1282–1290.

[88] Lahiri, R. 1973. Management Plan of Tiger Reserve in Sunderbanss, West Bengal, India. Department of Forests, West Bengal. 101 pp.

[89] Lapointe BE. 1997. Nutrient thresholds for bottom-up control of macroalgal blooms on coral reefs in Jamaica and southeast Florida. Limnol Oceanogr. 42: 1119–1131.

[90] Lee RE. 1999. Psychology. Cambridge University Press, New York, 614 pp.

[91] Leila J. Hamdan, and Robert B. Jonas 2007. Seasonal and interannual dynamics of free-living bacterioplankton and microbially labile organic carbon along the salinity gradient of the Potomac River. Estuaries and Coasts. 29(1):40-53.

[92] Levine J, Brewer S, Bertness M. 1998. Nutrients, competition and plant zonation in a New England salt marsh. J Ecol 86: 285–292.

[93] Mani, P., Krishnamurthy, K. and Palaniappan, R. 1986. Ecology of phytoplankton blooms in the Vellar Estuary, east coast of India. : Indian J. Mar. Sci. 15(1): 24-28.

[94] Manna S, Chaudhuri K, Bhattacharyya S, Bhattacharyya M. 2010. Dynamics of Sunderbans estuarine ecosystem: eutrophication induced threat to mangroves. Saline Systems. 6:8

[95] Margalef DR. 1951. Diversidad de especies en les communideades natural Public Institutte of Biologic Barcelonia . 1951; 9: 5 - 27.

[96] McLusky, D. S., Elliott, M., 2004. The Estuarine Ecosystem: Ecology, Threats and Management, third ed. Oxford University Press, Oxford, pp 216.
[97] Mishra, S., and Panigrahy, RC, 1995. Occurance of diatomblooms in Bahuda estuary, ast coast of India. Indian J. Mar. Sci., 24: 99-101.
[98] Morgan JAW, Winstanley C: Microbial biomarkers. In: van Elsas, JD, Trevors JT, Wellington EMH (Eds.), Modern Soil Microbiology. Marcel Dekker, New York 1997, 331–352.
[99] Nixon SW. 1995. Coastal marine eutrophication: a definition. social causes and future concerns. Ophelia. 31: 199-219.
[100] Nwankwo DI, Akinsoji A. 1992. Epiphyte community on water hyacinth Eichhornia crassipes (Mart.). Solms. in coastal waters of southwestern Nigeria. Arch. Hydrobiol. 124(4): 501-511.
[101] Oachs CA, Cole JL, Likens GE. 1995. Population dynamics of bacterioplankton in an oligotrophic lake. J Plank Res 17:265–391.
[102] Owens. NJP: 1986. Estuarine nitrification: a naturally occurring fluidized bed reaction". Est Coast Shelf Sci. 22: 31-44.
[103] Pace NR, Stahl DA, Lane DJ, and Olsen GJ: The analysis of natural microbial populations by ribosomal RNA sequences, Adv Microb Ecol 1986, 9, 1.
[104] Paerl HW. 1997. Coastal eutrophication and harmful algal blooms: Importance of atmospheric deposition and groundwater as "new" nitrogen and other nutrient sources. Limnol Oceanogr. 42: 1154–11.
[105] Pakulsht JD, Benner R, Amon R, Eadie B, Whitledge T. 1995. Community metabolism and nutrient cycling in the Mississippi River plume: evidence for intense nitrification at intermediate salinities, Mar Eco Prog Ser. 117: 207-218.
[106] Palmer M C. 1969. A Composite Rating of Algae Toleratng Organic Pollution. Journal of Phycology 5: 78–82.
[107] Pargiter, Frederick Eden, 1934 : A revenue history of the Sunderbanss, Volume-I (From 1765 to 1870) pp1-415, republished by Government of West Bengal in 2002.
[108] Patrick M, Ysebaert T, Damme SV, Bergh EV, Maris T & Struyf E. 2005. The Scheldt estuary: a description of a changing ecosystem Hydrobiologia. 540:1–11
[109] Paula, J. 1998. larval retention and dynamics of the prawns Palaemon longirostris H. Milne Edwards and Crangon crangon Linnaeus (Decapoda, Caridea) in the Mira estuary, Portugal, Inv. Reprod. Devel. 33:221-228
[110] Pennock JR. 1985. Chlorophyll distributions in the Delaware Estuary: Regulation by light-limitation. Estuar. st. Shelf Sci 21: 711-725.
[111] Perry R. 2003. A Guide to Marineplankton of southern California. Mar Sci Center, Ocean Globe 3rd edition, 1-17.
[112] Phytoplankton of Bhagirathi-Hooghly estuary: An illustrative account. Ind. Biol.,
[113] Pillay TVR. 1958. Biology of the Hilsa sp. Ind J Fish. 5:201–207.
[114] Pomeroy LR, Johannes RE. 1968. Occurrence and respiration of ultraplankton in the upper 500 meters of the ocean. Deep-Sea Res. 15: 381-391.
[115] Porter KG, Feig YS. 1980. The use of DAPI for identifying and counting aquatic microflora. Limnol Oceanogr. 25: 943-948.
[116] Prabhahar. C, Saleshrani. K and Enbarasan R. 2011. Studies on the ecology and distribution of phytoplankton biomass in Kadalur coastal zone Tamil nadu, India. Curr. Bot. 2(3):26-30, 2011.

[117] Rae R, Vincent WF. 1998. Phytoplankton production in subarctic lake and river ecosystems: development of a photosynthesis- temperature-irradiance model. J Plank Res 20:1293-1312.

[118] Ram ASP, Nair S and Chandramohan D. 2003. Seasonal Shift in Net Ecosystem Production in a Tropical Estuary. Limnol. & Oceanogra. 48(4): 1601-1607.

[119] Ray S. 2008. Comparative study of virgin and reclaimed islands of Sunderbans mangrove ecosystem through network analysis. Eecol Model. 215:207-216.

[120] Reservoir, Yala Province Songklanakarin J. Sci. Technol, 26(5) : 595-607.

[121] Reynolds CS. 1984. Phytoplankton periodicity: The interactions of form, function and environmental variability. Freshwater Biology. 14: 111-142.

[122] Reynolds, C. S. 1993. Scales of distribution and their role in plankton ecology, Hydrobiol., 249, 157 - 171.

[123] Richardson K. 1997. Harmful or exceptional phytoplankton blooms in the marine ecosystem. Advances in Marine Biology 31: 301- 385.

[124] Riley GA. 1967. The plankton of estuaries. In: Lauff, G. A. (ed.) Estuaries. AAAS Publ., Washington D. C., p. 316-328.

[125] Riley GA. 1967. The plankton of estuaries. In: Lauff, G. A. (ed.) Estuaries. AAAS Publ., Washington D. C., p. 316-328.

[126] Rissik D: Plankton: A Guide to their Ecology and Monitoring for Water Quality. Suthers IM, Rissik D (eds.). CSIRO Publishing, Collingwood, Australia; 2009. ISBN: 9780643090583.

[127] Robertson, A. I. and S. J. M. Blabber (1992). plankton, Epibenthos and fish communities. In: Tropical Mangrove Ecosystems (A. I. Robertson and D. M. Alongi Eds.). Coastal Estuar. Stud., 41: 173-224.

[128] Rosenberg R, Elmgren R, Fleischer S, Johnsson P, Persson G, Dahlin H. 1990. Marine eutrophication studies in Sweden. Ambio. 19:102-108.

[129] Ryther JH, 1969. Photosynthesis and fish production in the sea. Science 166:72-76. 969). Photosynthesis and fish production in the sea. Science. 166:72-76.

[130] Ryther JH, 1969. Photosynthesis and fish production in the sea. Science 166:72-76. 969). Photosynthesis and fish production in the sea. Science. 166:72-76.

[131] S. K. Mukhopadhyay, H. Biswas, T. K. De and T. K. Jana, 2006, Fluxes of nutrients from the tropical River Hooghly at the land-ocean boundary of Sunderbanss, NE Coast of Bay of Bengal, India, Journal of Marine System 62: 9 - 21.

[132] Santelli CM, Orcutt BN, Banning E, Bach W, Moyer CL, Sogin ML, Staudigel H, Edwards KJ: Abundance and diversity of microbial life in ocean crust. Nature 2008, 453: 653-657 doi: 10. 1038.

[133] Santoro, A. E. ;Francis, C. A. ;de Sieyes, N. R. ;Boehm, A. B. 2008. Shifts in the relative abundance of ammonia-oxidizing bacteria and archaea across physicochemical gradients in a subterranean estuary. Environmental Microbiology. 10(4): 1068-1079.

[134] Santra S C and Pal U C 1989 Phytoplankton of Bhagirathi- Hooghly estuary. An illustrative account. Indian Biology. 29: 1-27.

[135] Santra, S. C., Das, U. C., Sima Sen, T. M., Rita Saha, Dutta, S. & Gosh Dastidae, P. 1989. abundance of ammonia-oxidizing bacteria and archaea across

physicochemical gradients in a subterranean estuary. Environmental Microbiology. 10(4): 1068-1079.

[136] Sanyal, P. and Bal, A. R. 1986. Some observations on abnormal adaptations of mangrove in Indian Sunderbanss. Indian Soc. Coastal agric. Res. 4: 9-15.

[137] Sarkar A., S. Sengupta S, McArthur JM, Ravenscroft P, M. K. Bera, Ravi Bhushan, A. Samanta and S. Agrawal; 2009. Evolution of Ganges–Brahmaputra western delta plain: Clues from sedimentology and carbon isotopes. Quaternary Science Reviews., 28: 2564 - 2581.

[138] Scheffer M, Carpenter S, Foley JA, Folke C, Walker B. 2001. Catastrophic shifts in ecosystems. Nature. 413: 591–596.

[139] Sieburth, JMcN., SmatacekV, . and Lenz, J. 1978. Pelagic ecosystem structure: heterotrophic compartments of the plankton and their relationship to plankton size fractions. Limnol. Oceanogr., 23, 1256–1263.

[140] Sodhi NS, Brok BW, Bradshaw CJA. 2007. Diminishing Habitats in Regions of High Diversity. Trophial conservation Biology, . Blackwell Publishing Ltd, 1-32, ISBN 978-1-4051-5073-6.

[141] Sogin ML, Morrison HG, Huber JA, Welch DM, Huse SM, Neal PR, Arrieta JM, Herndl GJ. 2006. Microbial diversity in the deep sea and the underexplored "rare biosphere".

[142] Spalding, M. 1997. The global distribution and status of mangrove ecosystems. International Newsletter of Coastal Management-Intercoast Network, Special edition, 1 : 20-21.

[143] Spiers, A. G., 1999. Review of international/continental wetland resources. In: C. M. Finlayson and A. G. Spiers (eds.), Global Review of Wetland Resources and Priorities for Wetland Inventory. Wetlands Internation Publication 13. Wetlands International, Springer Science, pp 65-87.

[144] Stanley, D. J. ; Hait, A. K. ; 2000 Holocene depositional patterns, neotectonics and Sunderbans mangroves in the western Ganges-Brahmaputra delta. Journal of Coastal Research, 2000 – pp26-39

[145] Steele HL, Streit WR: Metagenomics. 2005. advances in ecology and biotechnology. FEMS Microbiol Lett 2005, 247: 105-111.

[146] Stockner, JG. and Antia, NJ. 1986. Algal picoplankton from marine and freshwater ecosystems: A multidisciplinary perspective. Can. J. Fish. Aquat. Sci. 43: 2472–2503.

[147] Stolte W, Mccollin T, Noordeloos A. and. Riegman R. 1994. Effect of nitrogen-source on the size distribution within marinephytoplankton population. Journal of Experimental Marine Biology and Ecology . 184: 83–97.

[148] Struyf E. 2005. The Scheldt estuary: a description of a changing ecosystem Hydrobiologia 540:1–11

[149] Sundh I, Bell RT. 1992. Extracellular dissolved organic carbon released from phytoplankton as a source of carbon for heterotrophic bacteria in lakes of different humic content. Hydrobiol 229:93–106.

[150] Teague, C. C. ; Barrick, D. E. ; Lilleboe, P. M. ; Cheng, R. T. ; Geoscience and Remote Sensing Symposium, 2007, IGARSS 2007, IEEE International 23-28 July 2007 p 2491 – 2494, Barcelona

[151] Tomas CR. 1997. Identifying marine phytoplankton. Academic Press, Harcourt Brace and Company, Torpnt. 858

[152] Tomlinson, B. P. 1986. The botany of mangrove. Cambridge University Press. London. UK.

[153] Tranvik LJ. 1988. Availability of dissolved organic carbon for planktonic bacteria in oligotrophic lakes of different humic content. Microb Ecol 16:311–322.

[154] Tringe SG, Rubin EM; Metagenomics. 2005. DNA sequencing of environmental samples. Nat Rev Gene. 6: 805- 814.

[155] Trivedi RK and Goel PK. 1980. Chemical and Biological methods for water pollution studies. Environmental Publication (India), pp. 250.

[156] Tseng LC, R Kumar, HU Dahms, QC Chen, JS Hwang. 2008a. Monsoon driven succession of copepod assemblages in coasta l waters of northeastern Taiwan strait. Zool. Stud. 47: 46-60

[157] Tulonen T, Kankaala P, Ojala A, Arvola L (1994) Factors controlling production of phytoplankton and bacteria under ice in a humic, boreal lake. J Plank Res 16:1411–1432.

[158] Turner RE, Rabalais NN (2003) Linking landscape and water quality in the Mississippi River basin for 200 years. Bioscience 53: 563-572.

[159] Turner RE, Rabalais NN. 1991. Changes in Mississippi River water quality this century, Bio Sci. 41:140-147.

[160] UNESCO. 1978. Monograph and Oceanographic methodology. Phytoplankton Manual. Sournia, A (eds.). In UNESCO Technicalpaper No. 28, 337.

[161] Valdes-Weaver, LM., Piehler MF. Pinckney J. L, Howe KE. Rossignol K and. Paerl HW. 2006. Long-term temporal and spatial trends in phytoplankton biomass and class-level taxonomic composition in the hydrologically variable Neuse-Pamlico estuarine continuum, North Carolina U. S. A. Limnology and Oceanography. 51(3): 1410–1420.

[162] Valiela I, McClelland J, Hauxwell J, Behr PJ, Hersh D, Foreman K. 1997. Macroalgal blooms in shallow estuaries: Controls and ecophysiological and ecosystem consequences. Limnol Oceanogr. 42:1105-1118.

[163] Ventrick EL, Hayward TL. 1984. Determining chlorophyll on the 1984 CalCOFI surveys. CalCOFI Rep. 25:74-79.

[164] Volunteer Estuary Monitoring: A Methods Manual. March 2006. EPA-842-B-06-003.

[165] Waniek, J. J. and N. P. Holliday: 2006 Large - scale physical control on phytoplankton growth in the Irminger Sea, Part II: Model study of the physical and meteorological

[166] Ward DM, Bateson MM, Weller R, Ruff-Roberts AL. 1992. Ribosomal RNA analysis of microorganisms as they occur in nature, Adv Microb Ecol. 12, 219.

[167] Weiss R. 1993. Dynamics of autotrophic picoplankton in marine and freshwater ecosystems. Adv. Microb. Ecol. 13: 327.

[168] Whitman WB, Coleman DC, Wiebe WJ. 1998. Prokaryotes: the unseen majority. Proc Natl Acad Sci USA. 95:6578-6583.

[169] Williams RB. 1972. Annual phytoplankton production in a system of shallow temperate estuaries. In: Chabreck, RH. (ed.) Proceedmgs of the Coastal Marsh and Estuary Management Synlposium. Louisiana State Univ. Press, Baton Rouge, p. 59-89.

[170] Willlams PJ. 1981. Microbial contribution to overall marine plankton metabolism: direct measurements of respiration. Ocranol Acta. 4: 359-364.
[171] Wofsy SC. 1983. A simple model to predict extinction coefficients and phytoplankton biomass in eutrophic waters. Limnol, Oceanogr. 28: 1144-1155.

Section 2

Water Treatment Technologies and Water Reuse

11

Water Reuse and Sustainability

Rouzbeh Nazari[1], Saeid Eslamian[2] and Reza Khanbilvardi[1]
[1]*City University of New York*
[2]*Isfahan University of Technology,*
[1]*USA*
[2]*Iran*

1. Introduction

Water reuse simply is the use of reclaimed water for a direct beneficial purpose in various sectors from home to industry and agriculture. For a number of semi-arid regions and islands, water reuse provides a major portion of the irrigation water. In addition, the reuse of treated wastewater for irrigation and industrial purposes can be used as strategy to release freshwater for domestic use, and to improve the quality of river waters used for abstraction of drinking water. Specific water reuse applications meet the water quality objectives. Water quality standards and guidelines which are related to irrigation and industrial water reuse are described in this chapter. Other reuse consumptions such as urban, recreational and environmental are also discussed.

Water quality is the most important issue in water reuse systems in ensuring sustainable and successful wastewater reuse applications. The main water quality factors that determine the suitability of recycled water for irrigation are pathogen content, salinity, specific ion toxicity, trace elements, and nutrients. It will be introduced the important criteria for evaluation water quality and World Health Organization guidelines (WHO, 1989) and the United States Environmental Protection Agency guidelines (USEPA, 1992, 2004) which are the two main guidelines that frequently used in many countries around the world. Finally, it will be discussed briefly about different treatment method selections; the degree of treatment required and the extent of monitoring necessary which depend on the specific application. Wastewater reuse can be applied for various beneficial purposes such as agricultural irrigation, industrial processes, groundwater recharge, and even for potable water supply after extended treatment. Water reuse allows the communities to become less dependent on groundwater and surface water resources and can decrease the diversion of water from sensitive ecosystems. Additionally, water reuse may reduce the nutrient loads from wastewater discharges into waterways, thereby reducing and preventing pollution. This "new" water resource may also be used to replenish overdrawn water resources and rejuvenate or reestablish those previously destroyed. Most common types of wastewater reuses are summarized in Table 1.

2. Agriculture reuse

The reuse of wastewater has been successful for irrigation of a wide array of crops, and increases in crop yields from 10-30% have been reported (Asano, 1998, 2004). For a number of semi-arid regions and islands, water recycling provides a major portion of the irrigation water. In addition, the reuse of treated wastewater for irrigation and industrial purposes can be used as strategy to release freshwater for domestic use, and to improve the quality of river waters used for abstraction of drinking water by reducing disposal of effluent into rivers (USEPA, 2003). By knowing that water for agriculture is critical for food security and also by understanding that agriculture remains the largest water user, with about 70% of the world's freshwater consumption, it can be understood that how important it is to have new source of water available for this section. According to recent Food and Agriculture Organization data (FAO Website), only 30 to 40% of the world's food comes from irrigated land comprising 17% of the total cultivated land. One of the broad strategies to address this challenge for satisfying irrigation demand under conditions of increasing water scarcity in both developed and emerging countries is to conserve water and improve the efficiency of water use through better water management and policy reforms. In this context, water reuse becomes a vital alternative resource and key element of the integrated water resource management at the catchment scale (Asano and Levine, 1996; Lazarova, 2000, 2001).

However, despite widespread irrigation with reclaimed wastewater, water-reuse programs are still faced with a number of technical, economic, social, regulatory, and institutional challenges. Some of the water-quality concerns and evaluation of long-term environmental, agronomic, and health impacts remain unanswered. But water quality is the most important issue in water reuse systems so to ensure sustainable and successful wastewater reuse applications, the potential public health risk associated with wastewater reuse should be evaluated and also the specific water reuse applications should meet water quality objectives. Water quality of the effluent which is going to be used as reuse water, is the most important issue related to water reuse systems that determines the acceptability and safety of the use of recycled water for a given reuse application. The options for sustainable reuse projects are related to the quality of the effluent, and the environmental risk associated with land application for a variety of crops and activities and irrigation type and even the quality standard can vary during irrigation and non-irrigation period (Eslamian et al., 2010). It might be higher during interim periods when irrigation is not practiced to ensure a relatively safe discharge to receiving water bodies. The main water quality factors that determine the suitability of recycled water for irrigation are pathogen content, salinity, sodicity (levels of sodium that affect soil stability), specific ion toxicity, trace elements, and nutrients. All modes of irrigation may be applied depending on the specific situation. If applicable, drip irrigation provides the highest level of health protection, as well as water conservation potential (Valentina and Akica, 2005). The most important criteria for evaluation of the treated wastewater are as follows (Kretzschmar et al., 2002):

- Salinity (especially important in arid zones)
- Heavy metals and harmful organic substances
- Pathogenic germs

Table 1 presents the most important water quality parameters and their significance in the case of municipal wastewater reuse.

The goal of each water reuse project is to protect public health without necessarily discouraging wastewater reclamation and reuse. The guidelines or standards required removing health risks from the use of wastewater and the amount and type of wastewater treatment needed to meet the guidelines are both contentious issues. The cost of treating wastewater to high microbiological standards can be so prohibitive that the use of untreated wastewater is allowed to occur unregulated.

Types of Reuse	Treatment	Reclaimed Water Quality	Reclaimed Water Monitoring	Setback Distances
Urban Reuse Landscape irrigation, vehicle washing, toilet flushing, fire protection, commercial air conditioners, and other uses with similar access or exposure to the water	Secondary[1] Filtration[2] Disinfection[3]	pH = 6–9 <10 mg/L biochemical oxygen demand (BOD) < 2 turbidity units (NTU)[5] No detectable fecal coliform/100 mL[4] 1 mg/L chlorine (Cl_2) residual (min.)	pH - weekly BOD - weekly Turbidity - continuous Coliform - daily Cl_2 residual - continuous	50 feet (15 m) to potable water supply wells
Agricultural Reuse For Non-Food Crops Pasture for milking animals; fodder, fiber and seed crops	Secondary Disinfection	pH = 6–9 < 30 mg/L BOD < 30 mg/L total suspended solids (TSS) < 200 fecal coliform/100 mL[5] 1 mg/L Cl_2 residual (min.)	pH - weekly BOD - weekly TSS - daily Coliform - daily Cl_2 residual - continuous	300 feet (90 m) to potable water supply wells
Indirect Potable Reuse Groundwater recharge by spreading into potable aquifers	Site specific Secondary and disinfection (min.) May also need filtration and/or advanced wastewater treatment	Site specific meet drinking water standards after percolation through vadose zone.	pH - daily Turbidity - continuous Coliform - daily Cl_2 residual - continuous Drinking water standards- quarterly Other - depends on Constituent	100 feet (30 m) to areas accessible to the public (if spray irrigation) site specific

[1] Secondary treatment processes include activated sludge processes, trickling filters, rotating biological contactors, and many stabilization pond systems. Secondary treatment should produce effluent in which both the BOD and TSS do no exceed 30 mg/L.

[2] Filtration means passing the effluent through natural undisturbed soil or filter media such as sand and anthracite.

[3] Disinfection means the destruction, inactivation or removal of pathogenic microorganisms. It may be accomplished by chlorination, or other chemical disinfectants, UV radiation or other processes.

[4] The number of fecal coliform organisms should not exceed 14/100 mL in any sample.

[5] The number of fecal coliform organisms should not exceed 800/100 mL in any sample.

Table 1. Reuse Chart (USEPA, 2004)

Parameter	Significance	Approximate Range in Treated Wastewater
Total Suspended solids (TSS)	TSS can lead to sludge deposits and anaerobic conditions. Excessive amounts caused clogging of irrigation systems. Measures of particles in wastewater can be related to microbial contamination, turbidity. Can interfere with disinfection effectiveness	< 1 to 30 mg/l
Organic indicators TOC Degradable Organics (COD, BOD)	Measure of organic carbon. Their biological decomposition can lead to depletion of oxygen. For irrigation only excessive amounts cause problems. Low to moderate concentrations are beneficial.	1 - 20 mg/l 10 - 30 mg/l
Nutrients N,P,K	When discharged into the aquatic environment they lead to eutrophication. In irrigation, they are beneficial, nutrient source. Nitrate in excessive amounts, however, may lead to groundwater contamination.	N: 10 to 30 mg/l P: 0.1 to 30 mg/l
Stable organics (e.g. phenols, pesticides, chlorinated hydrocarbons)	Some are toxic in the environment, accumulation processes in the soil.	
pH	Affects metal solubility and alkalinity and structure of soil, and plant growth.	
Heavy metals (Cd, Zn, Ni, etc.)	Accumulation processes in the soil, toxicity for plants	
Pathogenic organisms	Measure of microbial health risks due to enteric viruses, pathogenic bacteria and protozoa	Coliform organisms: < 1 to 104 /100 ml other pathogens: Controlled by treatment technology
Dissolved Inorganics (TDS, EC, SAR)	Excessive salinity may damage crops. Chloride, Sodium and Boron are toxic to some crops, extensive sodium may cause permeability problems	

Table 2. Water quality parameters for wastewater reuse and their significance (Asano, 1998)

Regulatory approaches stipulate water quality standards in conjunction with requirements for treatment, sampling and monitoring. These standards or guidelines are highly dependent on the kind of water use. Obviously, the landscape and forest irrigation has the lowest requirements concerning the treatment of effluent, compared to the potable reuse. But, the requirements of irrigation of limited crops (crops that need further processing) are not high and therefore it is applicable in economic terms.

The greatest health concern when using recycled water for irrigation is related to pathogens that could be present (Kretschmer et al., 2000). It is widely known that it is not practical to establish the presence or absence of all pathogenic organisms in wastewater or recycled water in a timely fashion. For this reason, the indicator organism, E-coli, was established many years ago to allow monitoring of a limited number of microbiological constituents.

Standards for wastewater reuse in many countries have been influenced by the WHO Health Guidelines (1989) (Table 3) and the USEPA Guidelines (2004) (Table 4). The Guidelines are set to minimize exposure to workers, crop handlers, field workers and consumers, and recommend treatment options to meet the guideline values. WHO's 1989 Guidelines which seems somehow old and there are no any newer WHO guidelines; for the safe use of wastewater in agriculture take into account all available epidemiological and microbiological data. The fecal coliform guideline (e.g. =1000 FC/100 ml for food crops eaten raw) was intended to protect against risks from bacterial infections, and the newly introduced intestinal nematode egg guideline was intended to protect against helminthes infections (and also serve as indicator organisms for all of the large settable pathogens, including amoebic cysts). The exposed group that each guideline was intended to protect and the wastewater treatment expected to achieve the required microbiological guideline was clearly stated. Waste stabilization ponds were advocated as being both effective at the removal of pathogens and the most cost-effective treatment technology in many circumstances.

Category	Reuse Conditions	Exposed Group	Intestinal Nematode (arithmetic mean no. eggs per liter)	Fecal Coliforms (geometric mean no. per 100 ml)	Wastewater treatment expected to achieve the required microbiological guideline
A	Irrigation of crops likely to be eaten uncooked, sports fields, public parks	Workers, consumers, publics	≤ 1	≤ 1000	A series of stabilization ponds designed to achieve the microbiological quality indicated, or equivalent treatment
B	Irrigation of cereal crops, industrial crops, fodder crops, pasture and trees	Workers	≤ 1	No standard recommendation	Retention in stabilization pond for 8-10 days or equivalent helminth and fecal coliform removal
C	Localized irrigation of crops in category B if exposure to workers and the public does not occur	None	Not applicable	Not applicable	Pretreatment as required by irrigation technology, but not less than primary sedimentation

Table 3. WHO (2001) guideline for use of treated wastewater in agriculture

Reuse Type	Treatment	Water Quality	Setbacks	Monitoring
……………… Public Contact ………………				
Irrigation for public areas: * Parks * Cemetery * Golf Courses * Other landscapes **Agricultural irrigation for:** * Food crops that will not be commercially processed * Any crops eaten raw	Secondary Filtration and Disinfection	* pH 6-9 * ≤ 10 mg/L BOD * ≤ 2 NTU * No detectable fecal coliforms/100 mL * at least 1 mg/L residual chlorine	* 50 feet to potable water	* Weekly: pH, BOD * Monthly: Coliforms * Continuously: Turbidity, Chlorine Residue
……………… Limited or No Public Contact ………………				
Irrigation of restricted access areas: * Sod farms * Silvicultures * Other areas with limited or no public access **Agricultural irrigation for:** * Food crops that will be commercially processed * Non-food crops and pastures	Secondary Disinfection	* pH 6-9 * ≤ 30 mg/L BOD * ≤ 30 mg/L TSS * ≤ 200 fecal coliforms/100 mL * at least 1 mg/L residual chlorine	* 300 feet to potable water * 100 feet to areas accessible to public (if spray irrigation is used)	* Weekly: pH, BOD * Monthly: Coliforms and TSS * Continuously: Chlorine Residue

Table 4. USEPA (2004) guideline for agricultural reuse of wastewater

In contrast, USEPA (2004) has recommended the use of much stricter guidelines for wastewater use in the USA. The USEPA (2004) has established guidelines to encourage states to develop their own regulations. The primary purpose of federal guidelines and state regulations is to protect human health and water quality. To reduce disease risks to acceptable levels, reclaimed water must meet certain disinfection standards by either reducing the concentrations of constituents that may affect public health and/or limiting human contact with reclaimed water. The elements of the guidelines applicable to reuse in agriculture are summarized in Table 4. For irrigation of crops likely to be eaten uncooked, no detectable fecal coliform/100 ml are allowed (compared to 1000 FC/100ml for WHO), and for irrigation of commercially processed crops, fodder crops, etc, the guideline is 200 FC/100 ml.

Much wastewater reuse in agriculture is indirect and that is, the wastewater is predisposed into rivers and the contaminated river water is used later on for irrigation. However, international guidelines for the microbiological quality of irrigation water used on a particular crop do not exist (Ayers and Westcott, 1985). The United States Environmental Protection Agency (USEPA) recommended that the acceptable guideline for irrigation with natural surface water, including river water containing wastewater discharges, be set at 1000 FC/10 ml (USEPA, 1981). This standard has been adopted in some other countries as an irrigation water quality standard, for example, Chile, in 1978 (Ayers and Westcott, 1985). This standard is also consistent with guidelines for unrestricted irrigation. FAO has now recommended that the WHO (1989) Guidelines be used interim irrigation water standards, until more epidemiological information is available. Eslamian and Tarkesh-Isfahani (2010b) evaluate the most efficient irrigation systems in wastewater reuse.

3. Industrial reuse

Reuse of reclaimed water for industrial proposes is developed in many industries of United State of America, Europeans and other developed countries. Reclaimed water reuse is one of the strategies for sustainable management. Industrial reuse has increased substantially since the early 1990s for many of the same reasons urban reuse has gained popularity, including water shortages and increased populations, particularly in drought areas, and legislation regarding water conservation and environmental compliance. Utility power plants are ideal facilities for reuse due to their large water requirements for cooling, ash sluicing, rad-waste dilution, and flue gas scrubber requirements (Metcalf and Eddy, 2003, 2007). Petroleum refineries, chemical plants, and metal working facilities are among other industrial facilities benefiting from reclaimed water not only for cooling, but for processing needs as well. For the majority of industries, cooling water is the largest use of reclaimed water because advancements in water treatment technologies have allowed industries to successfully use lesser quality waters. These advancements have enabled better control of deposits, corrosion, and biological problems often associated with the use of reclaimed water in a concentrated cooling water system. The most frequent water quality problems in cooling water systems are corrosion, biological growth, and scaling. These problems arise from contaminants in potable water as well as in reclaimed water, but the concentrations of some contaminants in reclaimed water may be higher than in potable water (EPA, 1981).

Industrial reuse can be explained and defined for a number of industries in the world, but if the most industrial water consumption, cooling towers, is considered to this subject, the industrial reuse is defined for each industry and it can be defined as a quality standard for reclaimed water reuse. Eslamian and Tarkesh-Isfahani (2010a) evaluate the urban reclaimed water for industrial reuses in North Isfahan, Iran. Based on this and other research projects results on eight various industries, and case studies, articles and books, reclaimed water quality parameter limitation for use in cooling towers are defined and shown in Table 5.

Parameter	Measured Standard Method	Unit	Selected Range of Concentration for IOR Consumed
Electrical conductivity (EC)	Platinum Electrode, number 2510 B of Standard methods	µmhos/cm	500-600
Hardness (as $CaCo_3$)	EDTA Titrimetric, number 2340 C of standard methods	mg\L	150-250
Alkalinity	Titrimetric, number 2320 B of standard methods	mg\L	100-150
Chloride	Argentometric, number 4500-Cl^- B of standard methods	mg\L	175-250
Orthophosphate (PO4)	Vanadomolybdophosphoric Acid Colorimetric, number 4500-P C of	mg\L	0-1
polyphosphate	Vanadomolybdophosphoric Acid Colorimetric, number 4500-P C of standard methods	mg\L	Good
NO_2^-	Colorimetric, number 4500-NO_2^- B of standard methods	mg\L	<1
NO_3^-	Ultraviolet Spectrophotometric, number 4500-NO_3^- B of standard methods	mg\L	<5
NH_3	Nesslerization, number D1426 of ASTM	mg\L	<1
TSS	Gravimetric, number 2540 D of standard methods	mg\L	5-10

Turbidity	Nephelometric, number 2130 B of standard methods	NTU	<2
TDS	Platinum Electrode, number 2510 B of Standard methods	mg\L	250-500
Ca	EDTA Titrimetric, number 3500-Ca B of standard methods	mg\L	50-75
BOD_5	Respirometric, number 5210 D of standard methods	mg\L	0-5
COD	Closed Reflux-Titrimetric, number 5220 C of standard methods	mg\L	20-40
pH	Electrometric, number 4500-H^+ B of standard methods	-	6-8
SO_4^{2-}	Gravimetric, number 4500-SO_4^{2-} C of standard methods	mg\L	0-250
Na^+ (as NaCl)	Direct Air-Acetylene Flame Atomic Absorption Spectrometric, number 3111 B of standard methods	mg\L	150.200
Cr	number 3111 B of standard methods	mg\L	<0.5
Cu	number 3111 B of standard methods	mg\L	<0.05-1
Se	number 3111 B of standard methods	mg\L	<1
Mn	number 3111 B of standard methods	mg\L	<0.3-1
Pb	number 3111 B of standard methods	mg\L	<1
Zn	number 3111 B of standard methods	mg\L	<1
Mg	number 3111 B of standard methods	mg\L	<20-30
Co	number 3111 B of standard methods	mg\L	<1
Cd	number 3111 B of standard methods	mg\L	<0.1-1
Fe	number 3111 B of standard methods	mg\L	0.1-0.3
Sr	number 3111 B of standard methods	mg\L	<1
As	number 3111 B of standard methods	mg\L	<1
Hg	number 3111 B of standard methods	mg\L	<1
SiO_2	Molybdosilicate Colorimetric, number 4500-SiO_2 C of standard methods	mg\L	<10-20
Oil and Greece	Partition- Gravimetric, number 5520 B of standard methods	mg\L	<1
Total Chlorine Residual	DPD Colorimetric, number 4500-Cl G of standard methods	mg\L	<4
Total coliform (as log)	-	Log MPN/100ml	<2.2 MPN/100ml
Fecal coliform (as log)	-	Log MPN/100ml	<2.2 MPN/100ml
SRB (sulfate reducing bacteria)	-	MPN/100ml	nil
PAHs	Gas Chromatographic-Flame Ionization Detector	µg/L	nil
THMs	Gas Chromatographic-Mass Spectrometry	µg/L	nil
MTBE	Gas Chromatographic-Flame Ionization Detector	µg/L	nil
OCP-Pesticide	Gas Chromatographic-Electron Capture Detector	µg/L	nil
OPP-Pesticide	Gas Chromatographic-Nitrogen Phosphorous Detector	µg/L	nil
2,4-D	High Performance Liquid Chromatographic	µg/L	nil

Table 5. Range of water quality parameters for reuse of reclaimed water in cooling towers

4. Urban reuse

Urban reuse systems are a crucial part of water recycling since it can provide the reclaimed water for various non-drinking purposes such as Irrigation of public parks and recreation centers, athletic fields, school yards and playing fields, highway medians and shoulders, and landscaped areas surrounding public buildings and facilities, Irrigation of landscaped areas surrounding single-family and multi-family residences, general wash down, and other maintenance activities. Urban reuse can be expanded to cover commercial uses such as vehicle washing facilities, laundry facilities, window washing and mixing water for pesticides, herbicides, liquid fertilizers, toilet and urinal flushing in commercial and industrial buildings. Reclaimed water can also help with human health and safety in dust control and concrete production for construction projects and control the expansion of suspended particles in the air and provide water for fire hydrants. A 2-year field demonstration/research garden compared the impacts of irrigation with reclaimed versus potable water for landscape plants, soils, and irrigation components. The comparison showed few significant differences; however, landscape plants grew faster with reclaimed water (Lindsey et al., 1996). But such results are not a given. Elevated chlorides in the reclaimed water provided by the City of St. Petersburg have limited the foliage that can be irrigated (Johnson, 1998). Dual distribution systems could be used to deliver the reclaimed water to customers through a parallel network of distribution completely separated and marked to distinguish from the community's drinking water line. Design considerations for urban water reuse systems should include two major components: water reclamation facilities and reclaimed water distribution system, including storage and pumping facilities. The reclaimed water distribution system has the potential to become a third water utility, along with drinking water and wastewater. Reclaimed water systems are operated, maintained, and managed in a manner similar to the drinking water system. One of the oldest municipal dual distribution systems in the U.S., in St. Petersburg, Florida, has been in operation since 1977. The system provides reclaimed water for a mix of residential properties, commercial developments, industrial parks, a resource recovery power plant, a baseball stadium, and schools. The City of Pomona, California, first began distributing reclaimed water in 1973 to California Polytechnic University and has since added two paper mills, roadway landscaping, a regional park and a landfill with an energy recovery facility. As part of planning of an urban reuse system, communities have the option of choosing continuous or interruptible reclaimed water system. In general, an interruptible source of reclaimed water can be used as long as reclaimed water will not be used as the only source of fire protection. For example, the City of St. Petersburg, Florida, decided that an interruptible source of reclaimed water would be acceptable, and that reclaimed water would provide only backup for fire protection. If a community determines that a non-interruptible source of reclaimed water is needed, then reliability, equal to that of a potable water system, must be provided to ensure a continuous flow of reclaimed water. This reliability could be ensured through a municipality having more than one water reclamation plant to supply the reclaimed water system, as well as additional storage to provide reclaimed water in the case of a plant upset. However, providing the reliability to produce a non-interruptible supply of reclaimed water will have an associated cost increase. In some cases, such as the City of Burbank, California, reclaimed water storage tanks are the only source of water serving an isolated fire system that is kept separate from the potable fire service. Retrofitting a developed urban area with a reclaimed water distribution system can be expensive. In some cases, however, the benefits of conserving potable water may justify the cost.

5. Environmental and recreational reuses

Water reuse provides a dependable, locally-controlled water supply and tremendous environmental benefits. Environmental reuse includes creating artificial wetlands, enhancing natural wetlands and sustaining stream flows. Uses of reclaimed water for recreational purposes range from landscape impoundments, water hazards on golf courses, to full-scale development of water-based recreational impoundments, incidental contact (fishing and boating) and full body contact (swimming and wading). As with any form of reuse, the development of recreational and environmental water reuse projects will be a function of a water demand coupled with a cost-effective source of suitable quality reclaimed water. In California, approximately 10 percent (47.6 mgd) (2080 l/s) of the total reclaimed water use within the state was associated with recreational and environmental reuse in 2000 (Leverenz et al., 2002). In Florida, approximately 6 percent (35 mgd or 1530 l/s) of the reclaimed water currently produced is being used for environmental enhancements, all for wetland enhancement and restoration (Florida Department of Environmental Protection, 2002). In Florida, from 1986 to 2001, there was a 53 percent increase (18.5 mgd to 35 mgd or 810 l/s to 1530 l/s) in the reuse flow used for environmental enhancements (wetland enhancement and restoration). Two examples of large-scale environmental and recreational reuse projects are the City of West Palm Beach, Florida, wetlands-based water reclamation project and the Eastern Municipal Water District multipurpose constructed wetlands in Riverside County, California. Other applications of environmental and recreational water reuse include creation of natural and man-made wetlands, recreational and aesthetic impoundments and stream augmentation. The objectives of these reuse projects are typically to create an environment in which wildlife can thrive and develop an area of enhanced recreational or aesthetic value to the community through the use of reclaimed water. Other benefits of environmental reuse include decreasing wastewater discharges and reducing and preventing pollution. Recycled water can also be used to create or enhance wetlands and riparian habitats.

6. Economic considerations

One the major aspects of water reuse is the socio economic impacts assessment of implementation of such resources. Wastewater can decrease impacts of water shortage in arid and semi-arid regions of the world and promote means of sustainable development in the world. However, this will be highly dependent of environmentally sound implementation and management for reuse systems. Poor planning and management could leave significant damages on health and environment by contaminating valuable drinking water supplies and bring unwanted socio economic losses. Economic sustainability and public reception depend on the usage of reclaimed water. Most researches and surveys (Angelakis et al., 2001; Mantovani et al., 2001), have concluded that the best practices are those that substitute reclaimed water in lieu of potable water for use in irrigation, environmental restoration, cleaning, toilet flushing, and industrial uses. The main benefits of using reclaimed water in these situations are conservation of water resources and pollution reduction. Treating and reusing wastewater is economically reasonable in terms of increasing the water availability and the benefits of saving the environment from discharge of wastewater into other systems and controlling the spread of contamination into water and soil. Demand for municipal, industrial and agriculture is on the rise and are expected to reach 37, 23, and 340 bcm; respectively. Provided the low consumptive use of the municipal and industrial sectors, most of the appropriated water can be recovered (Kretschmer et al., 2003). In the agricultural sector, the large size of withdrawals encourages the collection and reuse of irrigation water. Wastewater is already in use around the world. In China, Chile and Mexico, extensive agriculture lands around are irrigated by wastewater (Sadik et al., 1994; Xie et al., 1993). Arab regions have also practicing wastewater reuse. About 7 bcm of wastewater was reused in 1996 out of 191 bcm the total withdrawal that year; this implies less than 4% recovery. Reused agriculture drainage was about 5 bcm out of 168 bcm withdrawn for that sector, less than 3% recovery and 2 bcm of municipal and industrial wastewater out of 23 bcm withdrawn, about 9% recovery (El-Ghamam, 1997). Wastewater is a source real economic activity involving local and federal government along with private industries. Various entities invest in getting rid of it or suffer the environmental damage. Either practice has a pervasive impact on public health and the sustainability of development. If wastewater is properly treated and reused, solves two major of saving local and regional environment and resolving water shortage. Over all the economic viability of water reuse has to be studied individually and the required treatment and cost efficiency, depend on type of pollutants, concentration and type of reuse.

7. Public health concerns and acceptance

The major emphasis of wastewater reclamation and reuse has been on non-drinking applications so far, such as agricultural and landscape irrigation, industrial cooling and in-building applications, such as toilet flushing, in large commercial buildings. Indirect or direct potable reuse raises more public concern because of real or perceived perception of aesthetics and long-term health concerns. Regardless, the value of water reuse is weighed within a context of larger public issues of necessity and opportunity and will not be implemented until two major problems of public health concerns and public acceptance is resolved. Each of these problems involves various issues from scientific concerns to human psychology. In the case of public health concerns, which are extremely viable concerns, presence of pathogenic organism and inorganic micro pollutants should be carefully examined for their short and long term impacts. Pathogens could impose serious threat to human health. They are found in water as bacteria, protozoa, helminthes and ruses which some of them can be easily detected and removed (Dishman et al., 1989). However, others are more difficult to detect and removed and there are not enough studies to assign a safe concentration limits to them. Furthermore, the risk of viral infections and waterborne diseases in general is still an unresolved issue. The inorganic pollutants of concerns in water reuse are nitrates, other nitrogen compounds and heavy metals which are easy to detect and remove. Organic micro pollutants also represent a large problem in direct potable reuse mainly because of lack of sufficient information on the health significance of many the known or suspected carcinogenic, mutagenic, allergenic, teratogenic organic compounds found in water (Crook, 1985). It is also necessary to mention that there are thousands of organic compounds in water that are awaiting discovery (Golden, 1984; Dishman et. al., 1989). The second problem that the potable use of reclaimed water has to facing is public acceptance. This a major obstacle to reuse and it roots in educational and psychological barriers which have to overcome in order to obtain public support. Numerous researches have highlighted the fact that public is not welcoming in this regard and most of the polls revealed major opposition to direct potable use (Gallup, 1973; Kasperson et al., 1974; Carley, 1985). The general feeling about use of wastewater for drinking purposes is negative, regardless of the degree of treatment and these feelings embody the psychological factors in the public's rejection of direct potable use of reclaimed water.

8. Conclusions

The world's population is on the rise and is expected to increase dramatically between now and the year 2020 (United Nations, 2006). This growth will put more pressure on our already scarce and damaged water resources. Communities around the world will be faced with an increased level of wastewater production with no use. Water reclamation and reuse can offer significant help for conserving and extending available water supplies. Water reuse may also present communities with an alternate wastewater disposal method as well as providing pollution abatement by diverting effluent discharge away from sensitive surface waters. However, water reuse has its own advantages and disadvantages which have been summarized in Table 6.

Advantages	Disadvantages
This technology reduces the demands on drinkable sources of freshwater.	If implemented on a large scale, revenues to water supply and wastewater utilities may fall as discharge of wastewaters is reduced.
It may reduce the need for large wastewater treatment systems, if significant portions of the waste stream are reused or recycled.	Reuse of wastewater may be seasonal in nature, resulting in the overloading of treatment and disposal facilities during the rainy season.
The technology may diminish the volume of wastewater discharged, resulting in a beneficial impact on the aquatic environment.	Application of untreated wastewater as irrigation water or as injected recharge water may result in groundwater contamination.
Capital costs are low to medium for most systems and are recoverable in a very short time; this excludes systems designed for direct reuse of sewage water.	Health problems, such as water-borne diseases and skin irritations, may occur in people coming into direct contact with reused wastewater
Operation and maintenance are relatively simple except in direct reuse systems where more extensive technology and quality control are required	In some cases, reuse of wastewater is not economically feasible because of the requirement for an additional distribution system.
Provision of nutrient-rich wastewaters can increase agricultural production in water-poor areas.	Gases, such as sulfuric acid, produced during the treatment process can result in chronic health problems.

Table 6. Advantages and disadvantages of water reuse

9. References

Angelakis, A., Thairs, T. and Lazarova, V., 2001, Water Reuse in EU Countries: Necessity of Establishing EU-Guidelines, State of the Art Review, Report of the EUREAU Water Reuse Group EU2-01-26, 52p.

Asano, T., and Levine, A.D., 1996, Wastewater reclamation, recycling and reuse: past, present, and future, Water Science and Technology, 33(10-11), 1-14.

Asano, T., 1998, Wastewater Reclamation and Reuse, Water Quality Management Library, Vol. 10, Technomic Publishing Company, Lancaster, Pennsylvania.

Asano, T., 2004, Water and Wastewater Reuse, An Environmentally Sound Approach for Sustainable Urban Water Management (http://www.unep.or.jp/)

Ayers, R.S., and Westcott, D.W., 1985, Water Quality for Agriculture, Food and Agricultural Organization of the United Nations, FAO Irrigation and Drainage, Paper 29, Rome, Italy.

Carley, R. L., 1985, Wastwater reuse and Public Opinion, Journal of the American Water Works Association, 77(7), 72.

Crook, J., 1985, Water Reuse in California, Journal of the American Water Works Association, 77(7), 61.

Dishman, M., Sherrard, J., and Rebhun, M., 1989, Gaining Support for Direct Potable Water Reuse, Journal of Professional Issues in Engineering, 115 (2), 154-161.

El-Ghamam, A. R. I. H., 1997, The Future of Agriculture and Food Production in Saudi Arabia: A Briefing, Country Report.

Eslamian, S., Tarkesh-Isfahani, S. and Malekpour, I., 2010, Investigating heavy metals concentration of a wastewater treatment plant for agricultural and landscape reuses, Dryland Hydrology: Global Challenges Local Solutions, September 1-4, Westin La Paloma,Tucson, AZ, USA.

Eslamian, S., and Tarkesh-Isfahani, S., 2010a, Evaluating the urban reclaimed water for industrial reuses in North Isfahan, Iran, The 4th International Symposium on Water Resources and Sustainable Development, Algiers, Algeria.

Eslamian, S. and Tarkesh-Isfahani, S., 2010b, Evaluating the most efficient irrigation systems in wastewater reuse, Pakistan Agriculture: Challenges and Opportunities, Kashmir, Pakistan.

Florida Department of Environmental Protection (2002), (dep.state.fl.us/water/ wetlands/).

Food and Agriculture Organization (http://www.fao.org/corp/statistics/en/).

Gallup, G.J. ,1973, Water Quality and Public Opinion, Journal of the American Water Works Association, 65(8), 513.

Johnson, W.D., 1998, Innovative Augmentation of a Community's Water Supply - The St. Petersburg, Florida Experience, Proceedings of the Water Environment Federation, 71st Annual Conference and Exposition, October 3-7, Orlando, Florida.

Kasperson, R. E. et al., 1974, Community adoption water reuse system in the United States, U. S. Office of Water Resources Research.

Kretzschmar, R., 2002, Best Management Practices for Florida Marinas, Florida Department of Environmental Quality, Florida.

Kretschmer, N., Ribbe, L., and Gaese, H., 2000, Wastewater Reuse for Agriculture, Technology Resource Management and Development, Scientific Contributions for Sustainable Development, Special Issue, Water Management, Vol. 2, Cologne.

Lazarova, V., 2000, Wastewater Disinfection: Assessment of the Available Technologies for Water Reclamation, in Water Conservation, Vol. 3: Water Management, Purification and Conservation in Arid Climates, Goosen, M.F.A. and Shayya, W.H., eds., Technomic Publishing Co. Inc., 171.

Lazarova, V., 2001, Role of Water Reuse in Enhancing Integrated Water Resource Management, Final Report of the EU Project CatchWater, EU Commission.

Leverenz, H., Tchobanoglous, G. and Darby, J. L., 2002, Review of Technologies for the Onsite Treatment of Wastewater in California, Report No. 02-2, Prepared for California State Water Resources Control Board, Center for Environmental and Water Resources Engineering, UC Davis, Davis, CA.

Lindsey, P.R., Waters, K., Fell, G. and Setka Harivandi, A., 1996, The Design and Construction of a Demonstration/Research Garden Comparing the Impact of Recycled vs. Potable Irrigation Water on Landscape Plants, Soils and Irrigation Components, Water Reuse Conference Proceeding, American Water Works Association, Denver, Colorado.

Mantovani, P., Asano, T., Chang, A. and Okun, D.A., 2001, Management Practices for Nonpotable Water Reuse, WERF, Project Report 97-IRM-6.

Metcalf and Eddy, 2003, Wastewater Engineering: Treatment and Reuse, 4rd ed., Mc-Graw Hill Inc., New York.

Metcalf and Eddy, 2007, Water Reuse: Issues, Technologies, and Applications, McGraw-Hill, New York.

Sadik, A., and Shawki B., 1994, The Water Problems of the Arab World: Management of Scarce Resources." In Rogers, Peter, and Peter Lydon (eds.), Water in the Arab World: Perspectives and Prognoses, Ch 1, 1-37, the American University in Cairo Press, Cairo, Egypt.

United Nations, 2006, World Population, the World at Six Billion and World Prospects: The 2006 Revision, Department of Social and Economic Affairs, Population Division, New York.

USEPA, 1981, Process Design Manual: Land Treatment of Municipal Wastewater, EPA Center for Environmental Research Information, EPA 625/1-81-013, Cincinnati, Ohio.

USEPA, 1992, Guidelines for Water Reuse, United States Environmental Protection Agency and & USAID (United States Agency for International Development), Cincinnati, OH.

USEPA, 2003, National Primary Drinking Water Standards, United States Environmental Protection Agency, EPA 816-F-03-016, Washington, D.C.

USEPA, 2004, Guidelines for Water Reuse, United States Environmental Protection Agency, EPA 645-R-04-108, Washington, D.C.

Valentina L. and Akica B., 2005, Water Reuse for Irrigation: Agriculture, Landscapes, and Turfgrass, CRC Press.

Virginia Cooperative Extension Materials (http://pubs.ext.vt.edu/452/452-014/452-014.html).

WHO Website (http://www.who.int/bulletin/archives/78%289%291104.pdf)

World Bank 1995, From Scarcity to Security: Averting a Water Crisis in the Middle East and North Africa, Washington D.C.

World Health Organization (WHO), 1989, Health Guideline for the Use of Wastewater in Agriculture and Aquaculture, WHO Technical Report Series, No. 778, Geneva, Switzerland.

World Health Organization (WHO), 2001, Water Quality: Guidelines, Standards and Health Assessment of Risk and Risk Management for Water-related Infectious Diseases, Available from http://www.who.int/water_sanitation_health/dwq/whoiwa/en/

Xie, M., Kuffner, U. and Le Moignee, G., 1993, Using Water Efficiently: Technological Options, World Bank Technical Paper No. 205., Washington D.C.

12

In situ Remediation Technologies Associated with Sanitation Improvement: An Opportunity for Water Quality Recovering in Developing Countries

Davi Gasparini Fernandes Cunha[1], Maria do Carmo Calijuri[1], Doron Grull[1], Pedro Caetano Sanches Mancuso[1] and Daniel R. Thévenot[2]

[1]*Universidade de São Paulo,*
[2]*LEESU, Université Paris-Est,*
[1]*Brazil*
[2]*France*

1. Introduction

The access to safe water is of great importance to reduce the spread of diseases caused by water-related pathogens and to assure the life quality to the human-beings. According to the World Health Organization (WHO, 2011), diarrhea, for example, is responsible for two million deaths every year, mainly among children under the age of five. The environmental effects of some pollutants (e.g. endocrine disruptors, organic compounds) remain unclear and the harmful consequences of the exposure to contaminated water are certainly an important issue for the next decades. Moreover, many research have linked water quality to health problems, such as cancer (Rodrigues et al., 2003; Han et al., 2009), insufficient uptake of nutrients and trace-metals (Lind & Glynn 1999), diabetes, cerebrovascular and kidney disease (Meliker et al., 2007).

The costs and benefits of water quality have been the topic of stimulating discussion in the scientific community (Isaac, 1998; Hajkowicz et al., 2008; Saz-Salazar et al., 2009) because water quality decrease implies not only loss of lives, but also economic damages. The costs of the anthropogenic eutrophication reach US$2.2 billion in the United States (Dodds et al., 2009) and US$187.2 million in England and Wales every year (Pretty et al. 2002). The reduction of nutrient loading to the aquatic systems worldwide is the cornerstone of artificial eutrophication control (Smith et al., 1999), with repercussions in other fields like public health and economics.

The anthropogenic impacts on the quality of urban water bodies in developing countries are frequently exacerbated by poor levels of sanitation and inadequate water and wastewater management. Pressure from urban areas on the water quality was reported in Argentina (Almeida et al., 2007), Brazil (Jordão et al., 2007), India (Suthar et al., 2010) and Mexico (Bravo-Inclan et al., 2008). Rapid shifts in the land use patterns, unplanned urbanization and inefficient resources allocation are further aggravating environmental problems in such nations. Restrictions to the water uses are increasing as the pollution of rivers and lakes is offering more risks to the human health and to the maintenance of the ecological balance.

Within this context, the water resources management plays an important role in the conciliation of the water uses and the long-term sustainability. The *in situ* remediation of rivers, lakes and reservoirs is a decentralized alternative that may be convenient in some cases in comparison to off-site solutions. The main advantages of the *in situ* approach are, besides the relative small period of time required to its implementation, the suitability of the *in situ* facilities to the regions with lack of available areas to build off-site treatment plants (e.g. highly urbanized areas) and the lower expenses with pumping structures. Although it takes more time and requires more investments, the implementation of sanitation infrastructure is also necessary.

With the increase of the negative environmental impacts induced by the anthropogenic activities, the remediation of aquatic systems became an alternative to restore the ecological functions of the ecosystems and accelerate their recovery. The first and most important step in a remediation project is to define the remedial action aims to be accomplished at the site, involving the desirable mechanisms of treatment - biological (e.g. phytoremediation), physical and/or chemical (e.g. oxidation, air stripping, ion exchange, precipitation). Most of the current technologies for aquatic systems remediation were adapted from unitary processes used for drinking water production, industrial purposes or wastewater treatment. The flotation, for example, has been used in mining activities to separate the mineral of interest from the gangue since 1893 (Hoover, 1912). The technology was then adapted to treat water and wastewater through dissolved air flotation (e.g. Heinänen et al., 1995). Ultrafiltration membranes in turn have been mainly used for drinking water production (2 million m^3/day worldwide according to Laîné et al., 2000). According to the same authors, the oldest water industry with ultrafiltration plant started to operate in 1988 in France. The membranes are becoming cheaper over the years and the technology is more attractive for remediation of surface waters at the present time.

2. Water management in developing countries: Long and short term actions

The water management in developing countries would benefit from a well-weighted balance between long and short term actions (Fig. 1). The former actions should consider sanitation planning and infrastructure, whereas the latter ones should focus on the solutions for remediating the aquatic systems or attenuating their degradation level.

The main issues involved in both long and short term actions for water resources management are:

i. **Political commitment**. Sanitation and environmental recovery programs usually do not receive the same amount of investments in comparison to other areas (Varis et al., 2006). Local authorities willingness and specific government policies are necessary for meeting the health and environmental goals;
ii. **Institutional framework**. The effectiveness of the water management policies would be at risk with no solid institutions and skilled professionals. The technical and social challenges will just be overcome with trained planners, sector professionals and decision-makers;

iii. **Financing**. This is a complicated question because water quality recovery brings benefits to the health and to the environment, making it a public good. However, at the same time, water is also a private good (at the level of households);
iv. **Technology development and Innovation**. Sanitation and remediation of aquatic systems depend upon technology development, under a cost-benefit analysis. Moreover, the technology has to be adapted and optimized to the local peculiarities. In the case of the developing countries, technology transfer from developed countries might be necessary;
v. **Monitoring**. Monitoring programs play a pivotal role in assessing if the targets were met. Such programs must be able to provide feedback to improve the monitored system with a view to increasing efficiency and reducing costs;
vi. **Social acceptance**. It is desirable that people get involved with the decision-making process, increasing the chances of social acceptance of the water resources management policies or programs.

According to Calijuri et al. (2010), the water resources sustainability can be defined as a state of dynamic equilibrium between the disturbances imposed to the water bodies by the anthropogenic activities and the aquatic systems ability to self-regulation (i.e. their elasticity in response to a certain impact). When the impact is strong enough to prevent the self-recovery of the original condition, actions towards remediation, including palliative/temporary solutions, are required to avoid critical levels of degradation and severe impairment of different water uses.

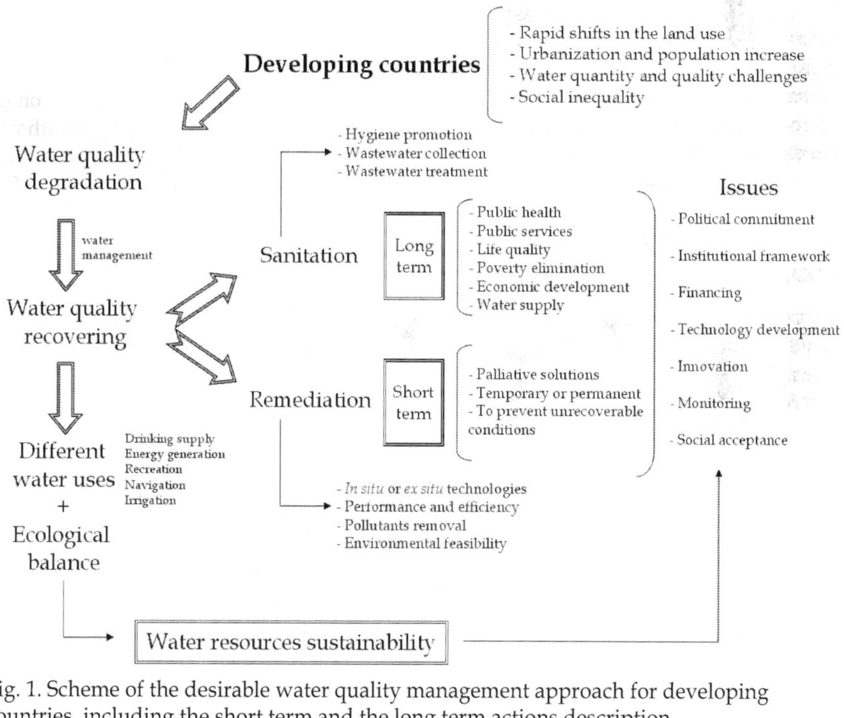

Fig. 1. Scheme of the desirable water quality management approach for developing countries, including the short term and the long term actions description.

2.1 Sanitation infrastructure

Approximately 1.1 billion people in the world do not have access to improved water supply sources and 2.4 billion people do not have access to any type of improved sanitation at all (WHO, 2011). It is clear that there is a need for additional water and sanitation services from government in partnership with other actors worldwide and especially in the megacities from the developing countries (Biswas et al., 2004). The implementation of sanitation infrastructure is a long term action that has to be continuously monitored and updated. Solid waste collection and disposal, water distribution and water and wastewater treatment are the main components of the sanitation in a country. Such items are related to life quality promotion, poverty elimination and economic development. The temporal evolution of sewage, water supply and solid waste collection services in Brazil is shown in Fig. 2.

According to the Brazilian Institute of Geography and Statistics (IBGE, 2011), there was an increase in the availability of the sewage system (i.e. wastewater collection) in the urban areas of the country from 1992 (46%) to 2009 (59%). Similar increase was observed for the availability of the solid waste collection services (62% in 1992 and 82% in 2009). The situation was worse in the rural areas, where the figures for sewage system and water supply systems reached only 5% and 33% in 2009, respectively (Fig. 2). In the year 2000, only 20% of the Brazilian municipalities treated the domestic wastewater (IBGE, 2000). The remaining loads ended up in the water bodies, contributing to water quality degradation (e.g. by increasing organic matter content and decreasing dissolved oxygen concentrations).

As shown for Brazil, the sanitation conditions in other developing countries are similar: higher levels of drinking water and sewage systems in urban areas, as compared to rural ones (Massoud et al., 2009). According to the WHO (2010), people living in low-income nations are least likely to have access to adequate sanitation infrastructure. The absence of sanitation and the access to unsafe water are therefore risk factors that are linked with increased mortality and morbidity worldwide.

2.2 *In situ* remediation technologies

Dissolved Air Flotation (DAF), Ultrafiltration Membranes (UM) and Enhanced Biological Removal (EBR) are examples of *in situ* technologies for water or wastewater treatment (Table 1). Such technologies can be used either individually or in association (Geraldes et al., 2008).

The DAF consists of the following steps:

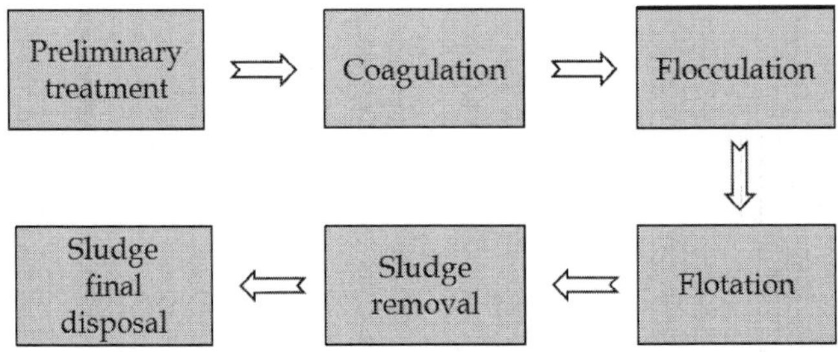

i. **Preliminary treatment** - barriers are installed in the river channel to remove coarse material;
ii. **Coagulation** - chemical compounds are added to the water to promote the coagulation (e.g. ferric chloride or aluminium hydroxide). Specific coagulation time and velocity gradients are needed;
iii. **Flocculation** - through certain flocculation time and velocity gradient, flakes are formed by aggregation of colloids;
iv. **Flotation** - small air bubbles are produced in the bottom of the river channel and their upward movement is able to bring the colloidal and particulate matter to the water surface, as a sludge;
v. **Sludge removal** - the sludge is removed from surface water through rotating blades or other mechanical device;
vi. **Sludge final disposal.**

The operation costs of a DAF system vary between 0.10 and US$0.20/m^3. One of the biggest concerns in the DAF plants is the significant consumption of chemicals for aggregation and flocculation of colloids and consequently the high production of sludge. Different processes have been used for thickening and dewatering the sludge produced by DAF (i.e. increasing the solids content) and reducing its volume (Dockko et al., 2006). However, the feasibility of the alternative for final sludge disposal (e.g. application in the agriculture or disposal in a landfill) depends upon its toxicity due to the presence of metals and other persistent pollutants (Mantis et al., 2005; Luz et al., 2009).

The treatment process with UM is based in the following steps:

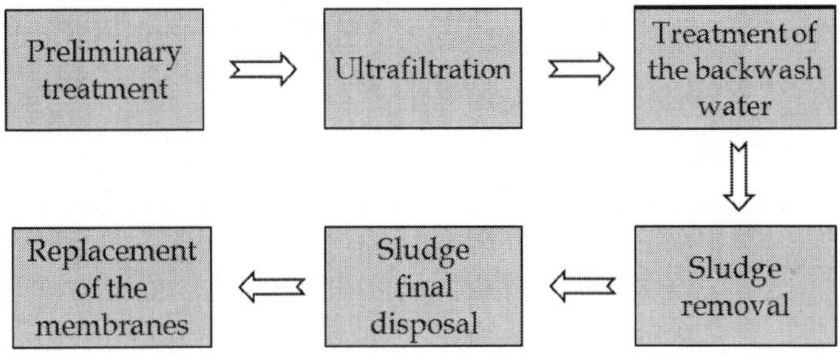

i. **Preliminary treatment** - removal of coarse material and sand;
ii. **Ultrafiltration** - the water passes through semipermeable membranes;
iii. **Treatment of the backwash water** - the contaminants rejected by the membranes accumulate on the membranes forming a fouling layer and requiring a periodical backwash to remove the debris. The water used for backwash has to be treated as it contains high concentrations of pollutants;
iv. **Sludge removal from backwash water;**
v. **Sludge final disposal**;
vi. **Replacement of the membranes** - the replacement must be achieved according to the membrane life, usually estimated as 5 to 8 years.

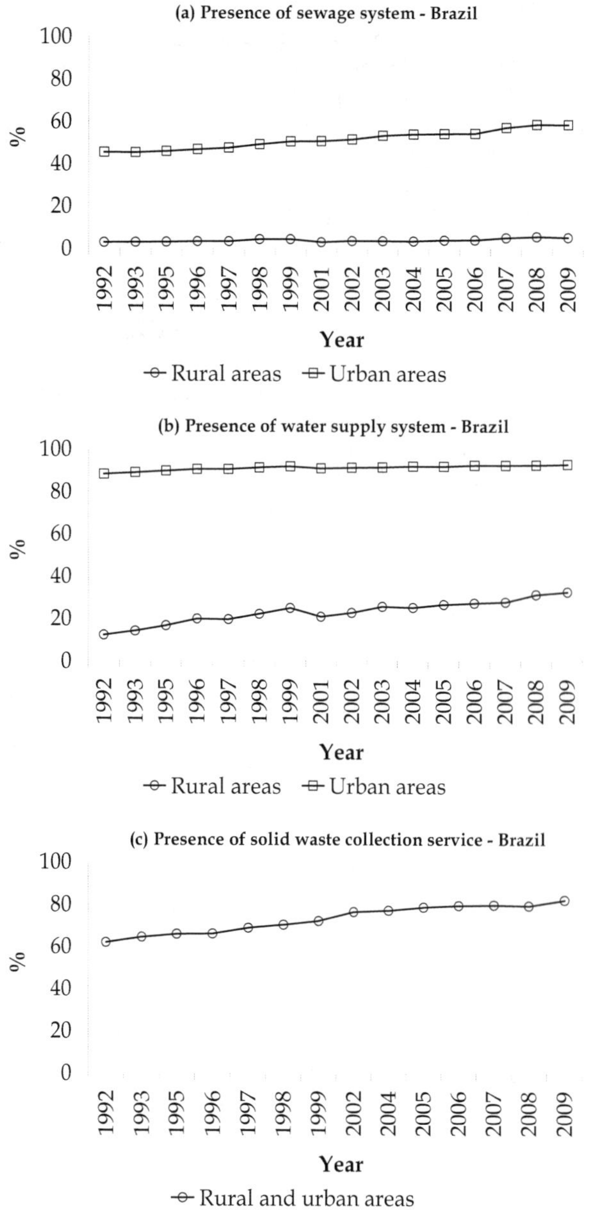

Fig. 2. Evolution (%) of the availability of the (a) sewage, (b) drinking water supply systems and (c) solid waste collection services in Brazil from 1992 to 2009. Reference: IBGE (2011).

In situ technology	Brief description	Total operating costs (US$/m³)	Benefits	Limitations
Dissolved Air Flotation (DAF)	Small air bubbles carry the impurities to the surface (as sludge) after previous coagulation and flocculation	0.10 - 0.20	- High efficiency for the removal of some nutrients (e.g. phosphorus), hydrocarbons and surfactants	- Significant amount of sludge to be managed - Chemicals consumption for coagulation and flocculation
Ultrafiltration Membranes (UM)	The water is forced against a semipermeable membrane	0.03 - 0.25	- No addition of chemicals - The membranes may be fed with the water course own pressure - Smaller footprint	- Preliminary treatment is required to remove coarse and sand material (to protect the membranes) - Treatment of the backwash water is necessary
Enhanced Biological Removal (EBR)	The growth of certain types of bacteria and the biological conversion are stimulated with the addition of specific enzymes or inoculums (commercially available: see Table 2).	0.01 - 0.15	- Minimum infrastructure is required - Some added reagents increase gas transfer rates and dissolved oxygen concentrations - Degradation of toxics - Rapid odor elimination	- Acceptability of the addition of inoculums to the aquatic systems, especially if they are exogenous and if the surface water is used for drinking water production

Table 1. Brief description of some *in situ* technologies: Dissolved Air Flotation, Ultrafiltration Membranes and Enhanced Biological Removal; comparison of their operating costs, efficiency and performance criteria.

The more stringent regulations regarding water quality associated with the reduction of costs of the ultrafiltration membranes have made this technology more attractive in the last years. UM can be used for drinking water production (Laîné et al., 2000; Xia et al., 2005) and river water treatment (Konieczny et al., 2009). The lower level of energy requirements and the lower consumption of chemicals are the main advantages of the UM systems. The need for frequent backwash preventing membrane fouling, especially when the organic matter content in the raw water is high, is an important issue to be managed.

However, the addition of a preliminary step with coagulation [e.g. with $FeCl_3$, $Fe_2(SO_4)_3$ or $Al_2(SO_4)_3$] and flocculation may increase the efficiency of the system and avoid too rapid membrane fouling (Konieczny et al., 2006; Babel & Takizawa, 2011; Bergamasco et al., 2011). The treatment of the backwash water is also an issue because the effluent from the backwash contains significant concentrations of the pollutants previously accumulated on the membranes. The backwash water is normally sent to a thickening tank, where the suspended matter settles to the bottom (as sludge). The clear water on top is then pumped to the upstream part of the treatment plant. Some recent studies consider the recycling of the backwash water as an interesting alternative, e.g. through a blending of 10% of backwash water and 90% of raw water (Gora & Walsh, 2011). The operating costs of the UM are expected to approximately range between 0.03 and US$0.25/$m^3$ and the main factors influencing the costs are the quality of the water to be treated, labor, energy and chemicals (if any) consumption as well as membrane replacement.

The EBR in turn is based in a single step, the **application of bioremediators** (Table 2) in the river or reservoir to be treated and subsequent monitoring of their efficiency in relation to the targeted pollutants.

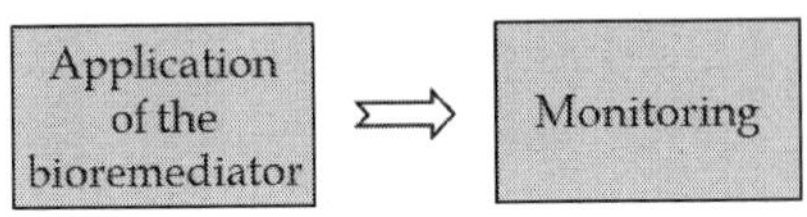

The bioremediators, which can be microorganisms (e.g. bacteria) or enzymes, are able to increase biodegradation and gas transfer rates. There is a considerable variety of products available in the market. Some examples are shown in Table 2 with the commercial name of the bioremediators, their definition, main applications and highlights according to their manufacturer. The bioremediators are expected to perform the removal, transformation or detoxification of pollutants from the aquatic environment into a less toxic form (Whiteley & Lee, 2006). The effect of the bioremediators is normally based in the combination of several processes, like solubilization (physical), oxidation (chemical) and catalysis (biological). Some recent studies have recognized the effectiveness of enzymatic processes of water remediation, mainly when associated with established technologies (Demarche, in press). However, some factors like costs and stability of the biocatalysts require further investigation.

There are two broad types of remediation through microorganisms: biostimulation (stimulation of the growth of indigenous microorganisms) and bioaugmentation (introduction of specific microorganisms to the local population). In both cases, predation, competition, adaptation to the environmental matrix (water, wastewater), possible adsorption on available solids as well as survival strategies play an important role in determining the overall efficiency of the remediation (Fantroussi & Agathos, 2005). The use of immersed biofilms (e.g. artificial plastic substrates) is also an alternative to perform water treament taking advantage of the nutrient uptake by algae (Jarvi et al. 2002) or the organic matter degradation by bacteria (Bishop, 2007). Although biofilms are usually unwanted in drinking water treatment stations due to the biofouling, they may be useful for water remediation and biodegradation of persistent compounds. Ammonia-nitrogen concentrations removal, for example, may be considerably improved with the growth of nitrifying bacteria (Jiao et al., 2011).

Company / Product	Definition	Highlights
Bio-Organic Catalyst™	"Fermentation supernatant derived from plants and minerals, which is blended synergistically in combination with a non-ionic surfactant to create a broad spectrum bio-organic catalyst"	- Dissolved oxygen levels increase - Biological nitrogen removal is enhanced - Solubilization rates of insoluble fats, oils and grease increase
Advanced BioCatalytics Corporation/ Accell3 Green™	"Refined, fermentation-derived, bioactive stress proteins that are formulated with surfactants"	- Bacterial metabolism is enhanced to accelerate the breakdown of organic material by oxidation - Organic contaminants are digested more rapidly and more completely to carbon dioxide and water
Enzilimp™	"Facultative bacteria are added to boost the nitrogen cycle and accelerate the organic matter degradation"	- Odors elimination - Reduction of Total and Fecal Coliforms, Biochemical and Chemical Oxygen Demand
Bioplus Biol2000™	"Facultative bacteria that are able to promote the degradation of organic compounds from industrial effluents"	- Biochemical and Chemical Oxygen Demand reduction - Phosphorus and nitrogen concentrations decrease
BioTC™ Rinenbac/Rinenzim	"Specific bacteria to stimulate degradation of fats and organic matter"	- Odors elimination - Reduction of Coliforms and Biochemical Oxygen Demand
Realco™ Realzyme	"Enzymes that are able to transform biofilms into water-soluble organic residues"	- Avoid the contamination from a biofilm

Table 2. Examples of some bioremediators available in the market, including their definition and major benefits, according to their manufacturer.

3. Case study with DAF: The Pinheiros River (São Paulo State, Brazil)

The Pinheiros River (23º42'S; 46º40'W) is located in the Southeast Region of Brazil (Fig. 3) in an extremely urbanized area in the Metropolitan Region of São Paulo, whose population is approximately 19.7 million inhabitants. This river links the Tietê River to the Billings Reservoir (storage volume: 995 million m^3; residence time: 30 months), a multipurpose water system used for drinking supply, energy generation (through Henry Borden Power Plant), navigation and recreation.

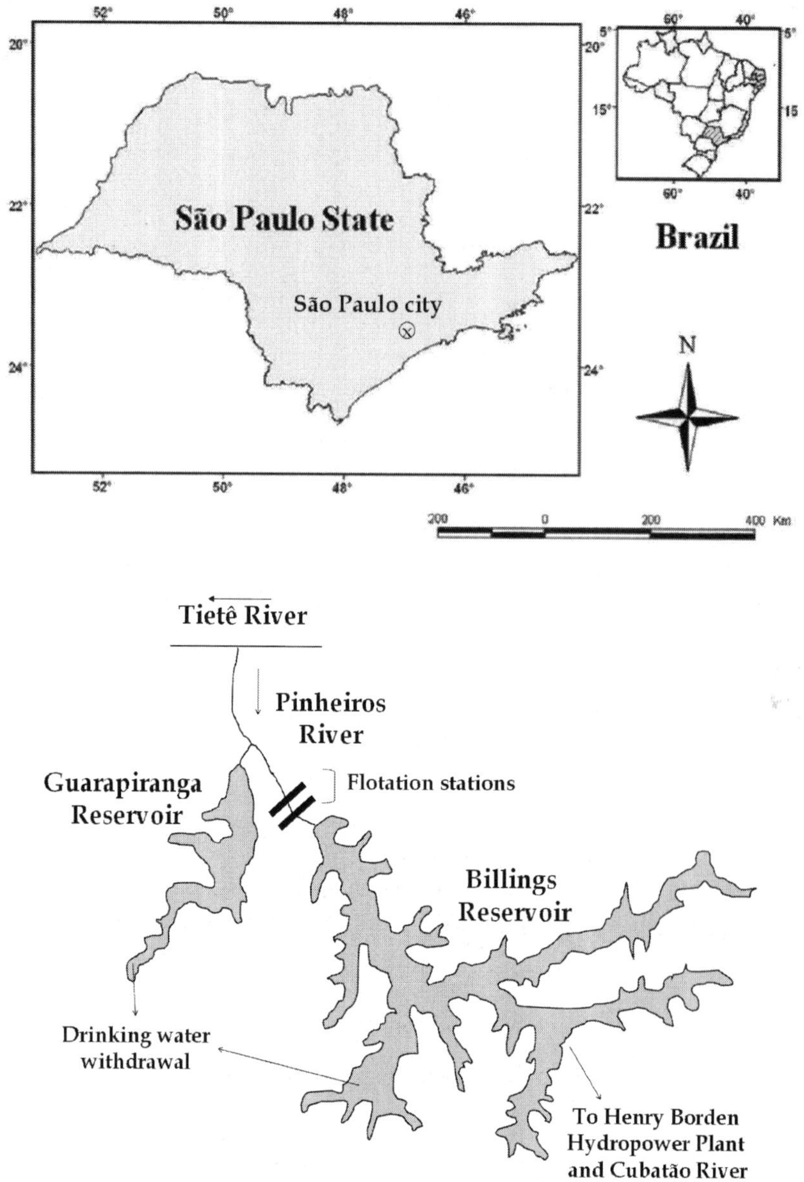

Fig. 3. Map of the São Paulo State, located in the Southeast Region of Brazil, and a scheme of the Tietê and Pinheiros Rivers and the Billings and Guarapiranga Reservoirs, two important multipurpose reservoirs in the area.

Considering the technology availability and the need for recovering the Pinheiros River water quality to assure no impacts to the Billings Reservoir, the DAF was tested. The *in situ* DAF pilot-scale system, composed by two treatment plants (Zavuvuz - 23°40′44″S; 46°41′53″W and Pedreira - 23°42′01″S; 46°40′56″W), was installed in the Pinheiros River channel to treat 10 m^3/s (Table 3) with subsequent pumping of the water to the Billings Reservoir (Fig. 4). Previous coarse material removal was necessary. Coagulation with ferric chloride and flocculation were followed by the flotation step (i.e. the production of tiny air bubbles in the bottom of the river). The upward movement of such bubbles was responsible for bringing the impurities to the surface, where they were collected through rotating blades for sludge removal. The treated flow (10 m^3/s) of the system installed in the Pinheiros River is significantly bigger than other similar treatment stations in Brazil: 0.05 m^3/s reported by Lopes & Oliveira (1999), 0.15 m^3/s (Oliveira et al., 2000) and 0.75 m^3/s (Coutinho & von Sperling, 2007).

Operational parameter or variable (unit)	**Value or range**
Tietê River mean flow (m^3/s)	120
Pinheiros River mean flow (m^3/s)	5*
Zavuvuz Stream mean flow (m^3/s)	0.7
Treated flow (m^3/s)	10
Recycle flow (m^3/s)	0.8 - 1.0
Hydraulic detention time (min)	25 - 30
Ferric chloride dosages (mg/L)	50 - 400
Rapid mixing (coagulation) time (min)	0.5
Rapid mixing velocity gradient (/s)	800
Slow mixing (flocculation) time (min)	27
Slow mixing velocity gradient (/s)	60
Sludge production in both plants (m^3/day)	150
Solids content in the sludge after centrifugation (%)	20
Energy consumption in both plants (kWh/day)	42,000

Table 3. Major operational parameters or variables of the *in situ* DAF pilot-scale system (both plants) in the Pinheiros River, São Paulo, Brazil. * When the pilot-scale system was operating, there was a contribution from the Tietê River waters to the total flow of the Pinheiros River. Reference: adapted from Cunha et al. (2010).

The system was operated from August 2007 to March 2010. A comprehensive monitoring program (148 water variables and more than 200,000 laboratory analyses) was delineated to assess the efficiency and feasibility of the pilot-scale prototype. Detailed information about the monitoring results in the Tietê and Pinheiros Rivers and in the Billings Reservoir may be found in some recent papers (Cunha et al., 2010; Cunha et al., 2011a and Cunha et al., 2011b). The efficiency achieved by each flotation station and the overall effect for some variables is shown in Table 4. The global removal efficiency achieved by both DAF treatment stations was 90% for total phosphorus, 54% for apparent color, 53% for chemical oxygen demand, 48% for turbidity, 40% for total suspended solids, 31% for soluble iron and only 2% for nitrogen-ammonia. The prototype promoted an increment of about 60% in the dissolved oxygen concentrations.

Through the operation and performance assessment of the DAF system, some positive aspects have been observed:

i. The technology was available and the local government was willing to promote the reclamation of the water quality of the Pinheiros River and to stimulate additional energy generation with the Billings Reservoir water. The conjunction of technology availability, environmental and economic issues (as shown in Fig. 1) proved to be important to boost actions towards effective water management;
ii. The treated flow that was transferred to the Billings Reservoir (10 m^3/s) was convenient for favouring the different water uses in the reservoir, such as energy and drinking water production;
iii. Considering the combined effect of both treatment stations, the pilot-scale system reached a significant percentage of removal of total phosphorus, one of the targeted nutrients whose loads to the Billings Reservoir must be reduced to help preventing the artificial eutrophication.

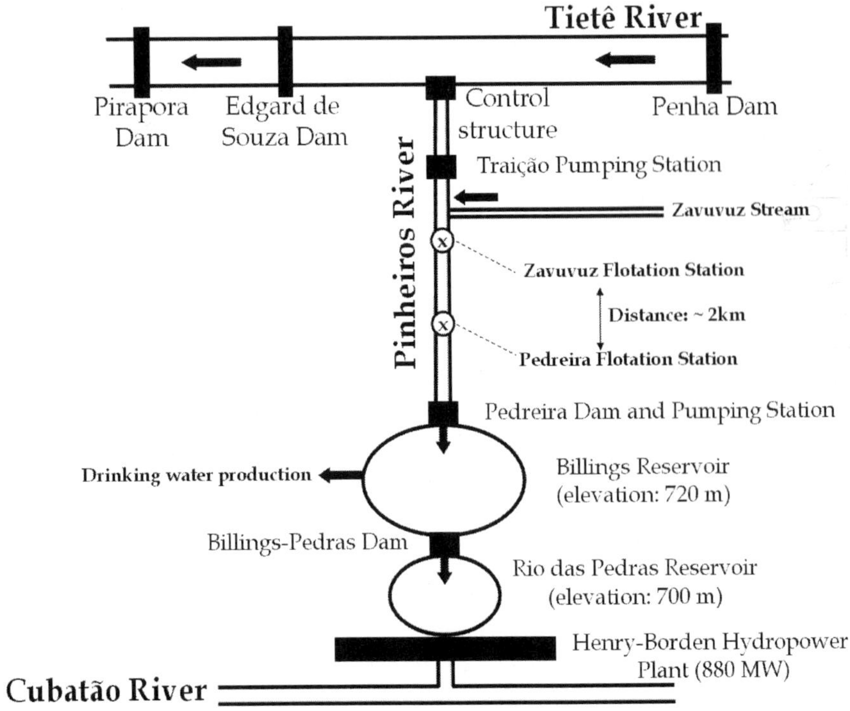

Fig. 4. Scheme of the *in situ* DAF pilot-scale systems placed in the Pinheiros River. Reference: adapted from Cunha et al. (2010).

However, our experience indicated the following negative factors and limitations:

i. The operation and maintenance of an "opened system" have to consider the potential influences of external factors, like sudden changes in the water flow and quality, both natural or human-induced;
ii. Sludge production and energy consumption were high. During the studied period, the sludge was sent to a landfill because no other alternative was considered safe due to the significant level of contamination (e.g. by heavy metals);
iii. As expected, ammonia-nitrogen was not removed by the DAF system and we assume that the high concentrations of this nutrient in the Billings waters may contribute to water quality decrease, affecting the aquatic life. A biological component with adequate residence time would probably be necessary for removing ammonia-nitrogen. Total suspended solids and metals (aluminium, chromium and iron) concentrations were also high in the treated water.

Variable	**Relative efficiency (%) of removal (−) or increase (+)**		
	Zavuvuz Flotation Station	**Pedreira Flotation Station**	**Overall effect**
Aluminium (soluble)	−17	null	null
Ammonia-Nitrogen	−1	−1	−2
Apparent color	−48	−27	−54
Cadmium (total)	−22	−17	−22
Chromium hexavalent	null	−33	null
Chemical oxygen demand	−41	−18	−53
Copper (total)	−13	−14	−60
Dissolved oxygen	+69	+24	+63
Ionic conductivity	null	null	null
Iron (soluble)	−19	null	−31
Lead (total)	−18	−42	−36
Total phosphorus	−48	−84	−90
Total suspended solids	−30	−10	−40
Turbidity	−37	−35	−48

Table 4. Efficiency (%) of removal (−) or increment of the *in situ* DAF pilot-system in the Pinheiros River. Reference: adapted from Cunha et al. (2010).

The DAF system in the Pinheiros River has focused on integrative approach, technology application, environmental quality and sustainability in the long-term. Nevertheless, since some inefficiencies and gaps were detected, further studies regarding sludge management and efficiency improvement for the removal of some variables (e.g. through complementary processes like those previously described in this chapter) are necessary. Urban waters in developing countries are a challenging issue and the operation of the DAF system in the Pinheiros River was an important contribution for the water resources management.

4. Conclusion

The term "developing countries" is often used to describe nations whose inhabitants have a standard of living between "low" and "medium", a growing industrial base and a rising Human Development Index. According to Kofi Annan (Secretary-General of the United Nations from 1997 to 2006), "developed country is one that allows all its citizens to enjoy a free and healthy life in a safe environment".

Therefore, environmental sustainability is an important step towards the full development of the developing countries, with a view to ensuring social equity, economic strength and environmental quality. Specifically regarding the management of water resources, two main issues should be considered. Our chapter has shown that the implementation of sanitation facilities (e.g. sewage collection and treatment systems) requires significant investments with long-term returns. On the other hand, the remediation of polluted rivers and reservoirs should be seen as a short-term palliative and emergency action to prevent these aquatic systems to reach levels of irreversible degradation, before necessary wastewater collection and treatment are available. The *in situ* approach for remediation may be desirable from the environmental point of view and also economically convenient. By analyzing the main benefits and limitations of three *in situ* technologies (Dissolved Air Flotation, Ultrafiltration Membranes and Enhanced Biological Removal), our investigation has suggested that the costs of remediation of aquatic systems ranged from US$ 0.01/m^3 (the cheapest cost in the range for treatment with bioremediators) to US$ 0.25/m^3 (the most expensive cost in the value range for ultrafiltration membranes). The technologies described in this chapter can be used simultaneously (for example, the DAF associated with the biological treatment with enzymes or with biofilms) to increase the efficiency and meet environmental standards. Although further research is required to find alternatives to solve the detected inefficiencies, our experience with the operation of a pilot-scale DAF system in the Pinheiros River (São Paulo, Brazil) was positive. It has indicated that integrated concepts of water management are necessary to explore urban waters as resources (and not risks) for human activities.

A well-balanced combination of actions, policies and programs for increasing the levels of sanitation coverage and promoting the remediation of impacted aquatic systems is a great opportunity to the developing countries. Political commitment, technology development or transfer from other country, comprehensive monitoring and involvement of the local citizens are factors that can legitimate the whole process and increase the probability of economic and environmental effectiveness and public acceptability.

5. References

Advanced BioCatalytics Corporation. (2011). Information about the bioremediator. In: *Advanced BioCatalytics Corporation*. 02 Aug 2011. Available from www.abiocat.com.

Almeida, C.A., Quintar, S., González, P. & Mallea, M.A. (2007). Influence of urbanization and tourist activities on the water quality of the Potrero de los Funes River (San Luis - Argentina). *Environmental Monitoring and Assessment*, Vol. 133, No. 1-3, (2007), pp. 459-465.

Babel, S. & Takizawa, S. (2011). Chemical pretreatment for reduction of membrane fouling caused by algae. *Desalination*, Vol. 274, No. 1-3, (2011), pp. 171-176.

Bergamasco, R., Konradt-Moraes, L.C., Vieira, M.F., Fagundes-Klen, M.R. & Vieira, A.M.S. (2001). Performance of a coagulation-ultrafiltration hybrid process for water supply treatment. *Chemical Engineering Journal*, Vol. 166, No. 2, (2001), pp. 483-489.

Bio-Organic Catalyst. (2011). Information about the bioremediator. In: *Bio-Organic Catalyst*. 02 Aug 2011. Available from <bio-organic.com>.

Bioplus. (2011). Information about the bioremediator. In: *Bioplus*. 02 Aug 2011. Available from <www.athcsm4.com.br/bioplus>.

BioTC.(2011). Information about the bioremediator. In: *BioTC*. 02 Aug 2011. Available from <biotecnicontrolambiental.yolasite.com>

Bishop, P.L. (2007). The role of biofilms in water reclamation and reuse. *Water Science and Technology* , Vol. 55, No. 1-2, (2007), pp. 19-26.

Biswas, A., Lundqvist, J., Tortajada, C. & Varis, O. (2004). Water management for megacities. *Stockholm Water Front*, Vol. 2, (2004) pp. 1–13.

Bravo-Inclan, L.A., Saldana-Fabela, M.P. & Sanchez-Chavez, J.J. (2008). Long-term eutrophication diagnosis of a high altitude body of water, Zimapan Reservoir, Mexico. *Water Science and Technology*, Vol. 57, No. 11, (2008), pp. 1843-1849.

Calijuri, M.C., Cunha, D.G.F. & Povinelli, J. (2010). *Sustentabilidade: um desafio na gestão dos recursos hídricos* (1st edition), EESC/USP, São Carlos, SP, Brazil.

Coutinho, W. & von Sperling, M. (2007). Avaliação de desempenho da estação de tratamento por flotação dos córregos afluentes à Represa da Pampulha - Belo Horizonte. *Proceedings of 24º Congresso Brasileiro de Engenharia Sanitária e Ambiental*, Belo Horizonte, MG, Brazil, 2007.

Cunha, D.G.F., Grull, D., Damato, M., Blum, J.R.C., Eiger, S., Lutti, J.E.I. & Mancuso, P.C.S. (2011a). Contiguous urban rivers should not be necessarily submitted to the same management plan: the case of Tietê and Pinheiros Rivers (São Paulo-Brazil). *Annals of the Brazilian Academy of Sciences* (in press).

Cunha, D.G.F., Grull, D., Damato, M., Blum, J.R.C., Lutti, J.E.I., Eiger, S. & Mancuso, P.C.S. (2010). On site flotation for recovering polluted aquatic systems: is it a feasible solution for a Brazilian urban river? *Water Science and Technology*, Vol. 62, No. 7, (2010), pp. 1603-1613.

Cunha, D.G.F., Grull, D., Damato, M., Blum, J.R.C., Lutti, J.E.I., Eiger, S. & Mancuso, P.C.S. (2011b). Trophic state evolution in a subtropical reservoir over years in response to different management procedures. Water Science and Technology, Vol. 64, No. 12, (2011), pp. 2338-2344.

Demarche, P., Junghanns, C., Nair, R.R. & Agathos, S.N. (in press). Harnessing the power of enzymes for environmental stewardship. *Biotechnology Advances* (in press).

Dockko, S., Park, S.C., Kwon, S.B. & Han, M.Y. (2006). Application of the flotation process to thicken the sludge from a DAF plant. *Water Science and Technology*, Vol. 53, No. 7, (2006), pp. 159-165.

Dodds, W.K., Bouska, W.W., Eitzmann, J.L., Pilger, T.J., Pitts, K.L., Riley, A.J., Schloesser, J.T. & Thornbrugh, D.J. (2009). Eutrophication of U.S. Freshwaters: Analysis of Potential Economic Damages. *Environmental Science & Technology*, Vol. 43, No. 1, (2009), pp. 12-19.

Enzilimp. (2011). Information about the bioremediator. In: *Enzilimp*. 02 Aug 2011. Available from <www.enzilimp.com.br/site/default.asp>.

Fantroussi, S.E. & Agathos, S.N. (2005). Is bioaugmentation a feasible strategy for pollutant removal and site remediation? *Current Opinion in Microbiology*, Vol. 8, No. 3, (2005), pp. 268-275.

Geraldes, V., Anil, A., Pinho, M.N. & Duarte, E. (2008). Dissolved air flotation of surface water for spiral-wound module nanofiltration pre-treatment. *Desalination*, Vol. 228, (2008), pp. 191-199

Gora, S.L. & Walsh, M.E. (2011). Recycle of waste backwash water in ultrafiltration drinking water treatment processes. *Journal of Water Supply and Technology*, Vol. 60, No. 4, (2011), pp. 185-196.

Hajkowicz, S., Spencer, R., Higgins, A. & Marinoni, O. (2008) Evaluating water quality investments using cost utility analysis. *Journal of Environmental Management*, Vol. 88, No. 4, (2008), pp. 1601-1610.

Han, Y.Y., Weissfeld, J.L., Davis, D.L. & Talbott, E.O. (2009). Arsenic levels in ground water and cancer incidence in Idaho: an ecologic study. *International Archives of Occupational and Environmental Health*, Vol. 82, No. 7, (2009), pp. 843-849.

Heinänen, J., Jokela, P. & Ala-Peijari, T. (1995). Use of dissolved air flotation in potable water treatment in Finland. *Water Science and Technology*, Vol. 31, No. 3, pp. 225-238.

Hoover, T.J. (1912). *Concentrating Ores by Flotation*. The Mining Magazine, London, 221 p.

IBGE. (2000, 2011). National Survey on Sanitation and Statistical Series on Sewage, Water Supply and Solid Waste Collection. In: *Brazilian Institute of Geography and Statistics*. 09 Jul 2011. Available from <www.ibge.gov.br/home>.

Isaac, R.A. (1998) Costs and benefits of water quality: Massachusetts as a case example. *Water Science and Technology*, Vol. 38, No. 11, (1998), pp. 15-21.

Jarvi, H.P., Neal, C., Warwick, A., White, J., Neal, M., Wickham, H.D., Hill, L.K., Andrews, M.C. (2002). Phosphorus uptake into algal biofilms in a lowland chalk river. *Science of the Total Environment*, Vol. 282-283, (2002), pp. 353-373.

Jiao, Y., Zhao, Q., Jin, W., Hao, X., You, S. (2011). Bioaugmentation of a biological contact oxidation ditch with indigenous nitrifying bacteria for in situ remediation of nitrogen-rich stream water. *Bioresource Technology*, Vol. 102, No. 2, (2011), pp. 990-995.

Jordão, C.P., Ribeiro, P.R., Matos, A.T., Bastos, R.K., Fernandes, R.B. & Fontes, R.L. (2007). Environmental assessment of water-courses of the Turvo Limpo River basin at the Minas Gerais State, Brazil. *Environmental Monitoring and Assessment*, Vol. 127, No. 1-3, (2007), pp. 315-326.

Konieczny, K., Sakol, D. & Bodzek, M. (2006). Efficiency of the hybrid coagulation–ultrafiltration water treatment process with the use of immersed hollow-fiber membranes. *Desalination*, Vol. 198, (2006), pp. 102-110.

Konieczny, K., Sakol, D., Plonka, J., Rajca, M. & Bodzek, M. (2009). Coagulation-ultrafiltration system for river water treatment. *Desalination*, Vol. 240, (2009), pp. 151-159.

Laîné, J.M., Vial, D. & Moulart, P. (2000). Status after 10 years of operation - overview of UF technology today. *Desalination*, Vol. 131, (2000), pp. 17-25.

Lind, Y. & Glynn, A.W. (1999) Intestinal absorption of copper from drinking water containing fulvic acids and an infant formula mixture studied in a suckling rat model. *Biometals*, Vol. 12, No. 2, (1999), pp. 181-187.

Lopes, M.A.B.N. & Oliveira, J.C.G. (1999). O processo de flotação em fluxo como alternativa de despoluição de lago urbano - caso Aclimação na cidade de São Paulo. *Proceedings of 20° Congresso Brasileiro de Engenharia Sanitária e Ambiental,* Rio de Janeiro, RJ, Brazil.

Luz, T.N., Tidona, S., Jesus, B., Morais, P.V. & Sousa, J.P. (2009). The use of sewage sludge as soil amendment. The need for an ecotoxicological evaluation. *Journal of Soils and Sediments,* Vol. 9, No. 3, (2009), pp. 246-260.

Mantis, I., Voutsa, D. & Samara, C. (2005). Assessment of the environmental hazard from municipal and industrial wastewater treatment sludge by employing chemical and biological methods. *Ecotoxicology and Environmental Safety,* Vol. 62, No. 3, (2005), pp. 397-407.

Massoud, M.A., Tarhini, A. & Nasr, J.A. (2009). Decentralized approaches to wastewater treatment and management: Applicability in developing countries. *Journal of Environmental Management,* Vol. 90, (2009), pp. 652-659.

Meliker, J.R., Wahl, R.L., Cameron, L.L. & Nriagu, J.O. (2007). Arsenic in drinking water and cerebrovascular disease, diabetes mellitus, and kidney disease in Michigan: a standardized mortality ratio analysis. *Environmental Health,* Vol. 6, (2007), pp. 1-11.

Oliveira, J.C.G., Netto, A.M., Angelis, J.A. & Barbosa, M.A. (2000). Estação de flotação e remoção de flutuantes do Parque do Ibirapuera - São Paulo - a aplicação do processo de tratamento por flotação em fluxo para a recuperação de lagos urbanos. *Proceedings of 21° Congresso Brasileiro de Engenharia Sanitária e Ambiental,* João Pessoa, PB, Brazil.

Varis, O., Biswas, A.K., Tortajada, C. & Lundqvist, J. (2006). Megacities and Water Management. *International Journal of Water Resources Development,* Vol. 22, No. 2, (2006), pp. 377-394

Pretty, J.N., Mason, C.F., Nedwell, D.B. & Hine, R.E. (2002). A Preliminary Assessment of the Environmental Costs of the Eutrophication of Fresh Waters in England and Wales. In: *University of Essex.* 19 Jun 2011. Available from <www.essex.ac.uk/ces/occasionalpapers/EAEutrophReport.pdf>.

Realco. (2011). Information about the bioremediator. In: *Realco.* 02 Aug 2011. Available from <www.realco.be>.

Rodriguez, M.J., Vinette, Y., Serodes, J.B. & Bouchard, C. (2003) Trihalomethanes in drinking water of greater Quebec region (Canada): Occurrence, variations and modelling. *Environmental Monitoring and Assessment,* Vol. 89, No. 1, (2003), pp. 69-93.

Saz-Salazar, S.D., Hernández-Sancho, F. & Sala-Garrido, R. (2009) The social benefits of restoring water quality in the context of the Water Framework Directive: A comparison of willingness to pay and willingness to accept. *Science of The Total Environment,* Vol. 407, No. 16, (2009), pp. 4574-4583.

Smith, V.H., Tilman, G.D. & Nekola, J.C. (1999) Eutrophication: impacts of excess nutrient inputs on freshwater, marine, and terrestrial ecosystems. *Environmental Pollution,* Vol. 100, No. 1-3, (1999), pp. 179-196.

Suthar, S., Sharma, J., Chabukdhara, M. & Nema, A.K. (2010). Water quality assessment of river Hindon at Ghaziabad, India: impact of industrial and urban wastewater. *Environmental Monitoring and Assessment,* Vol. 165, No. 1-4, (2010), pp. 103-112.

Whiteley, C.G. & Lee, D.J. (2006). Enzyme technology and biological remediation. *Enzyme and Microbial Technology,* Vol. 38, (2006), pp. 291-316.

WHO. (2010). World Health Statistics. In: *WHO Press, World Health Organization: Switzerland.* 10 Jun 2011. Available from <www.who.int/whosis/whostat/en>.

WHO. (2011). Water supply, sanitation and hygiene development. In: *Water sanitation and Health, World Health Association.* 10 Jun 2011. Available from <www.who.int/water_sanitation_health/hygiene/en>.

Xia, S., Li, X., Liu, R. & Li, G. (2005). Pilot study of drinking water production with ultrafiltration of water from the Songhuajiang River (China). *Desalination,* Vol. 179, (2005), pp. 369-374.

13

Evaluation of the Removal of Chlorine, THM and Natural Organic Matter from Drinking Water Using Microfiltration Membranes and Activated Carbon in a Gravitational System

Flávia Vieira da Silva-Medeiros[1], Flávia Sayuri Arakawa[1], Gilselaine Afonso Lovato[1], Célia Regina Granhen Tavares[1], Maria Teresa Pessoa Sousa de Amorim[2], Miria Hespanhol Miranda Reis[3] and Rosângela Bergamasco[1]

[1]*Universidade Estadual de Maringá, Maringá, PR, Brazil*
[2]*Universidade do Minho, Campus Azurém, Guimarães,*
[3]*Universidade Federal de Uberlândia, Uberlândia, MG,*
[1,3]*Brazil*
[2]*Portugal*

1. Introduction

Due to its low cost, stability, and effectiveness, adding chlorine to drinking water is one of the most common treatments to ensure its bacteriological quality (Al-Jasser, 2007). Chlorine inactivates various types of micro-organisms and its residual properties help to prevent micro-organism regrowth during water flow in pipes (Connell, 1997).

Beyond off-flavors development due to chlorination by-products, chlorine flavor by itself constitutes one of the major complaints against tap water. In 1996, chlorine taste was the third most reported taste default of tap water in the US (Suffet et al., 1996). Due to the unpleasant taste of tap water, consumers may prefer bottled water as drinking water, even if bottled drinking water consumption would be associated with a higher economic and ecological cost. (Rodriguez et al., 2004) showed that the perception of tap water quality is closely related to the residual chlorine level: people living near a treatment plant who may receive a higher chlorine level in their tap water were generally less satisfied by tap water quality and perceived more risks associated with it than people living far from the plant. It was reported that, in the US, bottled water drinkers have three main categories for decisions: safety of water; healthfulness of the water; and taste of the water (Mackey et al., 2004). Consumers supplied with tap water containing a residual chlorine level greater than 0.24 mg/L Cl_2 were less satisfied with tap water when compared to consumers receiving lower concentrations (Rodriguez et al., 2004). This value almost coincides with the free chlorine residual (0.2 mg/L) that must be maintained in the distribution system, reducing the likehood of further contamination (Clark & Coyle, 1990). When taken together, this studies underline that the consumers would reject tap water in safe conditions due the chlorine flavor.

The residual chlorine could be removed using an activated carbon filter at the time of consumption. Activated carbon has already been used to remove chlorine excess of tap water used in food industries (Jaguaribe et al., 2005). Due to its well-developed pore structure, activated carbon in either powder or granules has an excellent adsorbent capacity.

Beyond this concern with chlorine flavor, several studies reported that chlorination of organic matter in fresh water resulted in the formation of disinfection by-products (DBPs) (Richardson, 2003; Rook, 1976), especially trihalomethanes (THMs), which remain a human health concern. Trihalomethanes (THMs) are a group of volatile organic compounds (VOCs) classified as disinfection by-products (DBPs). They were first identified by (Rook, 1976) and are formed during the chlorination of water, when chlorine reacts with naturally occurring organic matter: mainly humic and fulvic acids. Their general formula is CHX_3, where X may be any halogen or a combination of halogens. However, generally speaking this term is used to refer only to those compounds containing either chlorine or bromide, because these are the ones most commonly detected in chlorinated water (chloroform, bromodichloromethane, dibromochloromethane and bromoform). Brominated trihalomethanes are formed when hypochlorous acid oxidizes bromide ion present in water to form hypobromous acid, which subsequently reacts with organic materials to form these compounds (Pavon et al., 2008; Richardson, 2003). Iodinated THMs have been identified in chlorinated drinking water; however, they are not widely measured and are not regulated, even though iodinated compounds may be more toxic than brominated and chlorinated compounds (Richardson, 2003).

The International Agency for Research on Cancer (IARC) has classified chloroform and bromodichloromethane as possible carcinogens for humans (Group 2B) based on limited evidence of carcinogenicity in humans but sufficient evidence of carcinogenicity in experimental animals. Dibromochloromethane and bromoform belong to Group 3 (not classifiable as regards their carcinogenicity to humans), based on inadequate carcinogenicity in humans and inadequate or limited carcinogenicity in experimental animals (Pavon et al., 2008; WHO, 2006). In the case of THMs, approximately equal contributions to total exposure come from four sources: the ingestion of drinking water, inhalation of indoor air, inhalation and dermal exposure during showering or bathing, and the ingestion of foods (WHO, 2006). Trihalomethanes have been detected in different aqueous matrixes: tap water, swimming pool water, distilled water, ultrapure water and even in water that has not been subjected to chlorination processes, such as ground water, mineral water, snow, rain water, sea, and river water. However, the concentrations of these compounds in unchlorinated water tend to be much lower than those usually found in tap water. The presence of these levels of THMs may be due to several causes. In cases in which the chloroform > bromodichloromethane > dibromochloromethane > bromoform pattern is conserved, the THMs are likely to have originated from the infiltration of chlorinated water. The sources of chlorinated water to ground water may include the irrigation of lawns, gardens and parks; leaking drinking water distribution and sewer pipes, and industrial spills, among others.

Adsorption in carbonaceous materials, as carbon nanotubes and carbon spheres was reported as an effective technique in removing THM from water (Lu et al., 2005; Morawski et al., 2000). However, the authors did not consider the application of simpler and cheaper technology, as granular activated carbon.

(Amy et al., 1990) found that the majority of THM formation potential is presented by small to medium organic compounds with a specific ultraviolet absorbance (SUVA) values less than 3.0. The SUVA parameter represents the ratio UV_{254}/DOC and constitutes an indicator of carbon aromacity in water (Uyak et al., 2008).

Numerous research studies involving microfiltration (MF) and ultrafiltration (UF) of surface waters in rivers and lakes have proved that NOM is the main source of fouling during membrane processes (Uyak et al., 2008). The application of activated carbon (AC) in conjunction with MF/UF membranes is a promising technology for the removal of organic compounds in drinking water treatment, which incorporates the adsorption capabilities of activated carbon and the microorganism and particle removal ability of the MF/UF membranes (Tsujimoto et al., 1998; Yuasa, 1998). (Ravanchi & Kargari, 2009) highlighted the importance to propose innovative integrated membrane processes in order to became this process more commercial.

Moreover, nanofiltration membranes also showed potential in removing THMs from drinking water (Uyak et al., 2008). The application of nanofiltration and reverse osmosis in drinking water treatment is increasing in developed countries (Clever et al., 2000). However, their application in developing countries is still limited due the high costs of the membranes and pumping (Glucina et al., 2000).

Alternatively, microfiltration membrane processes can be designed to operate with gravity as the driving force, in simple systems that could be operated by non-trained people. Besides this, gravitational systems present the following advantages: energy saving, once pumps are not necessary; simpler tubing is required because it operates at low pressures. However, this kind of researched is scarce in the scientific literature (Peter-Varbanets et al., 2009).

In this way, the objective of this publication is to show the efficiency of a microfiltration membrane working alone and associated with activated carbon for removing THM and their precursors from tap water in a gravitational module.

2. Process configuration

2.1 Materials

Acetate cellulose microfiltration membrane (pore diameter=3.0 μm) was purchased from ADVANTEC/MFS (Japan). Commercial Granular Activated Carbon (GAC) 20x40 mesh made from coconut was supplied by BAHIACARBON (Brazil). Table 1 presents the textural characteristics of this activated carbon.

Surface area ($m^2\,g^{-1}$)	Micropore area ($m^2\,g^{-1}$)	Total pore volume ($cm^3\,g^{-1}$)	Average pore diameter (Å)
715.46	677.77	0.3856	20.04

Table 1. Textural characteristics of the activated carbon used in this work.

Filtration tests were carried out with the membrane working alone and with GAC as a pretreatment. The gravitational module used in this study (Fig. 1) operated exclusively with gravity as driven force, which produced a pressure of approximately 0.36 bar, due the position of the lung tank in relation to the filtration module.

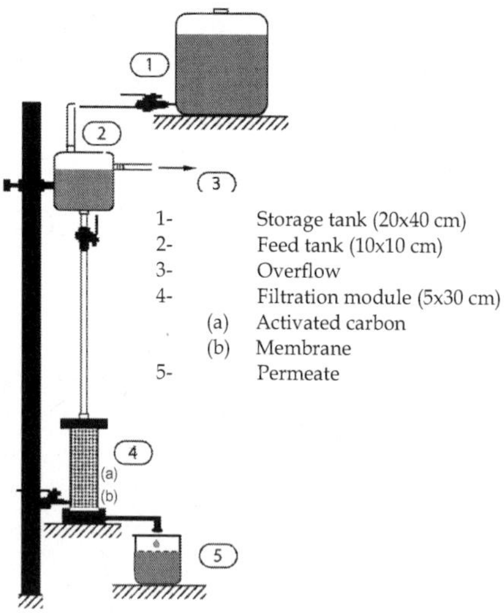

Fig. 1. Scheme of the gravitational filtration system.

The filtration module was composed of two parts: a flat-sheet membrane cell made of stainless steel and an acrylic cartridge to support the activated carbon. The effective membrane was 103 cm^2 and the activated carbon bed was 20 cm high. In each experiment, a new membrane was used, rinsed with ultrapure water and compacted by filtering ultrapure water during 2 h before starting a filtration test.

2.2 Raw water

The raw water used in this study was prepared using finished water of Pirapó River water (PRW), located in Maringá-Brazil. Quality parameters of this water are summarized in Table 2.

Parameters	Value
DOC (mg/L)	1.89
UV_{254} (L/cm)	0.032
SUVA (L/mg m)	1.69
Color (UC)	1.0
Turbidity (NTU)	0.67
pH	7.26

Table 2. Finished Pirapó River water (PRW) quality parameters

Two different types of water were prepared, and the details of each preparation are given below:

a. Chlorinated water

Chlorinated water samples were obtained by adding sodium hypoclorite (NaOCl ~2.5%, commercial grade) to tap water (PRW) in order to adjust the chlorine concentration to 2.0

mg/L, the limit regulated by Brazilian Government. The removal of chlorine, DOC, UV_{254} and SUVA were evaluated using the chlorinated water.

b. Water containing THMs

Trihalometanes mix at 2000 μg/L, produced by Supelco was diluted to 10 mg/L, and this final solution was used to adjust the THMs concentration in tap water (PRW), in order to maintain the total THMs concentration around 120 μg/L, slightly higher than the limit regulated by the Brazilian Government of 100 μg/L.

2.3 Analytical methods

Dissolved organic carbon (DOC) measurements were performed with a HACH DR2010 spectrophotometer, using the low range direct method. Besides, UV_{254} absorbance measurements were conducted in accordance with Standard Methods by a HACH DR 2010 UV spectrophotometer at a wavelength of 254 nm. Water samples for DOC and UV_{254} measurements were first filtered through a pre-washed 0.45 μm membrane filter to remove turbidity, which can interfere in these measurements, and distilled ultra filtered water was used as the background correction on the spectrophotometer.

Chlorine measurements were performed with a HACH DR/2010 spectrophotometer, using the DPD method for free chlorine determination, according to the Standard Methods for the Examination of Water and Wastewater (APHA, 1998). The samples were collected and immediately analyzed, using the HACH DPD free chlorine reagent.

THM concentrations were determined by Gas chromatography–mass spectrometry (GC-MS). For determination of THM was used in a chromatograph GC-MS with mass detector DSQ II with autosampler from Thermo triplus Red Space. The recovery of THMs was optimized using excess of KCl (4g per 12 mL of sample), according to Caro et al. (2007).

2.4 Flux and resistance measurements

Firstly, the permeated flux of deionized water was measured during 120 min in a clean membrane using new and clean membrane.

After the tests with raw water, final flux with deionized water was determined, also during 120 min. At the end of this measurement, the membrane was removed from the experimental module (Fig. 1) and mechanically cleaned in order to remove the cake formed on the membrane surface. After that, a new flux measurement with deionized water was done.

Resistances due to different fouling mechanisms were determined in order to investigate the fouling behavior. Resistances were calculated following the resistance-in-series model adapted from (A.I. Schafer, 2005) as presented in equation (1):

$$J = \frac{\Delta P}{\eta(R_m + R_p + R_c)} \tag{1}$$

where J is permeate flux [kg m^{-2} s^{-1}], ΔP is trans-membrane pressure [kg m^{-1} s^{-2}], η is dynamic viscosity [kg m^{-1} s^{-1}], and R denotes a resistance [m^2 kg^{-1}]: R_m is membrane hydraulic resistance, R_p is resistance due to pore blocking, and R_c is resistance due to cake formation.

Each resistance was experimentally measured in the gravitational system with and without GAC pretreatment. Membrane hydraulic resistance, R_m, was determined measuring the flux of deionized water through a clean membrane sheet. In this case, the others resistances are equals to zero.

The sum of resistances due to pore blocking and cake formation (R_p+R_c) was determined measuring the flux of deionized water with the fouled membrane, i.e. with the membrane that was used for raw water filtrations without clean procedures.

After mechanical cleaning, the flux measurements was carried out in order to obtain the resistance due to pore blocking, R_p.

Membrane fouling percentage (%F) was calculated according to equation (2), as proposed by (Balakrishnan et al., 2001). This percentage represents the drop in deionized water flux after the tests of filtration with raw water.

$$\%F = \frac{(J_i - J_f)}{J_i} \times 100 \tag{2}$$

where %F is membrane fouling percentage, and J_i and J_f are deionized water fluxes in clean and fouled membranes, respectively.

3. Process performance

3.1 Flux measurements with raw water

In general, the application of activated carbon as pretreatment to micro/ultrafiltration membranes affects the permeate flux of membranes and improves the removal of several parameters that may cause fouling (Fabris et al., 2007; Kim & Gai, 2008). However, there is a lack of information in the literature about this application in gravitational systems.

The permeate fluxes of the membrane operating alone and with activated carbon as pretreatment were determined during the evaluation of chlorine and THM removals from tap water in the gravitational module. Considering that the observed values of permeate fluxes were almost equals in both assays, Fig. 2 illustrates the average values of both assays during the 480 min of operation. The membrane working alone is referred as M3 and the membrane working with activated carbon as pretreatment is referred as M3C.

The permeate flux increased when using activated carbon as pretreatment. In the first 50 minutes of operation, it is observed a permeate flux between 2000 and 400 kg h^{-1} m^{-2} to the membrane M3 (without GAC pretreatment) and between 3000 and 700 kg h^{-1} m^{-2} to the membrane M3C (with GAC pretreatment). After 50 min of operation, the permeate flux was around 200 and 300 kg h^{-1} m^{-2} to the membranes M3 and M3C, respectively. However, in the last 100 minutes of operation, it is observed that the permeate flux was around 100 kg h^{-1} m^{-2} to the membrane M3, and around 200 kg h^{-1} m^{-2} to the membrane M3C. The better performance of the membrane M3C in comparison to the membrane M3 is probably due to the adsorption of organic matter in the activated carbon surface, which mitigates fouling effects (Kim & Gai, 2008).

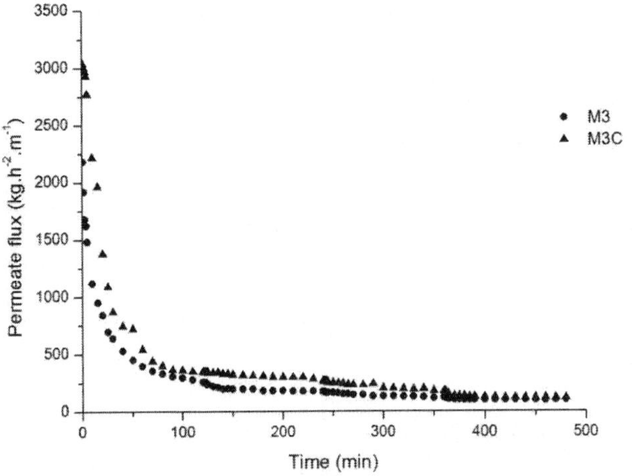

Fig. 2. Permeate fluxes of tap water to the membranes with and without GAC as pretreatment (M3C and M3, respectively).

Fig. 3 presents the results of total resistance against time to the 3.0 μm membrane working alone and associated with activated carbon, calculated using Equation 1. Total resistance is referred as the sum of R_m, R_p, and R_c.

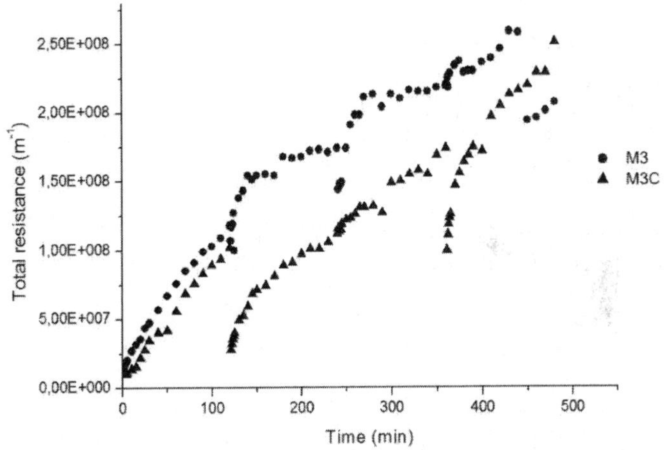

Fig. 3. Total resistance in tap water assays.

It is possible to notice from Fig. 3 that the application of GAC as pretreatment reduced the total resistance of the membrane during the operation with tap water. This decrease in the total resistance is probable the main cause of the flux increase observed during the filtration with GAC as pretreatment.

3.2 DOC and UV_{254} rejection capacity

Figs. 4 and 5 present the reduction in DOC and UV_{254} compounds, respectively, after microfiltrations of tap water with and without GAC pretreatment.

DOC rejection by the process with the membrane working alone (M3) was between 50 and 70% (Fig. 4) during the complete experiment. The range presented by UV_{254} rejection was between 35 and 55% (Fig. 5). These results are comparable to previous results found to nanofiltration membranes (Alborzfar et al., 1998), with the advantage of using microfiltration and a gravitational system, much more cheaper than a nanofiltration system using pumps.

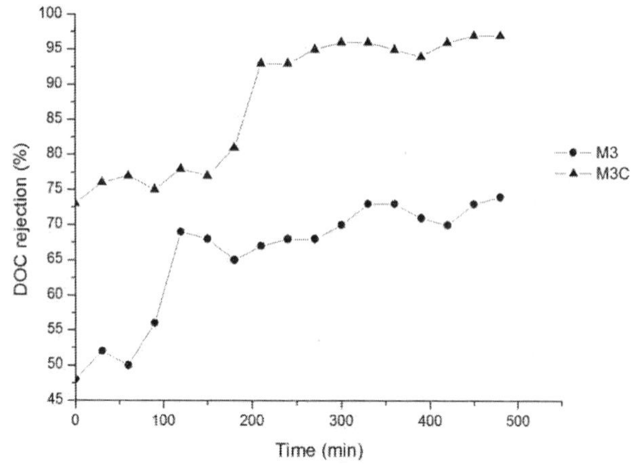

Fig. 4. DOC rejection in tap water after microfiltrations with the membrane working alone (M3) and with GAC as pretreatment (M3C).

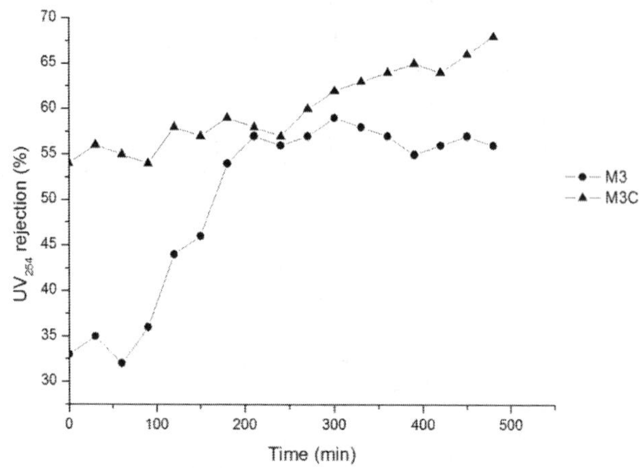

Fig. 5. UV_{254} rejection of in tap water after microfiltrations with the membrane working alone (M3) and with GAC as pretreatment (M3C).

Besides this, M3C presented a removal of 70 to 95% to DOC (Fig. 4) and a removal of 55 to 70% to UV254 (Fig. 5). The application of activated carbon as pretreatment significantly increased the removal of dissolved organic matter, as reported before (Bao et al., 1999).

3.3 Chlorine rejection capacity of the evaluated systems

Fig. 6 shows the rejection performances of free chlorine, in the initial concentration of 2 mg/L, by the two studied systems, M3 and M3C, in the gravitational module.

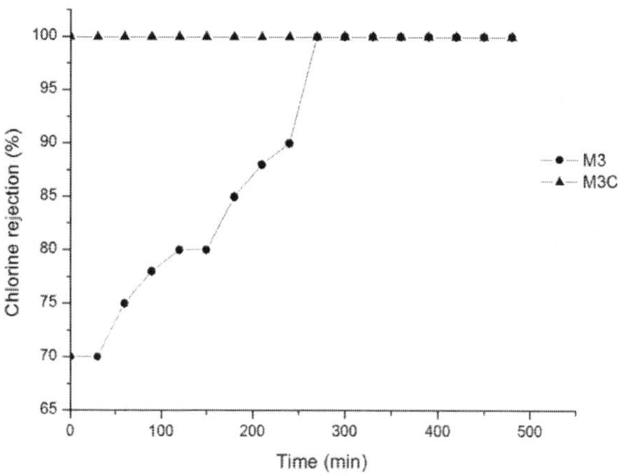

Fig. 6. Chlorine rejection in tap water after microfiltrations with the membrane working alone (M3) and with GAC as pretreatment (M3C).

Chlorine rejection increased gradually to the membrane working alone (M3), from 70% in the first minutes of operation to 100% after 270 min. It probably happened due to a formation of a cake fouling layer on the surface of the membrane, which improved the removal rate of contaminants (Kim & Gai, 2008).

In contrast to M3, the chlorine rejection of the system using activated carbon as pretreatment (M3C) was of 100% during the complete experiment. (Jaguaribe et al., 2005) reported that, due to its well-developed pore structure, coconut shells activated carbon (working alone) reduces around 40% of free chlorine presented in water.

3.4 THM rejection capacity of the evaluated systems

Total THM and its four compounds were chosen to evaluate the change in permeate concentration over the course of filtration tests using the membrane working alone (M3) and using activated carbon as pretreatment to the membrane (M3C). Fig. 7 illustrates the total THM rejection capacity of M3 and M3C processes.

Total THM rejection of M3 process was approximately equals to 74% during the complete experiment, which could be considered a suitable result, since similar results were obtained when using nanofiltration membranes at higher pressures (Uyak et al., 2008), and considering the fact that M3 is a microfiltration membrane operating only by gravity.

Moreover, M3C process presented a removal of 93 to 99% of THM compounds due to the considerable adsorption capacity of activated carbon toward various pollutants, especially THM, as reported before (Razvigorova et al., 1998).

Fig. 8 shows the rejection performances of three species of THM by the two studied systems, M3 and M3C during the 480 min of filtration. The THM species are chloroform (CFM), bromodichloromethane (BDCM) and dibromochloromethane (DBCM). Since bromoform (BFM) was not detected in the tested water, hence, three species were taken in account.

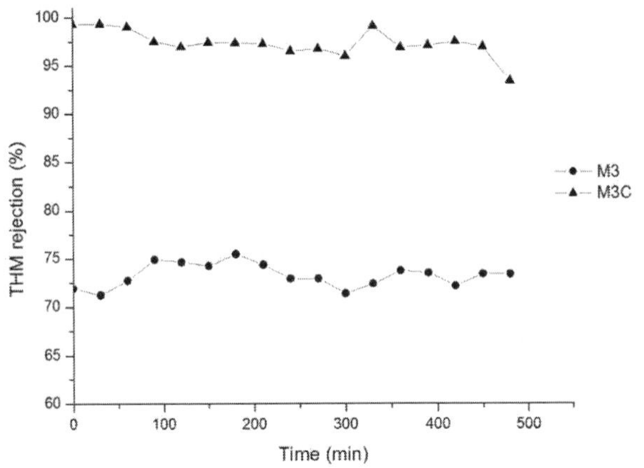

Fig. 7. Total THM rejection in tap water after microfiltrations with the membrane working alone (M3) and with GAC as pretreatment (M3C).

The selected microfiltration membrane working alone (M3) was effective in removing the THM compounds during the 480 min of filtration. It was depicted in Fig. 8 that the rejection efficiencies of CFM were found to be around 65 and 70% during the 480 min of operation. Moreover, the rejection efficiencies of BDCM and DBCM were practically 100%. As observed in previous similar studies (Uyak et al., 2008), the removal efficiency of M3 process was increased with increasing the molecular weight of THM species. As bromine atom replace the chlorine atoms, greatly increasing the molecular weight, resulted in higher removal efficiency. The higher removal efficiency of BDCM and DBCM was attributed to higher molecular weight and brominating characteristics (Uyak et al., 2008).

For the system using activated carbon as pretreatment to the microfiltration membrane (M3C), Fig. 8 illustrates the removal of CFM to be around 99% in the first minutes of operation and around 94% after 100 minutes. The rejection rates of BDCM and DBCM were also around 100%. The application of activated carbon as pretreatment enhanced the THM compounds removal, and this fact is especially notable to CFM, once the rejection rates of BDCM and DBCM were considerably high also when the gravitational module operated only with the microfiltration membrane (M3). It was found in the literature that magnitude of adsorption of these chlorinated compounds was in the following order: BDCM>DBCM>CFM. The molecules containing bromine were adsorbed with highest efficiency compared to the remainder of lower radius (chlorine). It means that not only the size of pores determines the adsorption but also surface chemical character may influence it (Razvigorova et al., 1998).

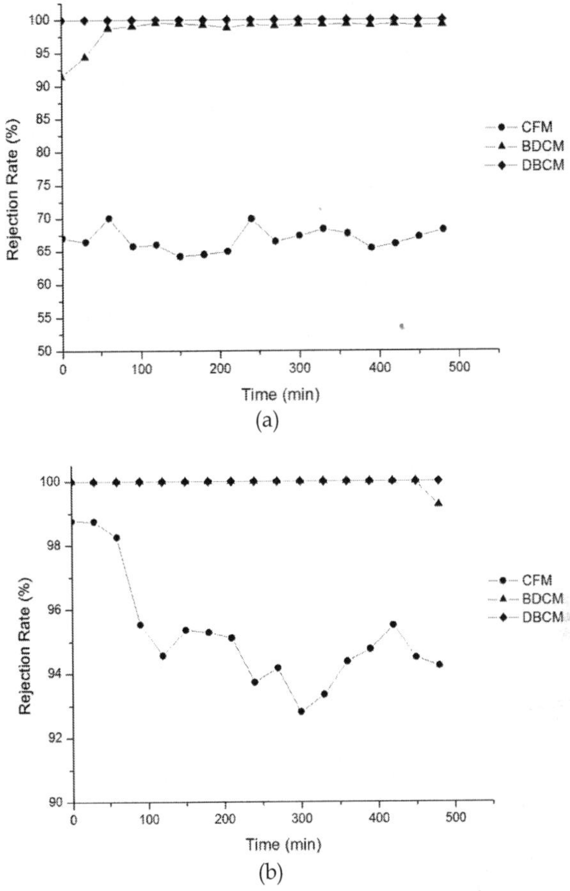

Fig. 8. THM compounds rejection of (a) M3 and (b) M3C processes.

3.5 Flux and resistance measurements

Fig. 9 illustrates the permeate flux values measured with deionized water to the membrane with and without the GAC pretreatment before and after the filtrations with tap water.

The association with activated carbon significantly improved the performance of the 3.0 μm membrane in relation to the initial and final flux values. The initial flux to the carbon+membrane system was approximately 2700 kg m^{-2} h^{-1}, while the membrane working alone presented initial flux around 2000 kg m^{-2} h^{-1}.

Considering final flux values, the membrane associated with GAC achieved values around 222 kg m^{-2} h^{-1} and the membrane working alone presented values not higher than 60 kg m^{-2} h^{-1}. These results could be related to the adsorption of organic matter by the activated carbon (Choo & Kang, 2010).

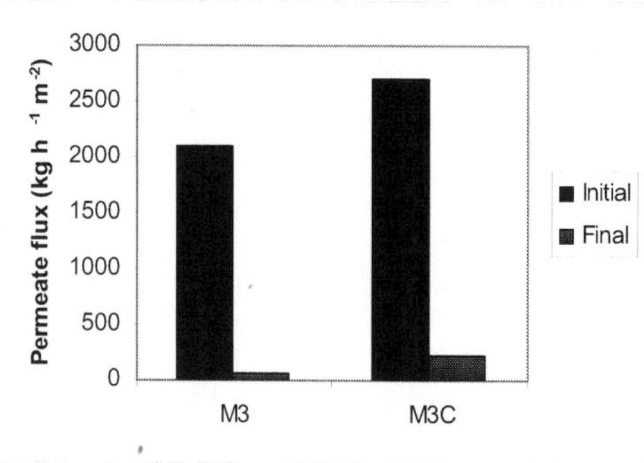

Fig. 9. Initial and final flux values of deionized water to the membrane with and without GAC as pretreatment (M3C and M3, respectively).

Table 3 shows the results of resistances and fouling percentage to the 3.0 μm membrane working alone and associated with activated carbon.

Process	R_m x10^8 (m^{-1})	R_c x10^8 (m^{-1})	R_p x10^8 (m^{-1})	R_t x10^8 (m^{-1})	%F
M3	0.141	1.31	4.17	5.621	97.3
M3C	-	1.13	0.156	1.395	91.9

Table 3. Results of resistance and fouling percentage.

It was observed a considerable decrease in the R_c, R_p, and R_t values when GAC was applied as pretreatment, indicating that the activated carbon reduced the fouling due the cake formation (R_c) and pore blocking (R_p). These results are in accordance with the previous reports (Kim & Gai, 2008). The hydraulic membrane resistance (R_m) was not determined to the membrane working with GAC as pretreatment because it is considered an inherent property of the membrane.

It was also observed a decrease in the fouling percentage with the application of activated carbon as pretreatment to the 3.0 μm membrane, which could also be expected, according to previous reports in the literature (Choo & Kang, 2010).

4. Conclusions

Microfiltration and its association with activated carbon are technologies that have potential for use in drinking water treatment. The conclusions that can be drawn from the results of this experimental investigation are as follows:

- Experimental results show that the microfiltration membrane evaluated in this study was effective in removing DOC, UV_{254}, chlorine, CFM, BDCM, and DBCM compounds. Further, brominated THM compounds were removed more significantly than chlorinated THM ones. The higher removal efficiency of DBCM was attributed to higher molecular weight and brominating characteristics.
- The application of activated carbon as pretreatment increased the permeate flux and also increased the rejection efficiency of DOC, UV_{254}, chlorine, CFM, BDCM, and DBCM compounds. It was attributed to the chemical interaction that probably happens between activated carbon surface and the studied compounds.

- The system using activated carbon and microfiltration membrane (M3C) applied in a gravitational module is a promising alternative to improve drinking water quality, due to its efficiency, simplicity and low cost.

5. Acknowledgments

The authors acknowledge the financial support of the CNPq and the FAPEMIG.

6. References

Al-Jasser, A. O. (2007). Chlorine decay in drinking-water transmission and distribution systems: Pipe service age effect. *Water Research,* Vol. 41, No. 2, (Jan), pp. 387-396, ISSN 0043-1354.

Alborzfar, M.;Jonsson, G. & Gron, C. (1998). Removal of natural organic matter from two types of humic ground waters by nanofiltration. *Water Research,* Vol. 32, No. 10, (Oct), pp. 2983-2994, ISSN 0043-1354.

Amy, G. L.;Alleman, B. C. & Cluff, C. B. (1990). Removal of Dissolved Organic-Matter by Nanofiltration. *Journal of Environmental Engineering-Asce,* Vol. 116, No. 1, (Jan-Feb), pp. 200-205, ISSN 0733-9372.

APHA. American Public Health Association (1998). Standard Methods for the Examination of Water and Wastewater, 20th ed. Washington.

Balakrishnan, M.;Dua, M. & Khairnar, P. N. (2001). Significance of membrane type and feed stream in the ultrafiltration of sugarcane juice. *Separation Science and Technology,* Vol. 36, No. 4, pp. 619-637, ISSN 0149-6395.

Bao, M. L.;Griffini, O.;Santianni, D.;Barbieri, K.;Burrini, D. & Pantani, F. (1999). Removal of bromate ion from water using granular activated carbon. *Water Research,* Vol. 33, No. 13, (Sep), pp. 2959-2970, ISSN 0043-1354.

Caro J., Serrano A., Gallego M. (2007), Sensitive headspace gas chromatography-mass spectrometry determination of trihalomethanes in urine. *Journal of Chromatography B, Analytical Technologies in the Biomedical and Life Sciences,* Vol. 848, No. 2, (Nov 7), pp. 277-282, ISSN: 1570-0232.

Choo, K. H. & Kang, S. K. (2010). Why does a mineral oxide adsorbent control fouling better than powdered activated carbon in hybrid ultrafiltration water treatment? *Journal of Membrane Science,* Vol. 355, No. 1-2, (Jun 15), pp. 69-77, ISSN 0376-7388.

Clark, R. M. & Coyle, J. A. (1990). Measuring and Modeling Variations in Distribution-System Water-Quality. *Journal American Water Works Association,* Vol. 82, No. 8, (Aug), pp. 46-53, ISSN 0003-150X.

Clever, M.;Jordt, F.;Knauf, R.;Rabiger, N.;Rudebusch, M. & Hilker-Scheibel, R. (2000). Process water production from river water by ultrafiltration and reverse osmosis. *Desalination,* Vol. 131, No. 1-3, (Dec 20), pp. 325-336, ISSN 0011-9164.

Connell, G. F. (Ed.). 1997. *The Chlorination/Chloramination Handbook.* Denver.

Fabris, R.;Lee, E. K.;Chow, C. W. K.;Chen, V. & Drikas, M. (2007). Pre-treatments to reduce fouling of low pressure micro-filtration (MF) membranes. *Journal of Membrane Science,* Vol. 289, No. 1-2, (Feb 15), pp. 231-240, ISSN 0376-7388.

Glucina, K.;Alvarez, A. & Laine, J. M. (2000). Assessment of an integrated membrane system for surface water treatment. *Desalination,* Vol. 132, No. 1-3, (Dec 20), pp. 73-82, ISSN 0011-9164.

Jaguaribe, E. F.;Medeiros, L. L.;Barreto, M. C. S. & Araujo, L. P. (2005). The performance of activated carbons from sugarcane bagasse, babassu, and coconut shells in removing residual chlorine. *Brazilian Journal of Chemical Engineering,* Vol. 22, No. 1, (Jan-Mar), pp. 41-47, ISSN 0104-6632.

Kim, H. S. & Gai, X. J. (2008). The role of powdered activated carbon in enhancing the performance of membrane systems for water treatment. *Desalination,* Vol. 225, No. 1-3, (May 1), pp. 288-300, ISSN 0011-9164.

Lu, C. S.;Chung, Y. L. & Chang, K. F. (2005). Adsorption of trihalomethanes from water with carbon nanotubes. *Water Research,* Vol. 39, No. 6, (Mar), pp. 1183-1189, ISSN 0043-1354.

Mackey, E. D.;Baribeau, H.;Crozes, G. F.;Suffet, I. H. & Piriou, P. (2004). Public thresholds for chlorinous flavors in US tap water. *Water Science and Technology,* Vol. 49, No. 9, pp. 335-340, ISSN 0273-1223.

Morawski, A. W.;Kalenczuk, R. & Inagaki, M. (2000). Adsorption of trihalomethanes (THMs) onto carbon spheres. *Desalination,* Vol. 130, No. 2, (Nov 1), pp. 107-112, ISSN 0011-9164.

Pavon, J. L. P.;Martin, S. H.;Pinto, C. G. & Cordero, B. M. (2008). Determination of trihalomethanes in water samples: A review. *Analytica Chimica Acta,* Vol. 629, No. 1-2, (Nov 23), pp. 6-23, ISSN 0003-2670.

Peter-Varbanets, M.;Zurbrugg, C.;Swartz, C. & Pronk, W. (2009). Decentralized systems for potable water and the potential of membrane technology. *Water Research,* Vol. 43, No. 2, (Feb), pp. 245-265, ISSN 0043-1354.

Ravanchi, M. T. & Kargari, A. (2009). *New Advances in Membrane Technology.* InTech, ISBN 978-953-307-009-4, Vienna, Austria.

Razvigorova, M.;Budinova, T.;Petrov, N. & Minkova, V. (1998). Purification of water by activated carbons from apricot stones, lignites and anthracite. *Water Research,* Vol. 32, No. 7, pp. 2135-2139, ISSN 0043-1354.

Richardson, S. D. (2003). Disinfection by-products and other emerging contaminants in drinking water. *Trac-Trends in Analytical Chemistry,* Vol. 22, No. 10, (Nov), pp. 666-684, ISSN 0165-9936.

Rodriguez, M. J.;Turgeon, S.;Theriault, M. & Levallois, P. (2004). Perception of drinking water in the Quebec City region (Canada): the influence of water quality and consumer location in the distribution system. *Journal of Environmental Management,* Vol. 70, No. 4, (Apr), pp. 363-373, ISSN 0301-4797.

Rook, J. J. (1976). Haloforms in Drinking-Water. *Journal American Water Works Association,* Vol. 68, No. 3, pp. 168-172, ISSN 0003-150X.

Schafer, A.; Fane, A. & Waite, T. (2004). *Nanofiltration – Principles and Applications,* Elsevier, ISBN 978-1-85617-405-3, Oxford.

Suffet, I. H. M.;Corado, A.;Chou, D.;McGuire, M. J. & Butterworth, S. (1996). AWWA taste and odor survey. *Journal American Water Works Association,* Vol. 88, No. 4, (Apr), pp. 168-180, ISSN 0003-150X.

Tsujimoto, W.;Kimura, H.;Izu, T. & Irie, T. (1998). Membrane filtration and pre-treatment by GAC. *Desalination,* Vol. 119, No. 1-3, (Sep 20), pp. 323-326, ISSN 0011-9164.

Uyak, V.;Koyuncu, I.;Oktem, I.;Cakmakci, M. & Toroz, I. (2008). Removal of trihalomethanes from drinking water by nanofiltration membranes. *Journal of Hazardous Materials,* Vol. 152, No. 2, (Apr 1), pp. 789-794, ISSN 0304-3894.

WHO. 2006. Guidelines for Drinking-Water Quality. Geneva.

Yuasa, A. (1998). Drinking water production by coagulation-microfiltration and adsorption-ultrafiltration. *Water Science and Technology,* Vol. 37, No. 10, pp. 135-146, ISSN 0273-1223.

14

Application of Hybrid Process of Coagulation/ Flocculation and Membrane Filtration to Water Treatment

Rosângela Bergamasco[1], Angélica Marquetotti Salcedo Vieira[1], Letícia Nishi[1], Álvaro Alberto de Araújo[2] and Gabriel Francisco da Silva[2]

[1]*Universidade Estadual de Maringá,*
[2]*Universidade Federal de Sergipe,*
Brazil

1. Introduction

Nowadays, the concern about contamination of aquatic environments has increased, especially when water is used for human consumption (Madrona et al., 2010). Contamination of water resources, especially in areas with inadequate sanitation and water supply, has become a risk factor for public health, with water playing a role as a vehicle for transmission of biological agents (viruses, bacteria, and parasites) as well as a source of contamination by chemicals (industrial effluents). Among the waterborne diseases, enteric diseases are most frequent. Approximately 19% of waterborne gastroenteritis outbreaks in the United States are attributed to protozoan parasites (Lindquist, 1999), particularly *Giardia* and *Cryptosporidium* species, due to their wide distribution in the environment, high incidence in the population, and resistance to conventional water treatment (Iacovski et al., 2004).

Giardia duodenalis is a flagellated protozoan parasite that causes giardiasis. The cysts are transmitted by fecal-oral route in a direct (person to person) or indirect way (contaminated water and food). In humans it can cause from self-limited to chronic enteritis, debilitating diarrhea with steatorrhea and weight loss (Rey, 2001). *Cryptosporidium* is a genus of protozoan parasites with several species that have been treated as a public health problem, especially after the advent of HIV/AIDS (Fayer, 2004). In immune depressed individuals it causes more serious infection; in some cases, the patient can die if untreated. It is considered an opportunistic parasite (Rey, 2001).

Despite regulations and control measures turning to be more and more stringent, outbreaks of waterborne *Cryptosporidium* spp. and *Giardia* spp. have been reported worldwide (United States Environmental Protection Agency [USEPA], 1996; Centers for Disease Control and Prevention [CDC], 2006). Therefore, the treatment applied to the collected water must ensure that it is free of pathogens and chemicals that pose health risks, when distributed by the water supply system. Furthermore, physicochemical parameters must meet the drinking water standards required by the laws of each country (Bergamasco et al., 2011). Thus, there is great importance in either the development of more sophisticated treatments or the improvement of the current ones.

In addition to these microorganisms, many other impurities can harm human health if not reduced or eliminated. They do not approach each other naturally during the coagulation/ flocculation process, being necessary the presence of a coagulant agent. Coagulation and flocculation processes are essential parts of water treatment, and the clarification of water using coagulants is practiced since ancient times (International Water Association [IWA], 2010). Alum has been the most widely used coagulant because of its proven performance, cost effectiveness, relative easy handling, and availability. Recently, much attention has been drawn on the extensive use of alum. According to Driscoll & Letterman (1995), the utilization of alum has raised a public health concern because of the large amount of sludge produced during the treatment and the high level of aluminum that remains in the treated water. McLachlan (1995) discovered that the intake of a large quantity of alum salt may cause Alzheimer disease.

Among the new techniques for drinking water treatment is the use of natural coagulants, aiming at a better quality of treated water by reducing the use of chemicals and also due to others advantages of natural coagulants. The biopolymers may be of great interest since they are natural low-cost products, characterized by their environmentally friendly behavior. Advantages of natural coagulant/flocculants have led some countries, like Japan, China, India, and United states, to adopt the use of natural polymers in surface water treatment do produce drinking water (Kawamura, 1991). The application of these coagulants in the coagulation/flocculation process has been successfully performed to remove turbidity, color, and natural organic matter (NOM) from natural water in order to produce clean potable water.

Thus, considering the coagulation step, the use of natural polyelectrolytes such as chitosan and *Moringa oleifera* (moringa) could be an option with many advantages over chemical agents, particularly the biodegradability, low toxicity, low residual sludge production, and less risks to health. Polyelectrolytes such as chitosan have a large number of surface charges that increase the efficiency of the coagulation process. Regarding the moringa, many researchers are currently seeking to identify the compound responsible for the coagulating characteristic of the seed of this plant, although there are still no definitive conclusions. It is known that it is a protein or polypeptide with coagulant properties (Ndabigengesere et al., 1995; Okuda et al., 2001; Ghebremichael et al., 2005). According to Davino (1976) the mechanism of coagulation/flocculation caused by the protein found in the seed of *Moringa oleifera* Lam resembles the mechanism caused by polyelectrolytes.

The use of coagulants for drinking water treatment, in spite of being efficient in the removal of most contaminants, is not able to generate water of high potability standards, which leads to the necessity of the simultaneous use of other techniques. Membrane filtration technique is already widely recognized and can be implemented in combination with coagulation processes.

This way, this chapter will look at the use of alternative techniques for water treatment based on the use of natural coagulants (chitosan and moringa seeds) associated with the membrane filtration process (micro and ultrafiltration) to obtain drinking water for human consumption.

1.1 Chitosan

Chitosan is a linear copolymer of d-glucosamine and N-acetyl-dglucosamine produced by the deacetylation of chitin, a natural polymer of major importance and the second most abundant natural polymer in the world, after cellulose (Rinaudo, 2006). It is described as a cationic polyelectrolyte and is expected to coagulate negatively charged suspended particles found in natural waters with increased turbidity (Divakaran & Pillai, 2002). Chitosan has been investigated as coagulant/flocculant for the removal of impurities from natural water and wastewater (Kawamura, 1991; Klopotek et al., 1994) because it can be conditioned and used for pollutant complexation in different forms, from water-soluble forms to solid forms (Renault et al., 2009). Some of the applications proposed for chitosan are: (i) removal of turbidity (Divakaran & Pillai, 2002; Bergamasco et al., 2011), natural organic matter (NOM) and color (Eikebrokk, 1999); (ii) inactivation of bacteria (Chung et al., 2003); and (iii) metal removal with higher efficiency when associated with ultrafiltration (Verbych et al., 2005).

The use of chitosan as coagulant in the coagulation/flocculation process (CFQ) for surface water treatment was studied by Bergamasco et al. (2011), achieving satisfactory results regarding color and turbidity removal, with values above 87%. The surface water sample in this case was from the Pirapó River at Maringá - PR, Brazil, with initial turbidity of 240 NTU and initial color of 1045 Hu. The chemical oxygen demand (COD) was initially of 19,3 mgO_2/L and it was reduced in 59.9% after treatment with chitosan, showing that the higher coagulation/flocculation of chitosan is related with the compounds that give color and turbidity to surface water.

The efficiency of the use of chitosan is more evident when compared with the process of coagulation/flocculation using aluminum sulfate as coagulant (CFS). The main difference in removal efficiency occurs when evaluating the removal of COD and TDS (Total dissolved solids). Comparing CFQ and CFS processes, COD removal efficiency was 59.9% using CFQ and 38.1% using CFS, and TDS removal was 41.9% for CFQ and 7.5% for CFS. In the same study, Bergamasco et al. (2011) observed that coagulation with aluminum sulfate generated sludge with SVI (sludge volume index) of 38.7 mg/mL, and the sludge formed by coagulation using chitosan showed SVI of 56.8 mg/mL. According to McLachlan (1995), the biggest advantage of chitosan over aluminum sulfate as coagulant is the fact that it is biodegradable, generating an easy to handle organic sludge that can be taken to a common landfill. Furthermore, chitosan improves the sedimentation step, as the flocs are more compact.

In a study by Eikebrokk & Saltnes (2001), it was verified that the fractions of color and organic carbon removal, with chitosan concentration of 4.0 mg/L, were 70 and 30%, respectively. In this work the authors also compared chitosan with metallic coagulants, otaining a reduction of 50% in the generated sludge when using chitosan, solving this way the problem of the concentration of trace metals in treated water. Still, the sludge disposal was simplified due to its biodegradability characteristics.

One can verify that chitosan presents good removal of turbidity and color, and is advantageous in terms of generated sludge, which can be disposed of in ordinary landfills, since it is biodegradable and has no trace metals.

1.2 *Moringa oleifera*

Moringa oleifera (moringa) is a tropical plant belonging to the family Moringaceae (Katayon et al., 2006), a single family of shrubs with 14 known species. Moringa is native of India but is now found throughout the tropics (Bhatia et al., 2007). Moringa seeds contain a non-toxic natural organic polymer which is an active agent with excellent activity and coagulating properties. The tree is generally known in the developing world as a vegetable, a medicinal plant, and a source of vegetable oil (Katayon et al., 2006). Its leaves, flowers, fruits, and roots are used locally as food ingredients. The medicinal and therapeutic properties of moringa have led to its utilization as a cure for different ailments and diseases, physiological disorders, and in Eastern allopathic medicine (Akhtar et al., 2007). Additionally, the coagulant is obtained at extremely low or zero net cost (Ghebremichael et al., 2005).

If moringa is proven to be active, safe, and inexpensive, it is possible to use it widely for drinking water and wastewater treatment. Besides, moringa may yet have financial advantages bringing more economic benefits for the developing countries (Okuda et al., 1999).

The moringa seed has a protein that when solubilized in water is able to promote coagulation and flocculation of compounds that cause color and turbidity in highly turbid water. Several studies have also shown their effective antimicrobial and antifungal capacity, thereby contributing to good water quality at low cost (Chuang et al., 2007; Coelho et al., 2009).

The process used to obtain the coagulant (which would be mainly constituted by the protein in the seed) is usually performed with the solubilization of the protein in water under stirring and filtration, but the use of salts is able to promote a greater solubilization of this protein in the medium, which would favor the coagulation/flocculation process.

Madrona et al. (2010) evaluated the extraction of the coagulant protein in the presence of potassium chloride (KCl) at different concentrations. The authors have shown that greatest coagulation efficiencies are achieved with KCl 1 mol/L, reaching nearly 100% removal of color and turbidity from water with initial turbidity of 850 NTU. In the same study, conducted by Madrona (2010), the author evaluated the effectiveness of other salts in the coagulant extraction process, such as magnesium chloride ($MgCl_2$) and sodium chloride (NaCl), compared with KCl, all at a concentration of 1 mol/L. A protein content of 4499 mg/L was achieved for the extraction using NaCl, and 4818 mg/L using KCl. The process of protein extraction with $MgCl_2$ was able to release only 950 mg/L of protein in solution, which is very close to the concentration released by the extraction with water (873 mg/L).

Moringa has been found to be ineffective as a natural coagulant for low turbidity drinking water but effective for high turbidity water in previous studies (Okuda et al., 2001). This was verified in a study by Nishi (2011). The authors obtained values of color and turbidity removal over 90% when the water to be treated showed high values of initial turbidity, between 350 and 450 NTU. A moringa concentration of 150 mg/L would have been sufficient to achieve this level of removal. The coagulant derived from moringa seeds, as it contains a certain amount of organic matter, can give color and turbidity to the treated water. Thus, when water with low initial turbidity undergoes coagulation/flocculation, depending on the concentration of moringa coagulant used, the effect of turbidity and color removal is not as satisfactory as the level of removal obtained for these parameters using this coagulant in the treatment of high turbidity water. In this case the effect of the coagulant protein overlaps the additional organic load and the removal of the evaluated parameters increases.

An important point to be considered when using moringa as a coagulant is related to the pH of the water to be treated. For chemical coagulants, water pH adjustment is necessary for the flakes to be properly formed. In the case of the moringa, there is no need for this adjustment, and this parameter is not changed after treatment, as evidenced by Vieira et al. (2010). Moringa is an efficient coagulant in a wide pH range (6-8), which is an advantage compared with other coagulants, as the pH adjustment step can be eliminated in the coagulation/flocculation processes.

1.3 Coagulation/flocculation and membrane filtration

In conventional water treatment plants, the coagulation/flocculation process is followed by filtration. However, nowadays, membrane separation has been widely studied for potable water production, since the MF/UF membranes are physical barriers that are able to efficiently remove suspended particles and colloids (Xia et al., 2007; Guo et al., 2009), turbidity, bacteria, algae, parasites, and viruses for clarification and disinfection purposes (Guo et al., 2009), as well as to control trihalomethane precursors (Bottino et al., 2001).

To overcome the problems caused by natural organic matter (NOM) in MF and UF applications, conjunctive use of coagulation and membranes is becoming more attractive for water treatment because the coagulation is an opportunity to join NOM with other particles present in water before NOM reaches the membrane surface. Application of the coagulation and ultrafiltration unit operations contributes to the improvement of treated water quality and the enhancement of the membrane performance. In comparison with conventional processes such as coagulation, flocculation, sedimentation and/or flotation, and rapid or slow sand filtration, MF/UF technology has many advantages such as superior quality of treated water, much greater compactness, easier control of operation and maintenance, use of fewer chemicals, and lower production of sludge (Bergamasco et al., 2009). The combined processes of coagulation/flocculation followed by micro or ultrafiltration usually result in better water quality, since the membrane processes often function as a polishing step in water treatment, being able to remove the impurities which are not removed by coagulation/flocculation processes. In addition, unlike the conventional filtration process, micro/ultrafiltration can retain bacteria and other microorganisms.

Bouchard et al. (2003) studied the processes of microfiltration and coagulation/ microfiltration, using ferric chloride and aluminum sulfate as coagulants in the combined process. Microfiltration tests were performed in a mini-module of submerged cross-fiber Zenon membranes, with porosity of 0.1 μm. For the combined process, the results obtained were 60% removal of TOC and a reduction of more than 80% of the compounds that absorb UV at 254nm. The coagulation/microfiltration process was shown to be beneficial, allowing a significant reduction in membrane fouling. These results were already expected, because when coagulation occurs, colloids are destabilized and cluster forming larger flocs, thus contributing to reduce membrane fouling. Comparing the processes of ultrafiltration and coagulation/ultrafiltration, using ferric chloride as a coagulant agent, removal of TOC and compounds that absorb UV at 254nm of 30% and 60%, respectively, was observed for the ultrafiltration process. For the combined coagulation/ultrafiltration, the removal of TOC and compounds that absorb UV at 254nm increased to 60 and 80%, respectively (Bouchard et al., 2003).

Konradt-Moraes (2004) studied the combined process of coagulation/flocculation/ ultrafiltration using ceramic membrane with pore size of 0.05μm and transmembrane pressure of 2 bar. Under optimum conditions of coagulation and flocculation for the biopolymer chitosan, removal of color, turbidity, compounds that absorb UV at 254nm, nitrite, phosphate, total coliforms, and *Escherichia coli* close to 100% and TOC removal of 75% were achieved. Thus, according to Konradt-Moraes (2004), there is great potential in the use of the combined coagulation/flocculation/membrane separation process. However, few research papers have been published to date, and therefore a wide range of study possibilities is open, that can lead to a deeper knowledge of the variables involved in the process as a whole, which in turn allows not only process scale-up, but also its transfer to the companies responsible for public drinking water supply.

2. Experimental results – Case studies

The case studies presented below used the processes of coagulation/flocculation with natural coagulants and membrane filtration for removing color, turbidity, *Giardia* and *Cryptosporidium*, to obtain drinking water for human consumption.

The first case describes the utilization of chitosan as natural coagulant and ceramic ultrafiltration membranes in pilot scale. The second case deals with moringa and polymeric microfiltration membranes in a bench-scale filtration module. Both cases used surface water from the Pirapó River, which serves a population of over 300,000 inhabitants in the city of Maringá, Brazil.

2.1 Process of coagulation/flocculation with chitosan followed by ultrafiltration for surface water treatment

A pilot plant, shown in Figure 1, has been used by the research group headed by the researcher Professor Rosângela Bergamasco, in the Environmental Preservation and Control Laboratory at the State University of Maringá, Brazil. This unit has been the basis for studies of water purification processes that are subsequent to coagulation/flocculation using natural coagulants such as chitosan and *Moringa oleifera*. These studies, based on the treatment of water from the Pirapó River, which is responsible for supplying the city of Maringá - PR, Brazil, have demonstrated the effectiveness of the evaluated coagulants, as well as the applicability of the combined processes, yielding high-quality water within the specifications required by Brazilian law. The surface water characterization is presented in Table 1.

In the study conducted by Bergamasco et al. (2011), the removal of UV-254nm absorbing compounds showed a significant increase when the hybrid process was used, changing from 85.8% with CFQ to 99.4% with CFQ-UF at 2 bar. The results achieved in the filtration with ceramic membranes of stainless steel with Al_2O_3/ZrO_2 (0.1μm) (TAMI, France), at transmembrane pressures of 1 and 2 bar are presented in Table 2. The UV absorbance of organic matter, in the range of 254–280 nm, reflects the presence of unsaturated double bonds and π-π electron interactions such as in aromatic compounds. However, it is known that natural organic matter (NOM) is a mixture of organic compounds called humic materials, but proteins, polysaccharides and other classes of biopolymers also contribute to NOM. This indicates that besides the compounds detected by UV-254nm absorption, other organic compounds may also be present in surface water, and therefore this parameter is not a suitable indicator of NOM removal. Other parameters should be considered for a better understanding of the process.

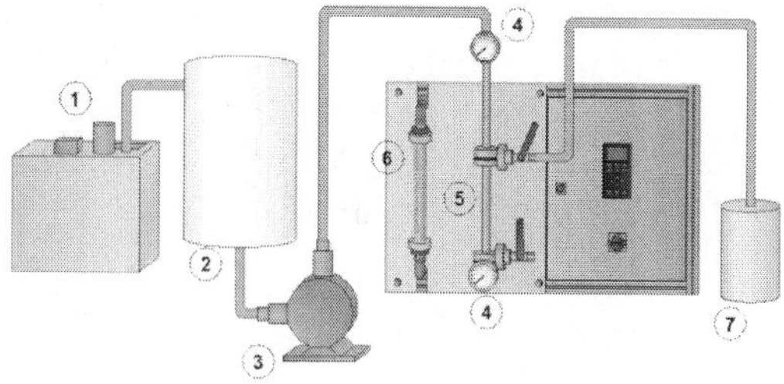

Fig. 1. Schematic diagram of the micro/ultrafiltration experimental unit. 1- thermostatic bath; 2- feed tank; 3- pump; 4- manometers; 5- membrane filtration module; 6- flowmeter (rotameter); 7- permeate.

Water quality parameter	Values
Apparent color (Hu)(1)	1695
True color (Hu)(1)	1045
Turbidity (NTU)(2)	240
Chemical oxigen demand (COD) (mgO_2/L)	19.3
Total organic carbon (TOC) (mgC/L)	6.4
UV-254nm absorbing compounds (UV-254nm) (cm^{-1})	0.923
pH	8.17
Total suspended solids (TSS) (mg/L)	1332
Total dissolved solids (TDS) (mg/L)	228
Total coliforms (CFU/100 mL)(3)	3955
Escherichia coli (CFU/100 mL)(3)	800

1. Hu = $mg_{Pt\text{-}Co}/L$
2. NTU = Nefelometric turbidity unit
3. CFU/100 mL = Colony forming units per 100 mL of water sample

Table 1. Characterization of the surface water from Pirapó River

Filtration tests were performed to characterize the flow of pure water through the membranes, using deionized water. Flux was calculated using Equation 1, where $f_{permeate}$ is the permeate flux, m is the mass of collected water, ρ25°C is water density at 25°C, Δt is the time interval during which water was collected, and Am is the filtering area of the membrane.

$$f_{permeate} = \frac{m}{\rho 25°C \times \Delta t \times Am} \tag{1}$$

The removal efficiency for each parameter analyzed in the different treatment processes was calculated from Equation 2, where Ci and Cf are the initial and final concentrations, respectively, for each parameter.

$$\%\ Removal\ efficiency = \left(\frac{Ci - Cf}{Ci}\right) \times 100 \tag{2}$$

Deionized water (DW) fluxes were determined before each experiment (Ji) and after the filtration of solutions SW, CFQ and CFS (Jf) to determine the fouling of the membrane. The percentage of fouling (%F) was calculated according to Equation 3, proposed by Balakrishnan et al. (2001), using the steady-state flux values, which assume that the flux tends to constant values. The percentage of fouling represents a decrease in the deionized water flux after tests with contaminated water.

$$\%F = \frac{(Ji - Jf)}{Ji} \times 100 \tag{3}$$

In Equation 3, %F is the percentage of membrane fouling, Ji is the initial water flux obtained in the first filtration with deionized water and Jf is the final water flux obtained by filtration of deionized water after the filtration of surface water.

The parameters apparent color, turbidity, and pH were evaluated according to the Standard Methods (American Public Health Association [APHA], 1995). Turbidity measurements were conducted using a turbidimeter (HACH, 2100P). A digital pH meter (Digimed DM-2) was used for pH measurements. Color measurements were conducted using HACH DR/2010 spectrophotometer - Method 8025. COD values were determined using HACH DR/2010 - Method 10129. TOC was determined using an Aurora 1030C TOC Analyzer with 1088 Rotary TOC Autosampler. Absorbance measurements at 254nm were performed using a Logen Scientific UV-Vis spectrophotometer. UV absorbance at 254nm was also used in this study as an indication of the removal of organics from water. UV absorbance is commonly used as an index of the aromatic level (Kim & Yu, 2005).

Comparing CFQ and CFS processes, the main differences in removal efficiency are observed with respect to COD removal (59.9% using CFQ and 38.1% using CFS) and TDS removal (41.9% for CFQ and 7.5% for CFS). It is also observed in Table 2 that the process of coagulation/flocculation with the natural coagulant chitosan was very effective in removing compounds responsible for color and turbidity, as well as UV-254nm absorbing compounds. Similar results were observed by other authors such as Roussy et al. (2005) and Rizzo et al. (2008).

The working conditions for the experiments presented in Table 2 were as follows: The pH was adjusted to 5.0 and concentration of natural chitosan coagulant was 1.0 mg/L. When using aluminum sulfate as coagulant (15 mg/L), pH was maintained at 7.0. In the rapid mixing step the speed was kept at 120rpm for 2.5 min, whereas the speed used in the slow mixing step was 20rpm for 20 min (Konradt et al., 2008). Temperature was maintained at 25.0±2°C during coagulation/flocculation. All experiments were performed at least in triplicate.

Parameter	CFQ (%)	CFQ - UF (%)		CFS (%)	CFS - UF (%)	
		P=1 bar	P=2 bar		P=1 bar	P=2 bar
Apparent color	98.1	99.4	99.1	99.8	99.2	100
True color	97.1	99.2	99.1	99.3	99.1	99.8
Turbidity	99.3	99.9	99.9	99.2	99.8	99.8
UV-254nm	85.8	91.8	99.4	83.4	96.3	88.5
COD	59.9	90.9	97.4	38.1	89.3	85.1
Total coliforms	62.5*	99.0	99.0	99.0	99.0	99.0
Escherichia coli	99.0	99.0	99.0	99.0	99.0	99.0
TSS	60.3	97.8	88.8	94.8	93.5	97.9
TDS	41.9	40.2	86.1	7.52	49.3	42.9

*reduction of 62.5% in the number of CFU/100mL initially present in the surface water. Source: Bergamasco et al., 2011.

Table 2. Removal efficiency (%) of coagulation/flocculation and coagulation/flocculation-ultrafiltration processes at 1 bar and 2 bar, using chitosan (CFQ) or aluminum sulfate (CFS) for the coagulation of surface water

Coagulation pretreatment allows a higher rejection of organics by microfiltration (MF) and UF and the cut-off criterion due to initial membrane pore size is no longer valid (Schafer et al., 2001). The most consistent system was CFQ-UF at the pressure of 1 bar, achieving removal efficiencies above 90% for all parameters assessed, except for TDS, which had 40.2% removal. But if the use of CFS-UF is considered, TDS removal reached a maximum of 49.3%. Another relevant parameter to be analyzed is COD, whose removal was higher when using CFQ-UF at 2 bar (97.4%) than when using CFS-UF at the same pressure (85.1%). An important point to be evaluated is the type of membrane used for filtration, because depending on the material the membrane is made of, fouling effects may be more or less pronounced, which will result in differences in permeate flux and percent removal of the assessed parameters.

The hybrid process of coagulation with chitosan followed by filtration in 0.1 µm pore size polysulfone membrane, resulted more effective for UV-254nm, TOC, and iron removal. Adding chitosan as coagulant, 70% of UV-254nm absorbing compounds and 47% of TOC (averaged values) were removed (Bergamasco et al., 2009). In this study the authors evaluated water from the Saint-Charles River in Quebec, Canada, with an initial UV-254nm absorbance value of 0.083-0.091.

One can see that the ceramic membranes can be more effective for water treatment than polysulfone membranes. Comparing the results with two membranes of different materials and same porosity, ceramic membranes led to a reduction of 91.8% in UV-254nm absorbance (Bergamasco et al., 2011), while the reduction obtained with polysulfone membranes was 70% (Bergamasco et al., 2009). Chitosan improves the sedimentation step, as the flocs are more compact, which is favorable when hybrid coagulation/flocculation/filtration systems are operated. Another fact that should be taken into consideration when applying UF for water treatment is the occurrence of membrane clogging, which causes a drop in permeate flux and is a result of a set of phenomena related to the solution nature and to the characteristics of the membrane (Bergamasco et al., 2011). This fact can be explained by two mechanisms, commonly attributed to the removal of organic matter by UF: sieve retention and adsorption sequestration. In sieve retention the UF membrane acts as a barrier for particle penetration. The particles are retained on the membrane surface and form a cake that grows in thickness as the filtration progresses. The second mechanism involves the entry and capture of the particles into the membrane matrix (Guo et al., 2009).

Bergamasco et al. (2011), using ceramic membrane in the combined process of coagulation/flocculation with chitosan and ultrafiltration, obtained higher permeate fluxes than with UF of surface water (SW) and CFS-UF under the same pressures. For the pressure of 2 bar the permeate flux of the CFQ-UF process was approximately twice that of the CFS-UF process, thus justifying the use of chitosan as a coagulant prior to the ultrafiltration step for surface water treatment. The authors presented the results obtained by the resistance-in-series model for the different types of resistance observed in the UF step using SW, CFQ, and CFS at pressures of 1 bar and 2 bar, as shown in Table 3. The coagulation/flocculation conditions to obtain water for use in the ultrafiltration experiments were the same as for the experiments presented in Table 2.

	Resistance x10¹¹ (m^{-1})*		
	R_f	R_{cp}	R_t
SW			
ΔP=1 bar	1.22	6.79	9.62
ΔP=2 bar	8.11	9.58	20.76
CFQ			
ΔP=1 bar	4.63	0.82	7,06
ΔP=2 bar	9.92	0.12	13.11
CFS			
ΔP=1 bar	0.42	5.78	8.11
ΔP=2 bar	5.89	12.89	21.86

*Rf: fouling resistance; Rcp: concentration polarization resistance; Rt: total resistance.
Source: Bergamasco et al. (2011).

Table 3. Resistances on the membrane during the ultrafiltration process with SW, CFQ, and CFS at 1 bar and 2 bar, using Al2O3/ZrO2 (0.1 μm) ceramic membranes and filtration time of 200 min.

It was observed by means of Table 3, that for the same type of water (without treatment, coagulated with chitosan, or coagulated with aluminum sulfate) the fouling resistance (R_f) due to solute adsorption into membrane pores and walls (Chang et al., 2001) increased with increasing transmembrane pressure, and this can be explained by the higher compression.

This type of resistance can be eliminated only by washing the membrane. It was also observed that R_f was greater for CFQ than for SW and CFS, but concentration polarization resistance (R_{cp}) and R_t were lower for CFQ than for SW and CFS at 1 bar and 2 bar. The floc cake resistance is lower than the resistance due to the unsettled floc and the uncoagulated organics, as reported by Guigui et al. (2002). The use of chitosan as a coagulant can lead to the formation of denser flocs. Thus, its negative impact on the filtration can be explained. The cake is formed by large aggregates, decreasing the average flux through and among these aggregates. However, the performance evaluation of the hybrid systems (CFS-UF and CFQ-UF) showed that the permeate quality was increased when compared with individually operated systems (CFS and CFQ). This is justified by the excellent ability of the UF process to remove particles and colloids.

2.2 Process of coagulation/flocculation with moringa followed by microfiltration for surface water treatment

The other study on water purification processes performed in the Environmental Preservation and Control Laboratory at the State University of Maringá, Brazil, evaluated the coagulation/flocculation process using the natural coagulant *Moringa oleifera,* followed by microfiltration (Nishi, 2011). As mentioned previously, surface water from the Pirapó River was used for this study. Samples of high and low turbidity were mixed to obtain water with different initial turbidity values. The samples used in this study had initial turbidity of 50, 150, 250, 350, and 450 NTU. The prepared samples were artificially contaminated with 10^6 cysts/L of *Giardia* spp. and 10^6 oocysts/L of *Cryptosporidium* spp. obtained from the positive control (suspension of cysts and oocysts) present in the commercial kit Merifluor (Meridian Bioscience, Cincinnati, OH, USA). After being prepared, the samples were subjected to the processes of coagulation/ flocculation with moringa seeds (CFM), microfiltration (MF), and the combined coagulation/flocculation with moringa seeds followed by microfiltration (CFM-MF).

Moringa coagulant solution was prepared and used the same day. Mature moringa seeds from the Federal University of Sergipe (UFS) were used as raw material. The seeds were manually removed from the dry pods and peeled. To prepare the 1% stock solution of moringa (concentration of 10,000 mg/L), 1 g of peeled seeds was crushed and added to 100 mL of distilled water. Subsequently, the solution was stirred for 30 min and vacuum filtered (Cardoso et al., 2008; Madrona et al., 2010). From the 1% stock solution, moringa solutions were prepared with different concentrations: 25, 50, 75, 100, 125, 150, 175, 200, 225, 250, 275, and 300 mg/L. CFM tests were performed on a simple jartest, under the following conditions: rapid mixing speed (RMS) of 100 rpm, coagulation time (CT) of 3 min, slow mixing speed (SMS) of 15 rpm, flocculation time (FT) of 15 min, and settling time (ST) of 60 min (Madrona et al. 2010).

To determine the coagulant concentrations which resulted in the highest removal of the evaluated parameters, 5x12 factorial experiment was applied. For the factor "A" five different levels of initial water turbidity were tested: 50, 150, 250, 350, and 450 NTU. Factor "B" consisted of twelve levels of concentration of moringa coagulant. The parameters analyzed in the experiments - color, turbidity, pH, and removal of *Giardia* and *Cryptosporidium* - were evaluated in triplicate for each combination of the factors "A" and "B". The results were analyzed by ANOVA using the F test and phase contrast (Nkurunziza et al., 2009) to obtain the optimum concentration of each coagulant for each water sample with initial turbidity of 50 to 450 NTU, to be later used in the combined process of coagulation/flocculation/membrane filtration. The software Statistica, version 8.0/2010, was used for the statistical analysis, and p values of less than 0.05 were considered significant.

In this study, the same methodologies described in section 2.1 were used to evaluate the removal efficiency of turbidity and color and the pH of treated water. The concentration of (oo)cysts of *Giardia* and *Cryptosporidium* was assessed by the membrane filtration technique with mechanical extraction and elution (Aldom & Chagla, 1995; Dawson et al., 1993; Franco et al., 2001).

The initial characteristics of water samples used in the study are presented in Table 4.

Turbidity (NTU)	Color (uH)	pH
50	350	7.80
150	902	7.81
250	1000	7.50
350	1849	7.64
450	1885	7.70

Source: Nishi, 2011

Table 4. Water sample parameters before treatment processes.

The results for the removal efficiency of turbidity, color, *Giardia* and *Cryptosporidium*, and pH values for water after treatment with moringa, under the aforementioned conditions, are presented in Table 5. Using moringa as coagulant agent, turbidity removal ranged from 3 to 97.4%. The lowest removal efficiencies, between 3 and 45.6%, were observed for water with low initial turbidity (50 NTU). Removals above 70% were observed for the samples with

		Moringa concentration in the water samples (mg/L)											
Initial turbidity (NTU)	Removal efficiency (%)	25	50	75	100	125	150	175	200	225	250	275	300
50	Turbidity	27.7	23.0	20.7	33.4	33.2	35.6	44.4	45.6	41.6	27.6	10.0	3.0
	Color	0.35	0.11	1.65	3.07	4.00	6.60	22.8	27.6	30.0	26.0	15.4	16.7
	Giardia	6.00	42.3	38.4	69.2	84.6	76.9	82.0	80.0	80.0	76.9	85.0	69.0
	Cryptosporidium	76.0	91.0	90.0	98.0	93.0	91.0	98.0	86.0	91.0	88.0	89.0	83.0
	pH	7.90	8.20	8.10	8.20	8.20	8.10	8.20	7.90	8.00	8.10	8.00	7.90
150	Turbidity	42.0	52.4	69.8	71.0	74.0	75.8	67.7	65.3	69.0	73.6	76.0	72.0
	Color	10.0	47.5	67.0	68.8	70.8	73.5	73.5	71.4	65.0	63.4	66.4	61.6
	Giardia	74.0	97.0	85.0	98.0	94.0	98.0	98.0	98.0	97.0	97.0	91.0	82.0
	Cryptosporidium	42.0	50.0	77.0	81.0	81.0	92.0	92.0	92.0	85.0	92.0	81.0	58.0
	pH	7.93	7.96	7.85	7.74	7.8	7.76	7.70	7.74	7.81	7.74	7.72	7.68
250	Turbidity	68.9	74.8	80.6	93.4	90.1	94.4	93.9	94.2	90.9	94.6	91.8	92.7
	Color	21.8	46.5	46.2	68.4	64.4	80.7	79.0	81.3	78.0	77.8	88.4	81.6
	Giardia	80.0	65.0	65.0	80.0	80.0	95.0	95.0	95.0	95.0	90.0	92.5	90.0
	Cryptosporidium	67.0	61.0	75.0	74,0	86.0	96.0	95.0	92.0	94.0	90.0	87.0	78.0
	pH	7.60	7.70	7.80	7.80	7.60	7.60	7.70	7.80	7.80	7.80	7.70	7.80
350	Turbidity	49.4	62.8	70.8	75.0	82.0	90.0	93.7	95.0	96.0	95.8	96.4	92.5
	Color	47.0	82.3	76.4	94.0	94.0	88.2	97.0	97.0	94.0	88.2	94.0	94.0
	Giardia	47.0	82.3	76.4	94.0	94.0	88.2	97.0	97.0	94.0	88.2	94.0	94.0
	Cryptosporidium	22.0	81.0	68.0	95.0	86.0	81.0	97.0	96.0	92.0	86.0	96.0	92.0
	pH	7.78	7.87	7.75	7.77	7.81	7.74	7.82	7.75	7.73	7.76	7.73	7.78
450	Turbidity	63.0	61.0	75.0	79.9	92.0	94.0	97.2	97.4	97.2	97.0	96.7	94.7
	Color	39.0	47.8	61.6	68.5	88.3	91.5	96.1	96.1	96.4	96.0	95.6	92.8
	Giardia	63.0	92.0	96.0	96.0	92.0	96.0	90.0	97.0	94.0	97.0	93.0	89.2
	Cryptosporidium	45.0	51.0	94.0	85.0	80.0	94.0	86.0	93.0	95.0	98.0	90.0	76.0
	pH	7.80	7.70	7.60	7.60	7.60	7.60	7.00	7.60	7.60	7.60	7.50	7.60

Source: Nishi, 2011.

Table 5. Percentage of removal efficiency of turbidity, color, *Giardia*, and *Cryptosporidium*, and pH values after the process of coagulation/flocculation with moringa.

turbidity of 250, 350, and 450 NTU (Nishi, 2011). The decrease in efficiency of turbidity removal from water with 50 NTU of initial turbidity, after the addition of moringa, can be explained by the increased organic load. This is justifiable as long as moringa is an oilseed which is rich in organic substances such as oil, protein, fat, and vitamins. This increase in turbidity and color in water treated with moringa was also observed in other studies, especially when the water had relatively low initial turbidity and color (Ramos, 2005).

Nkurunziza et al. (2009), using a 3% solution of moringa seeds, prepared with saline water, to treat water from the rivers in the province of Rwanda, obtained removal efficiency of 83.2% in samples with turbidity of 50 NTU and higher values (99.8%) in water with turbidity of 450 NTU. The optimum concentrations found in this study were 150 mg/L for 50 NTU and 125 mg/L for other turbidity levels tested by the researchers. The results of turbidity removal from water with low initial turbidity (50 NTU) were higher than those obtained in the present case study (45.6%) and for water with high initial turbidity (450 NTU) the results were similar (97.4%). The differences may be due to the different preparation procedures of the moringa solution, by aqueous or saline extraction, as well as the different concentrations of the stock solution of moringa. In both studies, the coagulant properties of the moringa appear to be more efficient in water of high initial turbidity, in agreement with other literature reports (Ndabigengesere et al., 1995; Madrona et al., 2010).

Ndabigengesere et al. (1995), applying an aqueous solution of 5% moringa seeds to synthetic turbid water (kaolin added to tap water) with initial turbidity of 426 NTU, obtained removals from 80 to 90% and reached the optimum concentration of 500 mg/L of coagulant solution. This concentration is higher than the optimum concentration for water of 450 NTU obtained in this case study, which was 275 mg/L. This difference between the optimum concentrations of the moringa solution may be due to the different water source: Ndabigengesere et al. (1995) used synthetic turbid water prepared with kaolin, and the present study used surface water. The different efficiencies of turbidity removal and optimum concentrations can be explained by the different compositions of water samples used in the studies (raw water, synthetic turbid water), that is, the substances present in water can influence the action of the coagulant agent and the formation of flocs, as well as by the preparation procedure of the moringa solution (aqueous or saline extraction), evaluated concentrations, and seed quality, among other factors.

Regarding color, the removal ranged from 0.11 to 30% for water with initial turbidity of 50 NTU. The highest removals for this sample were obtained with moringa concentration ranging from 175 to 250 mg/L. For water with higher initial turbidity (150 to 450 NTU), the removal efficiency ranged from 10 to 97%, the highest values being obtained for concentrations of 150 mg/L or higher (Nishi, 2011). It is observed that color removal by moringa is similar to its behavior with respect to turbidity: the lowest values of this parameter are obtained for water with high initial turbidity, which agrees with literature data (Cardoso et al., 2008; Nkurunziza et al., 2009; Madrona et al., 2010).

Concerning the pH of water samples after the coagulation process with different concentrations of moringa, it was observed that the average pH was 7.6, with variation of approximately 10% (Nishi, 2011). There was little variation among the samples regardless of the amount of moringa solution added, which consists of one of the benefits of moringa as a coagulant agent, that is, its addition does not significantly alter the pH of the water (Ndabigengesere et al., 1995; Nkurunziza et al., 2009), unlike the treatment with aluminum

The MF membrane was composed of hollow fibers made of polyimide, with porosity of 0.40 μm. The operating pressure was 1.0 bar. To maintain uniformity in the experiments, the initial volume was fixed in 5 L, and the test time was 60 min.

In this study, the same methodologies described in section 2.1 were used to evaluate membrane flux and fouling, as well as to analyze the removal efficiency of turbidity, color, *Giardia* and *Cryptosporidium*, and the pH of treated water.

The results obtained in the processes of microfiltration (MF) and coagulation/flocculation with moringa followed by microfiltration (CFM-MF) are presented below. These results are presented together to show if the pretreatment (coagulation/flocculation with moringa) had differences in relation to the MF process without pretreatment. The removal efficiencies and the pH of the water treated by the MF and CFM-MF processes are presented in Table 7.

		Initial turbidity (NTU)			
Treatment process	Removal efficiency (%)	150	250	350	450
MF	Turbidity	81.09	84.16	76.82	76.33
	Color	78.28	83.45	74.27	72.56
	Giardia	ND	ND	ND	ND
	Cryptosporidium	ND	ND	ND	ND
	pH	7.38	7.85	7.36	7.81
CFM-MF	Turbidity	93.54	92.28	84.78	99.39
	Color	96.15	92.19	88.96	100.0
	Giardia	ND	ND	ND	ND
	Cryptosporidium	ND	ND	ND	ND
	pH	7.33	7.72	7.34	7.51

ND - not detected. Source: Nishi, 2011

Table 7. Removal efficiencies of turbidity, color, *Giardia* and *Cryptosporidium*, and pH values of the water treated by the MF and CFM-MF processes.

It can be observed that the largest color and turbidity removals occurred with the combined CFM-MF process, compared with the MF process without pretreatment. There were no changes in the pH of the treated water. It is clear that the use of coagulation/flocculation with moringa prior to microfiltration improves the quality of treated water (Nishi, 2011).

A few studies were found in the literature regarding the CF/MF process using moringa as a coagulant for surface water treatment. Madrona (2010) evaluated the combined process of coagulation/flocculation with moringa and MF with ceramic membranes, and obtained 97 to 100% removal of turbidity and color in the treatment of surface water from the Pirapó River, in Maringá, Paraná. These results were similar to those obtained in the present study, which used a polymer membrane for the MF process. Parker et al. (1999), using hollow fiber MF membranes with 0.2 μm pores for the treatment of water that had been previously treated in settling tanks, obtained water with turbidity below 0.1 NTU, with average removal of 99.46%, similar to those obtained in this study.

Neither in the microfiltration (MF) process alone, nor in the combined (CFM-MF) processes, (oo)cysts of *Giardia* and *Cryptosporidium* were detected in the filtered water, being below the detection limit (<1 cyst or oocyst/L) (approximately 6 log removal), in agreement with literature data. Jacangelo et al. (1995), studying the application of three MF membranes with pore sizes between 0.08 and 0.22 μm for the treatment of water contaminated with *Giardia* and *Cryptosporidium,* found that the protozoa concentration was below detectable levels in the filtered water (<1 cyst or oocyst/L) from two of the membranes (corresponding to log removal> 4.7 to> 7.0 for *Giardia* and > 4.4 to> 6.9 for *Cryptosporidium*). They also concluded that the level of removal depends on the concentration of protozoa in the water to be treated and on membrane integrity. In another study, MF membranes with average pore size of 0.2 μm resulted in significant removal of particles that were the same size as *Giardia* cysts (5-15 μm). Log removal was, on average, 3.3 to 4.4. The removal of particles that were the same size as *Cryptosporidium* oocysts (2-5 μm) was lower, 2.3 to 3.5 log removal. These removals were obtained according to the concentration of (oo)cysts used for artificial contamination of water and proved to be independent of the membrane flux (114-170 L/hm^2) (Karimi et al., 1999).

Thus, one can say that MF may act as a barrier against protozoan (oo)cysts. The coagulation/flocculation with moringa associated with microfiltration resulted in high levels of removal of the evaluated parameters.

Figure 3 shows the permeate flux versus time for the microfiltration of deionized water (DW), raw water without coagulant (SW), and pretreated water (CFM).

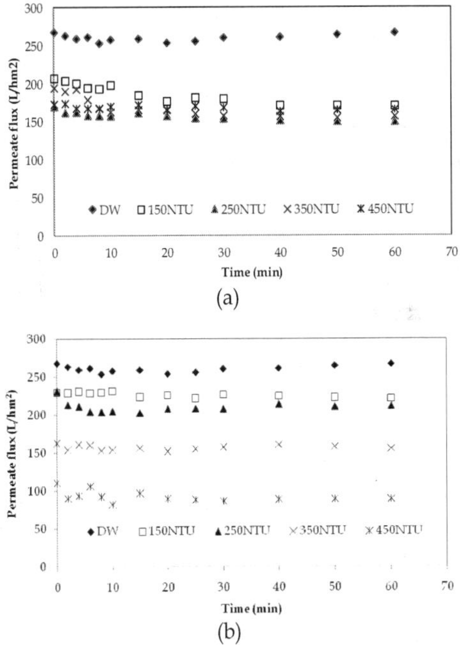

Source: Nishi, 2011.

Fig. 3. Permeate flux with deionized water (DW) and raw water with initial turbidity from 150 to 450 NTU in the MF (a) and CFM-MF (b) processes.

For the MF process with raw water, that is, without previous treatment (coagulation/ flocculation), permeate flux ranged from 157 to 187 L/hm² for water samples of turbidity from 150 to 450 NTU. In the combined process (CFM-MF), permeate flux ranged from 157 to 226 L/hm² for water samples with initial turbidity of 150 to 350 NTU. Samples of 450 NTU presented the lowest permeate flux, 91 L/hm², on average (Nishi, 2011). This may be due to the presence of a greater number of particles that can cause the process of concentration polarization and due to superposition of various fouling mechanisms in the membrane, which may cause the decrease of the permeate flux (Stopka et al., 2001).

The combined processes of coagulation/flocculation/microfiltration showed slightly higher fluxes when compared with the microfiltration process alone. The improvement in permeate flux using coagulation/flocculation prior to microfiltration was also observed in other studies (Katayon et al., 2007; Horčičková et al., 2009).

The percentage of fouling (%F) for the MF process with raw water (SW) and water coagulated/flocculated with moringa (CFM) with initial turbidity from 150 to 450 NTU is shown in Figure 4.

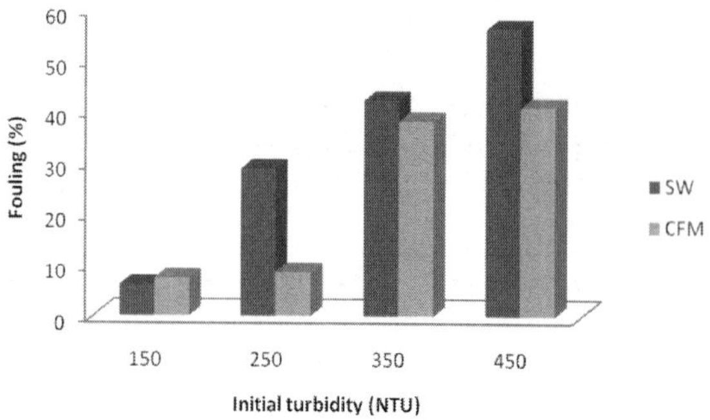

Fig. 4. Percentage of fouling for the MF process with raw water (SW) and water coagulated/flocculated with moringa (CFM) with initial turbidity from 150 to 450 NTU.

It is observed that the MF process with raw water showed higher percentages of fouling, ranging from 6.13 to 56.32% when compared with the combined process of coagulation/ flocculation with moringa followed by MF, which presented percentages of fouling from 7.48 to 40.9% (Nishi , 2011). This reduction in membrane fouling when using the process of coagulation/flocculation as pretreatment was also observed in other studies. Madrona (2010) used coagulation/ flocculation with moringa, followed by MF with ceramic membranes with porosity of 0.1 and 0.2 μm, for the treatment of surface water and observed fouling percentages of around 94% during the filtration of raw water and slightly lower values, around 88%, when water previously coagulated/flocculated with moringa was filtered. Carroll et al. (2000) used polypropylene hollow fiber MF membrane to filter surface water from the Moorabool River, Australia, and observed fouling percentages of 80% for water without pretreatment and 50% for water pretreated by coagulation with alum.

According to Cheryan (1998), the type and extent of fouling depend on the chemical nature of the membrane, the solute, and the solute-membrane interactions, as well as on the porosity of the membrane and the working pressure used in the process.

3. Conclusions

Performance evaluation of the hybrid systems (CFS-UF, CFQ-UF, and CFM-MF) showed that the permeate quality was increased when compared with individually operated systems. This is justified by the excellent ability of the MF/UF process to remove particles and colloids. The results also indicate that when applying CF-MF/UF at optimum conditions, a hygienic barrier effect was achieved for the treatment scheme, in which nearly 100% removal of total coliforms, *E. coli, Giardia* and *Cryptosporidium* was obtained at the end of the process. In addition, the combined processes CFQ-UF, CFM-MF, and CFS-UF produced drinking water in accordance with the legislation.

Given the above considerations, one can say that chitosan and *Moringa oleifera* have a potential application as natural coagulants in CF-MF/UF hybrid processes for treating drinking water with relatively high turbidity. This process can be used reliably to produce drinking water of excellent quality.

4. List of abbreviations

Coagulation/flocculation = CF
Coagulation/flocculation using chitosan as coagulant = CFQ
Coagulation/flocculation using aluminum sulfate as coagulant = CFS
Coagulation/flocculation using chitosan as coagulant followed by ultrafiltration = CFQ-UF
Coagulation/flocculation using aluminum sulfate as coagulant followed by ultrafiltration = CFS-UF
Coagulation/flocculation using moringa as coagulant = CFM
Coagulation/flocculation using moringa as coagulant followed by microfiltration = CFM-MF
Microfiltration = MF
Ultrafiltration = UF
Moringa oleifera = moringa
Natural organic matter = NOM

5. References

Akhtar, M., Moosa Hasany, S., Bhanger, M. I. & Iqbal, S. (2007). Sorption potential of *Moringa oleifera* pods for the removal of organic pollutants from aqueous solutions. *Journal of Hazardous Materials*, Vol. 141, No. 3, pp. 546–556.

Aldom, J.E. & Chagla, A.H. (1995). Recovery of *Cryptosporidium* oocysts from water by a membrane filter dissolution method. *Letters in Applied Microbiology*, Vol. 20, pp. 186-187.

American Public Health Association [APHA]. [1995]. *Standard Methods for the Examination for Water and Wastewater* (19th edition). Byrd Prepess Springfield, ISBN 0875532233, Washington.

Balakrishnan, M., Dua, M. & Khairnar, P.N. (2001). Significance of membrane type and feed stream in the ultrafiltration of sugarcane juice. *Journal of Separation Science and Technology*, Vol. 36, No. 4, pp. 619-637.

Bergamasco, R., Bouchard, C., da Silva, F.V., Reis, M.H.M., Fagundes-Klen, M.R. (2009). An application of chitosan as a coagulant/flocculant in a microfiltration process of natural water. *Desalination*, Vol. 245, pp. 205–213.

Bergamasco, R., Konradt-Moraes, L.C., Vieira, M.F., Fagundes-Klen, M.R. & Vieira, A.M.S. (2011). Performance of a coagulation–ultrafiltration hybrid process for water supply treatment. *Chemical Engineering Journal*, Vol. 166, pp. 483–489.

Bhatia, S., Othman, Z. & Ahmad, A.L. (2007). Pretreatment of palm oil mill effluent (POME) using *Moringa oleifera* seeds as natural coagulant. *Journal of Hazardous Materials*, Vol. 145, No. 1-2, pp. 120-126.

Bottino, A., Capannelli, C., Del Borghi, A., Colombino, M. & Conio, O. (2001). Water treatment for drinking purpose: ceramic microfiltration application. *Desalination*, Vol. 141, pp. 75–79.

Bouchard, C., Laflamme, E., Serodes, J., Ellis, D., Rahni, M. & Rodrigues, M. (2003). Étude en la laboratorie de I'ultrafiltration et de la coagulation-ultrafiltration d'une eau colorée. *Proceedings of 17éme Symposium de I'Est du Canada sur la Recherche Portant sur la Pollution de I'eau.*

Brown, T.J. & Emelko, M.B. (2009). Chitosan and metal salt coagulant impacts on *Cryptosporidium* and microsphere removal by filtration. *Water Research*, Vol. 43, No. 331–338.

Bustamante, H.A., Shanker, S.R., Pashley, R.M. & Karaman, M.E. (2001). Interaction between *Cryptosporidium* oocysts and water treatment coagulants. Water *Research*, Vol. 35, pp. 3179-3189.

Cardoso, K.C., Bergamasco, R., Cossich, E.S. & Konradt-Moraes, L.C. (2008). Otimização dos tempos de mistura e decantação no processo de coagulação/floculação da água bruta por meio da *Moringa oleifera* Lam. *Acta Scientiarum - Technology*, Vol. 30, pp. 193-198.

Carroll, T., King, S., Gray, S. R., Bolto, B. A. & Booker, N. A. (2000). The fouling of microfiltration membranes by nom after coagulation treatment. *Water Research*, Vol. 34, No. 11, pp. 2861 - 2868.

Centers for Disease Control and Prevention [CDC]. (2006). *Surveillance Summaries*, December 22. MMWR, 55 (No. SS-12).

Chang, I.S., Bag, S.O. & Lee, C.H. (2001). Effects of membrane fouling on solute rejection during membrane filtration of activated sludge. *Process Biochemistry*, Vol. 36, pp. 855–860.

Cheryan, M. (1998). *Ultrafiltration and microfiltration handbook*. Technomic Publishing CO, Illinois, Lancaster, USA.

Chuang, P.H., Lee, C.W., Chou, J.Y., Murugan, M., Shieh, B.J. & Chen, H.M. (2007). Anti-fungal activity of crude extracts and essential oil of *Moringa oleifera* Lam. *Bioresource Technology*, Vol. 98, pp. 232–236.

Chung, Y.C., Wang, H.L., Chen, Y.M. & Li, S.L. (2003). Effect of abiotic factors on the antibacterial activity of chitosan against waterborne pathogens. *Bioresource Tech.*, Vol. 88, pp. 179-184.

Coelho, J.S., Santos, N.D.L., Napoleão, T.H., Gomes, F.S., Ferreira, R.S., Zingali, R.B., Coelho, L.C.B.B., Leite, S.P., Navarro, D.M.A.F. & Paiva P.M.G. (2009). Effect of ***Moringa*** *oleifera* lectin on development and mortality of *Aedes aegypti* larvae. *Chemosphere*, Vol. 77, No. 7, pp. 934-938.

Davino, F. (1976). *Tecnologia de tratamento de água: água na indústria*. Almeida Neves, Rio de Janeiro, Brazil.

Dawson, D.J., Maddocks, M., Roberts, J. & Vidler, J.S. (1993). Evaluation of recovery of *Cryptosporidium parvum* oocysts using membrane filtration. *Letters in Applied Microbiology*, Vol. 17, pp. 276-279.

Divakaran, R. & Pillai, V.N.S. (2002). Flocculation of river silt using chitosan. *Water Research*, Vol. 36, No. 9, pp. 2414-2418.

Driscoll, C.T. & Letterman, R.D. (1995). Factors regulating residual aluminium concentrations in treated waters. *Environmetrics*, Vol. 6, pp. 287-309.

Eikebrokk, B. (1999). Coagulation-direct filtration of soft, low alkalinity humic waters. *Water Science and Technology*, Vol. 40, No. 9, pp. 55–62.

Eikebrokk, B. & Saltnes, T. (2001). Removal of natural organic matter (NOM) using different coagulants and lightweight expanded clay aggregate filters. *Water Science & Technology: Water Supply*, Vol. 1, No. 2, pp. 131–140.

Fayer, R. (2004). *Cryptosporidium*: a water-borne zoonotic parasite. *Veterinary Parasitology*, Vol. 126, pp. 37–56.

Franco, R.M.B., Rocha-Eberhardt, R. & Cantusio Neto, R. (2001). Occurrence of *Cryptosporidium* oocysts and *Giardia* cysts in raw water from the Atibaia river, Campinas, Brazil. *Rev. Inst. Med. Trop. S. Paulo*, Vol. 43, No. 2, pp. 109-111.

Ghebremichael, K.A., Gunaratna, K.R., Henriksson, H., Brumer, H. & Dalhammar, G. (2005). A simple purification and activity assay of the coagulant protein from Moringa *oleifera* seed. *Water Research*, Vol. 39, No. 11, pp. 2338-2344.

Guigui, C., Rouch, J.C., Durand-Bourlier, L., Bonnelye, V. & Aptel, P. (2002). Impact of coagulation conditions on the in-line coagulation/uf process for drinking water production. *Desalination*, Vol. 147, pp. 95-100.

Guo, X., Zhang, Z., Fang, L. & Su, L. (2009). Study on ultrafiltration for surface water by a polyvinylchloride hollow fiber membrane. *Desalination*, Vol. 238, pp. 183–191.

Horčičková, J., Mikulášek, P. & Dvořáková, J. (2009). The effect of pre-treatment on crossflow microfiltration of titanium dioxide dispersions. *Desalination*, Vol. 240, pp. 257-261.

Hsu, B.M. & Huang, C. (2002). Influence of ionic strength and pH on hydrophobicity and zeta potential of *Giardia* and *Cryptosporidium*. *Colloids and Surfaces A: Physicochemical and Engineering Aspects*, Vol. 201, pp. 201-206.

Iacovski, R.B., Barardi, C.R.M. & Simões, C.M.O. (2004). Detection and enumeration of *Cryptosporidium* sp. oocysts in sewage sludge samples from the city of Florianópolis (Brazil) by using immunomagnetic separation combined with indirect immunofluorescence assay. *Waste Manage Res.*, Vol. 22, pp. 171–176.

International Water Association [IWA]. (2010). Coagulation and Flocculation in Water and Wastewater Treatment. In: *Water Wiki*, april 28Th 2011, Available from: http://iwawaterwiki.org/xwiki/bin/view/Articles/CoagulationandFlocculationinWaterandWastewaterTreatment.

Jacangelo, J. G., Adham, S. S. & Laîné, J-M. (1995). Mechanism of *Cryptosporidium, Giardia* and MS2 virus removal by MF and UF. *Journal of the American Water Works Association*, Vol. 87, No. 9, pp. 107–121.

Karimi, A.A., Vickers, J.C. & Harasick, R.F. (1999). Microfiltration goes Hollywood: the Los Angeles experience. *Journal of the American Water Works Association*, Vol. 91, No. 6, pp. 90–103.

Katayon, S., Noor, M.J.M.M., Asma, M., Ghani, L.A.A., Thamer, A.M., Azni, I., Ahmad, J., Khor, B.C. & Suleyman, A.M. (2006). Effects of storage conditions of *Moringa oleifera* seeds on its performance in coagulation. *Bioresource Technology*, Vol. 97, No. 13, pp. 1455–1460.

Katayon, S., Noor, M.J.M.M., Tat, W.K, Halim, G.A., Thamer, A.M. & Badronisa, Y. (2007). Effect of natural coagulant application on microfiltration performance in treatment of secondary oxidation pond effluent. *Desalination*, Vol. 204, pp. 204-212.

Kawamura, S. (1991). Effectiveness of natural polyelectrolytes in water treatment. *Journal Awa*, Japan, Vol. 79, No. 6, pp. 88-91.

Kim, M.H. & Yu, M.J. (2005). Characterization of NOM in the Han River and evaluation of treatability using UF - NF membrane, Environmental Research, No. 97, pp. 116 - 123.

Klopotek, A.D., Wlaasiuky, D. & Klopotek, B.B. Compounds based on chitosan as coagulants and flocculants, *Proceedings of International Conference on Chitin and Chitosan*, Gydnia, Poland, August, 1994.

Konradt-Moraes, L.C. (2004). Estudo da coagulação-ultrafiltração para produção de água potável. Master of Science Thesis, Universidade Estadual de Maringá - Maringá, PR, Brasil, 135 pp. (in Portuguese).

Konradt, L.C.; Bergamasco, R.; Tavares, C.R.G.; Bongiovani, M.C. & Hennig, D. (2008). Utilization of the coagulation diagram in the evaluated of the natural organic matter (NOM) removal for obtaining potable water, International Journal of Chemical Reactor Engineering, No. 6, pp. 1 - 6.

Lechevallier, M.W. & Au, K. (2004). *Water treatment and pathogen control: process efficiency in achieving safe drinking water*. WHO Drinking Water Quality Series, UK, 136pp.

Lindquist, A. (1999). *Emerging pathogens of concern in drinking water*. United States Environmental Protection Agency, EPA, 600/R-99/070.

McLachlan, D.R.C. (1995). Aluminum and the risk for Alzheimer's Disease. *Environmetrics*, Vol. 6, pp. 233-275.

Madrona, G.S., Serpelloni, G.B., Vieira, A.M.S., Nishi, L., Cardoso, K.C & Bergamasco, R. (2010). Study of the effect of saline solution on the extraction of the *Moringa oleifera* seed's active component for water treatment. *Water, Air, & Soil Pollution*, Vol. 211, pp. 409–415.

Madrona, G.S. (2010). Extração/purificação do composto ativo da semente da *Moringa oleifera* Lam e sua utilização no tratamento de água para consumo humano. Doctoral Thesis, Universidade Estadual de Maringá - Maringá, PR, Brasil, 176 pp. (in Portuguese).

Muyibi, S.A. & Evison, L.M. (1995). *Moringa oleifera* seeds for softening hardwater. *Water Research*, Vol. 29, No. 4, pp. 1099–1105.

Ndabigengesere A., Narasiah, K.S. & Talbot, B.G. (1995). Active agents and mechanism of coagulation of turbid waters using *Moringa oleifera*. *Water Research*, Vol. 29, No. 2, pp. 703-710.

Nishi, L. (2011). Estudo dos processos de coagulação/floculação seguido de filtração com membranas para remoção de protozoários parasitas e células de cianobactérias. Doctoral Thesis, Universidade Estadual de Maringá - Maringá, PR, Brasil, 203 pp. (in Portuguese).

Nkurunziza, T., Nduwayezu, J.B., Banadda, E.N. & Nhapi, I. (2009). The effect of turbidity levels and *Moringa oleifera* concentration on the effectiveness of coagulation in water treatment. *Water Science and Technology*, Vol. 59, pp. 1551-1558.

Okuda, T., Baes, A.U., Nishijima, W. & Okada, M. (1999). Improvement of extraction method of coagulation active components from *Moringa oleifera* seed. *Water Research*, Vol. 33, No. 15, pp. 3373-3378.

Okuda, T., Baes, A.U., Nishijima, W. & Okada, M. (2001). Isolation and characterization of coagulant extracted from *Moringa oleifera* seed by salt solution. *Water Research*, Vol. 35, No. 2, pp. 405-410.

Parker, D.Y., Leonard, M.J., Barber, P., Bonic, G., Jones W. & Leavell, K.L. Microfiltration treatment of filter backwash recycle water from a drinking water treatment facility. *Proceedings of American Water Works Association Water Quality Technology Conference.* Denver, CO, 1999.

Ramos, R.O. (2005). Clarification of water with low turbulence and moderate color using seeds of *Moringa oleifera*. Doctoral Thesis. State University of Campinas - Campinas,SP, Brazil. 276 pages. (in Portuguese).

Reddy, V.; Urooj, A. & Kumar, A. (2005). Evaluation of antioxidant activity of some plant extracts and their application in biscuits. *Food Chemistry*, Vol. 90, pp. 317-321.

Renault, F., Sancey, B., Badot, P.M. & Crini, G. (2009). Chitosan for coagulation/flocculation processes-an eco-friendly approach. *European Polymer Journal*, Vol. 45, pp. 1337-1348.

Rey, L. (2001). *Parasitologia* (3th edition). Guanabara Koogan, ISBN 8527706776, Rio de Janeiro, Brazil.

Rinaudo, M. (2006). Chitin and chitosan: properties and applications. *Progress in Polymer Science*, Vol. 31, pp. 603-632.

Rizzo, L., Di Gennaro, A., Gallo, M. & Belgiorno, V. (2008). Coagulation/chlorination of surface water: a comparison between chitosan and metal salts. *Separation and Purification Technology*, Vol. 62, pp. 79-85.

Roussy, J., Van Vooren, M., Dempsey, B.A. & Guibal, E. (2005). Influence of chitosan characteristics on the coagulation and the flocculation of bentonite suspensions. *Water Research*, Vol. 39, pp. 3247-3258.

Schafer, A.I., Fane, A.G. & Waite, T.D. (2001). Cost factors and chemical pretreatment effects in the membrane filtration of waters containing natural organic matter,. *Water Research*, Vol. 35, pp. 1509-1517.

Stopka, J., Bugan, S.G., Broussous, L., Schlosser, S. & Larbot, A. (2001). Microfiltration of beer yeast suspensions through stamped ceramic membranes. *Separation and Purification Technology*, Vol. 25, pp. 535-543.

United States Environmental Protection Agency [USEPA]. (1996). *National Primary Drinking Water Regulations: Monitoring Requirements for Public Drinking Water Supplies*, Final Rule, 40CFR Part 141.

Verbych, S., Bryk, M., Alpatova, A. & Chornokur, G. (2005). Ground water treatment by enhanced ultrafiltration. *Desalination*, Vol. 179, No. 1-3, pp. 237-244.

Vieira, A.M.S., Vieira, M.F., Silva, G.F., Araújo, A.A., Fagundes-Klen, M.R., Veit, M.T. & Bergamasco, R. (2010). Use of *Moringa oleifera* Seed as a Natural Adsorbent for Wastewater Treatment. *Water, Air, & Soil Pollution*, Vol. 206, pp. 273-281.

Xagoraraki, I. & Harrington, G.W. (2004). Zeta potential, dissolved organic carbon, and removal of *Cryptosporidium* oocysts by coagulation and sedimentation. *Journal of Environmental Engineering*, Vol. 130, pp. 1424-1432.

Xia, S., Li, X., Zhang, Q., Xu, B. & Li, G. (2007). Ultrafiltration of surface water with coagulation pretreatment by streaming current control. *Desalination*, Vol. 204, pp. 351–358.

15

Elimination of Phenols on a Porous Material

Bachir Meghzili[1,2], Medjram Mohamed Salah[1],
Boussaa Zehou El-Fala Mohamed[1] and Michel Soulard[3]
[1]*Laboratoire LARMACS, Université de Skikda Algérie,*
[2]*Université de Biskra,*
[3]*Equipe Matériaux à Porosité Contrôlée, IS2M, LRC CNRS 7228, UHA, ENSCMu, France*

1. Introduction

The surface water which feeds the majority of the stations of treatment of drinking water is charged by organic composed, including one great part makes up of humic substances. It is thus important to eliminate them to avoid the formation of generally toxic organohalogen compounds (Dore, 1989; Meier, 1988; Boudhar, 1999). The coagulation and the flocculation followed by a clarification remain the most frequent treatment to withdraw colloids present in water, which they are of organic or mineral origin (Bersillon, 1983; Lefebvre and Legube, 1990).

Conventional treatment of the clarification which could indeed eliminate these macromolecules from the humic type is not always sufficient; it often requires optimal conditions not very compatible with the practical conditions of operation and of the treatment as with the criteria of potability. Processes based on adsorption often constitute a technique of choice, complementary to treatment basic. The adsorption is one of the processes of the separation which finds its application in several fields, such as extraction, purification and depollution.

Among the most recent progress in the water treatment, the advanced processes of oxidation (advanced oxidation process AOP) considered to be effective, allow mineralization in aqueous medium of the toxic organic molecules with respect to the man and of the environment. The advanced processes of oxidation are based on the generation and the use of a very strong oxidant which is the radical hydroxyl. This last can be produced by various processes chemical, photochemical biological, electrochemical (Andreozzi and Al, 1999; Chiron and Al, 2000; Galze and Al, 1992; Safarzadeh-Amiri and Al, 1996; Dussert, 1997). These methods rest on the formation of very reactive chemical entities which will biologically break up the most recalcitrant molecules into molecules degradable or in mineral compounds (Golich and Bahnemann, 1997). The reactions generally studied on this level can be classified in three categories (Hoang, 2009):

- Reactions with the reagents électrophiles (O_3, HOCl, ClO_2 and NH_2Cl),
- Reactions with the radicalizing species and initiating reactions of production of radicals (HO and inorganic radicals: CO_3^{-}, SO_4^{-}, Cl_2^{-}, catalyse homogeneous by Fe^{II} and Fe^{III}, radiolysis of water, photocatalysis, catalytic ozonization etc…)
- Reactions of phototransformation (UV and solar) with or without catalyst.

In Algeria, organic material can represent, with it only, a big part of the organic load of surface waters, in particular in the case of water of stopping. The presence of natural organic matter in a surface water east at the origin of many problems encountered during the various stages of treatment of potabilisation. Initially, the natural organic matter is undesirable because it reacts with chlorine during disinfection to form volatile organic compounds (trihalométhanes [THM], acid dichloroacetic [DCAA], etc), produced potentially carcinogenic (Lefebvre and Legube 1990; Hooper and Al, 1996; Stevens and Al, 1976; Najm and Al, 1993).

The natural organic matter is also known for its role in the transport and the trapping of organic and/or inorganic pollutants (Bartschat and Al, 1992; Tippinge, 1993). It represents also a potential substrate for the biological growth in the distribution network of drinking water. The weak dehydration of muds resulting from the treatment of drinking water and the filling of the membranes of filtration are also related to the presence of the natural organic matter in water (Dulin and Knocke 1989; Wiesner and Al, 1989; Bersillon and Al, 1999). Among the organic compounds, the phenols are regarded as harmful pollutants even with weak concentrations because of the potential dangers on health and environment (Dutta and Al, 1992, 1998).

The choice of a adsorbent material depends inter alia its type of porosity, its specific surface and nature of the element to be trapped. Our choice was made on a bentonite of M'zila (Mostaganem), rich in montmorillonite, because of its properties particular to fix many substances, of its availability in Algeria and its low costs (Essington, 1994; Amar and Gaid, 1987; Boufatit and Al, 2007). Indeed, some phyllosilicates have the property to easily adsorb water molecules or organics in interfoliar space. This phenomenon called swelling depends on the load of the layer, the localization of this one and the nature of the cations of compensation (Cailliere and Al, 1982). The bentonite is a material which contains approximately montmorillonite 75% and whose size of the particles is lower than 2 μm (Bergaya and Al, 2006). The argillaceous mineral term or phyllosilicate corresponds to hydrated aluminium silicates, of lamellate structures. These clays generally constitute a considerable fraction of grounds (Auerbach and Al, 2004). The layers of the phyllosilicates are consisted a stacking of octahedral layers (O) and tetrahedral (T). The tetrahedral layer generally consists of atoms of silicon surrounded by four oxygen atoms and bound between them by covalent bonds Si-O. The octahedral layer is formed by hexagonal units, composed of atoms of coordinate magnesium or aluminium with six oxygen atoms or with functions hydroxyls. The layers T and O are bound by covalent bonds and imply apical oxygens. The space located between the two layers is called interfoliar space. A layer and an interfoliar space form a structural unit. The phyllosilicates have the possibility of easily adsorbing water molecules in this interfoliar space.

During a isomorphic substitution of an element by another of lower oxidation step, in tetrahedral or octahedral layer, the deficit of load "+" of the layer is made up by interfoliar cations known as of compensation, exchangeable by mineral or organic cations (Cailliere and Al, 1982; Decarreau, 1990; Bouras, 2003). The most frequent substitution for a montmorillonite is that of Al^{3+} by Mg^{2+} in the octahedral layer. For this clay, the distance between the negative sites located at the level of the octahedral layer and the exchangeable cation located at the surface of the layer are such as the forces d' attraction are weak (Cailliere and Al, 1982). Substitutions of So by Al in the tetrahedral layer are also possible.

The interfoliar cations are in general exchangeable by organic and mineral cations being in solutions put in contact with the phyllosilicate. One then characterizes each phyllosilicate by his Capacity of Cation Exchange (CEC). In the case of montmorillonites, the values of CEC lie between 75 and 160 milliéquivalents for 100 grams d' clay (Viallis-Terrisse, 2000). In tables 1 and 2 we show the characteristics physicochemical of bentonite of M' Zila, Algeria (ENOF 1997).

Surface Spécifique (m^2/g)	Masse spécifique (g/cm^3)	pH	Capacité d'échange (meq/100g)	Cations échangeables (meq/100g)			Na/Ca
				Ca^{2+}	Na^{2+}	Mg^{2+}	
65,00	2,71	9,00	75,8	43,6	25,2	4,8	0,58

Table 1. Physico-chemical characteristics of bentonite

Montmorillonite	Quartz	Carbonates	Feldspaths	Biotites
45 à 60%	15 à 20%	8 à 10%	3 à 5%	8 à 10%

Table 2. Mineralogical characteristics of bentonite

2. Material and method

2.1 Procedure

For these tests, distilled water used has a pH ranging between 6 and 6,3. The initial solution of phenol equal to 100 mg/L is prepared starting from the dissolution of phenol crystallized in distilled water. The solutions are prepared by dilution in the water distilled according to the desired concentrations. For the tests of adsorption, we maintained the concentration of the constant aqueous solution (5mg/L) and varies the mass of the adsorbent m= 5,10,15,20,30,40 and 50 mg. After agitation during 5 hours with room temperature with magnetic stirrers, these solutions are centrifuged with 2000 revolutions per minute during 45 minutes for the analysis of phenol by spectrophotometer UV with a wavelength of 270nm by means of a spectrophotometer of the type SHIMADZU UV-1605.

For the kinetics of adsorption one proceeds in the same way, in beakers of 500mL containing distilled water, one adds the optimal bentonite amount (30 mg/L) as a optimal given by the jar-test. Sampling carried out during time make it possible to follow the evolution of the concentrations of phenol remaining in solution. Balance is reached after one 5 hours duration. For the analysis of the solids after adsorption of phenol sampling of bentonite (1 gram) are put in contact with phenol solutions (V=1L, C= 5 and 100 mg/L). After agitation throughout 6 hour to room temperature, the solids are separated from the liquids by filtration. After drying in a drying oven with 80°C the solids are collected for analysis by diffraction of x-rays, thermogravimetry and infra-red spectroscopy.

2.2.1 Diffraction of x-rays (DRX)

This technique allows inter alia, obtaining information on the interfoliar distance from material before and after adsorption of phenol. The apparatus X' PERT Pro of PANALYTICAL uses an assembly in reflexion θ-θ equipped with a tube with copper anode and with a detector RTMS (Real Time Multiple Strip) of type X' celerator. The recording is done uninterrupted during 90 minutes between 3 and 70 ° into 2 θ with the wavelength CuKα under a tension of 50 Kv and an intensity of 40 my.

2.2.2 Thermogravimetric analyses

The thermogravimetric analyses (TG) give indications on the variation of mass of a sample subjected to a linear rise in temperature. In this present study, we used an thermo-analyzer TG-DSC Sensys evo of SETARAM working between 25 and 750°C, under reconstituted air (mixture O_2 / N_2) and with a speed of rise of 5°C/minute. The analyses related to samples of clays saturated and unsaturated with phenol (C=100mg/L and 5mg) and of mass m=40mg. These thermogravimetric analyses under oxidizing atmosphere make it possible, indeed, to differentiate the organic part of the inorganic part, in particular water.

2.2.3 Analyzes by infra-red spectrophotometer

The Infra-red Spectroscopy with Transform of Fourier (IRTF) is based on l' absorption an infra-red radiation by analyzed material. It allows via the detection of the vibrations characteristic of the chemical bonds, to carry out l'analyzes chemical functions present in material. The bentonite samples, initially out of powder, are mixed in a small proportion with KBr (2.3 mg in 100 mg of KBr), then transformed in the form of a pastille. For the analyses, we used an apparatus BRUKER Equinox 55 in the area 400-4000 cm^{-1}, with the following conditions of recording: detector MCT, number of scans 8, resolution 4 cm^{-1}.

3. Results and discussion

3.1 Analyses of the solids

Diffractogram X of bentonite in a rough state (figure 1), indicates a basal distance (d001) from 12,50 Å. Moreover, it highlights the presence of several crystalline phases (quartz, feldspar). After contact with the phenol solution, the diffractogram shows that the basal distance increased, because (d001) is of 14,94 Å. Since d001 represents the thickness of the layer plus the interfoliar spacing, the spacing of the layers of initial bentonite would be equal to: 12,50 Å - 9,60 Å= 2,90 Å

The spacing of the layers of bentonite with exchange out of phenol would become equal to: 14,94 Å - 9,60 Å = 5,34 Å. A difference of 2,44 Å is thus measured between the spacing of the layers of initial bentonite and those of modified bentonite. It is thus possible that phenol molecules can be easily adsorbed in this interfoliar space.

The results obtained by thermal analysis are illustrated on figure 2 where three curves are presented: that of mass (TG), its derivative (dTG) and heat flow (HF). Curve TG presents several losses of mass according to the temperature. For temperatures going of 25 with 200°C, one observes a first loss from approximately 7%, which corresponds at the water beginning. It is followed of a weak loss (~ 0,9%) between 200 and 450°C, allotted to the oxidation of the organic matter imprisoned between the layers of clay and a third loss of mass beginning towards 450°C corresponding to the deshydroxylation from material. Dehydration as well as the deshydroxylation is accompanied by endothermic effects on the curve HF, while the oxidation of phenol results in two weak exothermic peaks. For a stronger phenol concentration (curve not represented), the organic matter loss is of approximately 1,8%. She thus increases with the concentration of phenol of departure.

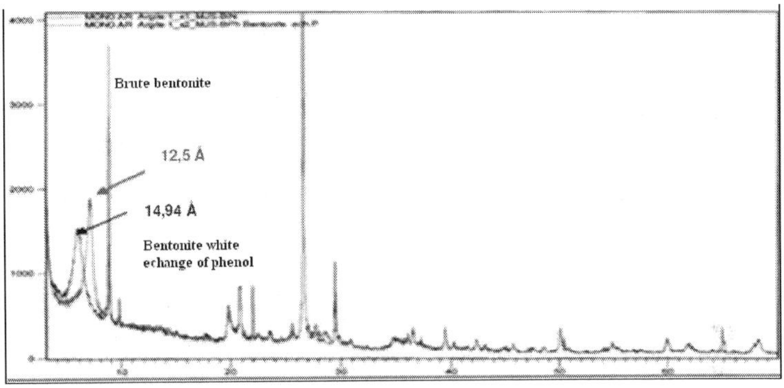

Fig. 1. Powder X-ray Diffractograms of initial bentonite and after phenol exchange

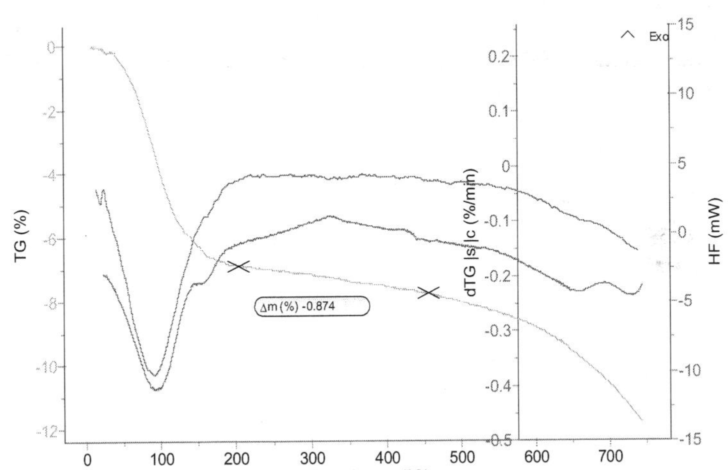

Fig. 2. TG Bentonite containing phenol (concentration at 5 mg/L)

On figure 3 are presented spectra FTIR of the various studied samples: bentonite brut, phenol and bentonite having adsorbed phenol. The presence of phenol should be evaluated by the presence of the characteristic bands, such as the connection OH (δO-H towards

3.3 Isotherms of adsorption of phenol

Several models were quoted in the literature to describe the isotherms adsorption. The models of Langmuir and Freundlich are most often used. Balance is described by curves with the room temperature, expressing the quantity of aqueous solution adsorbed per unit of mass of adsorbent according to the concentration of the aqueous solution to the state of balance.

It is a law in the form

$$\frac{x}{m} = f(ce)$$

X (mg): express the quantity of adsorbed aqueous solution
m: mass adsorbent (g)
Ce: concentration of the aqueous solution to the state of balance (mg/L)

From table 4 we determined the equations relating to the isotherms of adsorption (figures 6 and 7) as well as the parameters relating to the law of Freundlich (N and K) and to that of Langmuir (B and qm).

3.3.1 The equation of Freundlich

The equation of Freundlich is form:

$$\frac{x}{m} = KCe^{\frac{1}{n}}$$

The linearization of this expression in form logarithmic curve gives the following function.

Log (x/m) =1/n.logCe +logK

K and N are constants of balance

For our test:

Y=0,7352x +1,8468

Thus logK =1,8468 → K =70,27

1/n=0,7352 → n=1,36

R2=0,95

3.3.2 Equation to langmuir

$$q = \frac{x}{m} = qmX\frac{bCe}{1+bCe}$$

x/m: Quantity of adsorbed in mg/g to balance
qm = maximum capacity (ultimate) of adsorption
b= constant related to the energy of adsorption
Ce: Concentration with the balance (mg/l) of the organic compound

After passage to the opposite of function

$$\frac{x}{m},$$

the linearized form is the following one:

$$\frac{1}{q}=\frac{1}{qm}+\frac{1}{qmb}(\frac{1}{Ce})$$

Figure 7 shows the variation

$$\frac{1}{x/m}$$

according to

$$\frac{1}{Ce}$$

$$Y= 0{,}0134\ X +0{,}0018$$

$$R^2=0{,}96$$

$$\frac{1}{x/m} = 0{,}0134 \text{ thus } \frac{x}{m} = 74{,}62 \text{ mg/g}$$

$$b=7{,}44$$

The calculation

$$\frac{x}{m}$$

of watch which the bentonite could adsorb ~75 mg of phenol substances per gram of bentonite. We also note that the isotherms obtained follow best the law of Langmuir (R^2 =0,96). Table 5 presents the variation of the outputs of elimination of phenol according to time and figure 8, representing the abatement of phenol according to time, shows that the kinetics of adsorption slows down according to the reaction time. One can separate the phenomenon in two distinct stages:

- The first stage shows a fast increase in the yields of elimination during the first two hours of contact, which can be explained by the fixing of phenol on the surface of the adsorbent
- The second phase corresponds to the external mass transfer which is fast. shows a slow increase in the yield of elimination until the time of balance, which means that there is an internal mass transfer of the absorbable which generally corresponds to a phenomenon of diffusion in the internal porosity of the adsorbent.

Temps (min)	0	5	15	30	45	60	120	180	240	300	360
Co (mg/l)	5	5	5	5	5	5	5	5	5	5	5
Ct (mg/l)	5	4,14	3,98	3,70	3,5	2,96	2,59	2,10	1,95	1,8	1,8
R (%)	0	17,12	20,4	26	30	40,8	48,2	58	61	64	64
Ct/Co		0,828	0,796	0,74	0,70	0,592	0,518	0,42	0,39	0,36	0,36
$T^{1/2}$ ($min^{1/2}$)		2,236	3,872	5,477	6,708	7,745	10,954	13,416	15,491	17,320	18,973
1-(Ct/Co)		0,172	0,203	0,260	0,300	0,408	0,482	0,580	0,610	0,640	0,640

Table 5. Variation yield of elimination of phenol according to time

A few hours of contact with the adsorbent are enough to trap phenol by adsorption effectively. Indeed, the speed of adsorption is one kinetics of first order, function of the surface of the particles but inversely proportional to the diameter of these. The right obtained on figure 9, according to the square root of time, explains the diffusion (coefficient of correlation R^2 = 0,98).

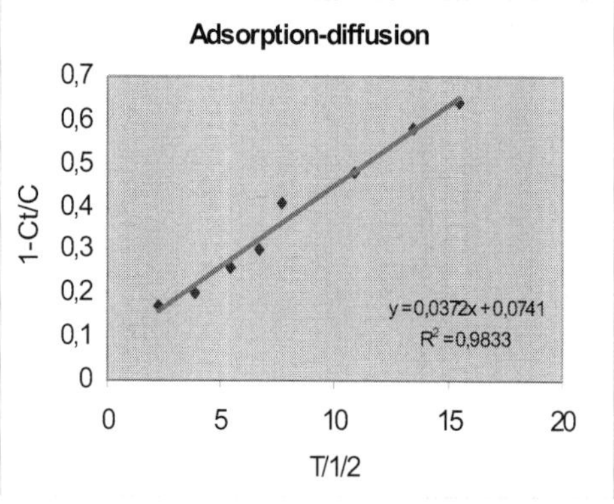

Fig. 8. Adsorption–diffusion on distilled water

Results can be compared with those of published diverse works. For example, Dali-Youcef and al., (2006) measured the adsorption of phenols on a local (bentonite) clay and a mud of dam. The results indeed confirm the capacities of the bentonite retaining more phenol with regard to the mud of dam and his capacity of adsorption is 32 mg / g. Banat and al., (2000) determined a weather of balance for the adsorption of the phenol about 6 hours and showed that the retention rate depended on the initial concentration and on the mass of bentonite used.

Al-Asheh and Al, (2003) examined the possibilities of using a bentonite for the retention of phenols and noticed that the increase in the mass of the adsorbent reduces the residual quantities of phenol in the final solution. Tests of adsorption of phenol on peat ashes and a bentonite were carried out by Viraraghavan and Alfaro, (1997). Their results indicate that the bentonite could retain phenol 46%. YU and Al, (2004) also showed that the adsorption of the phenol compounds by a montmorillonite increases with the initial phenol concentrations. In addition, more recently, of the tests of adsorption carried out by Nayak and Al, (2007) related to the retention of phenol on natural clay and modified clays. They concluded that it is possible to improve the capacity of adsorption of clays by specific treatments. Lastly, Boufatit and Al, (2007), which studied l' adsorption of organic compounds by Algerian clays rich in montmorillonite, show that the chlorinated phenols resulting from the degradation of the aromatic compounds (pesticides), are easily adsorbed on these clays.

Other materials were used for the retention of phenols in water. In particular, Guesbaya, (2005) used coagulation-flocculation with aluminium sulphate for water containing of phenol, but the retention remains very weak because the simple organic compounds tested are slightly eliminated whatever the amount from coagulant (sulphate of alumina) or the initial concentration of the compound. According to Al-Asheh and Al, (2003), in comparison with bentonite, the activated carbons have a higher adsorption capacity for the phenolic compounds. However, because of the relatively high cost of the activated carbon, the natural adsorbents, in spite of a lower performance, remain a solution interesting for the elimination of the contaminants of worn water.

4. Conclusion

The goal of this work was to study, experimentally, potential of bentonite gross of M' Zila in the adsorption of phenol pollutants. The study of the isotherms of adsorption enabled us to evaluate characteristics the characteristics adsorption of this bentonite. The results of the tests carried out in laboratory show the elimination of 64% of the concentrations (5 mg/L) of phenol and the bentonite can adsorb 75 mg/L phenol substances per gram and could be interesting in the elimination of the organic matter present in water. However, the optimum conditions for its use remain to be determined. The analysis of the isotherms makes it possible to determine if purification by adsorption can be carried out or not according to the initial phenol rate in water as well as an estimate of the mass of adsorbent making it possible to reduce in an important way the concentrations in pollutants. The preliminary reduction in the organic matter rate should limit the formation of potentially carcinogenic organochlorinated compounds likely to occur during a secondary treatment to chlorine.

5. References

Amar H. & A. Gaid (1987). Fixation on bentoniques clays of metallic ions in residual waters, *Rev. Sci. de l'Eau*, 3, 33-40.

Al-Ashehs., F. Banat & L. Abu-Aitah (2003). Adsorption of phenol using different types of activated bentonites. *Separation and Purification Technology* , 33, 1-10.

Andreozzi R., V. Caprio., A. Insola & R. Marotta (1999). Advanced oxidation process (AOP) for water purification and recovery. *Catal. Today*, 53, 51-59.

Auerbach S.M., K.A. Carrado & P.K. Sutta (2004). *Handbook of Layered Materials*, Marcel Dekker, New York, USA, 650 p.

Banat F. A., B. Al-Bashir., S. Al-Asheh & O. Hayajneh (2000). Adsorption of phenol by bentonite. *Environmental Pollution*, 107, 391-398.

Bartschat B.M., S.E. Cabanis & F.M. Morel (1992). Oligoelectrolyte model for cation binding by humic substances. *Environ. Sci. Technol.*, 26, 284-294.

Bergaya F., B.K.G. Theng &LAGALY G. (2006). *Handbook of clay science*, Developpement in clay science, series volume 1, 124 p.

Bersillon J.L. (1983). The mechanism of organic removal during coagulation. Thesis, Mc Master University, Hamilton Ontario, USA, 188 p.

Bersillon J.L., B. Lartiges, F. Thomas & L. Michot (1999). The transformation of water quality: from contribution to water treatment. Effect of mineral-organic-microorganism interactions on soil and Freshwater Environments. Kluwer Academic, Plenum Publisher, pp. 367-373.

Boudhar H. (1999). Improvement of the functioning of the small units of treatment of drinking water to eliminate néoformés elements, THM et turbidité. *Office International de l'Eau.*

Boufatit M., H. Ait-Amar & W.R. Mcwhinnie (2007). Development of Algerian material montmorillonite clay. Adsorption of phenol, 2-dichlorophenol and 2,4,6-trichlorophenol from aqueous solutions onto montmorillonite exchanged with transition metal complexes. *Desalination*, 206, 394-406.

Bouras O. (2003). Properties adsorbantes of pontées organophiles clays: synthesis and characterization, Doctoral thesis, Univ. Limoges, France.

Cailliere S., S. Henin & M. Rantureau (1982). *Minéralogie des argiles*. Masson, tome 1 et 2, pp. 184-189.

Chiron S., A. Fernandez-Alba, A. Rodriguez & E. Garcia-Calvo (2000). Pesticide chemical oxidation: State of the art. *Water Research*, 34, 366-377.

Dali-Youcef Z., H. Bouabdasselem & N. Bettahar (2006). Elimination of organic compounds by local clays. *Comptes Rendus de Chimie*, 9 (10), 1295-1300.

Decarreau A. (1990). Clayey materials: structure, properties and applications (collective work),Ed. Société Française de Minéralogie et de Cristallographie, pp. 8-21.

Dore, M. (1989). *Chimie des oxydants et traitement des eaux*, Ed.Tech. & Doc ; Lavoisier, Paris.

Dulin B.E. & W.R. Knocke (1989). The impact of incorporated organic matter on the dewatering characteristics of aluminium hydroxide Sludge. *Journal AWWA*, 81, 74-79.

Dussert B.W (1997). Advanced oxidation. UK: *Industrial wastewater*, pp.29-34.

Dutta n.N., G.S. Patil et S. Brothakur (1992). Phase transfer catalyzed extraction of phenolic substances from aqueous alkaline stream. *Separation Science and Technology*, 27, 1435.

Dutta N.N., S. Brothakur et R. Baruah (1998). A novel process for recovery of phenol from alkaline wastewater: laboratory study and predesign cost estimate. *Water Environ. Res.*, 70, 4-9.

ENOF (1997). Rapport d'analyse de la bentonite de M'Zila, Mostaghanem, Laboratoire Enof, Algérie.

Essington M.E. (1994). Adsorption of aniline and toluidines on montmorillonite, *Soil Science,* 158(3), 180-188.

Galze H., F. Beltran, T. Tuhkanen & J.W. Kang (1992). Chemical models of advanced oxidation processes. *Water Poll. Res. J. Canada* , 27, 23-42.

Guesbaya A. N. (2005). Coagulation-floculation by the sulfate of aluminum of phenolic organic compounds and humiques substances. *Larhyss Journal,* 4, 153-168.

Golich r. & D. Bahnemann (1997). Solar Water Treatment: Principles and Reactors, Intercalibration of Radical Sources and Water Quality Parameters, Swiss Federal Institute for *Environmental Science and Technology EAWAG,* Switzerland, pp. 137-148.

Hoang L. (2009). *Comparison of the energy returns on degradation of organic three by several processes of oxidation advanced(moved) in aqueous environment(middle), Doctoral thesis,* Univ. Poitiers, France.

Hooper S.M., R.S. Summers, S. Gabriele & D.M. OWEN (1996). Improving GAC performance by optimized coagulation. *Journal AWWA,* 88, 107-120.

Lefebvre e. & B. Legube (1990). Iron III Coagulation of humic substances extracted from surface waters: effect of pH and humic substances concentration. *Water Research,* 24, 591-606.

Meier, J.R. (1988). Genatoxic, Activity of organic chemicals in drinking water, *Mutation Res.,* 196, 211- 245.

Najm i.n., V.L. Snoeyink & Y. Richard (1993). Removal of 2, 4, 6-trichlorophénol and natural organic matter from water supplies using PAC in floc- blanket reactors. *Water Research,* 27, 551-560.

Nayak P. S. & B. K. Singh (2007). Removal of phenol from aqueous solutions by sorption on low cost clay. *Desalination*, 207, 71-79.

Safarzadeh-Amiri A., J.R. Bolton & S.R. Caster (1996). The use of iron in advanced oxidation processes. *J. Adv. Oxid. Technol.*, 1, 18-26.

Terrisse V. (2000). Interaction of constituent hydrated, main Silicates of Calcium of the cement, with the chlorides of alkaline. Analogy with clays, Doctoral thesis, Univ. de Bourgogne, Dijon, France.

Tippinge. (1993). Modelling the competition alkaline earth cations and trace species for binding by humic substances. *Environ.Sci.Technol.*, 27, 520-529.

Stevens A.A., C.J. Slocum, D. SeegeR & G.G. Robeck (1976). Chlorination of organics in drinking water. *Journal AWWA,* 68, 615-620.

Viraraghavan T. & F. Alfaro (1997). Adsorption of phenol from wastewater by peat, fly ash and bentonite. *Journal of Hazardous Materials,* 57, 59-70.

Wiesner M.R., M.M. Clark & J. Malleviale (1989). Membrane filtration of coagulated suspensions. *J. Environ. Engin.,* 115, 20-40. .

YU J., M. Shin, J. Noh & J. Seo (2004). Adsorption of phenol and chlorophenols on Ca-montmorillonite in aqueous solutions. *Geosciences Journal*, 8(2), 185-189.

16

Water Quality Improvement Through an Integrated Approach to Point and Non-Point Sources Pollution and Management of River Floodplain Wetlands

Edyta Kiedrzyńska[1,2] and Maciej Zalewski[1,2]
[1]International Institute of the Polish Academy of Sciences, European Regional Centre for Ecohydrology Under the Auspices of UNESCO, Lodz,
[2]University of Lodz, Department of Applied Ecology, Lodz,
Poland

1. Introduction

The world is faced with problems related to quality and quantity of water resources due to extensive industrialization, increasing population density and a highly urbanized society. Global scenarios suggest that almost two-thirds of the world's population will experience some water stress by 2025, which will accelerate the water environmental degradation to a unimaginable crisis scale (Momba, 2010).

Wetland are among the most important ecosystems on the Earth. The extent of the world's wetlands is now thought to be from 7 to 10 million km^2, or about 5 to 8 % of the land surface of the Earth (Mitsch and Gosselink, 2007). Wetlands include swamps, bogs, marshes, mires, fens, and also river floodplain wetlands.

River floodplain wetlands are very important hydrosystems that retain a significant part of the global freshwater bodies, and because of their location at lower elevations in the landscape, they are also highly exposed to accumulation of large loads of nutrients and other pollutants. This results in eutrophication, which in turn leads to degradation of biological diversity and the appearance of toxic cyanobacterial blooms, which pose threats to human and animal health.

This chapter will try to answer the frequently asked question "What exactly is a wetland?" and "What is the hydrological and biological characteristics of wetlands?" and "What are point and non-point sources pollution?". A section will also be presented on the role of river floodplain wetlands as key ecosystems important for regulation of the water, sediments and nutrients retention, and as a natural buffering system that can be considered as a tool for the reduction of nutrients and other pollutants transport by a river to downstream water ecosystems, and thus contributing to freshwater quality improvement. Part of the chapter will be devoted to application of the ecohydrological sustainable management of floodplain-wetland ecosystems, which is based on the restoration of natural mechanisms determining these ecosystems and functioning of the landscape for the increasing efficiency of water purification, and reducing the negative impact of pollution on the freshwater resources. The third part of the chapter will present a general assumption of the crucial international document "The Declaration on Sustainable Floodplain Management".

2. What is a wetland?

Wetlands sometimes are described as „the kidneys of the landscape" because they function as the downstream receivers of water and waste from both national and human sources. Furthermore, wetlands stabilize water supplies and water balance of the catchment area, thus ameliorating both floods and drought, and they have been found to clean polluted waters, protect shorelines, and recharge groundwater aquifers (Mitsch et al., 2009).

These ecosystems also have been called „ecological supermarkets" due to the extensive food chain and rich biodiversity they support. They play major roles in the landscape by providing unique habitats for a wide variety of flora and fauna. Now that we have focused our attention on the health of our entire planet, wetlands are being described by some as important carbon sinks and climate stabilisation on a global scale (Mitsch and Gosselink, 2007).

Wetland definitions and terms are many and are often confusing or even contradictory. Nevertheless, definitions are important both for the scientific understanding of these systems and for their proper management (Mitsch and Gosselink, 2007), and above all for using the wetlands for water quality improvement .

The Ramsar Convention on Wetlands (signed in Ramsar, Iran 1971) defines wetlands as areas of marsh, fen, peatland or water, whether natural or artificial, permanent or temporary, with water that is static or flowing, fresh, brackish or saline, including areas of marine water, the depth of which at low tide does not exceed six meters.

According to the U.S. Environmental Protection Agency wetlands are areas where water covers the soil, or is present either at or near the surface of the soil all year or for varying periods of time during the year, including the growing season. Wetlands vary widely because of regional and local differences in soils, topography, climate, hydrology, water chemistry, vegetation and other factors, including human disturbance. Indeed, wetlands are found from the tundra to the tropics and on every continent except Antarctica.

According to the wetland definition given by Mitsch and Gosselink (2007), it should include three main components: (i) wetlands are distinguished by the presence of water, either at the surface or within the root zone; (ii) wetlands often have unique soil conditions that differ from adjacent uplands; (iii) wetlands support biota such as vegetation adapted to wet conditions (hydrophytes) and, conversely, are characterized by the absence of flooding-intolerant biota.

Floodplain wetlands are one of the types of natural wetlands and are transitional between terrestrial of the river valley and open water river ecosystems (Fig. 1). Factors such as climate and geomorphology define the degree to which wetlands can exist, however the starting point is the hydrology, which, in turn, affects the physiochemical environment, including the soils, which in turn, together with the hydrology, determines what and how much of the biota, including vegetation, is found in a wetland (Mitsch et al., 2009).

Fig. 1. The Pilica River floodplain, upstream of the Sulejów Reservoir (central Poland); A - situation of high discharge (Q=83.2 m^3 s^{-1}) in spring 2006 (Photo by Piotr Wysocki); B - low discharge (Q= 6.7 m^3 s^{-1}) in summer 2006 (Photo by Mariusz Koch).

3. Wetland hydrology

Hydrologic conditions are extremely important for the maintenance of a floodplain wetland's structure and function, because they affect many abiotic factors, including soil anaerobiosis, nutrient availability (Mitsch and Gosselink, 2007; Vorosmarty and Sahagian, 2000). The hydrology of a river wetland creates unique physiochemical conditions that make such an ecosystem different from both well-drained floodplain systems and deeper old river bed systems.

The major components of river wetland's water budget include precipitation, evapotranspiration, surface flow, ground water fluxes, and other overbank flooding in floodplain wetlands. Water depth, flow patterns, and duration and frequency of flooding, sediments and nutrients transport (Kadlec and Knight, 1996; Magnuszewski et al., 2007; Altinakar et al., 2006; Kiedrzyńska et al., 2008a; Kiedrzyńska et al., 2008b), which result from all hydrologic inputs and outputs, influence the biochemistry of the soils and are major factors in the ultimate selection of the biota of wetlands (Mitsch and Gosselink, 2007; Kiedrzyńska et al., 2008a). The water status of a wetland defines its extent and determines the species composition in a natural floodplain wetland (Mitsch and Gosselink, 1993). However, biota components are active in altering the wetland hydrology and other physiochemical conditions (Zalewski 2000; 2006; Mitsch and Gosselink, 2007; Kiedrzyńska et al., 2008a).

manifest themselves (Zalewski and Naiman, 1985). The quantification of hydrological pulses along the river continuum (Junk et al., 1989; Vannote et al., 1980; Agostinho et al., 2004; Altinakar et al., 2006; Magnuszewski et al., 2007; Kiedrzyńska et al., 2008b) and monitoring of threats (Wagner and Zalewski, 2000; Mankiewicz-Boczek et al., 2006; Bednarek and Zalewski, 2007a, 2007b; Kiedrzyńska et al., 2008b; Urbaniak et al., in press), such as point and nonpoint source pollution (Takeda et al., 1997; Borah and Bera, 2003; Tian et al., 2010; Kiedrzyńska et al., 2010), are necessary for optimal regulation of processes towards the sustainable water and ecosystems management.

2. Increasing the carrying capacity of ecosystems that is their evolutionarily established resistance and resilience to absorb human-induced impacts (Target aspect). This aspect of ecohydrology expresses the rationale for a proactive approach to the sustainable management of freshwater resources. It assumes that it is not enough to simply protect the ecosystems, but in the face of increasing global changes, which are manifested in the growth of the population, energy consumption, material and human aspirations, it is necessary to increase the capacity of ecosystems. This can be achieved by regulation the interplay between hydrology and biota; analysis of dynamic oscillations of an ecosystem and its productivity and succession (as reflected by nutrient/pollutant absorbing capacity versus human impacts) should be the solution to process regulation (Bednarek and Zalewski, 2007a, 2007b; Kiedrzyńska et al., 2008a, Zalewski 2011).
3. Application of "dual regulation" in shaping and management of processes in river floodplain wetlands for purification and water quality improvement, biodiversity and ecosystem services for society (Methodology aspect). This means that a biotic component (macrophytes, bacteria) of a floodplain ecosystem can control and shape the chemical parameters of water and hydrological processes through effects on shaping the substrate roughness. These relationships also occur in the opposite direction - *vice versa*, what means using hydrology to regulate the biota (Zalewski, 2006, Zalewski and Kiedrzyńska, 2010). Great potential of the knowledge, which has been generated by dynamically developing ecological engineering (Mitsch 1993; Jorgensen 1996; Chicharo, 2009), should to a large extent accelerate the implementation of the above concept.

Sustainable management of the river floodplain wetlands gives a number of positive implications on the global ecosystem by improving the water quality, which depends on the development, dissemination and implementation of these principles and interdisciplinary knowledge, based on the latest achievements in environmental protection (Fig. 2).

The success of these actions depends on the profound understanding of the whole range of multi-dimensional processes involved. The first dimension is temporal: spanning a time frame from the past, paleohydrological conditions till the present, with a due consideration of future, global change scenarios. The second dimension is spatial: understanding the dynamic role of river and floodplain biota over a range of scales, from the molecular- to the valley-scale. Both dimensions should serve as a reference system for enhancing the buffering capacity of floodplain wetlands as key ecosystems important for the regulation of water, sediments and nutrients retention, and reduction of nutrients and other pollutants transport by a river to downstream water ecosystems, and thus contributing to freshwater quality improvement.

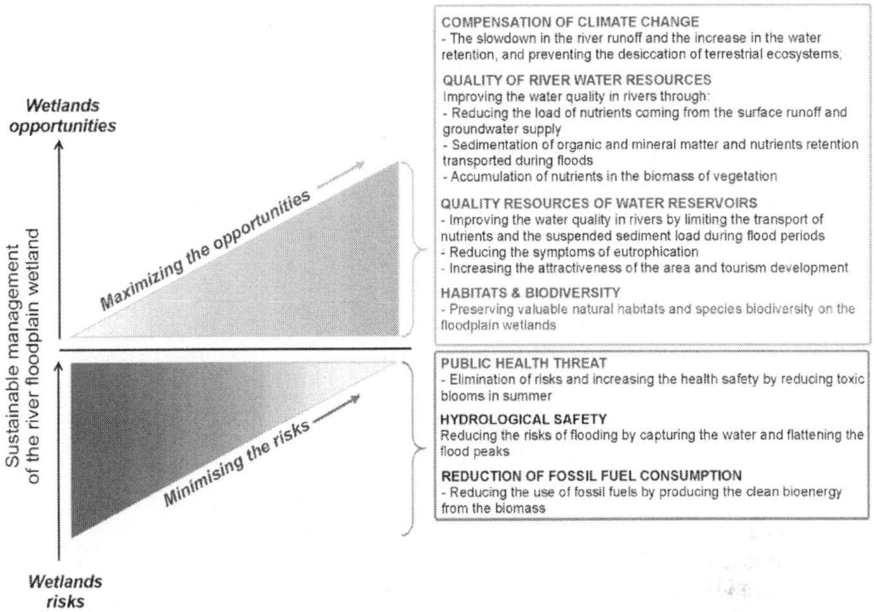

Fig. 2. Implications of the sustainable management of the floodplain wetlands.

6. Wetland water quality improvement – A new way of thinking

6.1 Water pollution – Point and non-point sources pollution

Water pollution is a crucial global problem, which requires ongoing evaluation and revision of water resource policy at all levels (from the international one down to individual aquifers and wells). It has been suggested that it is the leading worldwide cause of deaths and diseases, and that it accounts for the deaths of more than 14,000 people daily (West, 2006).

Water resources are usually referred to as polluted when they are impaired by anthropogenic contaminants and either do not support a human use, such as drinking water, and/or undergo a considerable shift in their ability to support their constituent biotic communities, such as fish. Natural phenomena, such as algae blooms, storms, and earthquakes, also cause major changes in the water quality and the ecological status of the water.

Surface water and groundwater have often been studied and managed as separate resources, although they are interrelated. Surface water seeps through the soil and becomes groundwater. Conversely, groundwater can also feed surface water sources. Sources of surface water pollution are generally grouped into two categories based on their origin (Winter, 1998).

Point source (PS) water pollution refers to contaminants that enter a waterway from a single, identifiable source, such as a pipe or ditch. Examples of sources in this category include discharges from a sewage treatment plant, a factory, or a city storm drain. Non–

point source (NPS) pollution refers to diffuse contamination that does not originate from a single discrete source. NPS pollution is often the cumulative effect of small amounts of contaminants gathered from a large area. A common example is the leaching of phosphorus and nitrogen compounds from fertilized agricultural lands. Nutrients runoff in stormwater from "sheet flow" over an agricultural field or forest are also cited as examples of NPS pollution. Excessive export of nutrients from PS and NPS pollution are the leading causes of eutrophication in lakes, reservoirs and rivers, and coastal water bodies worldwide (Alexander et al., 2008; Diaz and Rosenberg, 2008; Tian et al., 2010).

Eutrophication is a shift in the trophic status of a given water body in the direction of increasing plant biomass, by adding some artificial or natural substances, such as nitrates and phosphates, through e.g. fertilizers or sewage, to an aquatic system.

In other terms, it is a water bloom resulting from a great increase of phytoplankton in a water body. Negative environmental effects include hypoxia, the depletion of oxygen in the water, which induces reductions in specific fish and other animal populations. Thus, eutrophication of water resources leads to degradation of biological diversity and the appearance of toxic cyanobacterial blooms, which pose threats to human health and animals (Tarczyńska et al., 2001; Mankiewicz et al., 2001, 2005; Jurczak et al. 2004).

River wetlands are altered by the runoff of pollutants from point and diffuse sources of pollution flowing from the upper catchment areas and thus are purified. The effects of polluted water on wetlands have not received yet enough attention.

6.2 Wetlands as key ecosystems improving the water quality

Rivers and floodplain wetlands are the ecosystems that are particularly exposed to eutrophication and high anthropogenic stress (Meybeck 2003, Zalewski and Kiedrzyńska 2010). This is because they are situated in landscape depressions, into which the whole range of catchment anthropogenic modifications and impacts are transferred and accumulated (Altinakar et al., 2006; Zalewski, 2006; Magnuszewki et al., 2007), e.g. sediments and nutrients (Kiedrzyńska et al., 2008a; Kiedrzyńska et al., 2008b), dioxins (Urbaniak et al., 2009; Urbaniak et al., in press), microbial contamination (Gągała et al., 2009). These dramatically progressing disturbances are sometimes negatively amplified by degradation of the hydrological cycle and the loss of integrity between fluvial ecosystems and floodplains, which can result in the increased eutrophication (Tarczyńska et al., 2001; Izydorczyk et al., 2005; Izydorczyk et al., 2008) and the reduction of biodiversity and ecosystem services for societies (Zalewski 2008; Zalewski and Kiedrzyńska, 2010). However, the river valley with natural floodplain wetlands are areas that may be used in water purification.

Water quality improvement by the use of wetlands has been broadly discussed (Bastian and Hammer, 1993; Raisin and Mitchell, 1995; Nairn and Mitsch, 2000; Trepel and Kluge, 2002; Mitsch et al., 2005; Mitsch and Gosselink, 2007; Mitsch et al., 2009), especially the importance of natural floodplains for river self-purification and freshwater quality protection (Bayley, 1995; Loeb and Lamers, 2003; Zalewski 2006; Kiedrzyńska et al. 2008a; Kiedrzyńska et al. 2008b). An example can be the area of 24 km^2 of wetlands that collected the water from the Zala River catchment, and which has been reconstructed within the confines of a multidisciplinary research programme on the protection of the Lake Balaton (Hungary).

According to Pomogyi (1993), 96% of PO_4-P, 87% of NO_3-N and 58% of TP were retained in this area in 1990. Interesting studies conducted by Wassen (1995) in the Biebrza Valley in Poland reported that the floodplain vegetation is an important sink for nutrients, especially for N and P. Wetlands are also used in other European countries, e.g. in the Netherlands, Germany, Finland (Wassen et al., 2002; Olde Venterink et al., 2002) and in the United States, and around the world (Weller et al., 1996; Mitsch et al., 2005; Thullen et al., 2005; Mitsch et al. 2009).

Floodplains can optimize nutrient retention in the river ecosystem, especially in catchments with large areas of agriculture and can be considered as a tool for the reduction of nutrient transport by a river to downstream reservoirs and estuaries (Kiedrzyńska et al., 2008a; Kiedrzyńska et al., 2008b).

The highest nutrients' loads transported by rivers usually occurred during rising water stages of floods and they should be directed to floodplain areas upstream the reservoir at the very initial stages of floods, in order to diminish the load in a reservoir. The research on the Pilica River floodplain (central Poland) looked into the possibilities of enhancing this process, both through sedimentation and assimilation in the vegetation biomass. The research that was based on the DTM and hydraulic models demonstrated that sedimentation of flood sediments in the floodplain essentially reduces the transport to the reservoir. During floods, the sediment is effectively deposited and phosphorus is retained in the 30-kilometer section of the Pilica River floodplain. In the flooding area of 1007 ha, fine-grained flood sediments reached 500 t and the retention of P was 1.5 t. Furthermore, the efficiency in the assimilation of nutrients and the biomass production by autochthonous plant communities, with special emphasis on willow patches, was examined against a background of a hydroperiod. The potential of vegetation in the Pilica River floodplain (26.6 ha) for summer phosphorus accumulation was estimated at 255 kg P y^{-1}, however, a conversion of 24% or 48% of the area into fast-growing managed willow patches can increase the phosphorus retention up to 332 kg P y^{-1} or 399 kg P y^{-1}, respectively (Kiedrzyńska et al., 2008a). Theoretically, 1 kg of P can lead to some 1-2 t of algal biomass in a reservoir (Zalewski, 2005). Therefore, floodplain wetlands are mostly enriched with the riverine material and, at the same time, river water is purified by deposition of this material. Floodplains can, therefore, serve as natural, cleaning and biofiltering systems for reducing the concentrations of sediments, nutrients, micropollutants and, other pollutants coming from upper sections of the catchment area.

7. Wetland management - The Declaration on Sustainable Floodplain Management

In the 21st century, wetlands management should focus not only on the conservative protection of these valuable ecosystems, but also on the sustainable use and optimization of abiotic-biotic processes for problem solving and improving the water quality.

Floodplain wetlands are an integral part of river systems and therefore they play a fundamental role in the exchange of water masses and matter between a river and terrestrial ecosystems (Mitsch et al., 1979; Junk et al., 1989; Tockner et al., 1999; Mitsch et al., 2008; Kiedrzyńska et al., 2008b). Floodplains are "dynamic spatial mosaics", where water acts as a connector between various components (Thoms 2003; Kiedrzyńska et al., 2008a). This specific connection is crucial for maintaining the function and integrity of floodplain-river systems (Tockner et al., 1999; Amoros and Bornette, 2002; Thoms 2003). They are the hot spots of terrestrial and aquatic biodiversity in the catchment landscape due to a mosaic of plant communities and their spatio-temporal dynamics (Zalewski, 2008). Sustainable development of the river and floodplain environment needs to take into account the fact that biological structures and fundamental ecological processes, such as water and nutrients cycles, are to a large extent, suffering from deterioration (Zalewski, 2009).

The sustainable management of floodplains, which are the most diversified ecosystems and most resilient to human impact due to their hydrological pulse-driven self-regenerative capacity, is obviously very important. Therefore, there is still a further need for insights into these and other processes, whereas "engineering harmony" between river floodplain ecosystems and societies (UN MDGs) requires solutions from integrative, interdisciplinary science such as ecohydrology, a subdiscipline of sustainability science focused on ecological aspects of the hydrological cycle (Zalewski and Kiedrzyńska, 2010).

Such an integrated ecohydrological approach to sustainable management of wetlands is contained in the presented below Floodplain Declaration "Declaration on Sustainable Floodplain Management", which was elaborated based on presentations and discussions at the International Conference under the auspices of IHP of UNESCO "Ecohydrological Processes and Sustainable Floodplain Management: Opportunities and Concepts for Water Hazard Mitigation, and Ecological and Socioeconomic Sustainability in the Face of Global Changes" (19th - 23rd of May 2008, Lodz, Poland).

7.1 Declaration on Sustainable Floodplain Management

7.1.1 Recognition: Properties and values of floodplains

Floodplains are dynamic wetlands, an integral part of river basins with a high potential for biological productivity, biodiversity, flood mitigation, groundwater recharge, river purification and regulation of exchanges of nutrients between land and water, and other ecosystem services, all maintained by the pulse-regulated hydrology of running waters.

Floodplains are threatened by increasing population and improper management. Development of floodplains without consideration of the specifics of their ecological structure and dynamics thus diminishes biodiversity, reduces benefits to society related to water quality, cultural aesthetic values and - in consequence - causes economic losses.

7.1.2 Floodplains and global climate change

Floodplains are an important component of global environmental security and resilience because of their high compensatory potential to mitigate environmental change due to their capacity for water retention, food production, CO_2 sequestration, production of bio-fuels, and the diversity of habitats that they support.

7.1.3 Integrative science for problem solving

Understanding the functioning of floodplains and their potential for socio-economic benefits, requires integration of recent knowledge of:

- geomorphological and paleohydrological evolution of river valleys,
- hydrological processes and patterns of ecological succession,
- societal interactions and learning alliances,
- climate scenarios,
- strategic forecasts based on integrative modelling and adaptive management

In order to reverse floodplain degradation and increase ecological resilience and economic benefits, a shift in strategy from floodplain exploitation to floodplain sustainable use is necessary. Accordingly we need a change of public perception from sectoral, structural and reactive responses to an integrated, process-regulation-oriented and proactive approach.

7.1.4 Methodology for provisioning sustainable ecological services of floodplains

- *Ecohydrological management* of floodplains, will require "dual regulation" - a framework for harmonisation of biodiversity conservation with such human needs as flood mitigation, food and energy production, transport and recreation.
- *Hydrotechnical infrastructure* harmonised on the basis of integrative science and best management practices incorporating catchment scale ecosystem processes, will be a powerful tool for reversing degradation of biodiversity, and enhancing sustainable development and compensation of global changes
- *Cultural heritage* of the catchment should become an important element for spatial reconnection of floodplains to the adjacent landscape, as well as restoration of links to social, economic and cultural values.
- *People's perception* and attitudes to the changing environment can only be shaped by new solutions based on integrative science, which depend upon development of programs and methodologies for education and communication.

7.1.5 Tools for implementation

Policies by national and international institutions for water resources, energy, transportation, and environmental management must elevate the protection of pristine sections of the floodplains and promote sustainable use, and restoration of degraded floodplains on rivers, lakes and coastal zones.

Land use integrated planning, financial incentives, economic instruments, and environmental regulatory frameworks are essential tools for implementing the ecohydrological standards and criteria. In case of "novel floodplains", created by secondary succession after human impact, floodplain loss due to essential new development of e.g. transport systems should be mitigated through restoration of at least twice the area of degraded floodplain.

A network of long-term ecological processes, research sites, responsible institutions, and data bases is needed for improving progress and transfer of knowledge, and transfer and sharing of technology.

Public participation, facilitated by modern communication approaches, is fundamental to accommodating conflicting interests and uses of floodplains.

7.1.6 Recommendations for action plan

- Classification of different types of floodplains with special consideration of catchment perspective and ecosystem services;
- Development of methodology to assess rate and type of flood pulses necessary to maintain floodplain functions and structures and to reconcile protection and social needs;
- Formulation of principles for floodplain management, sustainable food and renewable energy production based on integrative science and the relevant science/policy interface.

8. Conclusion

Floodplain wetlands can purify and improve the water quality because they have a significant role in the water retention, sedimentation of mineral and organic matter, nutrients and pollutants. Furthermore, the floodplain wetland vegetation has a great biological potential for the assimilation and accumulation of nutrients in biomass and especially for the uptake of phosphorus.

Therefore well-managed river wetlands can serve as natural cleaning and biofiltering systems for reducing the concentrations of sediments, nutrients, micropollutants and, other serious pollutants.

On the one hand, in the 21st century, the floodplain wetlands management should focus on the protection of biodiversity and values of these important ecosystems, but on the other hand, also on the sustainable use and optimization of abiotic-biotic processes for problem solving and improving the water quality.

In accordance with the conclusions of the Floodplain Declaration, the successful reversal of degradation of floodplain ecosystems should become the objective for the development of a sound vision of co-evolution of Ecosphere and Anthroposphere, by engineering harmony between three dynamic and evolving components: catchment areas, water resources and a society, with an emphasis on the change from exploitative to participatory environmental consciousness. For this purpose, it is necessary to continue the integration of studies of highly specialized disciplines of environmental and social sciences into the framework of Ecohydrology - a holistic problem-solving concept. The system approach, foresight methodology and learning alliances are these important new components of the trans-disciplinary sustainability science that should be used for sustainable water management in the catchment area, and also the ecological and socio-economic potential of the basin should be used for the improvement of human health and the quality of life following the UN MDGs.

9. Acknowledgment

Part of these researches was developed within the framework of the following projects: 1) Project of the Polish Ministry of Science and Higher Education: NN 305 365738 "Analysis of point sources pollution of nutrients, dioxins and dioxin-like compounds in the Pilica River catchment area and drawing up the reclamation methods"; 2) LIFE+EKOROB project: Ecotones for reduction of diffuse pollutions (LIFE08 ENV/PL/000519); 3) The Pilica River Demonstration Project under the auspices of UNESCO and UNEP.

We are particularly grateful to the most active Members of the Steering Committee and the Advisory Committee of the International Conference "Ecohydrological Processes and Sustainable Floodplain Management: Opportunities and Concepts for Water Hazard Mitigation, and Ecological and Socioeconomic Sustainability in the Face of Global Changes" (19th - 23rd of May 2008, Lodz, Poland) for their methodological contribution to the Declaration on Sustainable Floodplain Management. The Floodplain Declaration is available from the following link on the II PAS ERCE under the auspice of UNESCO website at: http://www.erce.unesco.lodz.pl.

10. References

Agostinho, A.A.; Gomes, L.C. & Verissimo, S. (2004). Flood regime, dam regulation and fish in the upper Parana´ river: effects on assemblage attributes, reproduction and recruitment. *Reviews of Fish Biology and Fishery*, 14(1): 11–19.

Alexander, R.B.; Smith, R.A.; Schwarz, G.E.; Boyer, E.W.; Nolan, J.V. & Brakebill, J.W. (2008). Differences in phosphorus and nitrogen Delivery to The Gulf of Mexico from the Mississippi River Basin. *Environmental Science and Technology*, 42(3): 822-830.

Altinakar, M.; Kiedrzyńska, E. & Magnuszewski, A. (2006). Modeling of inundation pattern at Pilica river floodplain, Poland. In: *Climate Variability and Change Hydrological Impacts*. Demuth, S.; Gustard, A.; Planos, E.; Scatena, F. & Servat, E. (Eds). *IAHS Publ.* 308. 579-585. *Proceedings of the Fifth FRIEND World Conference*, Havana, Cuba, November, 2006.

Amoros, C. & Bornette, G. (2002). Connectivity and biocomplexity in waterbodies of riverine floodplains. *Freshwater Biology* 47: 517-539.

Athie, D. & Cerri, C.C. (1987). The use of macrophytes in water pollution control. *Water Sci. Technol.* 19, 10.

Bastian, R.K. & Hammer, D.A. (1993). The use of constructed wetlands for wastewater treatment and recycling. In: *Constructed wetlands for water quality improvement*. G.A. Moshiri (Ed.), , CRC Press, Florida, USA, pp.59-68.

Bayley, P.B. (1995). Understanding large river-floodplain ecosystems. *BioScience* 45: 153-158.

Bednarek, A. & Zalewski, M. (2007a). Potential effects of enhancing denitrification rates in sediments of the Sulejów Reservoir. *Environment Protection Engineering*. 33(2): 35–43.

Bednarek, A. & Zalewski, M. (2007b). Management of lowland reservoir littoral zone for enhancement of nitrogen removal via denitrification. *Proceedings of International Conference. W3M Wetlands: Monitoring, Modelling and Management*. Wierzba. Warsaw Press, pp. 293–299.

Borah, D.K. & Bera, M. (2003). Watershed-scale hydrologic and nonpoint-source pollution models: Review of mathematical bases. *American Society of Agricultural Engineers*, 46(3): 1553-1566.

Chicharo, L.; Wagner, I.; Chicharo, M.; Łapinska, M. & Zalewski, M. (2009). Practical experiments guide for Ecohydrology. UNESCO, Venice, Paris. 122pp, (ISBN 978-989-20-1702-0).

Compton, J.E. & Cole, D.W. (1998). Phosphorus cycling and soil phosphorus fractions in Douglas-fir and red alder stands. *Forest Ecol. and Manag.*110, 101-112.

Copper, J.R. (1994). Riparian wetlands and water quality. *J. of Environ. Qual.* 23, 896-900.

Diaz, R.J. & Rosenberg, R. (2008). Spreading dead zones and consequences for marine ecosystems. *Science*, 321(5891): 926-929.

Fink, J. (1963). Introduction to the plants phosphorus biochemistry. *National Publishing of Agriculture and Forestry*. Warsaw, Poland, pp. 241.

Gągała, I.; Izydorczyk, K.; Skowron, A.; Kamecka-Plaskota, D.; Stefaniak, K.; Kokociński, M. & Mankiewicz-Boczek, J. (2009). Appearance of toxigenic cyanobacteria in two Polish lakes dominated by *Microcystis aeruginosa* and *Planktothrix agardhii* and environmental factors influence. *Ecohydrology & Hydrobiology* 9 (2).

Gołdyn, R. & Grabia, J. (1996). A conception of using natural methods for additional wastewater treatment in Jarocin. In: *Constructed wetlands for wastewater treatment, II Scientific-Technical Conference*. Kraska, M. & Błażejewski, R. (Eds.), Poznań, Poland.

Izydorczyk, K.; Tarczynska, M.; Jurczak, T.; Mrowczynski, J. & Zalewski, M. (2005). Measurement of phycocyanin fluorescence as an online Early warning system for cyanobacteria in reservoir intake water. *Environmental Toxicology* 20: 425-430.

Izydorczyk, K.; Jurczak, T.; Wojtal-Frankiewicz, A.; Skowron, A.; Mankiewicz-Boczek, J. & Tarczyńska, M. (2008). Influence of abiotic and biotic factors on microcystin content in Microcystis aeruginosa cells in a eutrophic temperate reservoir. *Journal of Plankton Research* 30 (4): 393-400.

Jørgensen, S. E. (1996). The application of ecosystem theory in limnology. *Verh. Int. Verein Limnol.* Vol. 26, 181–192.

Junk, W.J.; Bayley, P.B. & Sparks, R.E. (1989). The flood pulse concept in river-floodplain systems. In: *Proceedings of the international large river symposium*. Dodge, D.P. (Ed.). *Can. Spec. Publ. Fish. Aquat. Sci*, 106, 110-127.

Jurczak, T.; Tarczynska, M.; Karlsson, K. & Meriluoto, J. (2004). Characterization and diversity of cyano- bacterial hepatotoxins (microcystins) in blooms from Polish freshwaters identified by liquid chromatography-electrospray ionisation mass spectrometry. *Chromatographia* 59, 571-578.

Kadlec, R.H. & Knight, R.L. (1996). *Treatment Wetlands*. Lewis Publishers, CPR Press. USA. 893 pp.

Keedy, P.A. (2010). *Wetland Ecology, Principles and conservation*. Second Edition. Cambridge University Press. UK. pp. 497.

Kiedrzyńska, E., Wagner-Łotkowska, I. & Zalewski, M. (2008a). Quantification of phosphorus retention efficiency by floodplain vegetation and a management strategy for a eutrophic reservoir restoration. *Ecological Engineering* 33, 15-25.

Kiedrzyńska, E.; Kiedrzyński, M. & Zalewski, M. (2008b). Flood sediment deposition and phosphorus retention in a lowland river floodplain: impact on water quality of a reservoir, Sulejów, Poland. *Ecohydrology & Hydrobiology* 8: 2-4.

Kiedrzyńska, E., Macherzyński, A., Skłodowski, M., Kiedrzyński, M., Zalewski M. (2010). Analysis of point sources of pollution of nutrients in the Pilica River catchment and use of ecohydrological approach for their reduction (in polish). In: *Hydrology in Environmental Protection and Management*. A. Magnuszewski (Ed.). Monograph of the Committee for Environmental Sciences PAS, 69, pp. 285 - 295.

Klopatek, J.M. (1978). Nutrient dynamic of freshwater riverine marshes and the role of emergent macrophytes. In: *Freshwater wetlands: Ecological processes and management potential*. Good, R.E.; Whigham, D.F. & Simpson, R.L. (Eds). Academic Press, New York, pp. 195-216.

Loeb, R. & Lamers, L. (2003). The effects of river water quality on the development of wet floodplain vegetations in the Netherlands. *Proceedings of the International conference "Towards natural flood reduction strategies"*. September 2003. Warsaw, Poland.

Magnuszewski, A.; Kiedrzyńska, E.; Wagner-Łotkowska, I. & Zalewski, M. (2007). Numerical modelling of material fluxes on the floodplain wetland of the Pilica River, Poland. In: *Wetlands: Monitoring, Modelling and Management*. Okruszko, T.; Szatyłowicz, J.; Mirosław - Świątek, D.; Kotowski, W. & Maltby, E. (Eds). A.A. Balkema Publishers - Taylor & Francis Group. pp. 205-210.

Mainstone, C.P. & Parr, W. (2002). Phosphorus in rivers - ecology and management. *The Science of the Total Environ.* 282/283, 25-47.

Mankiewicz, J., Walter, Z., Tarczynska, M., Fladmark, K.E., Doskeland, S.O., Zalewski, M. (2001). Apoptotic effect of cyanobacterial extract on rat hepatocytes and human lymphocytes. *Environmental Toxicology* 3 (16), 225-233.

Mankiewicz, J., Komarkova, J., Izydorczyk, K., Jurczak, T., Tarczynska, M., Zalewski, M. (2005). Hepatotoxic cyanobacterial blooms in the lakes of northern Poland. *Environmental Toxicology* 20, 499-506.

Mankiewicz-Boczek, J., Izydorczyk, K., Jurczak, T. (2006). Risk assessment of toxic *Cyanobacteria* in Polish water bodies. In: Kungolos, A.G., Brebbia, C.A., Samaras, C.P., Popov, V. (Eds) *Environmental Toxicology*. WIT press, Southampton, Boston, *WIT Transactions on Biomedicine and Health* 10, 49-58.

Meybeck, M. (2003). Global analysis of river systems: from Earth system controls to Anthropocene syndromes. Philosophical Transactions of the Royal Society of London, [B] 358 (1440), 1935-1955.

Mitsch, W. J. (1993). Ecological Engineering - a co-operative role with planetary life support system. *Environmental Science Technology* Vol. 27, 438-445.

Mitsch, W.J & Gosselink, J.G. (2007). *Wetlands*. Fourth Edition. John Wiley & Sons, Inc. USA.

Mitsch, W.J.; Dorge C.L. & Wiemhoff, J.W. (1979). Ecosystems dynamics and a phosphorus budget of an alluvial cypress swamp in southern Illinois. *Ecology* 60: 1116-1124.

Mitsch W.J. & Gosselink, J.G. (1993). *Wetlands*. Second Edition. John Wiley & Sons, Inc. USA. 722 pp.

Mitsch, W.J.; Zhang, L.; Anderson, C.J.; Altor, A.E. & Hernández, M.E. (2005). Creating riverine wetlands: Ecological succession, nutrient retention, and pulsing effects. *Ecological Engineering* 25, 510-527.

Mitsch, W.J.; Zhang, L.; Fink, D.F.; Hernandez, M.E.; Altor, A.E.; Tuttle, C.L. & Nahlik, A.M. (2008). Ecological Engineering of floodplain. *Ecohydrology & Hydrobiology* Vol.8, No. 2-4: 139-147.

Mitsch, W.J.; Gosselink, J.G.; Anderson, C.J. & Zhang, L. (2009). *Wetland Ecosystem*. John Wiley & Sons, Inc. USA. 295 pp.

Momba, MNB. 2010. Wastewater Protozoan-Driven Environmental Processes for the Protection of Water Sources. In: Momba M. and Bux F. Eds. Biomass. Croatia, downloaded from sciyo.com., pp.202.

Nairn, R.W. & Mitsch, W. J. (2000). Phosphorus removal in created wetland ponds receiving river overflow. *Ecological Engineering* 14: 107-126.

Olde Venterink, H.; Pieterse, N.M.; Belgers, J.D.M.; Wassen, M.J. & De Ruiter, P.C. (2002). N, P, and K budgets along nutrient availability and productivity gradients in wetlands. *Ecological Applications* 12, 1010-1026.

Olde Venterink, H.; Wassen, M.J.; Verkroost, A.W.M. & De Ruiter, P.C. (2003). Species richness-productivity patterns differ between N-, P-, and K- limited wetlands. *Ecology* 84, 2191-2199.

Ozimek, W. & Renman, G. (1996). The role of emergent macrophytes in the constructed wetlands for wastewater treatment. In: *Constructed wetlands for wastewater treatment. II Scientific-Technical Conference.* Kraska, M. & Błażejewski, R. (Eds). Poznań, Poland, pp. 9-17.

Pomogyi, P. (1993). Nutrient retention by the Kis-Balaton Water Protection System. *Hydrobiologia* 251, 309-320.

Raisin, G. W. & Mitchell, D. S. (1995). The use of wetlands for the control of non-point source pollution. *Water Science Technology* 32 (3): 177-186.

Sumorok, B. & Kiedrzyńska, E. (2007). Mycorrhizal status of native willow species at the Pilica River floodplain along moist gradient. In: *Wetlands: Monitoring, Modeling and Management.* Okruszko, T.; Maltby, E.; Szatyłowicz, J.; Świątek, D. & Kotowski, W. (Eds). A.A. Balkema Publishers - Taylor & Francis Group, pp. 281-286.

Surrency, D. (1993). Evaluation of aquatic plants for constructed wetlands. In: *Constructed wetlands for water quality improvement.* Moshiri, G.A. (Ed.), CRC Press. Florida, pp. 349-357.

Takeda, I.; Fukushima, A.& Tanaka, R. (1997). Non-point pollutant reduction in a paddy-field watershed using a circular irrigation system. *Water Research,* 31(11): 2685-2692.

Tarczyńska, M.; Romanowska-Duda, Z.; Jurczak, T. & Zalewski, M., (2001). Toxic cyanobacterial blooms in a drinking water reservoir-causes, consequences and management strategy. *Water Science Technology*: Water Supply 1 (2), 237–246.

Thoms, M.C. (2003). Floodplain-river ecosystems: lateral connections and the implications of human interference. *Geomorphology* 56: 335-349.

Thullen, J.S.; Sartoris, J.J. & Nelson, S.M. (2005). Managing vegetation in surface-flow wastewater-treatment wetlands for optimal treatment performance. *Ecological Engineering* 25, 583-593.

Tian, Y.W.; Huang, Z.L. & Xiao, W.F. (2010). Reductions in non-point source pollution through different management practices for an agricultural watershed in the Three Gorges Reservoir Area. *Journal of Environmental Sciences,* 22(2): 184-191.

Tockner, K.; Pennetzdorfer, D.; Reiner, N.; Schiemer F. & Ward, J.V. (1999). Hydrological connectivity, and the exchange of organic matter and nutrients in a dynamic river-floodplain system (Danube, Austria). *Freshwater Biology* 41: 521-535.

Trepel, M. & Kluge, W. (2002). Ecohydrological characterisation of a degenerated valley peatland in Northern Germany for use in restoration. *Journal for Nature Conservation* 10:155-169.

Urbaniak, M.; Zieliński, M.; Wesołowski, W. & Zalewski, M. (2009). Sources and distribution of polychlorinated dibenzo-para-dioxins and dibenzofurans in sediments of urban cascade reservoirs, Central Poland. *Environ. Protec. Engineer.,* No. 3, vol. 35, 93-103.

Urbaniak ,M.; Kiedrzyńska, E. & Zalewski, M. (in press). The role of a lowland reservoir in the transport of micropollutants, nutrients and the suspended particulate matter along the river continuum. *Hydrology Research* 00-00.

Vannote, R.L.; Minshall, G.W.; Cummings, K.W.; Sedell, J.R. & Cushing, C. E. (1980). The River Continuum Concept. *Canad. J. Fish. Aquati. Sci.,* 37, 130-137.

Vorosmarty, C.J. & Sahagian, D. (2000). Antropogenic disturbance of the terrestrial water cycle. *Bioscience* 50(9): 753–765.

Wagner, I. & Zalewski, M. (2000). Effect of hydrological patterns of tributaries on biotic processes in lowland reservoir - consequences for restoration. *Ecological Engineering*. Special Issue 16, 79-90.

Wassen, M.J. (1995). Hydrology, water chemistry and nutrient accumulation in the Biebrza fens and floodplains (Poland). *Wetlands Ecology and Management* 3, 125-137.

Wassen, M.J.; Peeters, W.H.M. & Olde Venterink, H. (2002). Patterns in vegetation, hydrology, and nutrient availability in an undisturbed river floodplain in Poland. *Plant Ecology* 165, 27-43.

Weller, C.M.; Watzin, M.C. & Wang, D. (1996). Role of wetlands in reducing phosphorus loading to surface water in eight watersheds in the Lake Champlain Basin. *Environmental Management* 20, 731-739.

West, L. (2006). World Water Day: A Billion People Worldwide Lack Safe Drinking Water. http://environment.about.com/od/environmentalevents/a/waterdayqa.htm. About. (March 26, 2006).

Winter, T.C.; Harvey, J.W.; Franke, O.L. & Alley, W.M. (1998). Ground Water and Surface Water: A Single Resource. *United States Geological Survey* (USGS) - Circular 1139. Denver, USA.

Zalewski, M. (2011). Ecohydrology for implementation of the EU water framework directive. Proceedings of the Institution of Civil Engineers. *Water Management* 8, 16 Issue, 375-385.

Zalewski, M. (2000). Ecohydrology - the scientific background to use ecosystem properties as management tools toward sustainability of water resources. In: Zalewski, M. (Ed.). *Ecological Engineering. Journal on Ecotechnology* 16: 1-8.

Zalewski, M., (2002). Ecohydrology - the use of ecological and hydrological processes for sustainable management of water resources. *Hydrological Sciences Journal* 47(5), 825-834.

Zalewski, M. (2005). Engineering Harmony. *Academia* 1(5), 4-7.

Zalewski, M. (2006). Flood pulses and river ecosystem robustness. In: *Frontiers in Flood Research.* Tchiguirinskaia, I.; Thein, K.N.N., K. & Hubert, P. (Eds). Kovacs Colloquium. June/July 2006. UNESCO, Paris. *IAHS Publication* 305. 212 pp.

Zalewski, M. (2007). Ecohydrology as a Concept and Management Tool. [In:] *Water and Ecosystems Managing Water in Diverse Ecosystems to Ensure Human Well-being*. King, C., Ramkinssoon, J., Clü sener-Godt, M., Adeel, Z. (Eds). UNU-INWEH UNESCO MAB, Canada: 39.53.

Zalewski, M. (2008). Rationale for the "Floodplain Declaration" from environmental conservation toward sustainability science. *Ecohydrology &Hydrobiology* Vol. 8, No. 2-4, 107-113.

Zalewski, M. (2009). Ecohydrology for engineering harmony between environment and society. *Danube News.* May 2009. No. 19. Volume 11.

Zalewski, M.; Janauer, G. S. & Jolankai, G. (1997). Ecohydrology - A new Paradigm for the Sustainable Use of Aquatic Resources. International Hydrological Program UNESCO. *Technical Document on Hydrology* No 7, Paris.

Zalewski, M. & Naiman, R.J. (1985). The regulation of riverine fish communities by a continuum of abiotic-biotic factors. In: *Habitat Modifications and Freshwater Fisheries*. Alabaster, J.S. (Ed.). FAO UN. Butterworths, London. 3-9.

Zalewski, M. & Kiedrzyńska, E. (2010). System approach to sustainable management of inland floodplains - declaration on sustainable floodplain management. *CAB Reviews: Perspectives in Agriculture, Veterinary Science, Nutrition and Natural Resources* 5, No. 056, 1-8.

17

Water Quality in the Agronomic Context: Flood Irrigation Impacts on Summer In-Stream Temperature Extremes in the Interior Pacific Northwest (USA)

Chad S. Boyd, Tony J. Svejcar and Jose J. Zamora
USDA-Agricultural Research Service, Eastern Oregon Agricultural Research Center, Burns, OR USA

1. Introduction

European arrival to the Pacific Northwest (U.S.A.) in the late 1800's signalled the beginning of an era of change for watershed use and water quality of free-flowing stream systems. Historic uses centered around trapping of beaver, utilization of livestock forage, and harvesting of anadramous fish and timber resources (Beschta, 2000). Impacts of these land uses on stream systems were severe; impaired flood plain development with beaver removal, interrupted nutrient cycling with overharvest of anadramous fish, streambank degradation due to overgrazing, and severe flooding associated with abusive logging practices (Trefethen, 1985; Meehan, 1991; Beschta, 2000). Perhaps most dramatic was the conversion of ecosystem type from stream system to impoundment associated with dam construction for hydroelectric power production (Beschta, 2000). The US Clean Water Act (CWA) of 1965 was crafted and passed into law in response to these and other issues, providing a legal and regulatory framework for managing land use practices that impact water quality (Adams 2007). Subsequent amendments of the CWA in 1972 and 1977 helped expand this framework to include impacts of non-point source pollution and mandated that states develop water protection programs (National Research Council, 1995; Adams, 2007).

While the forgoing issues continue to attract popular and political attention, legislative and regulatory mandates have dramatically reduced the acute effects of these practices and attention has now shifted to less dramatic, but nonetheless important associations between topical land use practices and water quality. There is an established and growing awareness of the impact of agriculturally-related non-point source pollution on stream systems. In this chapter we will 1) examine the scope of non-point source pollutant issues relating to the water quality/agriculture interface for streams in the Interior Pacific Northwest (PNW), 2) present a case-study examining the relationship between flood irrigation and in-stream temperature, and 3) make the case that stream temperature and non-point source water quality issues are complex problems that are best addressed by considering variation of the problem in both time and space.

1.1 Modern agriculture and water quality of free-flowing streams in the Pacific Northwest

Excessive erosion associated with row-crop agriculture has reduced water quality over much of the PNW, particularly in the region dominated by the Columbia Basin and Columbia Plateau of Eastern Washington, North-central Oregon, and Northern Idaho (Schillinger et al., 2010). Similarly, pathogens and nutrients from agricultural operations can and have disrupted PNW stream ecology. Pathogens now rank second and nutrients fourth on the list of top pollutants in U.S. water bodies and agricultural lands are a significant contributor to both pathogen and nutrient loading (Richter et al., 1997; Parajuli et al., 2008; USEPA, 2009). Nutrient by-products from agricultural operations, particularly nitrogen, can cause eutrophication of stream systems and associated reduction of dissolved oxygen and reduced biodiversity (Vitousek et al., 1997). From a management standpoint the issues of sediment, nutrient and pathogen contaminants have collectively been addressed using stream-side vegetation buffer strips in an attempt to attenuate pollutant entry into stream systems (Castelle et al., 1994; Schmitt et al., 1999; Dosskey 2002; Dorioz et al., 2006; Mayer et al., 2007). Vegetation buffers have been effective at reducing nutrient (Yates and Sheridan, 1983; Lowrance et al., 1985), pathogen (Tate et al., 2004; Knox et al., 2007) and sediment loading (Lyons et al., 2000; Lee et al., 2003) in streams. However, development of effective policy concerning the use of buffer strips has been complicated by the fact that the efficacy of buffers in reducing pollutant loading varies strongly in accordance with a number of design and environmental factors including buffer width, cover and height of plant material, slope, and soil attributes (Pearce et al., 1997, Atwill et al., 2005, George et al., 2011). For example, a review by Dorioz et al. (2006) reported variation in sediment retention by vegetation buffers of 2.5 orders of magnitude across studies. Ultimately, the use of riparian buffer strips will need to be paired with spatially explicit models (e.g., Tim and Jolly, 1994) to assist managers in integrating land-use practices with conservation measures to reduce pollutant yield at the watershed scale.

Livestock operations can have both direct and indirect effects on water quality of PNW stream systems. Direct effects can include increased nutrient and pathogen loading through fecal or urine additions (Nader et al., 1998; George et al. 2011); excessive nutrient loading can stimulate algal blooms leading to reduced dissolved oxygen concentrations (Belsky et al., 1999). Additionally, increased turbidity associated with hoof action can negatively impact aquatic organisms by decreasing primary production (Henley et al., 2000; Line, 2003). Strategic location of livestock attractants (e.g., salt, artificial water sources) can reduce livestock densities near riparian areas and associated nutrient inputs into stream systems (Tate et al., 2003; George et al. 2011). Indirect effects of livestock on water quality include impacts to streamside plant communities and physical damage to streambanks. Riparian vegetation plays a critical role in sustaining the biotic integrity of the stream ecosystem by anchoring bank material in place during high flow events (Kleinfelder et al., 1992; Clary and Leininger, 2000). Livestock grazing of this resource is generally sustainable under conditions of moderate utilization (e.g., graze to 10cm stubble height; Boyd and Svejcar, 2004; Volesky et al., 2011), but timing of grazing can also influence livestock impacts on water quality and riparian vegetation (Boyd and Svejcar, 2004). Severe utilization of streamside plants by livestock can lead to reduced plant biomass and altered stream channel conditions (e.g., wider and more shallow channel) that are associated with decreased water storage capacity, reduced filtration effects of vegetation buffers, and increased water temperature (Kauffman and Kruger, 1984; Winward, 1994; Toledo and Kauffman, 2001).

1.2 Agricultural impacts on stream temperature

Recently, concern has developed regarding summer in-stream temperature dynamics and agricultural practices that may be associated with elevated water temperature. Water temperature is an important attribute of the aquatic environment that has the potential to affect basic ecological processes such as nutrient cycling and can also modulate biotic ecology (Poole and Berman, 2001; Isaak and Hubert, 2001). Water temperature is a key driver of invertebrate demography and impacts fish species via temperature-dependent fluxes in dissolved oxygen content (Young and Huryn, 1998). In the Mountain and Northwest United States, concern over water temperature extremes and their impact on red band (*Oncorhynchus mykiss newberri*) and bull (*Salvelinus confluentus*) trout is an important driver of regulatory mechanisms governing agronomic and other land use practices.

Much of the interest surrounding the influence of agriculture on water temperature has focused on practices that reduce woody plants in the riparian area. A growing body of empirical evidence suggests that shade from woody plants can reduce daily maximum water temperature (Poole and Berman, 2001; Tate et al., 2005). Livestock grazing (as well as wildlife-associated herbivory) can reduce woody plant cover, particular when grazing occurs subsequent to the senescence of herbaceous vegetation (Holland et al., 2005; Clary et al., 1996; Matney et al., 2005). However, shade from woody plants is only one of a myriad of factor, including air mass characteristics, elevation, stream flow, stream gradient, adiabatic rate, channel width/depth, and groundwater inputs that can influence water temperature maxima in streams (Larson and Larson, 1996; George et al., IN PRESS). Diversion of stream water for purposes of irrigating agricultural crops can potentially impact a number of these processes including stream flow and groundwater dynamics.

2. Flood irrigation impacts on stream temperature: A case study

Flood irrigation is a common agricultural practice on interior PNW meadows used for hay and/or livestock forage. These meadows are typically low-to-mid elevation and contain, or are influenced by, a seasonal or perennial natural stream system. Water from the stream is diverted at the upstream end of the meadow into smaller irrigation ditches that parallel the stream at a distance on one or both sides. Water then disperses from the irrigation ditch across the meadow either passively through sub-surface flow, or actively via overland spillage from the irrigation ditch. Specific water management practices vary by and within drainage, however, the irrigation season generally begins in spring following the onset of mountain snowmelt. Efficiency and control of water usage is obviously low with flood irrigation, but it remains a popular form of irrigation due to low costs associated with the absence of power (i.e., electric or fossil fuel) and few major infrastructural requirements.

Excessive in-stream water temperature is the most frequent water quality impairment for streams in southeastern Oregon (Oregon Department of Environmental Quality 2002). Stream temperature regulations are associated mainly with concern over thermal requirements for cold water fish stocks (Boyd and Strudevant 1996, Gamperl et. al. 2002).

Many streams in this region lack or have minimal amounts of woody cover (i.e. shade); under such conditions, fluctuations in water temperature are strongly influenced by parallel fluctuations in air temperature (Stefan and Preud'homme 1993, McRae and Edwards 1994). Flood irrigation can potentially increase in-stream temperature through two mechanisms: 1) decreases in in-stream discharge, and 2) overland flow of warmer flood waters re-entering the stream. Conversely, flood irrigation could act to cool in-stream water temperature if sufficient flood water returns to the stream via groundwater input.

Previous research has suggested that flood irrigation may act to moderate summer in-stream temperature extremes by elevating the groundwater table in the surrounding meadow and increasing groundwater inputs into the stream (Stringham et al. 1998). If such a moderating effect were to occur, one prediction would be that daily in-stream temperature maximums would be less sensitive to the controlling influence of air temperature. Our objective in this case-study was to determine the impact of flood irrigation on seasonal temperature dynamics of a meadow stream in southeastern Oregon. Specifically, we characterized depth to groundwater and discharge patterns in irrigated and non-irrigated years, and tested the hypothesis that air temperature would have less effect on stream temperature in an irrigated compared to non-irrigated stream reach.

2.1 Study site

Our study took place in the Lake Creek drainage in Grant Co., OR, U.S.A. (11T0370683 UTM4890874) at an elevation of approximately 1500 m. Lake Creek is a perennial Rosgen C class (Rosgen 1994) stream. Its position near the base of the Strawberry Mountains causes strong seasonal discharge fluctuations associated with snowmelt in April through early June. The local area receives approximately 330 mm of annual precipitation, most falling as snow during the winter months; maximum air temperatures (approx. 28°C) occur in late July (Oregon Climate Service 2005). Soils were sandy clay loam underlain by a clay lens at depths ranging from 75 - 90 cm. We established a 3.5 km upper study reach and a 1.0 km lower study reach approximately 1.3 km downstream from the upper reach (Figure 1). The upper reach was within the irrigated portion of the meadow and the lower reach was not. Two tributaries entered the main channel of Lake Creek within the irrigated portion of the meadow. The beginning of the upper study reach was immediately downstream from the farthest downstream tributary. No tributaries entered within the lower study reach. Shade from woody plants was non-existent in either reach but the lower reach had limited topographic shading.

The meadow surrounding Lake Creek is seasonally flooded by two irrigation ditches flowing north to south along the east and west sides of the meadow (Figure 1). The West Ditch (0.08 m^3/sec^{-1}) and East Ditch (0.06 m^3/sec^{-1}) diversions were opened in 2004 from April 13 to June 30 and from April 21 to June 30 in 2005. Approximately 70% of the flow in West Ditch #1 was diverted into West Ditch #2. A lateral slope downward from the irrigation ditches to streamside elevations allowed for subsurface flow from irrigation ditches to the surrounding meadow. Diversions within the West Ditch were used to augment flood irrigation from subsurface flow. Water spreading from the East Ditch relied entirely on subsurface flow from the irrigation ditch.

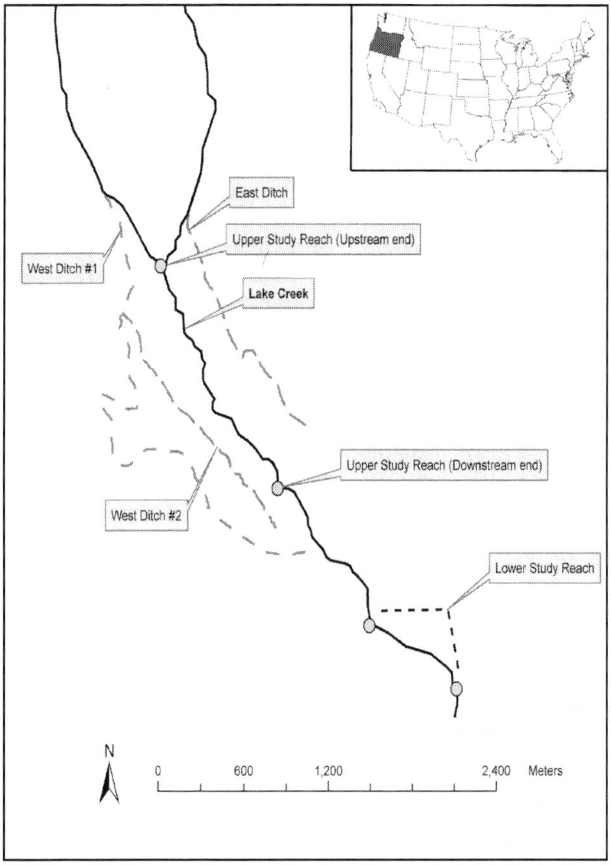

Fig. 1. Overview of study site on Lake Creek, southeast OR, U.S.A. depicting irrigation ditches and locations for upper and lower study reaches (inset shows location of Oregon within U.S.A.). The meadow surrounding the upper reach was flood irrigated from mid-April to June 30 of 2004-2005.

2.2 Methods and materials

Data for a non-irrigated year were collected in 2002 and for flood irrigated years in 2004 - 2005. We measured point-in-time stream discharge at approximately 10-day intervals in June and July using permanent main channel cross-sections located at the upstream ends of the upper and lower study reaches. Discharge was estimated using a magnetic-head pygmy flow meter with a top set wading rod and an Aquacalc 5000 discharge recorder (JBS Energy, Inc; West Sacramento, CA). Measurements were made at 60% of total depth from the stream water surface. Regression analysis was used to estimate mid-month values for June and July. Depth to groundwater within the upper reach was measured at 10-day intervals in June and July using shallow PVC wells spaced at 15-m intervals along 4 transects run

shut-off was much less as compared to the irrigated years (Figure 2). These data suggested a higher meadow water table with irrigation and a greater potential for groundwater inputs into the stream. Maximum daily groundwater temperatures increased from June to July and ranged from approximately 9 to 12°C across years and months (Figure 3). Increasing groundwater temperature in July was probably associated with increasing air temperature (Ward 1985). Groundwater temperatures were generally 8 to 14°C cooler than in-stream temperature maxima suggesting that any groundwater return flow into the main channel would have a buffering effect on maximum stream temperatures.

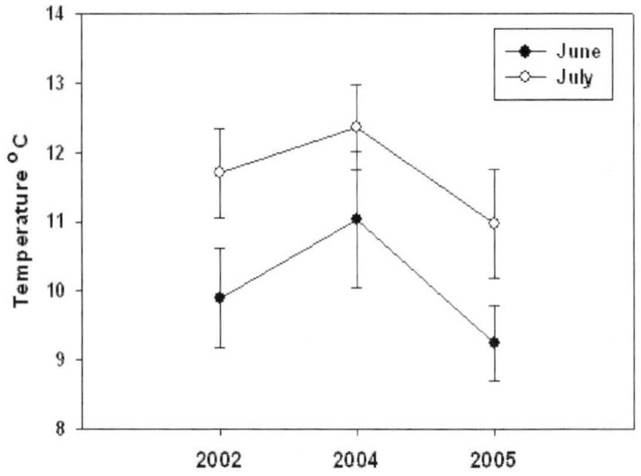

Fig. 3. Daily maximum groundwater temperature means and associated standard errors for the upper reach of Lake Creek, OR, U.S.A. The meadow surrounding the upper reach was flood irrigated from mid-April to June 30 of 2004-2005.

Maximum daily air temperature was associated positively with maximum daily water temperature in both irrigated and non-irrigated years (Figure 4). R^2 values ranged from 0.55 to 0.81, with the exception of July, 2002 which had values of 0.30 and 0.32 for the upper and lower reaches, respectively (Table 2). Explanatory power during this time period was decreased by several outlying points that were associated with cool nights followed by warm, but overcast days. If these days are excluded (n = 3), R^2 values for the upper and lower reaches increase to 0.48. In the non-irrigated year (2002) maximum daily water temperature in both study reaches responded similarly to air temperature although the intercept was slightly higher for the lower reach in June ($P = 0.003$, Figure 4). In the irrigated years intercepts differed ($P < 0.001$) between the upper and lower reaches and slopes were different ($P < 0.071$) for all but July of 2004 ($P = 0.159$, Figure 4). At a given maximum air temperature, water temperature was slightly less in the upper reach suggesting that flood irrigation helped moderate maximum daily water temperature. Although the y intercept was slightly higher for the lower reach in June of the non-irrigated year (2002), the lines of best fit for irrigated and non-irrigated reaches converge at higher air temperatures. In contrast, lines of best fit for the study reaches diverge at higher air temperatures during irrigated years (2004-2005), providing further evidence for an irrigation-related buffering effect.

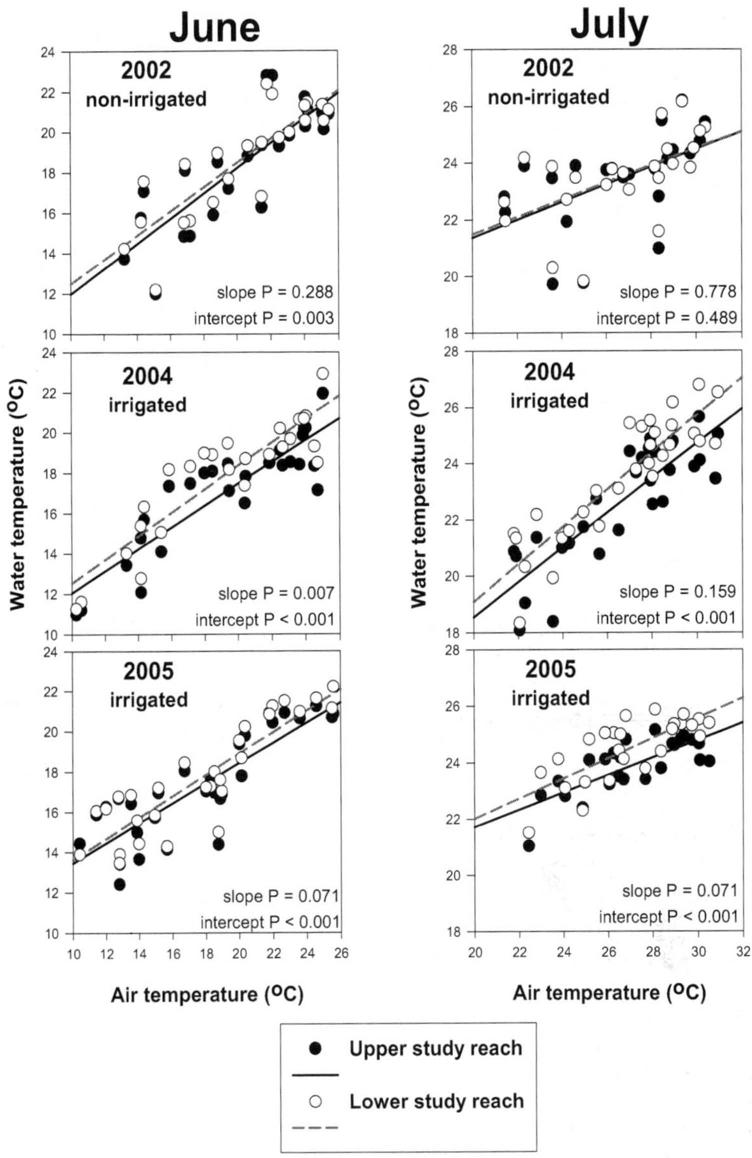

Fig. 4. The relationship between daily maximum air and stream temperature values for June and July within the upper and lower study reaches of Lake Creek, OR, U.S.A. The meadow surrounding the upper reach was flood irrigated from mid-April to June 30 of 2004-2005. Probability values for differences in slope and intercept values between irrigated and non-irrigated reaches are reported within graphs.

Year	Month	Reach	Slope	y intercept	R^2	Predicted maximum water temperature °C
2002	June	Upper	0.62	5.77	0.66	21.92
		Lower	0.60	6.51	0.71	22.03
2002	July	Upper	0.31	15.16	0.3	24.80
		Lower	0.30	15.45	0.32	24.82
2004	June	Upper	0.54	6.64	0.79	20.71
		Lower	0.58	6.71	0.81	21.84
2004	July	Upper	0.62	6.20	0.73	25.33
		Lower	0.66	5.82	0.79	26.37
2005	June	Upper	0.50	8.47	0.72	21.44
		Lower	0.53	8.31	0.78	22.11
2005	July	Upper	0.31	15.60	0.57	25.09
		Lower	0.36	14.93	0.55	25.93

Table 2. Predicted monthly maximum water temperature values for the upper and lower study reaches of Lake Creek, OR, U.S.A. The meadow surrounding the upper reach was flood irrigated from mid-April to June 30 of 2004-2005. Equations were in the form of *max water* = *max air(x)* + *b* and were generated for each year/month/location combination based on data for regression equations in Figure 4. Air temperatures of 26°C and 31°C were used to approximate observed maximum temperatures for June and July, respectively.

Without irrigation (2002) predicted maximum water temperatures in the 2 reaches were very similar. However, with irrigation (2004 and 2005) there was divergence with cooler temperatures in the upper reach. For irrigated years predicted water temperature maximums were approximately 0.9°C warmer in the lower reach as compared to the upper reach (Table 2). Temperature reduction in irrigated years was recorded under varying discharge levels including the late snow-melt run-off period in June and lower flow conditions in July (Table 1). Stringham et al. (1998) reported maximum stream temperatures of 1 to 3°C cooler for a flood irrigated vs. non-irrigated reach. The relatively smaller thermal response in our study may be associated with the presence of old irrigation ditches between the currently used ditches (Figure 1) and the active stream channel. These ditches intercepted some of the sub-surface flow and provided additional water surface area for evaporative losses. Our results are in contrast to those of Tate et al. (2005) who reported increased stream temperatures with flood irrigation. This discrepancy may be explained by the relatively low percentage of total stream discharge (about 20%) that we diverted for irrigation as opposed to that of Tate et al. (30 to 70%).

In summary, we found that air temperature exerted a measurable and strong influence on water temperature maximums, explaining up to 80% of the variation in this variable (Table 2). For the stream we studied, flood irrigation appeared to moderate the influence of air temperature on daily maximum water temperature; this effect was observed during irrigation and continued for at least 1 month after cessation of irrigation. Our results agree with earlier work from Stringham et al. (1998) and suggest that flood irrigation can help buffer daily maximum water temperature during summer air temperature extremes. In the present study, the magnitude of the predicted temperature reduction was generally < 1°C. However, stream size, percentage of water diverted, and proportion of surface/subsurface flows may all influence the effect of flood irrigation on water temperature.

3. Stream temperature as a complex problem

Boyd and Svejcar (2009) presented the notion that those problems facing natural resources managers today differ, in fundamental ways, when compared to those of previous generations. Specifically these authors argued that many of the issues facing managers today are "complex", in that the environmental factors associated with these issues vary in both space and time, making it difficult to generalize management prescriptions or characterize ecosystem responses. These same hindrances make it challenging to formulate biologically-realistic regulatory statues for stream temperature and other water quality parameters. In terrestrial systems, simultaneous consideration of space and time is difficult given that space has x, y and z (i.e., elevation) dimensions. Measurement of water temperature (and water quality in general) is somewhat different in that, practically speaking, space has only one dominant dimension (i.e., upstream or downstream). Thus the riparian ecosystem offers a unique opportunity for simultaneously considering environmental characteristics in space and time.

Our preceding examination of water temperature dynamics on Lake Creek is a good example of how variation in space and time can interact to in influence water quality parameters. Our approach in this case was to monitor water temperature over both space and time, in irrigated and non-irrigated years and to use this information to make inferences regarding the influence of flood irrigation on water temperature maxima. To further characterize variation in water temperature over both space and time, we plotted daily water temperature maximums for a non-irrigated year (2002) from June 1 (Julian day 152) to September 30 (Julian day 273; Figure 5). Data were interpolated using negative exponential smoothing and displayed in a contour plot (SigmaPlot 12.0; Systat Software Inc.; San Jose, CA) that allows the user to view change in temperature over time at a given point in the stream and change over space for a given day.

When viewed in this way, daily maximum water temperature is set within a background of strong intra-annual and spatial variation. From a management standpoint, these data imply that capturing variation in space and time should be an integral part of a water quality assessment strategy. For example, if we were to have measured temperature at one point within the irrigated reach (stream distance 0 - 3.5 km), and one location within the non-irrigated reach (3.5 to 5.9 km) then our conclusions regarding the influence of flood irrigation on stream temperature would have been strongly influenced by choice of sampling location, as compared to the multiple location sampling strategy that we employed. Similarly, our use of a two-month sampling window helped overcome undue influence of time at shorter sampling intervals. This same logic applies equally to determining compliance within a regulatory framework. For example, the Oregon Department of Environmental Quality stipulates that average daily maximum water temperature for cold-water fisheries should not exceed 20°C for any seven-day period, and that number drops to 12°C for streams containing bull trout spawning and juvenile rearing habitat (ODEQ, 2008). Determination of compliance with either standard for the section of Lake Creek we studied would depend strongly on when and where samples were taken (see Figure 5). The spatial and temporal variability of maximum water temperature for Lake Creek could also have strong relevance to the ecology of aquatic organisms. For example, while water temperatures at a given point/time on Lake Creek may exceed tolerance levels of cold water fishes, variability in temperature over space may act to provide thermal refugia that allow affected species to escape potentially harmful temperatures (Ebersole et al., 2001).

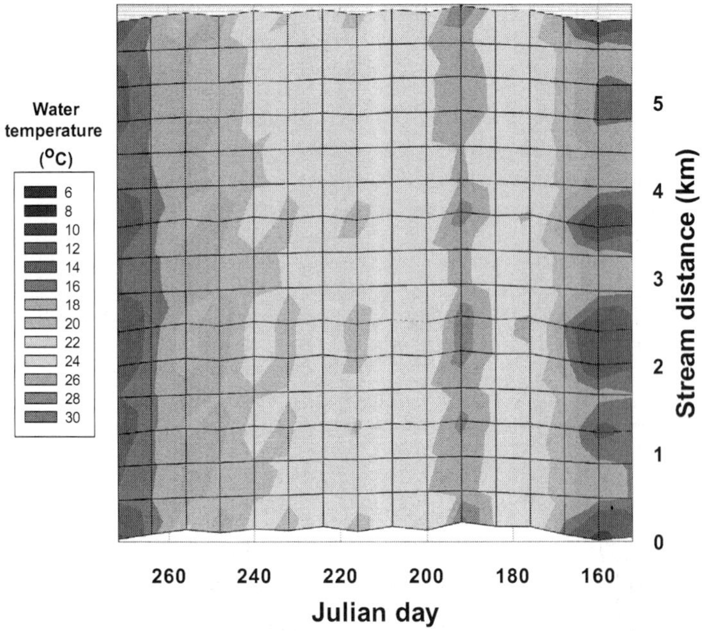

Fig. 5. Daily maximum water temperature values over space and time for Lake Creek, OR, U.S.A.

4. Conclusions

In conclusion, we suggest that complex problems such as water temperature management may not be solved with the broad, sweeping regulatory statues (e.g., the Clean Water Act and various amendments) historically used to solve more easily identifiable problems, but instead require an understanding of the inherent ecological uniqueness that defines complex problems on a case-by-case basis. We suggest that management and regulation of in-stream water temperature is a complex problem that will vary across both space and time and that natural resources professionals must recognize the dynamic nature of this relationship in designing management plans, regulatory policy, and future research. Such recognition is often at odds with contemporary policies and paradigms that focus attention on discreet temperature (or other water quality) values, thus creating tension between managers and regulators. Incorporating spatial and temporal variation into the concept of stream temperature will allow for a more representative characterization of water temperature regime and promote a more comprehensive understanding of the potential impacts of water temperature on the stream ecosystem and its inhabitants.

5. References

Adams, P.W. (2007). Policy and management for headwater streams in the Pacific Northwest: synthesis and reflection. *Forest Science*, Vol.53, pp. 104-118.

Atwill, E.R.; Tate, K.W.; Das Gracas, M.; Pereira, C.; Bartolome, J.W. & Nader, G.A. (2005). Efficacy of natural grass buffers for removal of *Cryptosporidium parvum* in rangeland runoff. *Journal of Food Protection*, Vol.69, pp. 177-184.

Belsky, A.J.; Matzke, A. & Uselman, S. (1999). Survey of livestock influences on stream and riparian ecosystems in the western United States. *Journal of Soil and Water Conservation*, Vol.54, pp. 419-431.

Beschta, R.L. (2000). Watershed management in the Pacific Northwest; the historical legacy. In: Land Stewardship in the 21st Century: Contributions of Watershed Management. USDA Forest Service Rocky Mountain Forest & Range Experiment Station General Technical Report.

Boyd, M. & Strudevant, D. (1996). The scientific basis for Oregon's stream temperature standard: common questions and straight answers. Oregon Department of Environmental Quality. 30 pp.

Boyd, C.S. & Svejcar, T.J. (2004). Regrowth and production of herbaceous riparian vegetation following defoliation. *Journal of Range Management*, Vol. 57, pp. 448-454.

Boyd, C.S. & Svejcar, T.J. (2009). Managing complex problems in rangeland ecosystems. *Rangeland Ecology and Management*, Vol. 62, pp. 491-499.

Castelle, A.J.; Johnson, A.W. & Conolly, C. (1994). Wetland and stream buffer requirements: a review. *Journal of Environmental Quality*, Vol.23, pp. 878-882.

Clary, W.P & Leininger, W.C. (2000). Stubble height as a tool for management of riparian areas. *Journal of Range Management*, Vol. 53, pp. 562-573.

Clary, W.P.; Shaw,N.L; Dudley, J.G.; Saab, V.A.; Kinney, J.W.& Smithman, L.C. (1996). Response of a depleted sagebrush steppe riparian system to grazing control and woody plantings. USDA, Forest Service, Intermountain Research Station Research Paper INT-RP-492.

Dorioz, J.M.; Wang, D.; Poulenard, J. & Trevisan, D. (2006). The effect of grass buffer strips on phosphorous dynamics - A critical review and synthesis as a basis for application in agricultural landscapes in France. *Agriculture, Ecosystems and Environment*. Vol.117, pp. 4-21.

Dosskey, M.G. (2002). Setting priorities for research on pollution reduction functions of agricultural buffers. *Environmental Management*, Vol.30, pp. 641-650.

Ebersole, J.L.; Liss, W.J. & Frissell, C.A. (2001). Relationship between stream temperature, thermal refugia and rainbow trout *Oncorhynchus mykiss* abundance in arid-land streams in the northwestern United States. *Ecology of freshwater fish*, Vol.10, pp. 1-10.

Gamperl, A. K., Rodnick, K. J., Faust, H. A., Venn, E. C., Byington, J. D., Bennett, M. T., Crawshaw, L. I., Kelley, E. R., Powell, M. S. & Li, H. (2002). Metabolism, swimming performance and tissue biochemistry of high desert redband trout (Oncorhynchus mykiss ssp.): Evidence of phenotypic differences in physiological function. Physiological and Biochemical Zoology. Vol.75, pp. 413-431.

George, W.F.; Jackson, R.D.; Boyd, C.S. & Tate, K.W. (2011). A scientific assessment of the effectiveness of riparian management practices. pp. 213-252. Briske, D.D. (*editor*). Conservation benefits of rangeland practices: Assessment, recommendations, and knowledge gaps. United States Department of Agricultur - Natural Resources Conservation Service.

Henley, W.F.; Patterson, M.A.; Neves, R.J. & Dennis Lemly, A. (2000). *Reviews in Fisheries Science*, Vol.8, pp. 125-139.

Holland, K.A; Leininger, W.C. & Trilica, M.J. (2005). Grazing history affects willow communities in a montane riparian ecosystem. *Rangeland Ecology and Management*, Vol.58, pp. 148-154.

Isaak, D.J.& Hubert, W.A. (2001). A hypothesis about factors that affect maximum summer stream temperatures across montane landscapes. *Journal of the American Water Resources Association*, Vol. 37, pp. 351-366.

Kauffman, J.B. & Krueger, W.C. (1984). Livestock impacts on riparian ecosystems and streamside management implications...A review. *Journal of Range Management*, Vol. 37, pp. 430-438.

Kleinfelder, D.; Swandon, S.; Norris, G. & Clary, W. (1992). Unconfined compressive strength of some streambank soils with herbaceous roots. *Journal of the Soil Science Society of America*, Vol. 56, pp. 920-924.

Knox, A.K.; Tate, K.W.; Dahlgren, R.A. & Atwill, E.R. (2007). Management reduces E. coli in irrigated pasture runoff. *California Agriculture*, Vol.61, pp. 159-165.

Larson, L.L. & Larson, S.L. (1996). Riparian shade and stream temperature: A perspective. *Rangelands*, Vol.18, pp. 149-152.

Lee, K.H.; Isenhart, T.M. & Shultz, R.C. (2003). Sediment and nutrient removal in an established multi-species riparian buffer. *Journal of Soil and Water Conservation*, Vol. 58, pp. 1-8.

Line, D.E. (2003). Changes in a stream's physical and biological conditions following livestock exclusion. *American Society of Agricultural Engineers*, Vol. 46, pp. 287-293.

Lowrance, R.R.; Leonard, R.A., Asmussen, L.E. & Todd, R.L. (1985). Nutrient budgets for agricultural watersheds in the southeastern coastal plan. *Ecology*, Vol.66, pp. 287-296.

Lyons, J.; Trimble, S.W. & Paine, L.K. (2000). Grass versus trees: Managing riparian areas to benefit streams of central North America. *Journal of the American Water Resources Association*, Vol.36, pp. 919-930.

Matney, C.A.; Boyd, C.S. & Stringham, T.K. (2005). Use of felled junipers to protect streamside willows from browsing. *Rangeland Ecology and Management*, Vol.58, pp. 652- 655.

Mayer, P.M.; Reynolds, S.K.; Steven, J., Jr.; McCutchen, M.D.; Marshall, D. & Canfield, T.J. (2007). Meta-analysis of nitrogen removal in riparian buffers. *Journal of Environmental Quality*, Vol.36, pp. 1172-1180.

McRae, G. & Edwards, C.J. (1994). Thermal characteristics of Wisconsin headwater streams occupied by beaver: Implications for brook trout habitat. *Transactions of the American Fisheries Society*, Vol.123, pp. 641-656.

Meehan, W.R. (1991). Influences of forest and rangeland management on salmonid fishes and their habitats. American Fisheries Society, Bethesda, MD. 751p.

Nader, G.; Tate, K.W.; Atwill, R. & Bushnell, J. (1998). Water quality effect of rangeland beef cattle excrement. *Rangelands*, Vol.20, pp. 19-25.

National Research Council (1995). Wetlands: characteristics and boundaries. Committee of the Characterization of Wetlands, National Academy of Sciences, Washington, D.C. 307p.

Oregon Climate Service. (2005). Climatological data for Seneca, Oregon. Oregon State university, Corvallis, OR.

Oregon Department of Environmental Quality. (2002). Oregon's Final 2002 303(d) List. http://www.deq.state.or.us/wq/WQLData/View303dList02.asp.

Oregon Department of Environmental Quality. (2008). Temperature Water Quality Standard Implementation - A DEQ Internal Management Directive. http://www.deq.state.or.us/wq/pubs/imds/Temperature.pdf

Parajuli, P.B., Mankin, K.R. & Barnes, P.L. (2008). Applicability of targeting vegetative filter strips to abate fecal bacteria and sediment yield using SWAT. *Agricultural Water Management*, Vol.95, pp. 1189-1200.

Pearce, R.A.; Trlica, M.J.; Leininger, W.C.; Smith, J.L. & Frasier, G.W. (1997). Efficiency of grass buffer strips and vegetation height on sediment filtration in laboratory rainfall simulation. *Journal of Environmental Quality*, Vol.26, pp. 139-144.

Poole, G.C. & Berman, C.H. (2001). An ecological perspective on in-stream temperature: natural heat dynamics and mechanisms of human-caused thermal degradation. *Environmental Management*, Vol. 27, pp. 787-802.

Rosgen, D.L. (1994). A classification of natural rivers. *Catena*, Vol.22, pp. 169-199.

Richter, B.D.; Braun, D.P.; Mendelson, M.A. & Master, L.L. (1997). Threats to imperilled freshwater fauna. *Conservation Biology*, Vol.11, pp. 1081-1093.

SAS Institute Inc. (1999). SAS procedures guide, release 8.0. Cary, NC.

Schillinger, W.F.; Papendick, R.I. & McCool, D.K. (2010). Soil and Water Challenges for Pacific Northwest Agriculture. Soil and Water Conservation Advances in the United States. pp. 47-79. Zobeck, T.M. & Schillinger, W.F. (*editors*). Soil Science Society of America Special Publication 60, Madison, WI.

Schmitt, T.J.; Dosskey, M.G & Hoagland, K.D. (1999). Filter strip performance and processes for different vegetation, widths, and contaminants. *Journal of Environmental Quality*, Vol.28, pp. 1479-1489.

Stefan, H.G. & E.B. Preud'homme. (1993). Stream temperature estimation from air temperature. *Water Resources Bulletin*, Vol.29, pp. 27-45.

Stringham, T.K., J.C. Buckhouse & Krueger, W.C. (1998). Stream temperatures as related to subsurface waterflows originating from irrigation. *Journal of Range Management*, Vol.51, pp. 88-90.

Tate, K.W.; Atwill, E.R., McDougald, N.K. & George, M.R. (2003). Spatial and temporal patterns of cattle feces deposition on rangeland. *Journal of Range Management*, Vol.56, pp. 432-438.

Tate, K.W.; Das Gracas, M; Pereira, C & Atwill, E.R. (2004). Efficacy of vegetated buffer strips for retaining *Cryptosporidium parvum*. *Journal of Environmental Quality*, Vol.33, pp. 2343-2251.

Tate, K.W.; Lile, D.F.; Lancaster, D.L.; Porath, M.L.; Morrison, J.A. & Sado, U. (2005). Statistical analysis of monitoring data to evaluate stream temperature - a watershed scale case study. *California Agriculture*, Vol.59, pp. 161-167.

Tim, U.S. & Jolly, R. (1994). Evaluating agricultural nonpoint-source pollution using integrated geographic information systems and hydrologic/water quality models. *Journal of Environmental Quality*, Vol.23, pp. 25-35.

Toledo, Z.O. & Kauffman, J.B. (2001). Root biomass in relation to channel morphology of headwater streams. *Journal of the American Water Resources Association*, Vol. 37, pp. 1653-1663.

Trefethen, J.B. (1985). An American Crusade for Wildlife. Boone and Crockett Club, Alexandria, VA. 409p.

United States Environmental Protection Agency (2009). Fact Sheet: Introduction to Clean Water Act (CWA) Section 303(d) Impaired Waters Lists. U.S. Environmental Protection Agency, Washington, D.C. Available at http://www.epa.gov/owow/tmdl/results/pdf/aug_7_introduction_to_clean.pdf (accessed September 26, 2011).

Vitousek, P.M.; Aber, J.; Howarth, R.W.; Likens, G.E.; Matson, P.A.; Schindler, D.W.; Schlesinger, W.H. & Tilman, G.D. (1997). Human alteration of the global nitrogen cycle: Causes and consequences. *Issues in Ecology*, Vol.61, pp. 1225-1232.

Voleksy, J.D; Schacht, W.H.; Koehler, A.E.; Blankenship, E. & Reece, P.E. (2011). Defoliation effects on herbage production and root growth of wet meadow forage species. *Rangeland Ecology and Management*, Vol.64, pp. 506-513.

Ward, J.V. (1985). Thermal characteristics of running waters. *Hydrobiologia*, Vol.125, pp. 31-46.

Winward, A.H. (1994). Management of livestock in riparian areas, p. 49–52. *In* : G . A. Rasmussen and J.P. Dobrowolski (*editors*), Riparian resources: a symposium on the disturbances, management, economics, and conflicts associated with riparian ecosystems. Coll. of Natur. Resour., Utah State University, Logan, UT.

Yates, P. & Sheridan, J.M. (1983). Estimating the effectiveness of vegetated floodplains/wetlands as nitrate-nitrite and orthophosphorous filters. *Agriculture, Ecosystems and Environment*, Vol.9, pp. 303-314.

Young, R.G. & Huryn, A.D. (1998). Comment: Improvements to the diurnal upstream-downstream dissolved oxygen change technique for determining whole-stream metabolism in small streams. *Canadian Journal of Fisheries and Aquatic Sciences*, Vol.55, pp. 1784-1785.

18

Effects of Discharge Characteristics on Aqueous Pollutant Concentration at Jebel Ali Harbor, Dubai-UAE

Munjed A. Maraqa[1], Ayub Ali[2], Hassan D. Imran[1], Waleed Hamza[1] and Saed Al Awadi[3]

[1]*United Arab Emirates University*
[2]*Griffith University*
[3]*Ports, Customs and Free Zone Corporation*
[1,3]*United Arab Emirates*
[2]*Australia*

1. Introduction

The Arabian Gulf is an important geographical location. The Gulf has been extensively used for transport purposes. Meanwhile, countries in the region benefit from the Gulf's diverse marine habitats and utilize its water for desalination or some industrial needs. Several pollutants are induced into the Gulf including those resulting from oil spill accidents, offshore exploration processes, ballast water discharge, reject brine discharge, dredging activities, and coastal construction projects. Meanwhile, some of the Gulf countries are developing new coastal industrial facilities or expanding existing ones. These facilities are not without an adverse impact on the marine environment.

Most of the work related to the quality of Arabian Gulf water has focused on understanding the flow dynamics and the impact of oil spills. A number of researchers, for example, used numerical modeling to investigate residual circulation and flow pattern of the Arabian Gulf (Hughes and Hunter, 1979; Lardner et al. 1987, 1993; Chao et al. 1992; Horton et al., 1994; Elshorbagy et al., 2006; Azam et al., 2006a, 2006b; Thoppil & Hogan, 2010). Other researchers assessed the Gulf water quality as affected by oil spills (El Samra et al., 1986; Lardner et al., 1988; Al-Rabeh et al., 1992; Spaulding et al., 1993).

Little attention, however, has been directed to investigate the impact of discharges from coastal industrial facilities on water quality in the Arabian Gulf. In this study, we will consider the case of Jebel Ali Harbor to numerically assess the harbor's water quality as affected by discharges from industrial facilities located at Jebel Ali Free Zone (JAFZ) area in Dubai, United Arab Emirates (UAE). The harbor at JAFZ area (Fig. 1) is one of the largest man-made ports in the world. The harbor receives discharge consisting of several treated industrial effluents. Water discharged into the harbor must adhere to the effluent quality criteria set forth in the Environmental Requirements established by the Ports, Customs and Free Zone Corporation (PCFC, 2003) at JAFZ area. PCFC has also established harbor water quality objective limits (PCFC, 2003) in order to protect marine life and to minimize the impact of industrial activities on the surrounding ecosystem. Regular monitoring of discharged treated wastewater as well as harbor water and sediments is conducted by the PCFC to assure adherence to effluent standard and quality objective limits.

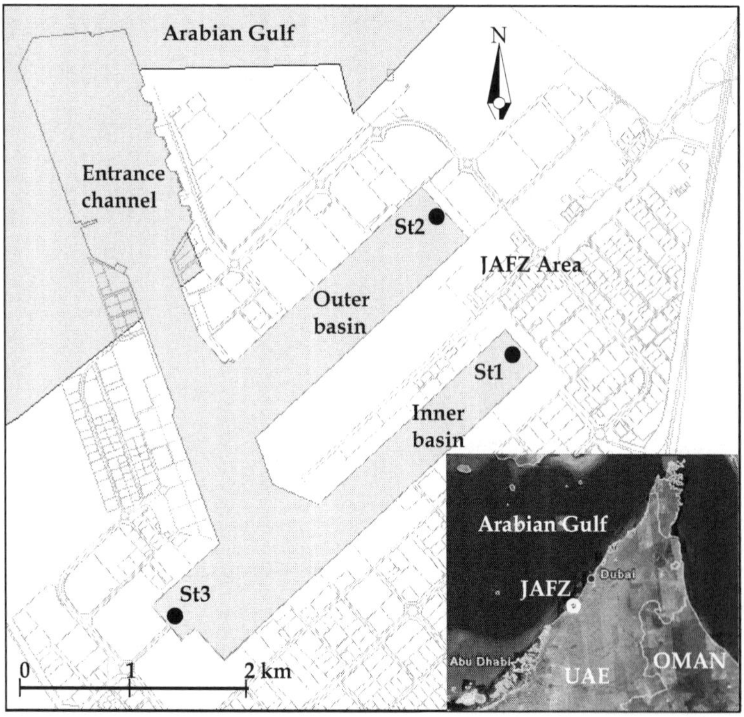

Fig. 1. Map of Jebel Ali Harbor. The circle in the bottom map is the location of JAFZ area.

Future expansion of industrial activities at JAFZ area, in addition to port activities and on-going as well as planned coastal construction projects in the vicinity of the harbor, may increase pollutant loading to the receiving water body. Limited work, however, has been conducted to assess the water quality of Jebel Ali Harbor. Maraqa et al. (2007) studied the fate of selected pollutants in the harbor and concluded that induced pollutants tend to accumulate in the harbor due to its limited flushing capacity. Furthermore, Maraqa et al. (2008) found that the main flow regime in the harbor follows alternate paths during flooding and ebbing, which creates eddy-like circulations in net flow distribution (see Fig. 2). Maraqa et al. (2008) also showed that dead-end locations at Jebel Ali Harbor have low water circulation and that flushing of a conservative pollutant discharged into the harbor takes a few months to several years depending on the discharge location.

This study expands on the work of Maraqa et al. (2008) to investigate variations in the concentration of pollutants induced into the harbor due to variations in the loading rate, discharge location and discharge concentration. A similar approach was used by Kashefipour et al. (2002; 2006) to assess the impact of various bacterial input rates on the receiving water in coastal basins in the UK. This study further explores the relationship between a continuous pollutant loading rate and average pollutant concentration in Jebel Ali Harbor water. As such, the study is of importance to modeling experts and managers interested in the hydrodynamic and transport properties of this harbor. The outcome of this study could further assist managers of the harbor decide on proper input rates and discharge locations so that water quality objective limits are not exceeded. Meanwhile, the general approach presented here may be of value in application to other systems.

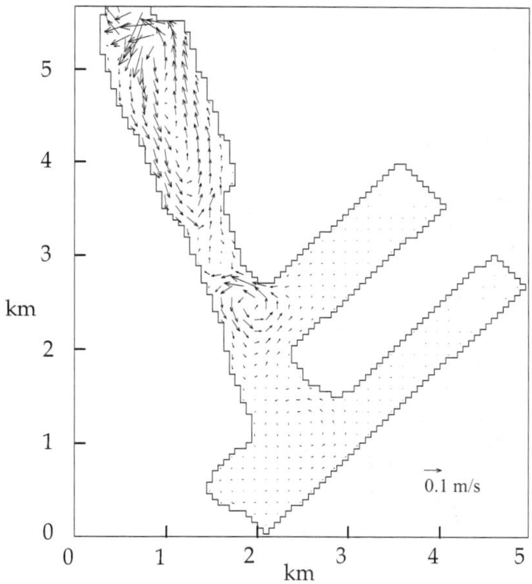

Fig. 2. Net flow over a tidal cycle of Jebel Ali Harbor (adopted from Maraqa et al., 2008)

2. Governing equations

Numerical modeling can be used to help achieve discharge conditions that meet pre-set environmental limits. Numerical modeling has been extensively applied to simulate water circulation and contaminant transport in harbors and semi-enclosed coastal areas. For example, efforts were made to better understand the hydrodynamic regimes (Estacio et al., 1997; Vethamnoy et al., 2005; Azam et al., 2006a; Dias and Lopes, 2006a,b; Maraqa et al., 2008; Montano-Ley et al., 2008), to investigate pollutant dispersion (Gesteira-Gomez et al., 1999; Das et al. 2000), and to assess water quality (Tao et al., 2001; Copeland et al., 2003; Fiandrino et al., 2003; Lopes et al., 2005; Cerejo and Dias, 2007). Other efforts were made to establish surveillance procedures (Lopes et al., 2005) and to quantify the impact of effluent discharge (Ganoulis, 1991; Kashefipour et al., 2002; Gupta et al., 2004, Kashefipour et al., 2006; Rucinski et al., 2007; Brewer et al., 2008).

Modeling of fluid flow is based on the principles of continuity of mass and conservation of momentum. For flows, which show little variation in the vertical dimension, it is acceptable to integrate these equations over the depth of water, resulting in two-dimensional (2D) equations of motion. In a 2D hydrodynamic (HD) model, the continuity equation is:

$$\frac{\partial \zeta}{\partial t}+\frac{\partial p}{\partial x}+\frac{\partial q}{dy}=0 \tag{1}$$

where, ζ is the water level (m); p and q are flux densities in x and y directions ($m^3/s/m$); t is time (s); x and y are space coordinates (m). The x-and y-momentum are given by Eq. (2) and (3), respectively:

$$\begin{aligned}&\frac{\partial p}{\partial t}+\frac{\partial}{\partial x}\left(\frac{p^2}{h}\right)+\frac{\partial}{dy}\left(\frac{pq}{h}\right)+gh\frac{\partial \zeta}{\partial x}+\\&\frac{gp\sqrt{p^2+q^2}}{C^2h^2}-\frac{1}{\rho_w}\left[\frac{\partial}{\partial x}(h\tau_{xx})+\frac{\partial}{\partial y}\left(h\tau_{xy}\right)\right]-\\&\Omega q-fVV_x\frac{h}{\rho_w}\frac{\partial}{\partial x}(p_a)=0\end{aligned} \tag{2}$$

$$\begin{aligned}&\frac{\partial q}{\partial t}+\frac{\partial}{\partial y}\left(\frac{q^2}{h}\right)+\frac{\partial}{dx}\left(\frac{pq}{h}\right)+gh\frac{\partial \zeta}{\partial y}+\\&\frac{gq\sqrt{p^2+q^2}}{C^2h^2}-\frac{1}{\rho_w}\left[\frac{\partial}{\partial x}\left(h\tau_{yy}\right)+\frac{\partial}{\partial x}\left(h\tau_{xy}\right)\right]-\\&\Omega p-fVV_y\frac{h}{\rho_w}\frac{\partial}{\partial y}(p_a)=0\end{aligned} \tag{3}$$

where, h is water depth (m); C is Chezy resistance ($m^{1/2}/s$); f is the wind friction factor (dimensionless), V, V_x, and V_y are wind speed and components in x and y directions (m/s), respectively; Ω is Coriolis parameter (s^{-1}); p_a is atmospheric pressure ($kg/m/s^2$); ρ_w is the density of water (kg/m^3); and τ_{xx}, τ_{xy} and τ_{yy} are components of effective shear stress (N/m^2).

The advection-dispersion (AD) model simulates the spreading of a substance in an aquatic environment under the influence of fluid transport and dispersion processes. The substance may be treated conservatively or with decay. The governing equation for a 2D AD model is given as (Adams and Baptista, 1986):

$$\begin{aligned}&\frac{\partial}{\partial t}(hc)+\frac{\partial}{\partial x}(uhc)+\frac{\partial}{\partial y}(vhc)=\frac{\partial}{\partial x}\left(hD_x\frac{\partial c}{\partial x}\right)\\&+\frac{\partial}{\partial y}\left(hD_y\frac{\partial c}{\partial y}\right)-khC+Q_s(c_s-c)\end{aligned} \tag{4}$$

where, c is substance concentration (mg/l); u and v are horizontal velocity components in x and y directions (m/s), respectively; D_x and D_y are dispersion coefficients in x and y directions (m^2/s), respectively; k is the linear decay rate coefficient (s^{-1}); Q_s is the source/sink discharge per unit horizontal area ($m^3/s/m^2$); c_s is substance concentration in the source/sink discharge (mg/l).

3. Methodology

3.1 Model description

Jebel Ali Harbor has an approach channel that starts 15 km offshore. The approach channel has a depth of 14-15 m and a width of 280 m reducing to 235 m. It bends after 10 km and becomes the entrance channel. It widens to 300 m at the bend and to 340 m at the entrance channel. There are two basins within the port. The outer 14-m deep basin is 2.3 km long and 600 m wide. The inner basin is 3.7 km long and 425 m wide, with a depth of 11.5 m. All channel and basin bottoms are sandstone. The surface area of the harbor is about 5.3 million m^2 and the total water volume is about 75 million m^3.

Maraqa et al. (2008) developed a 2D model using the MIKE21 modeling system of the Danish Hydraulic Institute (DHI, 2003a; 2003b) to simulate the HD and the AD processes within the Jebel Ali Harbor. Justification of the use of a depth-integrated 2D model was based on the nearly uniform temperature and salinity profiles found at different locations in the harbor (Maraqa et al., 2008). Since the AD model developed by Maraqa et al. (2008) was used in this study, a brief description of the model setup is presented below. For more details about the model setup the readers are referred to Maraqa et al. (2008).

The HD model of Jebel Ali Harbor is the basis for the AD model. The model was constructed with a rectangular grid system of 60×60 m^2. The dimensions of the grid were selected as a compromise between resolution and computational time. The origin of the model was 24°58′03″ latitude and 55°01′28″ longitude, taking east-west and north-south directions as the x and y directions, respectively. The entrance to the harbor was selected as the open boundary and the flow direction was considered perpendicular to the boundary. The closed side and bottom boundaries were considered as no flow boundaries. A constant water level and zero velocities were used as initial conditions at all grid points. Tide level was used as the boundary condition and the flow direction was considered perpendicular to the boundary. Latest topographical description of the harbor area was incorporated in the model (Jan de Nul Dredging Ltd., 2004). Although, the hydraulic regime of Jebel Ali Harbor is mainly dependent on the tide (Maraqa et al., 2008), meteorological forces were incorporated to improve the accuracy of the model. Observed meteorological conditions during January to December 2004 at the site were applied to the model in the first simulation year and similar meteorological conditions were used for a simulation period of 12 successive years.

The predicted tide level at the entrance of the harbor was used as the boundary condition for the HD model. The prediction was carried out using the Admiralty method (DHI, 2003a) facilitated in MIKE21 tools using major tidal constituents (see Table 1) with necessary seasonal corrections of -0.1 during February, March and April and +0.1 during July and August (ATT, 2003). Predicted tide levels were referenced to mean sea level datum and converted to local chart datum (CD) adding 1.02 m (ATT, 2003). The HD model was calibrated against tide level and flow data measured in December 2004. Through a rigorous calibration process, constant Chezy number and eddy viscosity were selected as 40 and 1.0 m^2/s, respectively. Simulated tide levels compared quite well with measured levels at three locations within the harbor. Also, simulated flow values through the entrance channel matched quite well with the measured ones at the same location (Maraqa et al., 2008).

In the development of the AD model of Jebel Ali Harbor, Maraqa et al. (2008) used a spatially varied dispersion coefficient determined by a formula suggested by Fischer et al. (1979):

$$D = 0.011\frac{\overline{u}^2 W^2}{du^*} \tag{5}$$

where, D is the dispersion coefficient (m^2/s); $\bar{u}$ is the average velocity (m/s); W is the width of the channel (m); d is the depth of the channel (m); u^* is the shear velocity (m/s) which is expressed as $(ghS)^{0.5}$; g is the gravitational acceleration = 9.81 (m/s^2); h is the hydraulic radius ≈ depth of the channel (m); and S is the water surface slope.

Constituent name	Amplitude (m)	Phase (°)
Principal lunar semidiurnal (M_2)	0.43	359
Principal solar semidiurnal (S_2)	0.17	49
Luni-solar declinational diurnal (K_1)	0.25	155
Lunar declinational diurnal (O_1)	0.17	100
First overtide of M_2 (F_4)	0.0	0
Second overtide of M_2 (F_6)	0.0	0

Table 1. Tidal constituents at Jebel Ali Harbor (ATT, 2003).

Values of the dispersion coefficient in the x and y directions were calculated using Eq. (5) based on a channel width of 300 m, an estimated average water surface slope of 2×10^{-5} m/m, and a hydraulic radius (considered as the average depth of flow) of 12.0 m (Maraqa et al., 2008). Furthermore, the mean velocity at each grid point during an ebb tide in spring was calculated from the HD model and was used as an average velocity for calculating the dispersion coefficient. Estimated dispersion coefficients at the dead-end locations within the inner and outer basins were found to be much lower (<0.0001 m^2/s) than those in the main channel (>0.01 m^2/s).

3.2 Model applications

In this study, the effect of variation in the loading rate on the average pollutant concentration in the harbor (C_{avg}) was numerically investigated by simulating pollutant concentration in harbor water subject to different continuous loading rates at specified locations. Three discharge points were selected (Fig. 1); one in the corner of the inner basin (St1), another in the corner of the outer basin (St2), and a third point at the west corner of the inner basin (St3). These locations are currently used to discharge treated industrial wastewater into the harbor (Maraqa et al., 2007). Conservative and degradable pollutants were considered with a loading rate (*LR*) that varied from 0.01-1.0 g/s. The lower limit of loading rates nearly corresponds to the current total rate of discharge of phosphate, while the upper limit is close to the rate of discharge of nitrate or BOD_5 (Maraqa et al., 2007). It should be noted that conservative pollutants discharged into Jebel Ali Harbor could include reject brine from desalination plants or any pollutant that does not undergo transfer and transform reactions. On the other hand, degradable pollutants could include BOD, coliform bacteria, or any other pollutant that undergoes transformation, not transfer, reactions.

Three different sets of simulations were conducted in this study. In the first set, the loading rate of a conservative pollutant varied while fixing the discharge concentration at 20 mg/l. This discharge concentration was chosen based on the current discharge concentration of some contaminants (Maraqa et al., 2007). Since the concentration at the source was fixed in this set of simulations, variations in the loading rates are due to variations in the discharge flow rates. The second set of simulations was carried out to investigate the impact of changing the discharge concentration, with fixed loading rates, on the value of C_{avg}. Discharge concentrations of 10, 20, and 40 mg/l at St1 were simulated for discharge rates of 0.01, 0.1 and 1.0 g/s of a conservative pollutant. The third set of simulations was conducted to study the impact of pollutant degradation on C_{avg}. Simulations of the latter cases were accomplished using the ECO Lab module of MIKE21 (DHI, 2003c) along with the AD model. A decay rate constant of 0.1, 0.2 and 0.5/yr were used with pollutant input rates of 0.01, 0.1 and 1.0 g/s at St1.

All simulated cases in this study are summarized in Table 2. In all simulated cases, a pollutant concentration background value of zero in harbor water was used as the initial and boundary conditions. For each simulated case, the concentration level at different locations and the total mass of the pollutant within the harbor were numerically estimated using the developed model by Maraqa et al. (2008) to find out C_{avg}. For a continuous and constant loading rate, steady-state conditions were assumed to be reached when the mean pollutant concentration over a tidal cycle at any point in the harbor did not change over time. This definition of steady-state concentration is similar to the definition of stationary-state concentration used by Edinger et al. (1998).

The value of C_{avg} at steady-state conditions was calculated by averaging the spatial concentration values over the entire harbor-modeling area when steady-state conditions prevailed. As indicated by the California Regional Water Quality Control Board (CRWQCB, 2007), an average concentration value of the main water mass is typically used when comparison with a water quality objective limit is intended. The CRWQCB (2007) also indicated that objective limits cannot be applied at or immediately adjacent to zones of initial dilution within which higher concentration can be tolerated.

4. Results and discussion

4.1 Time of steady-state conditions

The time to reach steady-state conditions due to a continuous discharge was almost the same for a certain discharge location independent of the loading rate. However, the time to reach steady-state was dependent on the discharge location with a value of about 12 yrs for discharge at St1, 7 yrs for discharge at St2, and 4 yrs for discharge at St3 (Table 2). It should be noted that the time to reach steady-state is not directly comparable to flushing time or residence time. Generally, flushing time is defined as "the ratio of the scalar in a reservoir to the rate of renewal of the scalar" (Geyer et al., 2000). Flushing time describes the exchange characteristics of a waterbody without identifying the underlying physical processes or their spatial distribution (Monsen et al., 2002). Residence time, on the other hand, is the time it takes a waterparcel to leave a semi-enclosed waterbody through its outlet (Monsen et al., 2002). Residence time is measured from an arbitrary start location within the waterbody, whereas the time to reach steady-state concentration used in this study depends primarily on the mixing characteristics of the entire waterbody. However, steady-state times were found quite similar to residence times simulated by Maraqa et al. (2008) for these stations since both of these time scales depend on the same physical processes.

Case	Location	Concentration (mg/l)	Discharge flow (m^3/s)	Loading rate (g/s)	Degradation rate constant (yr^{-1})	Time to reach steady-state (yr)
1	St1	20	0.0005	0.01	0	12
2	St1	20	0.0025	0.05	0	12
3	St1	20	0.0050	0.10	0	12
4	St1	20	0.0250	0.50	0	12
5	St1	20	0.0500	1.00	0	12
6	St2	20	0.0005	0.01	0	7
7	St2	20	0.0050	0.10	0	7
8	St2	20	0.0500	1.00	0	7
9	St3	20	0.0005	0.01	0	4
10	St3	20	0.0050	0.10	0	4
11	St3	20	0.0500	1.00	0	4
12	St1	10	0.0010	0.01	0	12
13	St1	10	0.0100	0.10	0	12
14	St1	10	0.1000	1.00	0	12
15	St1	40	0.00025	0.01	0	12
16	St1	40	0.0025	0.10	0	12
17	St1	40	0.0250	1.00	0	12
18	St1	20	0.0005	0.01	0.1	12
19	St1	20	0.0050	0.10	0.2	12
20	St1	20	0.0500	1.00	0.5	12
21	St1	20	0.0005	0.01	0.1	12
22	St1	20	0.0050	0.10	0.2	12
23	St1	20	0.0500	1.00	0.5	12
24	St1	20	0.0005	0.01	0.1	12
25	St1	20	0.0050	0.10	0.2	12
26	St1	20	0.0500	1.00	0.5	12

Table 2. Description of the simulated cases.

4.2 Spatial variations

Spatial variations of pollutant concentration in harbor water due to a loading rate of 1.0 g/s at St1 is shown in Fig. 3. Simulation shows that the concentration distribution changes significantly from year 2 to year 8, whereas it increases slightly after year 8 until it reaches steady-state conditions. From the circulation pattern of the harbor, as presented by Maraqa et al. (2008), pollutant distribution in the inner and outer basin is dominated by diffusion while that in the entrance channel is greatly affected by advection.

To examine the effect of pollutant source location, the simulations were repeated relocating the point source at St2 (Fig. 1). It was found (Fig. 4) that the concentration distributions in the inner and outer basins are different than the distribution with the source location at St1. But, the concentration distributions in the entrance channel were almost similar for the two cases (see Fig. 3 and Fig. 4). The concentration in the entrance channel is mostly influenced by the Gulf water rather than the inside basins because of the dominant advection processes. In any case, the highest concentration occurs at the source location, while the lowest concentration generally occurs at the west side of the entrance channel due to the inward net flow conditions (Maraqa et al., 2008). For a loading rate of 1.0 g/s, the average pollutant concentration in the harbor water reached 0.35 mg/l with the discharge point located at St2 and 1.3 mg/l with the discharge point located at St1. The reasons behind reaching steady-state conditions with lower concentration at St2 are faster transport due to advection and more dilution with Gulf water. Thus, it is necessary to examine the impact of pollutant loading using modeling technique before selecting the discharge location.

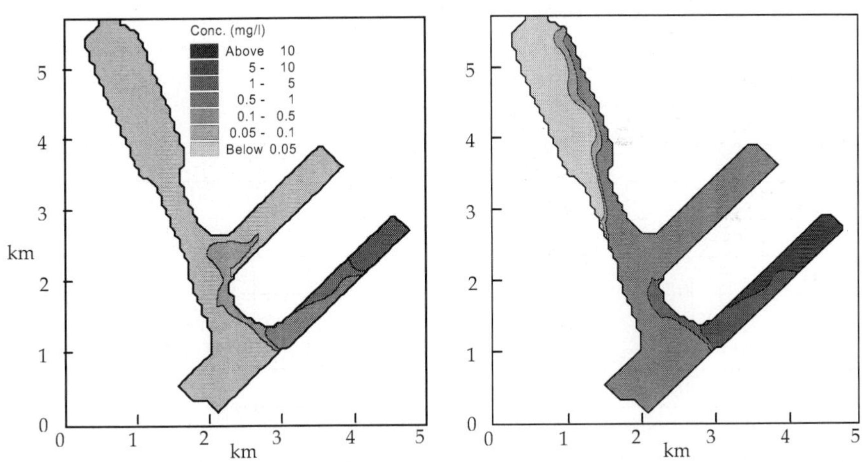

Fig. 3. Concentration map after 4 yrs (left) and 12 yrs (right) from start of simulation with a continuous loading rate of a conservative pollutant of 1.0 g/s at St1.

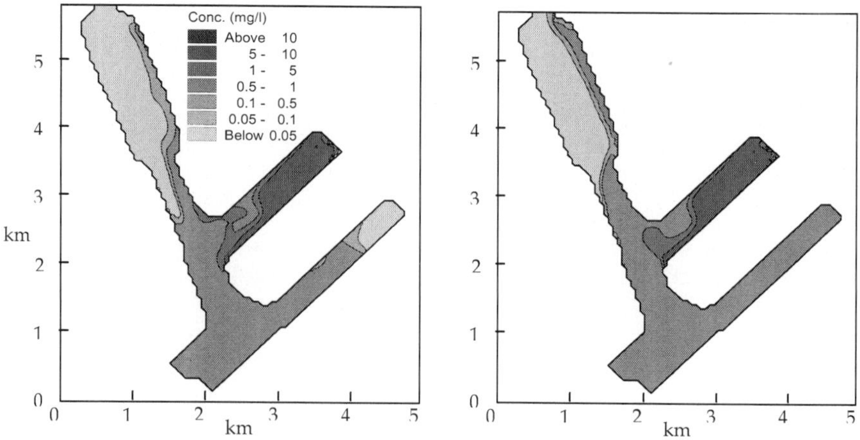

Fig. 4. Concentration map after 2 yrs (left), and 7 yrs (right) from start of simulation with a continuous loading rate of 1.0 g/s at St2.

4.3 Temporal variations

Temporal variation of pollutant concentration at St3 due to pollutant loading of 1.0 and 0.1 g/s at St1 is shown in Fig. 5. The figure shows that the concentration at St3 increases sharply at initial times and levels of at later times until it reaches a plateau value. For a conservative pollutant, the mechanisms of solute transport within the harbor are associated with the advection-dispersion processes. Advection driven by the tide was the principal transporting process within Jebel Ali Harbor. Winds and waves play a minor role in mixing and transport of pollutants because of the bottle-neck shape of the harbor.

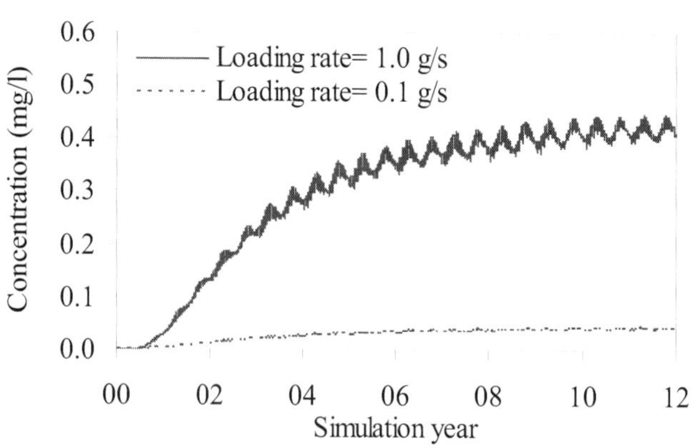

Fig. 5. Long-term variation of concentration at St3 subject to a loading rate of a conservative pollutant of 0.1 and 1.0 g/s at St1.

Further inspection of Fig. 5 shows that there are seasonal fluctuations in the concentration level. Such fluctuation occurs at hourly levels due to variations of tidal levels as demonstrated in Fig. 6. In general, the average concentration reduces during summer season when the tide levels are high and the concentration increases during winter season when the tide levels are low. Also, the concentration reduces during high water and increases during low water because of hourly tide level changes. This indicates that short-term monitoring of water quality in the harbor may not reflect on the long-term changes. For example, the concentration of a conservative pollutant at St3 reaches 0.46 mg/l in April of year 11 as a result of a loading rate of 1.0 g/s at St1. The concentration drops in July (of that year) to about 0.41 mg/l for the same loading rate at St1.

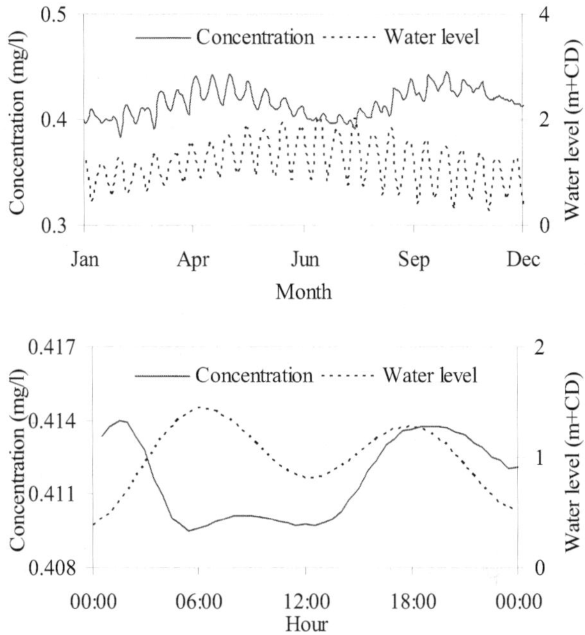

Fig. 6. Concentration at St3 due to a loading rate of 1.0 g/s at St1 showing seasonal variations (top) and daily variations (bottom) at the end of the 10th yr along with the tide levels.

4.4 Steady-state concentration of conservative pollutants

For conservative pollutants, the values of C_{avg} resulting from different loading rates at St1, St2, and St3 are presented in Fig. 7. As the figure shows, C_{avg} varies with both the loading rate and the discharge location. Higher concentration was observed when the source was located at St1. On the other hand, relatively lower concentrations were observed when the source was located at St2 or St3. Discharging at St2 and St3 produces almost the same average concentration in harbor water for similar loading rates. Similar concentration distributions were also observed in the entrance channel whether the source was located at St2 or St3, but different concentration distributions were observed in the inner and outer basins.

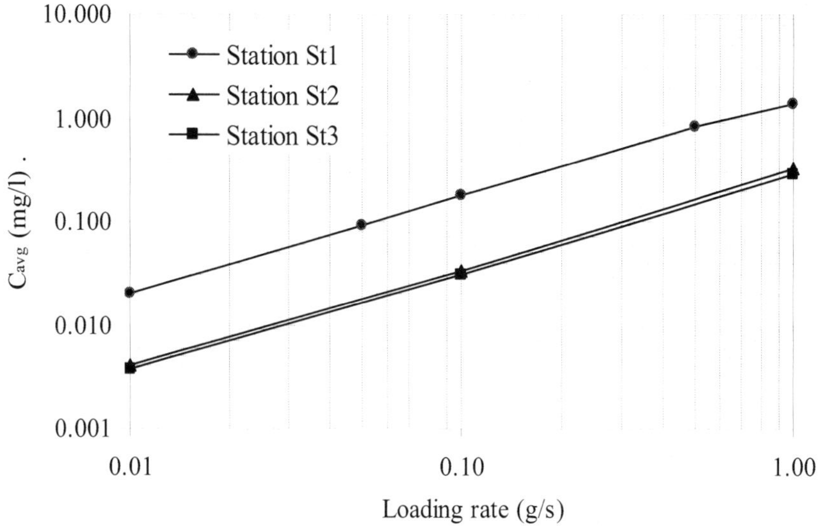

Fig. 7. Average concentration in harbor water for different discharge locations and loading rates of a conservative pollutant.

Based on the data presented in Fig. 7, the following best fit equations were produced relating the average concentration (mg/l) in harbor water to the loading rate (g/s):

$$\text{Discharge at St1: } C_{avg} = 1.468\,(LR)^{0.926} \tag{6}$$

$$\text{Discharge at St2: } C_{avg} = 0.322\,(LR)^{0.956} \tag{7}$$

$$\text{Discharge at St3: } C_{avg} = 0.286\,(LR)^{0.953} \tag{8}$$

Equations 6-8 have a coefficient of determination (r^2) of 0.999. From Eqs. (6)-(8), C_{avg} correlates almost linearly with the pollutant loading rate. At a given loading rate, discharge at St1 results in values of C_{avg} that are 4-5 times higher that those due to discharge at either St2 or St3. However, the maximum concentration in the harbor always occurred close to the discharge location. This is consistent with the findings of Kumar et al. (2000) who reported higher bacterial pollution near diffuser locations (discharge points). Also, the maximum concentration in this work was almost an order of magnitude higher than the average concentration in the harbor. For example, the maximum concentration due to a discharge of 1.0 g/s at St1 was 13.26 mg/l compared to an average concentration of 1.31 mg/l for this case.

The effect of changes in the discharge concentration on the value of C_{avg} is shown in Fig. 8. The figure shows that C_{avg} is more dependent on the loading rate and less dependent on the discharge concentration. In other words, it is the mass input rate, rather than the discharge concentration itself, that influences the concentration of the pollutant in the harbor. Thus, Fig. 7 (or Eqs. 6-8) can be used to determine the allowable discharge rate of a pollutant such that C_{avg} does not exceed a pre-set harbor objective limit. Further inspection of Fig. 8 shows that C_{avg} increases with the increase in the input concentration at a loading rate of 1 g/s, while it maintains nearly the same value regardless of the input concentration at lower loading rates. Such observation could be due to the high volume of discharge water associated with a high loading rate and a relatively low input concentration.

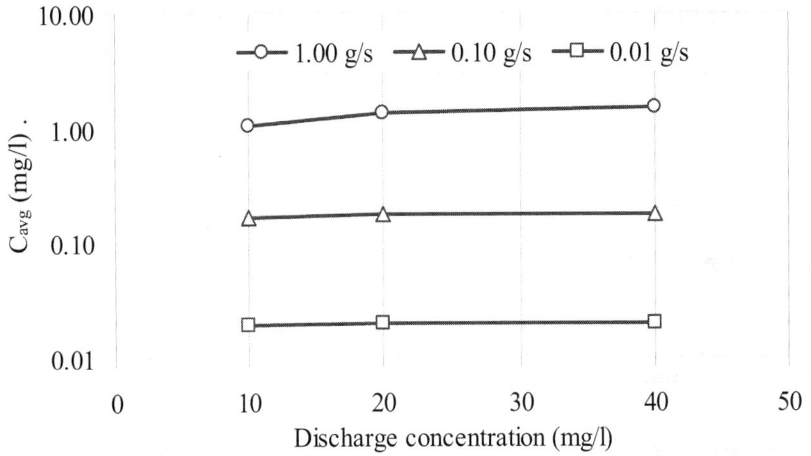

Fig. 8. Average concentration in harbor water for different discharge concentrations and loading rates of a conservative pollutant at St1.

4.5 Steady-state concentration of degradable pollutants

The simulations carried out under the previous cases were limited to pollutants that are conservative. The effect of degradation on the average concentration in harbor water is presented in Fig. 9 for degradable pollutants discharged at St1. As expected, an increase in the decay rate constant (k) results in a reduction in C_{avg}. This reduction is almost independent of the loading rate and averages 18%, 36% and 62% with a decay rate of 0.1, 0.2 and 0.5/yr, respectively. Meanwhile, the average concentration of a degrading pollutant introduced at St1 could be well predicted from that of a conservative pollutant using a decreasing exponential function with an average time (t) of 2 yrs as shown in Eq. (9):

$$(C_{avg})_k = (C_{avg})_{k=0}\, e^{-2k} \qquad (9)$$

where, $(C_{avg})_k$ is the average concentration of a degrading pollutant in harbor water and $(C_{avg})_{k=0}$ is the average concentration of a conservative pollutant and k is in units of yr^{-1}.

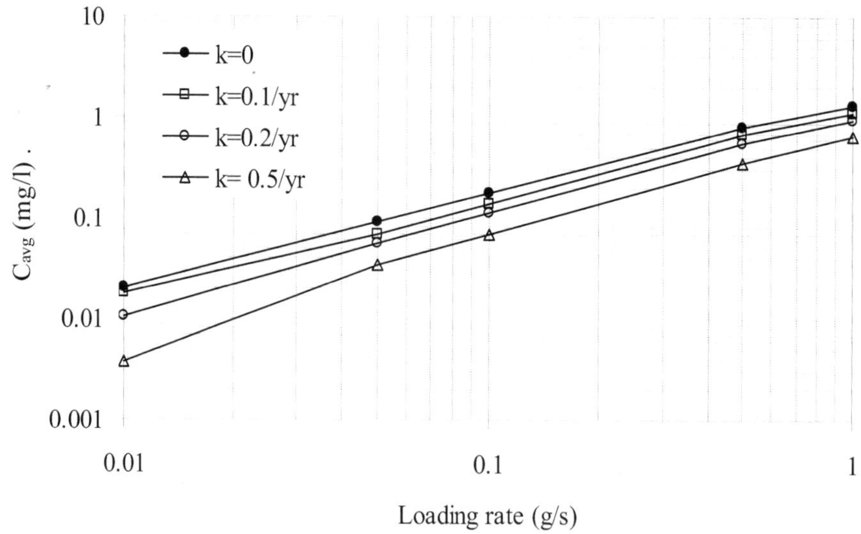

Fig. 9. Average concentration of pollutants with different degradation rates as a result of discharge at St1.

5. Conclusion

With a continuous pollutant discharge, the aqueous concentration at any point in Jebel Ali Harbor reaches steady-state conditions with duration that depends on the discharge location. The longest duration (about 12 yrs) occurs with discharge in the east corner of the inner basin. Results of this study show that the average steady-state pollutant concentration in harbor water varies with both the loading rate and the discharge location, but is independent of the discharge concentration. For a conservative pollutant discharged in Jebel Ali Harbor, developed relationships of average pollutant concentration in harbor water were found to correlate almost linearly with the discharge loading rate. It was also observed that discharging in the east corner of the inner basin results in an average steady-state concentration of about 4-5 times higher than values associated with discharge at the east corner of the outer basin or the west corner of the inner basin. For a degrading pollutant, the reduction in the steady-state average concentration is almost independent of the loading rate, but could be adequately predicted from that of a conservative pollutant using a decreasing exponential time function. Derived relationships of average aqueous pollutant concentration in the harbor versus the discharge loading rate will be useful for better management of harbor water quality.

6. Acknowledgement

This project was funded by the Ports, Customs and Free Zone Corporation at Jebel Ali, Dubai and the UAE University.

7. References

Adams, E. E. & Baptista, A.M. (1986). Ocean Dispersion Modeling. In: *Encyclopedia of Fluid Mechanics, Vol. 6, Complex Flow Phenomena and Modeling*, N.P. Cheremisinoff (ed.), pp. 865–895, Gulf Publishing, ISBN: 978-0872015180, Houston, Texas.

Al-Rabeh, A. H.; Cekirge, H. M. & Gunay, N. (1992). Modeling the fate and transport of Al-Ahmadi oil spill, *Water, Air, and Soil Pollution*, Vol. 65, No. 3-4, pp. 257-279, ISSN: 0049-6979.

ATT (2003). Admiralty Tide Tables. Admiralty Charts and Publications, the United Kingdom Hydrographic Office, Vol.3, NP 203-01.

Azam, M. H.; Elshorbagy W. & Nakata, K. (2006a). Three-dimensional modeling of the Ruwais Coastal Area of United Arab Emirates, *J. Waterway, Port, Coastal and Ocean Engineering*, Vol. 132, No. 6, pp. 487-495, ISSN: 0733-950X.

Azam, M. H.; Elshorbagy, W.; Ichikawa, T.; Terasawa, T. & Taguchi, K. (2006b). 3D model application to study residual flow in the Arabian Gulf, *J. Waterway, Port, Coastal and Ocean Engineering*, Vol. 132, No. 5, pp. 388-400, ISSN: 0733-950X.

Brewer, H. M.; Godin, G.; Cunderlik, J. M.; Houghton, E.; Firman, M. & Green, D. (2008). Assimilative capacity modelling of Collingwood Harbour: A case study on the selection of effluent limits for municipal wastewater treatment plant expansion, Proceedings of the Water Environment Federation, ISSN: 1938-6478, pp. 2172-2185(14).

Cerejo, M. & Dias, J. M. (2007). Tidal transport and dispersal of marine toxic microalgae in a shallow, temperate coastal lagoon, *Marine Environmental Research*, Vol. 63, No. 4, pp. 313–340, ISSN: 0141-1136.

Chao, S.-Y.; Kao, T. W. & Al-Hajri, K. R. (1992). A numerical investigation of circulation in the Arabian Gulf, *J. Geophys. Res.*, Vol. 97, No. C7, pp. 11219–11236, ISSN 0148–0227.

Copeland, G.; Monteiro, T.; Couch, S. & Borthwick, A. (2003).Water quality in Sepetiba Bay, Brazil, *Marine Environmental Research*, Vol. 55, No. 5, pp. 385–408, ISSN: 0141-1136.

CRWQCB (2007). San Francisco Bay Basin (Region 2) Water Quality Control Plan. California Regional Water Quality Control Board, Oakland, CA.

Das, P.; Marchesiello, P. & Middleton, J. H. (2000). Numerical modelling of tide-induced residual circulation in Sydney Harbour, *Marine and Freshwater Research*, Vol. 51, No. 2, pp. 97-112, ISSN 1323-1650.

Dias, J. M. & Lopes, J. F. (2006a). Implementation and assessment of hydrodynamic, salt and heat transport models: the case of Ria de Aveiro lagoon (Portugal), *Environmental Modelling and Software*, Vol. 21, No. 1, pp. 1–15, ISSN 1364-8152.

Dias, J. M. & Lopes, J. F. (2006b). Calibration and validation of hydrodynamic, salt and heat transport models for Ria de Aveiro lagoon (Portugal), *J. Coastal Research*, Vol. SI 39, pp. 1680–1684, ISSN 0749-0208.

DHI (2003a). Scientific Documentation, MIKE 21 Hydrodynamic Module, DHI Water and Environment, DHI Software, Denmark.

DHI (2003b). Scientific Documentation, MIKE 21 Advection-Dispersion Module, DHI Water and Environment, DHI Software, Denmark.

DHI (2003c). Scientific Documentation, MIKE 21 ECO Lab Module, DHI Water and Environment, DHI Software, Denmark.

Edinger, J. E.; Buchak, E. M. & Kollubru, V. S. (1998). Modeling flushing and mixing in a deep estuary, *Water, Air, and Soil Pollution*, Vol. 102, No. 3-4, pp. 345–353, ISSN: 0049-6979.

El Samra, M.I.; Emara, H.I. & Shunbo, F. (1986). Dissolved petroleum hydrocarbon in the northwestern Arabian Gulf, *Marine Pollution Bulletin*, Vol. 17, No. 2, pp. 65-68, ISSN: 0025-326X.

Elshorbagy, W.; Azam M. H. & Taguchi, K. (2006). Hydrodynamic characterization and modeling of the Arabian Gulf. *J. Waterway, Port, Coastal and Ocean Engineering*, Vol. 132, No. 1, pp. 47-56, ISSN: 0733-950X.

Estacio, F. J.; Garcia-adiego, E. M.; Fa, D. A.; Garcia-Gomez, J. C.; Daza, J. L.; Hortas F. & Gomez-Ariza, J. L. (1997). Ecological analysis in a polluted area of Algeciras Bay (southern Spain): External 'versus' internal outfalls and environmental implications, *Marine Pollution Bulletin*, Vol. 34, No. 10, pp. 780-793, ISSN: 0025-326X.

Fiandrino, A.; Martin, Y.; Got, P.; Bonnefont, J. L. & Troussellier, M. (2003). Bacterial contamination of Mediterranean coastal seawater as affected by riverine inputs: simulation approach applied to a shellfish breeding area (Thaulagoon, France), *Water Research*, Vol. 37, No. 8, pp. 1711–1722, ISSN: 0043-1354.

Fischer, H. B.; List, E. J.; Koh, R. C. Y.; Imberger, J. & Brooks, N. H. (1979). *Mixing in Inland and Coastal Waters*, Academic Press Inc. ISBN: 9780122581502, London, UK.

Ganoulis, J. G. (1991). Water quality assessment and protection measures of a semi-enclosed coastal area: The Bay of Thermaikos (NE Mediterranean sea), *Marine Pollution Bulletin*, Vol. 23, pp. 83-87, ISSN: 0025-326X.

Gesteira-Gomez, M.; Montero, P.; Prego, R.; Taboada, J. J.; Leitao, P.; Ruiz-Villarreal, M.; Neves, R. & Perez-Villar, V. (1999). A two-dimensional particle tracking model for pollution dispersion in A Coruna and Vigo Rias (NW Spain), *Oceanologica Acta*, Vol. 22, No. 2, pp. 167–177, ISSN: 0399-1784.

Geyer, W. R.; Morris, J. T.; Prahl, F. G. & Jay, D. A. (2000). Interaction between physical processes and ecosystem structure: A comparative approach. In: *Estuarine Science: A synthetic approach to research and practice*, J. E. Hobbie (ed.), pp. 177-210, Island Press, ISBN: 9781559637008, Washington, D.C.

Gupta, I.; Dhage, S.; Chandorkar, A. A. & Srivastav, A. (2004). Numerical modeling for Thane creek, *Environmental Modelling and Software*, Vol. 19, No. 6, pp. 571–579, ISSN 1364-8152.

Horton, C.; Clifford, M.; Schmitz, J. & Hester, B. (1994). SWAFS: Shallow water analysis and forecast system. Overview and Status Rep., Naval Oceanographic Office, Stennis Space Center, Miss.

Hughes, P. & Hunter, J. R. (1979). A Proposal for a physical oceanography program and numerical modeling of the KAP Region. UNESCO, Div. Mar. Sci., Paris, MARINF/27, 16 Oct. 1979. 102 pp.

Jan de Nul Dredging Ltd. (2004). Dredging and Land Reclamation, A company commissioned for dredging Jebel Ali Harbor, Personal Communication, June 2004.

Kashefipour, S. M.; Lin, B.; Harris, E. & Falconer, R. A. (2002). Hydro-environmental modelling for bathing water compliance of an estuarine basin, *Water Research*, Vol. 36, No. 7, pp. 1854–1868, ISSN: 0043-1354.

Kashefipour, S. M.; Lin, B. & Falconer, R. A. (2006). Modelling the fate of faecal indicators in a coastal basin, *Water Research*, Vol. 40, No. 7, pp. 1413–1425, ISSN: 0043-1354.

Kumar, R.; Subramanium, J. & Patil, D. (2000). Water quality modeling of municipal discharge from sea outfalls, Mumbai, *Environmental Monitoring and Assessment*, Vol. 62, No. 2, pp. 119–132, ISSN: 0167-6369.

Lardner, R. W.; Lehr, W. J.; Fraga, R. J. & Sarhan, M. A. (1987). Residual circulation in the Arabian Gulf. I: Density-driven flow. *Arabian J. Sci. Eng.*, Vol. 12, No. 3, pp. 341–354, ISSN: 1319-8025.

Lardner, R.W.; Cekirge, H.M.; Al-Rabeh, A.H. & Gunay, N. (1988). Passive pollutant transport in the Arabian Gulf, *Advances in Water Resources*, Vol. 11, No. 4, pp. 158-161, ISSN: 0309-1708.

Lardner, R. W.; Lehr, W. J.; Fraga, R. J. & Sarhan, M. A. (1989). A model of residual currents and pollutant transport in the Arabian Gulf, *Mathematical and Computer Modelling*, Vol. 12, No. 9, pp. 1185-1186, ISSN: 0895-7177.

Lardner, R. W.; Al-Rabeh, W.; Gunay, N.; Hossain, M.; Reynolds, R. M. & Lehr, W. J. (1993). Computation of residual flow in the Gulf using the Mt. Mitchell data and the KFUPM/RI hydrodynamic model. *Mar. Pollution Bull.*, Vol. 27, pp. 61–70, ISSN: 0025-326X.

Lopes, J. F.; Dias, J. M.; Cardoso, A. C. & Silva, C. I. V. (2005). The water quality of the Ria de Aveiro lagoon, Portugal: From the observations to the implementation of a numerical model, *Marine Environmental Research*, Vol. 60, No. 5, pp. 594–628, ISSN: 0141-1136.

Maraqa, M. A.; Ali, A. & Khan, N. (2007). Modeling selected water quality parameters at Jebel Ali Harbor, Dubai- UAE, J. *Coastal Research*, SI 50, pp. 794-799, ISSN 0749-0208.

Maraqa, M. A.; Ali, A.; Imran, H. D.; Hamza, W. & Al Awadi, S. (2008). Simulation of the hydrodynamic regime of Jebel Ali Harbor, Dubai-UAE, *Aquatic Ecosystem Health and Management*, Vol. 11, No. 1, pp. 105-115, ISSN: 1463-4988.

Monsen, N. E.; Cloern, J. E.; Lucas, L. V. & Monismith, S. G. (2002). A comment on the use of flushing time, residence time and age as transport time scales, *Limnology and Oceanography*, Vol. 47, No. 5, pp. 1545-1553, ISSN 0024-3590.

Montano-Ley, Y.; Peraza-Vizcarra, R. & Paez-Osuna, F. (2008). Tidal hydrodynamics and their implications for the dispersion of effluents in Mazatlan Harbor: An urbanized shallow coastal lagoon, *Water, Air, and Soil Pollution*, Vol. 194, No. 1-4, pp. 343-357, ISSN: 0049-6979.

PCFC (2003). Environmental Control Rules and Requirements (3rd edition), Environment, Health and Safety Department, Ports, Customs and Free Zone Corporation, Dubai, UAE.

Rucinski, D.K.; Auer, M.T.; Watkins, D.W.; Effler, S.W.; Doerr, S.M. & Gelda R.K. (2007). Accessing assimilative capacity through a dual discharge approach, *Water Resources Planning and Management*, Vol. 133, No. 6, pp. 474-485, ISSN: 0733-9496.

Spaulding, M.L.; Anderson, E. L.; Isaji, T. & Howlett, E. (1993). Simulation of the oil trajectory and fate in the Arabian Gulf from the Mina Al Ahmadi spill, *Marine Environmental Research*, Vol. 36, No. 2, pp. 79-115, ISSN: 0141-1136.

Tao, J.; Li, Q.; Falconer, R.A. & Lin, B. (2001). Modelling and assessment of water quality indicators in a semi-enclosed shallow bay, *Hydraulic Research*, Vol. 39, No. 6, pp. 611-617, ISSN: 0022-1686.

Thoppil, P.G. & Hogan, P.J. (2010). A modeling study of circulation and eddies in the Persian Gulf, Physical Oceanography, Vol. 40, No. 9, pp. 2122-2134, ISSN: 0022-3670.

Vethamony, P.; Reddy, G. S.; Babu, M. T.; Desa, E. & Sudheesh, K. (2005). Tidal eddies in a semi-enclosed basin: a model study, *Marine Environmental Research*, Vol. 59, No. 5, pp. 519–532, ISSN: 0141-1136.

19

The Effect of Wastes Discharge on the Quality of Samaru Stream, Zaria, Nigeria

Y.O. Yusuf and M.I. Shuaib
Department of Geography, Ahmadu Bello University, Zaria, Nigeria

1. Introduction

Water is the most important natural resource in the world. Since without it life cannot exist and most industries would not function or operate (Tebbut, 1998). It is essential to life and in fact the basis of life, being so it was almost inevitable that the development of water resources preceded any real understanding of their origin and formation (Ward, 1975). However, water is a unique resource having no substitute with its quality and quantity varying over space and time, hence is finite. Water is equally one of the remarkable substance known which is found in vast quantities in nature; it could be in gaseous, liquid and solid state. There are nearly $14x10^8$ cubic kilometers of water on the planet but 97.5% of this is salty water, fresh water account for only 2.35% of the total. Many people depend on fresh water from lakes, rivers and streams for their water supplies, and these sources contain respectively only 0.26 and 0.006% of the total volume of fresh water (UNEP, 1994)

Water plays a vital role in the development of a stable community, since human being can exist for days without food but absence of water for a few days may lead to death. Water is an essential pre-requisite for the establishment of a stable community. In the absence of which nomadic lifestyle becomes necessary and communities move from one area to another as demand for water exceed its availability. Hence, it is therefore not surprising that sources of water are often zealously guarded over the century. Many skirmishes have taken place over water right (Tebbut, 1998). Water is used for drinking, cooking, sanitation, agricultural purpose, industrial purposes and also used for generating hydro-electric power e.t.c. The amount of water used for other purposes, apart from food preparation and cooking, vary widely and are greatly influenced by the type and availability of water supply.

Furthermore, these facts show how important water is to man, so if water is contaminated it poses a great health hazard to man causing various diseases, and one of the greatest avenue for the spread of diseases is through water. Therefore the presence of water in the environment does not suffice, rather how useful it is to man is what qualifies it as a resource to man. Considering that the utility of water is limited, evaluating in terms of quality, quantity and reliability of all the possible water sources becomes expedient. World Health organization (WHO, 1976) reported that 1.5 billion people worldwide drink filthy water, and this is thought to be increasing for about 20 million each year, up to 25,000 people die each day in the world from water related diseases. Waste water is a complex mixture of natural inorganic and organic material mixed with man-made substances. It contains everything discharged to the sewer, including material washed from roads and roots, and of course where the sewer is damaged groundwater will also gain entry (Edmund *et al*, 1976 and Gray, 2000).

The contamination of water can affect aquatic animals whose survival depend on the quality of water in which they live presently, aquatic life existence is being threatened by man's activities such as industrial waste, domestic waste, waste of animals and human discharge into the stream. These activities have been affecting river quality and in turn the living organisms in the stream (Vega *et al*, 1996).

Samaru stream, which takes its source from Samaru, Zaria, Kaduna state (Nigeria) is a headwater tributary of river Kubanni. The most noticeable problem in Samaru is the inadequate or lack of pipe sewer system. This obviously creates some inefficiency in the disposal of liquid and solid wastes. Refuse disposal is by open dump system. Soak away pit are used as a means of disposal of domestic wastes. All the aforementioned constitute a major pollution in the study area.

There is a network of open gutter which convey storm water away to the major drain running parallel to Zaria-Sokoto road. This drain discharge directly into the stream, this wastes water carries with it faeces, solid materials, ashes, and several other assorted materials which if not deposited along the course of the stream end up in Kubanni dam. These problems are aggravated during the rainy season and these wastes are transported and deposited in the river channel and subsequently into the Kubanni reservoir which may increase the incidence of water borne disease.

The aim of this research is to assess the effect of wastes discharge on the quality of Samaru stream, Zaria, Kaduna state, Nigeria. In order to achieve the aim, the following objectives are imperative, to:

1. Determine the effect of pollution on the quality of the Samaru stream.
2. Compare upstream and downstream water quality of the Samaru stream.
3. Give possible suggestions on measures to improve the quality of water in the Samaru stream.

2. Background to the study area

Zaria is located at approximately 11°3′ north of the equator and on longitude 7° 42′ East of the Greenwich meridian and at about 660m above sea level. It is the second largest city in Kaduna state after Kaduna town north ward along Kaduna-Kano highway (see fig.1 and 2). It has four (4) geopolitical zones (local government areas) including Zaria city local government, Sabon Gari local government area to the north, Soba local government to the east and Giwa local government area to the west.

Samaru is located in Zaria and situated in Sabon-Gari local government area of Kaduna state. It is located at latitude 11° 10′ north and longitude 7° 37′ east. The area is an extension of urban Zaria to the north along Zaria-Sokoto and Zaria Kaura Namoda railway line. It is also one of the major settlements that make up the urban Zaria. It is an educational and administrative settlement which brought about the establishment of new settlement for non-residents of Zaria City. It grows as a result of the establishment of Ahmadu Bello University, which was established in 1962 and other institutes like Federal Institute for Chemical and Leather Research, Federal College of Aviation and Industrial Development Corporation.

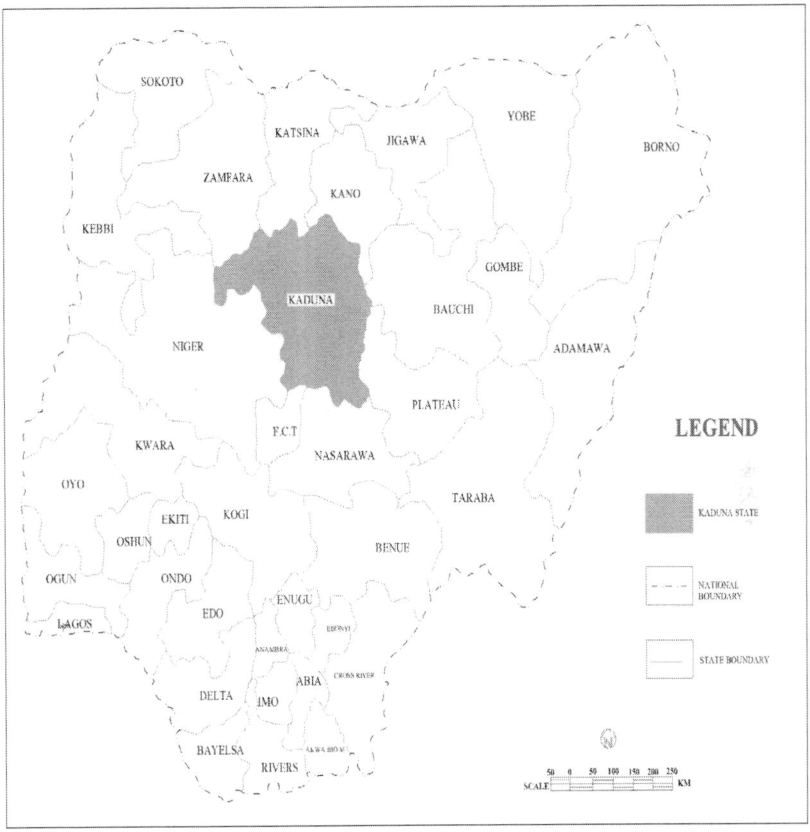

Fig. 1. Map of Nigeria showing Kaduna State

The Kubanni River has its source from the Kampagi hill in Shika near Zaria. It flows in a Southeast direction through the premises of Ahmadu Bello University. The Samaru stream which is one of the tributaries of Kubanni River has a stream length of 1.05km within an area of 2.28km^2 and has drainage density of 0.4605 km/km^2.

The climate of the study area is a tropical savanna climate, with distinct wet and dry seasons (Aw climate Koppens classification). Zaria experiences six (6) months of rainy season and six (6) months of dry season. The rainy season is from May to late October, while the dry season is from early November to April, this is as a result of the interplay of the two dominant air masses within the region i.e. the tropical continental air messes (cT) and Tropical maritime air masses (mT) (Iguisi and Abubakar, 1998).

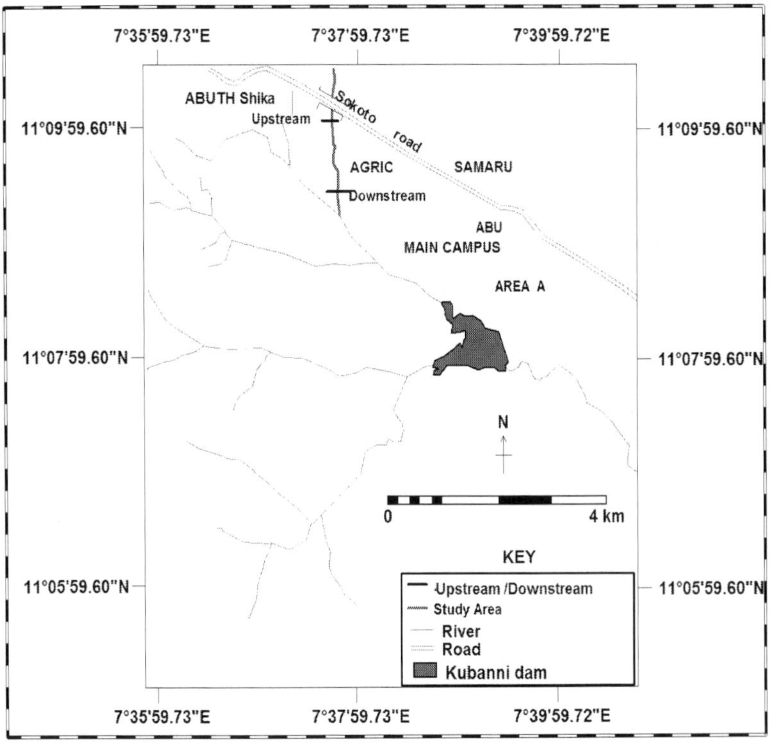

Fig. 3. Sumaru stream showing the upstream / downstream points of data collection.
Source: Adapted and modifed from Zaria SW topographivc map

4. Results and discussion

4.1 Upstream variation in water quality

Water samples were collected and tested at the upstream section of the channel using several parameters including pH, colour, Dissolved Oxygen (DO), Biological Oxygen Demand (BOD). These samples were collected both at the early part of the day (7am) and at the later part of the day (6pm). The mean values for each of the parameters at the different times period of the day over 30 days are summarized in table 1.

Parameter	Mean Upstream	
	7am	6pm
pH	7.25	7.23
Colour	15.97	22.9
DO	6.98	6.76
BOD	2.22	2.25

Table 1. Upstream mean values of parameter

4.1.1 Upstream variation in pH

From the result obtained the upstream mean pH value for the early part of the day was 7.25 while the later part of the day recorded a mean value of 7.23. This indicate that there is no much difference in upstream pH value for both morning and evening period, they are within the range of neutrality. This low variation in both time periods is further exemplified by the trend line graph shown in figure 4a.

4.1.2 Upstream variation in colour

From the derived result the upstream mean colour value for the early part of the day was 15.97, while the later part of the day recorded the mean colour value of 22.9, giving a range of 6.95, which indicate a wide variation for both time periods (morning and evening). This wide variation is further revealed by the trend line graph in figure 4b, which shows a wide and haphazard distribution over the time period.

4.1.3 Upstream variation in dissolved oxygen (DO)

The DO has a mean upstream value of 6.98(mg/l) recorded in the morning while 6.76(mg/l) was the mean value recorded for the evening period, this gave a low range of 0.22(mg/l), this means that there is no significant variation in DO upstream for both morning and evening. These trends are confirmed in figure 4c.

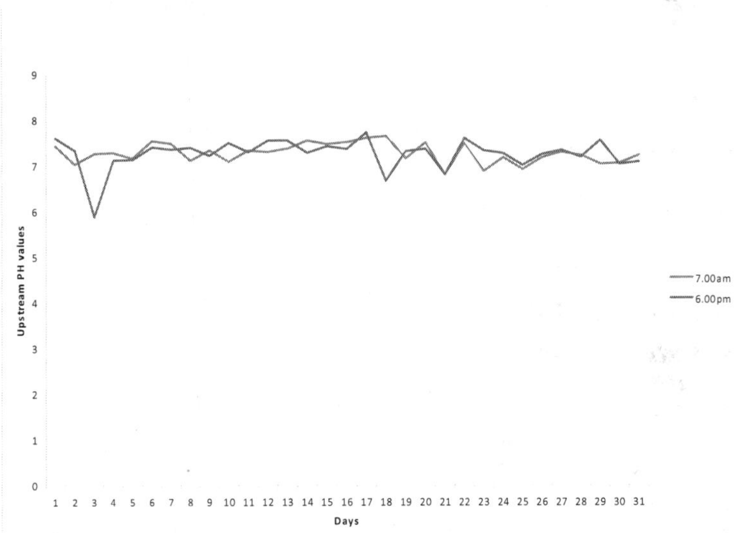

Fig. 4a. Upstream trend in pH for 30 days

4.1.4 Upstream variation in BOD

The result obtained reveals that upstream mean value for Biological oxygen demand is lower in the morning with 2.25(mg/l) than that recorded for the evening period with

2.5(mg/l), giving a range of 0.03mg/l indicating that there is low variation in BOD in both morning and evening. This low variation is depicted in the trend line for both morning and evening; this is shown in figure 4d.

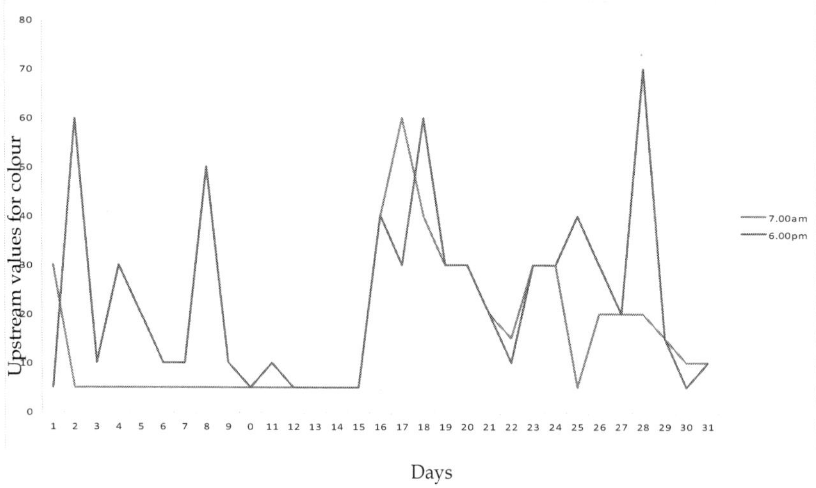

Fig. 4b. Upstream trend in colour for 30 days

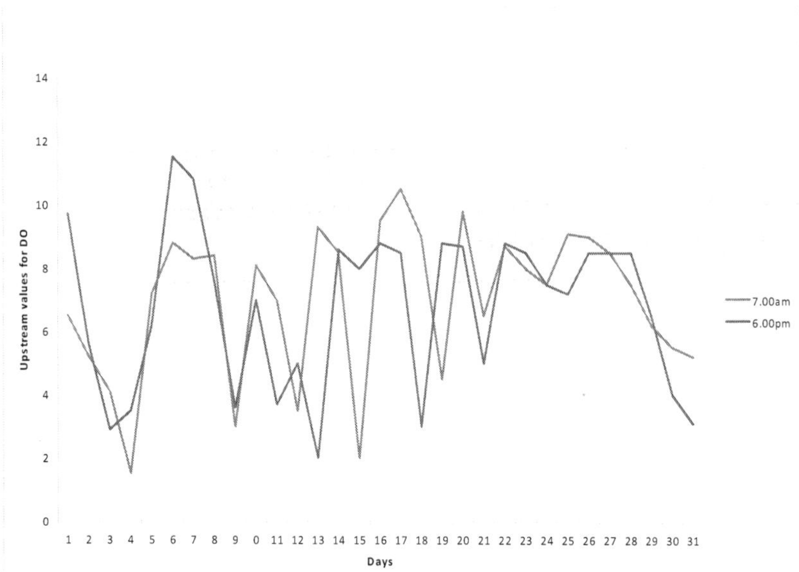

Fig. 4c. Upstream trend in Dissolved Oxygen for 30 days

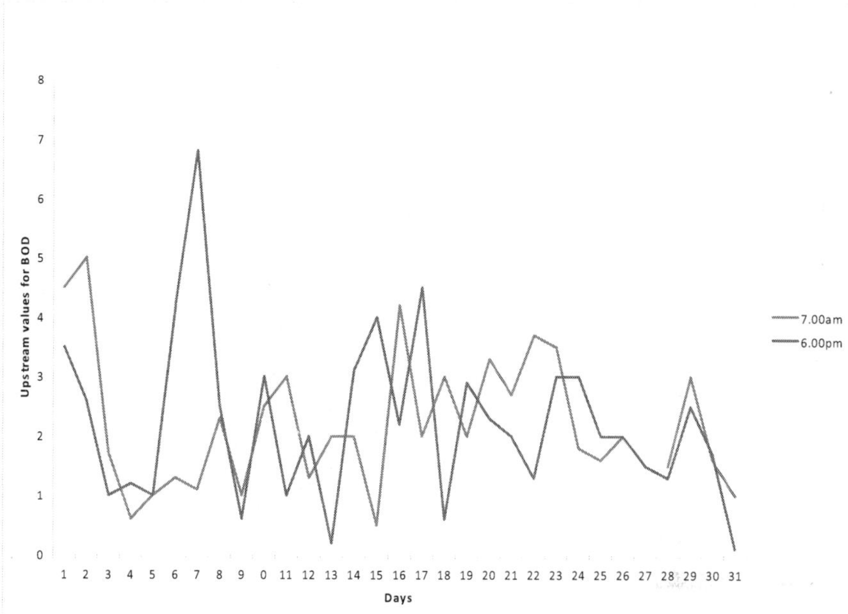

Fig. 4d. Upstream trend in BOD for 30 days

4.2 Downstream variation in water quality

Water samples were collected and tested at the downstream section of the channel using same parameters. These samples were collected both at the early part of the day and at the late part of the day. The mean values for each of the parameter at different time period of the day over 30 days are summarized in table 2.

Parameter	Downstream Mean	
	7am	6pm
pH	7.20	7.28
Colour	16.61	18.06
TDS	162.42	154.09
DO	6.71	6.83
BOD	2.15	2.69
COD	568.45	567.74

Table 2. Downstream mean values of parameters

4.2.1 Downstream variation in pH

From the result obtained, the downstream mean pH value for the early part of the day was 7.20 while in the later part of the day it was recorded as 7.20; this shows that there is no significance difference between the downstream (morning and evening), they fall within the neutrality level, giving a range of 0.08 which is relatively low. This low variation is further revealed by the trend line graph in fig 5a.

4.2.2 Downstream variation in colour

The downstream mean value of colour in the early part of the day was 16.61 while in the later part of the day it was recorded as 18.06, this shows that it was higher in the later part of the day, giving a range of 1.45 which is relatively high. This variation is depicted in the close trend line graph for both morning and evening as shown in fig 5b.

4.2.3 Downstream variation in dissolved oxygen (DO) mg/l)

From the result obtained it was observed that the mean DO was higher in the later part of the day i.e. 6.83 mg/l, while in the early part of the day it was 6.71 mg/l. Given a range of 0.12 which is relatively low, means there is no significant variation in dissolved oxygen downstream for both morning and evening. This low variation is further revealed in the trend line graph shown in fig 5c.

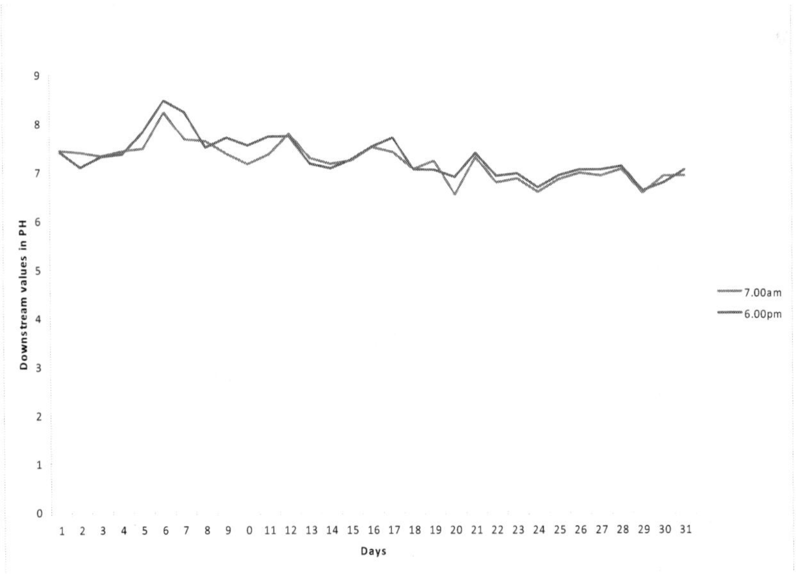

Fig. 5a. Downstream trend in pH for 30 days

4.2.4 Downstream variation in biological oxygen demand (BOD) (mg/l)

Mean BOD downstream value was 2.15 in the morning while 2.69 was the mean value recorded for the evening. This gave a range value of 0.54 which is relatively low. The trend line graph is presented in fig 5d.

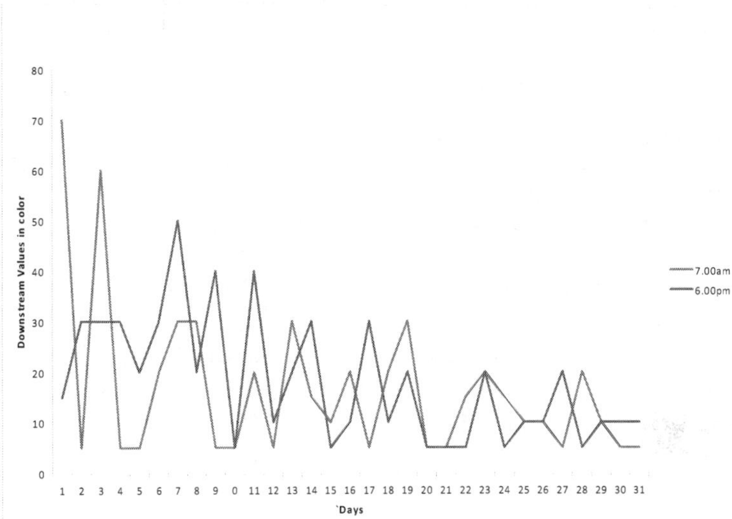

Fig. 5b. Downstream trend in color for 30 days

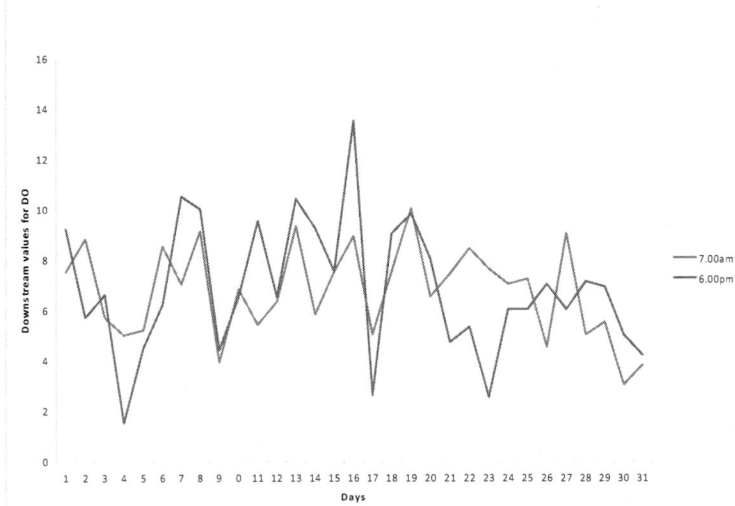

Fig. 5c. Downstream trend in DO for 30 days

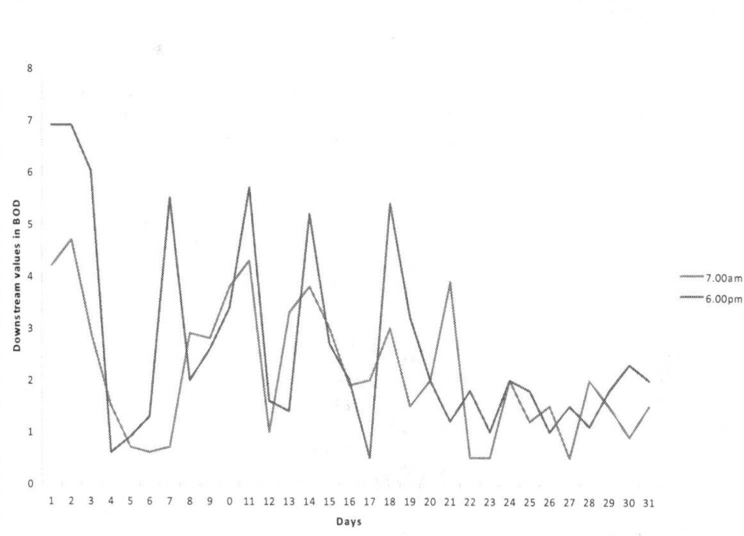

Fig. 5d. Downstream trend in BOD for 30 days

4.3 Upstream-downstream trend for the morning and evening

The upstream and downstream values for the early part of the day were correlated for the various parameters to examine the relationship between the upstream and downstream trend for each parameters at the early part of the day. From the correlation carried out, it was observed that the upstream and downstream trend for all the parameters were all positively correlated but weak with the exception of total dissolved solid (TDS) and BOD which have correlation co-efficient r values of 0.7219 and 0.5915. This shows that for an upstream increase in any of the parameters there is also a downstream increase for such parameters and vice-versa (see table 3).

Parameters	r^2	R
pH	0.0827	0.2876
Colour	0.0041	0.0640
TDS	0.5211	0.7219
DO	0.0736	0.2713
BOD	0.3499	0.05915
COD	0.1212	0.3481

Table 3. r^2 and r value for upstream - downstream relationship in the morning.

The upstream and downstream values for the later part of the day were correlated for the various parameters to examine the relationship between the upstream and downstream trend for each of the parameters at the later part of the day. From the correlation carried out, it was observed that the upstream and downstream trend for all the parameters were positively correlated but weak with the exception of TDS which has a correlation coefficient (r) value of 0.7198 (Table 4).

Parameter	r^2	R
pH	0.0149	0.1221
COLOUR	0.0261	0.1616
DO	0.0447	0.2114
BOD	0.0283	0.1682

Table 4. r^2 and r values for upstream - downstream relationship in the evening

5. Conclusion and recommendation

The result obtained from this study infers that the Samaru stream is still well oxygenated and can be said to be safe. However, the safety of this stream is being threatened by the continuous deposition of wastes into it from Samaru. This is why it can be classified as Class 2 as presented by Audu (2002) which can be said to be of doubtful water quality and needing improvement especially as it is found to be of low aesthetic quality.

Similarly the pH values observed from the study falls within the maximum permitted limit of 6.5 - 8.5 as specified by the Nigeria standard for drinking water quality. On the other hand the colour levels observed in the study is above the maximum permitted level of 15 as specified by the Nigerian standard for drinking water (NIS, 2007).

Parameters used in this study which were found to be positively and strongly correlated upstream and downstream suggest that there is no significant difference in the upstream and downstream water quality of the Samaru stream.

Based on the results obtained from the study the following recommendations became imperative.

1. Other water quality parameters, in addition to those observed in this study should be tested during the dry season to check the pollution, such as nitrate, ammonia, calcium and sulphate because of human habitation due to the discharge of waste water and agricultural practises.
2. It will be of significance to carry out another study collecting water sample both in dry season and rainy season's so as to critically examine the variation between the two seasons.
3. Location of domestic, industrial and agricultural wastes should be made far away from water bodies, this will greatly reduce the amount of wastes reaching the water bodies.
4. There is need for public enlightenment on the need to protect our environment and the benefit to be derived therefrom through the mass media such as radio, television and newspaper.
5. Control of the farming activities within the catchment should be fully implemented to avoid concentration in flow of organic and inorganic materials into the rivers.

6. References

Audu, C. (2002): The Effect of Sabon-Gari Discharge on the Quality of a Portion of River Galma, Zaria. *Unpublished M.Sc theses,* Department of Water Resources and Environmental Design, A.B.U Zaria.

Ayoade, J.O. (1988): *Tropical Hydrology and Water Resources*: Macmillan Publishers Ltd, London Pp.275

Edmund, B., Besselievre, P.E. and Schwartz, M. (1976) *The Treatment of Industrial Waste.* Second Edition, Mcgraw Hill Inc. New York.

Gray, N.F. (2000): *Water Technology.* First Edition Vinod Vasistha Publishers London.

Iguisi, E.O. and Abubakar, S.M. (1998). "The Effect of Land use on Dam Siltation" Paper presented at 41st Annual Conference of Nigeria Geographical Association, University of Uyo, Akwa Ibom.

Mortimore, M.J. (ed.) (1970): *Zaria and Its Region.* No. 4 Department of Geography, A.B.U. Zaria.

NIS (2007) *Nigerian Standard for Drinking Water.* Standards Organization of Nigeria, Lagos.

Nyagba, L. (1986) *Soils, Climate and Vegetation.* Institute of Education Press, A.B.U. Zaria

Ojo, O. (1982) *The Climate of West Africa.* HEB, London.

Tebbut, T.H.Y. (1998) *The Principle of Water Quality Control.* Fifth Edition. Butter Worth, Heineman, Linacre House Jordon Hills Oxford.

Vega, M.M., Fernández, C., Blázquez, T., Tarazona, J.V. and Castaño, A. (1996) *Biological and Chemical Tools in the Toxicological Risk.* Assessment of Jarama River Spain Environmental Pollution, 135 - 139 pp.

Ward, R.C. (1975) *Principles of Hydrology.* McGraw-Hill, London.

WHO (1971): *International Standard for Drinking Water* 3rd Edition Geneva, www.rivernetwork.org.

WHO (1976): *International Standard for Drinking Water* 3rd Edition Geneva, www.rivernetwork.org.

Wright, J. B. and McCury, P (1970) *Land Forms in Zaria and the Surrounding Regions.* Third World Planning Review, Liverpool University Press.

20

Water Quality in Hydroelectric Sites

Florentina Bunea[1], Diana Maria Bucur[2],
Gabriela Elena Dumitran[2] and Gabriel Dan Ciocan[3]
[1]*National Institute for R&D in Electrical Engineering ICPE-CA,*
[2]*Politehnica University of Bucharest,*
[3]*Université Laval, Laboratoire de Machines Hydrauliques,*
[1,2]*Romania*
[3]*Canada*

1. Introduction

The most widely form of renewable energy is the hydropower, which produces electrical power using the gravitational force of falling or flowing water. Comparing to fossil fuel powered energy plants, hydropower plants are considered „green" energy source, because they do not produce direct waste and have almost no output level of greenhouse gas carbon dioxide. Hydropower is the most important source of renewable electricity generation - 86.3 %, and essential to operate the others sources of renewable energy that are random generation.

Hydro energy importance also comes from its own source - the water, an essential life resource. Therefore, to maintain the water quality is a main concern from ecological, economical and sustainable development point of view (Bunea et al., 2010).

Hydroelectric sites use the available head and flow rate of a water course. Sometimes, this is made using more natural configuration, but often it involves important construction works and arrangements. The most common hydroelectric sites are based on:

a. local rise of the water level by means of a dam, which creates a reservoir upstream the dam. The hydropower plant is usually placed next to the dam;
b. deviation of water course through a free surface channel or tunnel. At the downstream end of the channel or tunnel, the water is put under pressure and driven to the turbines;
c. other mixed arrangements, with surface or underground hydropower plant, specific to mountain areas. These sites have high dams and large reservoirs.

The water quality used in a hydropower plant depends on many elements such as: the size and depth of the lake, placement along water course, intake depth, hydropower releases, temperature variations, rain intensity and frequency, reservoir thermal stratification and hydropower plant operation regimes.

The ecological impact of hydropower plants will be presented. In the first part is analyzed the processes in the reservoir (fig. 1), then, the influence of water realized on the downstream river. Also, some methods and their efficiency for improving water quality will be presented. Finally a review of chemical and quality water evolution in time will be presented for a hydroelectric site of Romania.

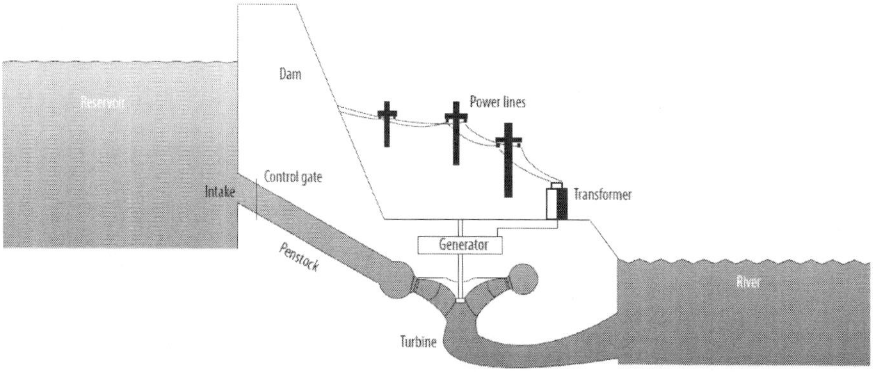

Fig. 1. Cross section of a hydroelectric power plant

During warm seasons the large reservoirs become subject of thermal stratification. Because the upper layers are close to the free water surface, they have a higher level of dissolved oxygen (DO). On the opposite side, the lower layers have a low level of DO, mainly because of the organic sediments at the bottom of the reservoir. When DO level goes under 5.0 mg/l, the aquatic life is endangered and large quantities of fish can die if the DO remains at 1÷2 mg/l for a few hours.

In hydropower plants, the water that goes to the turbines is taken from the lower layers of the reservoirs, sometimes with low DO content, which can affect the downstream water quality. The DO level from downstream water depends also on water head, periodic temperature variations, intensity and frequency of rain, hydropower plant design and its operation regimes.

Recently, the number of studies concerning water quality from hydropower releases increased. Many environment or ecological issues were reported, in different types of hydroelectric schemes. Scientist and engineers try to find solutions and mechanisms which will improve water quality, especially DO level. Generally, the low DO level is caused by organic sediments left on the reservoir bottom floor from the initial filling. When these organic sediments decompose, they absorb the oxygen from water, producing sulphuretted hydrogen, carbon dioxide and methane (like greenhouse gas). This pollution alters the local flora and fauna, even causing total extermination of some aquatic species.

Low DO level happens when the reservoir has a depth greater than 15 m and a volume bigger than $61 \cdot 10^6$ m^3, the power output is more than 10 MW, and the retention time is longer than 10 days.

Romania has about 170 hydroelectric sites, a quarter of them having reservoirs larger than $61 \cdot 10^6$ m^3 and deeper than 15 m, so they are susceptible for a low DO level (Bucur et al., 2010).

The usual methods used to increase the DO level in the hydropower plants downstream waters include selective intakes, air diffusers, hub and draft tube deflectors. These equipments are used in the hydropower plants with different success rates in the aeration process. Generally, in order to increase the DO with 1mg/l, an air quantity of 1% from water volume is necessary (March, et al, 1992). A bibliographical revue is presented in this paper and recommendations are done for the implementation of aeration devices.

There is no legal support for DO level control downstream hydropower plants, but there are intense concerns regarding this issue. Usually, turbine aeration is made only in order to reduce turbine central vortex, so to increase the efficiency and reduce unsuitable pressure fluctuations and structure vibrations. The aeration made to increase the DO level downstream hydropower plant must be more consistent. Injection of a bigger air quantity can decrease the turbine efficiency; therefore air injection becomes an important factor for the balance between power output and ecology.

2. Hydroelectric reservoirs

Regarding the mean multiannual water flow, the surface water sources in Romania are much higher than ground water sources. Each type of water source has its own physic-chemical and biological characteristics, varying from region to region, depending on the mineralogical composition of the crossed areas, by the contact time, temperature, weather conditions, etc.

Water accumulated in reservoirs has the physical - chemical qualities significantly different from water flowing in the river, before the dam construction and the hydropower development. Thus, the processes occurring in lakes can have an important impact on the water quality. On one hand the stagnation of water leads to a natural settling of suspended materials which determinate a good transparency of the water and less sensitive to weather conditions. On the other hand, the stagnation of water leads to thermal and chemical stratification which excludes the water circulation on vertically direction.

2.1 Seasonal stratification of water in hydroelectrical reservoirs

Thermal structure of lakes varies by climate, by configuration of the lake basin, by the water intake surface and by the total mineralization of water. The most common structure is the direct stratification, which involves the higher temperatures at the water surface and lower temperatures to the bottom. For this kind of stratification the decrease of water temperature is not uniform with depth. Temperate regions are characterized by dimictic lake ecosystems. Most lakes in Romania are considered dimictic, meaning they mix twice a year - spring and fall. In the winter season, reverse stratification will be installed, while in the summer period a direct stratification will appear.

The lakes dynamic is characterized by energy and mass exchange processes. Dominant energy flow comes from the kinetic energy of wind and thermal energy produced by solar radiation. The vertical profile of temperature/density established in a lake results by superposing these two energy contributions (Dumitran and Vuta, 2011).

Thermal stratification consists in the existence of a vertical thermal gradient in the water mass. The low thermal conductivity of water contributes also, assuring that thermal energy is very slowly transferred to the bottom layers of the water. This transfer is accelerated by vertical turbulent mixing and convective cooling of the water body. In time, the cumulative effect of heat loss and convective cooling can be felt throughout the water column, reducing the lake water temperature and causing a full mixing between the water layers (Pourriot and Meybeck, 1995). The cumulative effect of heat loss and convective cooling can be felt in the entire water column, thus reducing the lake water temperature and producing a full mixing between the water near the surface and deeper layers. Turbulent mixing is a process that precludes stratification, tends to destabilize the water column and is caused by shear induced by wind action (Stevens and Imberger, 1996).

Convective cooling occurs only if the net heat flow from the lake surface is negative. Thus the lake is losing heat to the atmosphere and the water layers near the surface are cooling, becoming denser than deeper waters.

At this point thermal stratification becomes unstable, and the volumes of water near the surface descend to a water layer with same temperature. Because of friction, running water entails other volumes of water, producing a new vertical mixing. The movement is done without wind energy contribution and there is a destratification tendency of superior layer to the equilibrium depth (Fig. 2). In summer, a typical temperature/density profile for a temperate lake is composed from two layers of small temperature/density gradient (epilimnion, hipolimnion) divided by a layer of high temperature/density gradient (metalimnion).

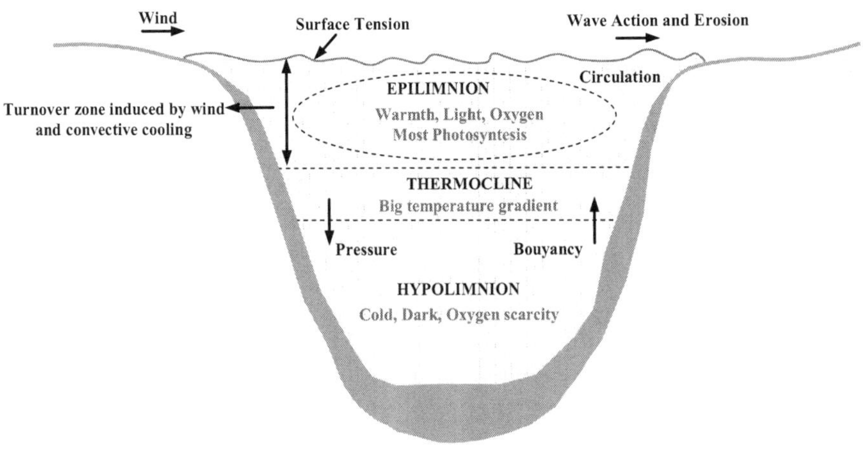

Fig. 2. Cross section of thermally stratified lake

Over the year, the lake water follows the cycle. In spring the ice melts into the lake, the wind picks up and the lake mixes. This is called spring turnover. Oxygen and nutrients get distributed throughout the water column as the water mixes. Then, as the weather becomes warmer, the surface water warms again and sets up summer stratification. During the summer the lake has a barocline structure, so at the surface a stable warmer layer of water overlies a colder water layer. The water movement due to wind and convection currents produces a mixing process which homogenize just the epilimnion, while the water temperature in the hypolimnion is kept at around 4 °C. In the fall the sunlight is not as strong as during summer and the nights become cooler. This change in season allows the epilimnion to cool off. As the water in the epilimnion cools, the density difference between the epilimnion and hypolimnion is not as great. Wind can then mix the layers. In addition, when the epilimnion cools it becomes denser and sinks to the hypolimnion, mixing the layers. This mixing allows oxygen and nutrients to be distributed across the whole water column. In winter, when surface water temperature drops below 4 °C, circulation ceases again and winter stagnation appear, characterized by an inverse thermal stratification. During this period the water mass is characterized by lower temperature at the surface and higher to bottom.

The lake stratification entails lower dissolved oxygen concentration in the bottom and the emergence of anaerobic oxidation processes. The stratification of lakes has a negative impact on trophic evolution of these ecosystems. Thus, the organic matter content and nutrients concentration will be increasing and sometimes even the hydrogen sulfide will appear at the bottom of lakes.

2.2 Day/night stratification of water in hydroelectrical reservoirs

In temperate regions the temperature differences between day and night are significant, so the water cooled during the night, goes down in a deeper layer. This depth is direct correlate with the reservoir size, so it can vary from 5 m up to 20 m (Read et al., 2011). In these conditions, the thermoclin layer appears which is characterized by a temperature drop of 10 to 15 °C (Fig.3).

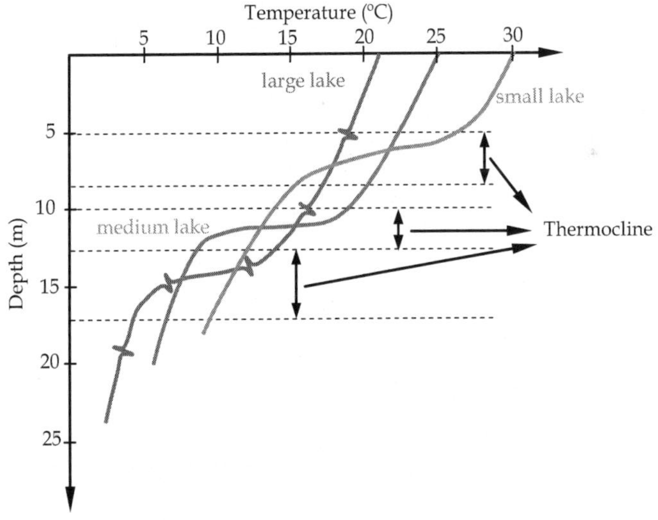

Fig. 3. The batimetric zones of the lake

2.3 Eutrophication of hydroelectrical reservoires

Biological quality of water is essentially affected by eutrophication, a phenomenon favored by building a hydropower plant. Eutrophication has proved to be one of the most widespread and serious anthropogenic disturbances to aquatic ecosystems. The major cause for eutrophication is the increased loading of nutrients, especially phosphorous. Increasing wastewaters, introduction of phosphorous containing detergents, use of fertilisers, and erosion in the watershed are the major reasons for increased loading of nutrients. The effects of the eutrophication phenomenon are negatively reflected on water quality, for reservoir ecosystem and also for river ecosystem (Fig. 4). Thus, eutrophication may lead in some cases even to the impossibility of using the water for certain uses.

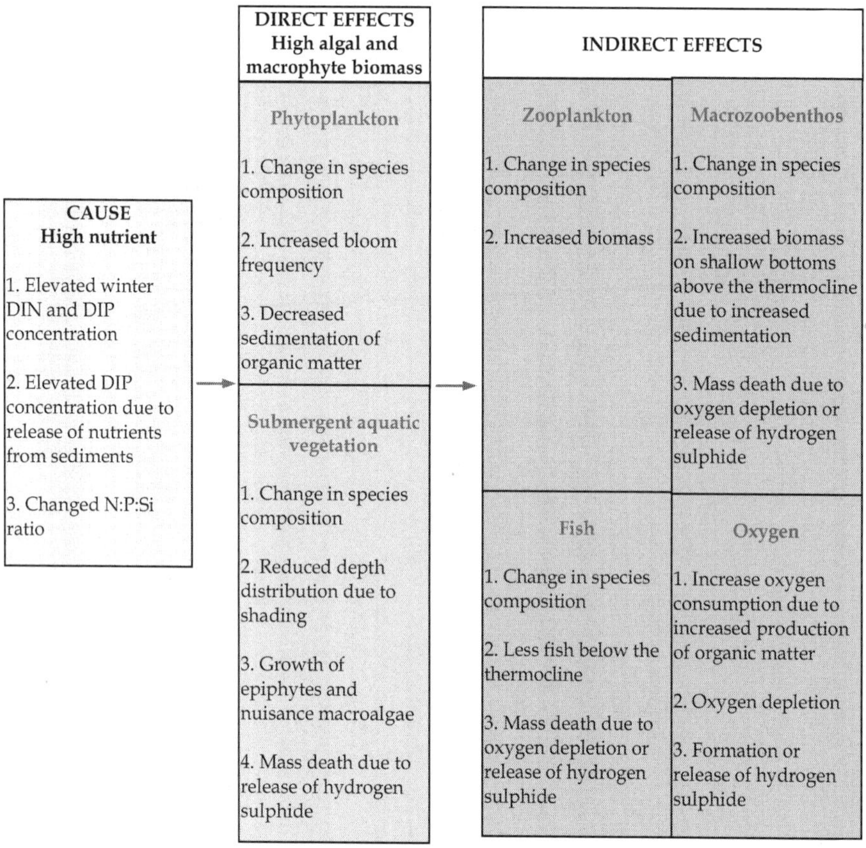

Fig. 4. Causes and effects of eutrophication

Effects of eutrophication emergence affect the lake ecosystem by:

- organoleptic changes of water (color, taste, odor and turbidity) by increasing the biomass of planktonic algae. Water may have a green color due to high content of green algae or diatoms, a red color in blue-green algae species presence, or even brown. This effect gives an unaesthetic aspect of water and leads to additional costs when water is used as a source of drinking water;
- premature clogging of filters and grids of treatment plants which are supply directly from the lake, due to increased phytoplankton biomass;
- biological clogging of the lake and therefore a reduction of its volume due to growth of organic matter content and organic detritus at the lake bottom;
- inability to release and to transform the organic matter due to its excessive quantity;
- pronounced decrease in the dissolved oxygen content, especially at the bottom of the lake, due to increased organic matter decomposition reactions;
- pronounced increase in the concentration of carbon dioxide, iron, manganese, ammonia and hydrogen sulfide due to occurrence of anaerobic decomposition conditions when dissolved oxygen is depleted;
- corrosion of water storage facilities due to the occurrence of precipitation reactions of iron and manganese. The same effect occurs in water in the presence of some *Cyanophyceae* species (*Oscillatoria*), which corrodes steel tanks in the presence of light;
- the appearance of toxic substances in water disposed by some *Cyanophyceae* species (*Microcystis* and *Anabaena flos-aquae aeroginosa*), causing human gastrointestinal disease;
- replacement of special fish species by common species due to changes in water quality.

From all the effects of water eutrophication, the most important consequence is the decrease of oxygen availability. During the day, plants produce oxygen, through photosynthesis, using sunlight. In the night, all organisms consume the oxygen dissolved in water by endogenous breathing.

When excessive amounts of biomass exist in the water body, decomposing organic matter will lead to higher oxygen consumption. Thus the oxygen in the water will be depleted, leading on the one hand to the impossibility of aquatic organisms breathing and on the other hand to the occurrence of anaerobic decomposition. Therefore all the biotic components of the aquatic ecosystem will suffer.

The heterotrophic organisms will be the first affected (fish and shellfish) because of their increased sensitivity to changes taking place in the chemical composition of water, like excessive alkalinity that occurs during intense photosynthesis processes and the lack of dissolved oxygen.

Eutrophication leads to changes in the populations of organisms that live in water. This is done through changes in ecological factors, which are becoming limiting factors for development of the aquatic organisms. Mostly are considered abiotic ecological factors, such as light, temperature, water movement or quantity of certain nutrients in water.

Since eutrophication involves a high input of allochthonous nutrients, often the light is the limiting factor for algae flourishing. Thus, due to changes in optical properties of water through cover of the surface water by vegetation, a high mortality of zoobenthos, nekton and zooplankton appear.

The nutrients demand vary widely from one species to another, both in terms of type and nutrient intake, so that a deterioration of the relationship between nutrients (nitrogen, phosphorus, silicon and iron) determine changes in qualitative and quantitative composition of the phytoplankton. From all the nutrients in the aquatic ecosystems, phosphorus is given the most attention, because is essential for all phytoplankton species. From the 5000 phytoplankton species with high abundance and wide geographical distribution, only 300 produce algae flourishing. Among the species that produce large biomass are many *Cyanobacteria,* which have the capacity to produce toxic substances in the water with effects over health. Changes in the phytoplankton composition in an aquatic ecosystem will cause major changes in the entire trophic chain. Thus, the composition of primary consumers (zooplankton and fish) will change (Cooke et al., 2005).

2.4 Accidental pollution of hydroelectrical reservoirs

The death of fish is mainly caused by the level of dissolved oxygen. There are also some situations of water pollution with toxic substances, but they will not be detailed in this paper. As an example, in river Târnava Mare (downstream Odorheiul Secuiesc, Romania) an historical pollution incident happened in 2002 (table 1) [Serban, 2005]. It caused fish morbidity, because of high temperature (over 20°C), water flow lower than annually average, low water velocity (0.3 ÷ 0.6 m/s) and overloading with some organic substances from an upstream wastewater treatment. All these made the level of DO lower than 4mg/l.

River	F [km²]	Q_{ma} [m³/s]	v [m/s]	Q [m³/s]	OD [mg/l]	T_{apa} [°C]	CCO-n [mg/l]	NH_4 [mg/l]
Crasna	1702	5.56	0.31	1.40	4.2	25	16.6	8.82
Someş	9753	82.60	0.60	33.9	4.5	28.4	15.8	4.2
Zalău	54	0.80	0.55	0.32	3.8	25	18.2	5.2
Someşul Mic	2954	17.50	0.38	6.88	1.03	28	32.4	6.2
Bistriţa	650	8.38	0.40	4.70	4.2	21	4.49	2.6
Târnava Mare (Od. Scuiesc)	646	5.70	0.35	1.31	4	23.9	32.0	4
Târnava Mare (Copşa Mică)	2960	12.10	0.33	3.12	2.6	22.4	13.9	0.8

Table 1. Fish morbidity on water courses in Romania in 2002 after water pollution incident (Q_{ma} - average multi annual water flow, Q – water flow at the incident time)

If the water pollution is limited, the reconstruction of original water quality is possible only by eliminating the accidental pollution sources.

2.5 Management of water quality in hydro electrical reservoirs

Because the lake stratification has a negative effect on water quality, the depth of reservoirs is a great disadvantage from water quality point of view. Therefore very deep lakes are not desired. For deeper reservoirs, one measure to combat the stratification is to locate water intakes at different depths.

The main effect of the stratification in the hypolimnion and sediments is the increased consumption of oxygen and appearance of anoxic conditions which impoverish the deep water fauna. This condition may also lead to a series of chemical and microbial processes like nitrate ammonification, denitrification, desulphurication and methane formation. The release of phosphorous from the sediments is extremely important as it accelerates eutrophication.

The following actions are required to maintain water quality in the reservoirs used by the hydroelectric development:

- the watershed management by river bed erosion control works, which will reduce the intake of silt in the lake;
- discharge of the effluent downstream the reservoir section;
- reducing water pollution;
- setting up sanitary protection perimeters around the lake and adjacent control of tourist areas;
- prevention of lake stratification and insurance of water vertical circulation through water intake at various depths and periodic use of bottom discharge system at a flow able to ensure hygiene riverbed downstream and hipolimnion renewal.
- discharges in reservoir mass is advantageous to be submerged, perpendicular to the surface of the lake for a maximum effect of aeration and movement of water layers.
- flows discharged from tailrace must have a minimum impact for downstream environment. Discontinuous discharge destroys the river bed, erodes the banks and can even break the roads and bridges. Such flows also have a stressful effect on fish.

In this way the water quality can be maintained in reservoirs without negative effects on water quality.

3. Aeration methods inside hydraulic turbines

As presented before, the water quality downstream hydropower plant depends mainly on the quality of water from the upstream reservoir. After water passes through the turbines, a supplementary degasification of water takes place, because of the low pressure in turbine draft tube, which lowers the DO level. This process happens mostly in Francis turbines at partial load operating regimes.

This is the main reason for developing and installing new aeration methods to increase DO level of turbined water. From hydraulic point of view, an air quantity injected downstream turbine runner, could affect turbine efficiency. For this reason, it is recommended that air inflow to be maximum 1÷3% of turbine water flow.

3.1 Existing solutions for aeration of hydraulic turbines

Measurement data are available for different technical solutions for water aeration:

- auto ventilation turbines, developed by Voith Hydro and Tennessee Valley Authority. The aeration can be made central, distributed or peripheral (through the outlet edge of blade) (Figure 5). Test were made for each aeration system individual and combined with the others. The air injection is made through new or existing passages (vacuum braking system and snorkel tubes), using air compressors or natural air admission (proffered for a lower cost).

- a system with air injection in turbine and another one with oxygen injection through porous hoses in penstock were installed at Tims Ford Dam (Harshbarger et al., 1999 and 1995).

The autoventing turbines (central, peripheral or distributed) were implemented for the first time at Norris Dam. These aeration systems can be used individually or combined. The justification for any solution depends on many parameters, characterizing each hydroelectric site. For autoventing turbines all combination were tested. When a group operates with all aeration systems the DO increased up to 5.5 mg/l. In this case the air absorbed in turbine is twice than the air absorbed by the original runners.

Depending on the operational conditions and the aeration system, the energetic efficiency decreased with 0 ÷ 4 %.

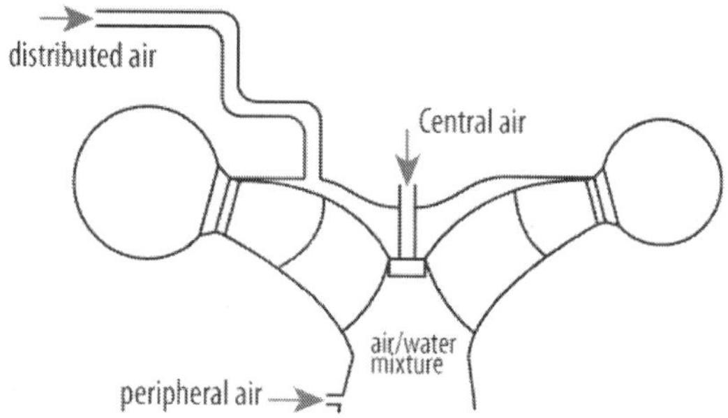

Fig. 5. Aeration methods for autoventing turbines

In other researches (March et al., 1992) the DO level in downstream water was up to 6 mg/l, trying to affect as little as possible the energetic efficiency. A few solutions were tested, like injecting air through runner, the design of a new deflector, low pressure edge blades, coaxial diffuser, injection in the conical part of the draft tube, or combination of them.

Aeration performance can be evaluated by measuring the DO level upstream and downstream turbine,

$$E = OD_u - OD_d \tag{1}$$

where DO_u and DO_d are the OD concentration upstream and downstream turbine.

The effect of aeration over the hydraulic efficiency of the turbine is

$$\Delta\eta = \eta_a - \eta_0 \tag{2}$$

where η_a is turbine efficiency with aeration system and η_0 is turbine efficiency without aeration system.

Two aeration systems were tested at Tims Ford Dam (Harshharger et al., 1995), in order to achieve a DO level of 6 mg/l, by injecting air in turbine and oxygen in penstock, through porous rubber pipes. For an upstream DO level of maxim 1 mg/l, if both aeration systems were operational, the DO level got to 5.2 mg/l and if only air system was on the DO level was 4.2 mg/l.

The air was injected with high pressure compressors under runner cover or in the draft tube. Also, porous line diffusers were installed in penstock, for oxygen injection, in case the desired DO level (6 mg/l) is not reached only with the air injection system. The cost of this oxygen injection system in 1995 was of 300'000 $.

Both systems were used during low DO level periods. In order to evaluate the DO level increase and turbine efficiency, the air, oxygen and water discharge were modified during tests. In any of the cases, the turbine efficiency decreased with maxim 1%, so the aeration didn't affect it too much. But this technology was rejected because of the rate between initial installing cost, long time operation and service cost.

Another research study, developed during two years, was made at la Bagnell Dam (Sullivan et al., 2006), over two turbines (with runner aeration orifices), an old one and a new one, with some changes. The tests were made for 51 combinations of water discharge, downstream water depth and aeration orifices diameters and were determined the water discharge through orifices, DO level and water temperature in sections upstream and downstream turbine. As a general conclusion, the oxygen transfer efficiency increases with the increase of air discharge and of downstream water level.

For the older turbine, it was noticed that for smaller openings the runner orifices are more efficient than draft tube orifices, and that when both aeration systems were operational and the water discharge was small, the DO level was over 5 mg/l.

Some researchers made studies concerning DO, temperature and fish growth downstream the hydro plants (Boring, 2005). For the Saluda River (U.S.) a model based on the historical data from 1990-2005 had been developed. In accordance with Environmental Protection Agency (EPA), the criteria established for 2006 are: for survival of trout - min. 3 mg/l, for growth protection - 6.5 mg/l for an average of 30 days, and for sensitive cold-water invertebrates - min. 4 mg/l.

The studies and researches continued with mathematical modelling of the flow (Rohland et al., 2010) for the three classical aeration methods in a Francis turbine. Each classical injection method has different characteristics and influence over the dimension and distribution of bubbles that flow through the draft tube and over the operating efficiencies.

The parameters that influence the efficiency are: the shape and length of draft tube, bubble retention time, quantity of air or the void fraction, air admission intake, bubble size and distribution. From quantity of air or the void fraction point of view, central aeration is the most efficient. The calculations for turbine efficiency and the aeration methods were used for optimization of aeration solution at Bridgewater plant, one of the first power plants designed in respect to aeration.

In Romania, environmental impact and water quality are main concerns, but the aspects of DO level are not taken in consideration. Even if the hydropower operators are preoccupied about environmental issues, there is no legal support for DO level. Turbine aeration (especially central zone) is made only in a few sites, but with hydraulic purposes (to reduce central vortex at partial load) in order to reduce pressure fluctuations.

3.2 Aeration efficiency and main parameters

As mentioned before, two main parameters must be considered in the aeration process: the DO transfer through air injection and the total energetic consumption to realize it.

The DO transfer necessary for water quality improvement, depends on many physical parameters: the quantity of injected air, the gas - liquid contact surface, time of contact, temperature, pressure gradient flow, DO level gradient, turbulence level of flow.

The energetic consumption necessary to introduce the air in liquid depends on the following operational parameters: the quantity of injected air, injection method (natural or induced), air influence over turbine efficiency (flow changes).

In order to obtain a good global balance the DO transfer must be done with a minim energetic consumption. For this purpose is necessary to generate an optimal dimension for the gas bubble - as small as possible (the positive effect is double, because of increasing contact surface and retention time), but with low hydraulic losses for decreasing energetic consumption.

The main parameters that must be considered to find the best compromise between the improvement of water quality and modification of turbine efficiency - see table no 2.

Physical aeration parameters	⇔	Energetic operational parameters
- Air – water contact surface, - Bubble shape coefficient, - Retention time, - Pressure gradient, - Flow turbulence - Air void fraction, - Air/water temperature.		- Energetic consumtion for air injection (natural injection is more advantageous), - Efficiency looses due to changes in internal flow.

Table 2. Parameters considered for aeration efficiency balance

The air injected in turbine affects its efficiency in two ways: because of the flow perturbation caused by air introduced in water flow and because of energetic consumption necessary to introduce the air. Considering the above, the injected air flow is limited to 1÷3% of water inflow. Usually, this air quantity is not enough to have a good aeration and to obtain the minim reference DO level (5÷6 mg/l) in downstream waters. The natural absorption of atmospheric air is preferred, because it uses the existent turbine depression.

The hydrodynamics and the mass transfer of bubble columns are the subject of many research studies. The matter is analyzed from different aspects: the velocity induced by the bubbles to the liquid (Krishna et al., 1999), (Wiemann and Mewes, 2005), (Ekambara and Dhotre, 2010), bubbles generation regimes (uniform, transient and heterogeneous) (León-Becerril et al, 2002), (Pincovschi et al, 2007), size distribution of gas bubble gas (Katerina et al, 2004), (Polli et al, 2002), Laser Doppler Anemometry and Particle Image Velocimetry measurements (Mayur et al, 2010) (Laakkonen et al, 2005), (Becke et al., 1999), (Ciocan et al., 2011), void fraction and volumetric transfer coefficient (Krishna and van Baten, 2003), numerical simulation of mass transfer (Painmanakul et al. 2009), (Connie et al, 2003).

Mathematical models for turbulent two phase flow in complex configurations is one of the most difficult part in gas – liquid flow simulations, not solved until now. A more detailed representation of bubble movement and of their interaction with the liquid phase can get to development of turbulence models. In spite of consistent efforts for a correct description of closing rules for drag forces, lift forces and mass forces, the precisely model of interfacial forces remains an open matter in this kind of numerical simulations.

This paragraph is focused on correlation of aeration quality (oxygen transfer for a constant injected air flow), with the energetic efficiency of the transfer (energy consumption for introduction of the air flow).

Further on is presented the air bubble dimension influence on DO transfer, in relation to aerator required pressure, for different air discharges. Five metallic perforated plates were used (Bunea et al., 2010) with different orifice sizes (*d*) and identical geometries (Figure 6).

The total perforated surface area is equal in all configurations (12 mm²). The tests were made in same hydrodynamic conditions, in a tank with 79.2 l of water. For each plate were determined the DO level (C), water temperature (*t*) and pressure losses (Δp) on aeration system. The experimental results concerning DO level in time, are determined according ASCE regulations 2-91/1993.

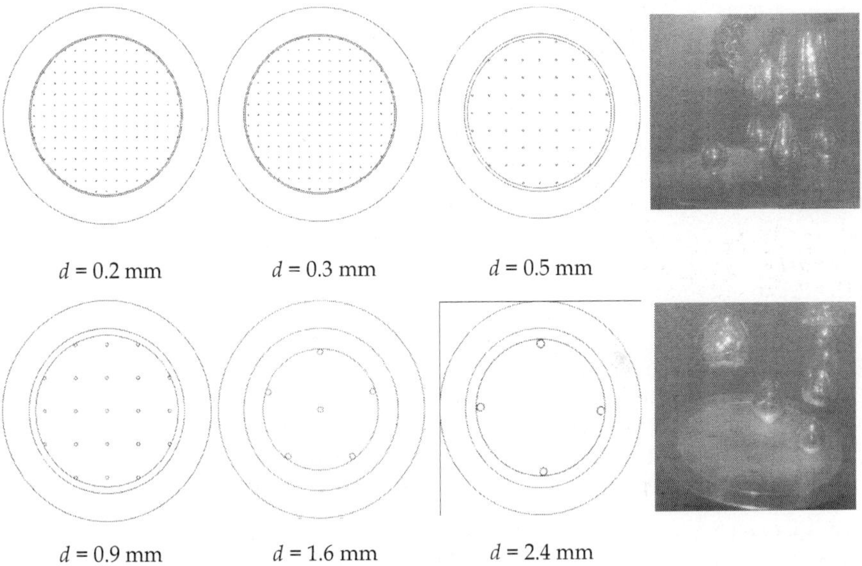

Fig. 6. The layout of the orifices on metallic plates

By keeping constant the following parameters: air flow rate Q_{air}, water volume and total area of perforations, the increasing of air-water interfacial area for the first layer of bubbles (*A*), standard oxygen transfer rate (*SOTR*) and standard aeration efficiency *(SAE)* with the decreasing of orifice diameter can be observed, while the pressure loss on perforated metallic plate (Δp) increases with maximum 27 mmH_2O (Table 3).

d (mm)	A (mm^2)	Δp (mmH_2O)	SOTR (mg/min)	SAE (kg/kWh)
1.6	326	16	33.37	2.50
0.9	781	28	33.35	2.48
0.5	1524	30	49.74	3.67
0.3	3150	34	67.45	4.95
0.2	5233	43	77.08	5.59

Table 3. Evolution of mass transfer parameters with the orifice diameter for constant air discharge Q_{air}=360 l/h

Figure 7 shows the influence of orifice diameters on standard aeration efficiency provided by the metallic plates for different air flow rates (Bunea et al, 2010). Also, by increasing orifices diameter is necessary to increase the air flow rate, otherwise the plates generate bubbles from half of their surface. A bigger air flow rate leads to increased pressure losses on the aeration device and diminishes the standard aeration efficiency. Finally is obtained the efficiency of the oxygen transfer rate related to power needed to inject the air.

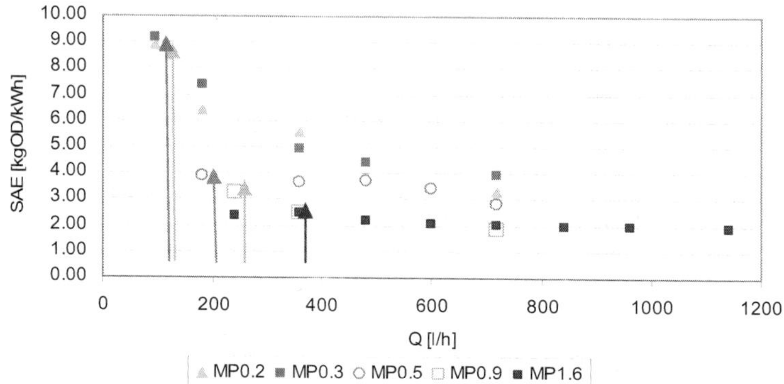

Fig. 7. Variation of SAE with the air flow rate and the orifice diameter of plates

A basic study of the air column in a water tank shows the importance of the quality of air injection, meaning: bubble size, pressure loss on the aeration device etc. for the oxygenation purpose. Different types of bubble aeration systems have been tested and compared. Starting with the experimental results, the influence of air bubbles dimension upon mass transfer is pointed out and the results are correlated with the pressure needed for the aeration device operation (pressure loss) at different air flow rates (Q = 180÷1160 l/h). Thus, efficiency SAE increases while air flow rate and the size of orifices in plates decreases.

4. Water chemical parameters in reservoirs of low head hydropower plants

It is important to study in time the water chemical parameters in large reservoirs. Big hydroelectric projects change the environment, creating retention lakes, the interaction between water and equipment materials causing different corrosion stages. In the same time, the equipments and their operation can affect the water quality (oil leaks, water degasification the through the turbines etc.).

The Lower Olt cascade from Romania (fig. 8) is an important and unique hydroelectric site, made of five identical plants (HPP). Each HPP has four bulb turbine-pump units, with a total installed discharge of 500 m^3/s and 13.5 m net head. The corresponding reservoirs have a volume to normal retention level between $62 \cdot 10^6$ m^3 and $99 \cdot 10^6$ m^3.

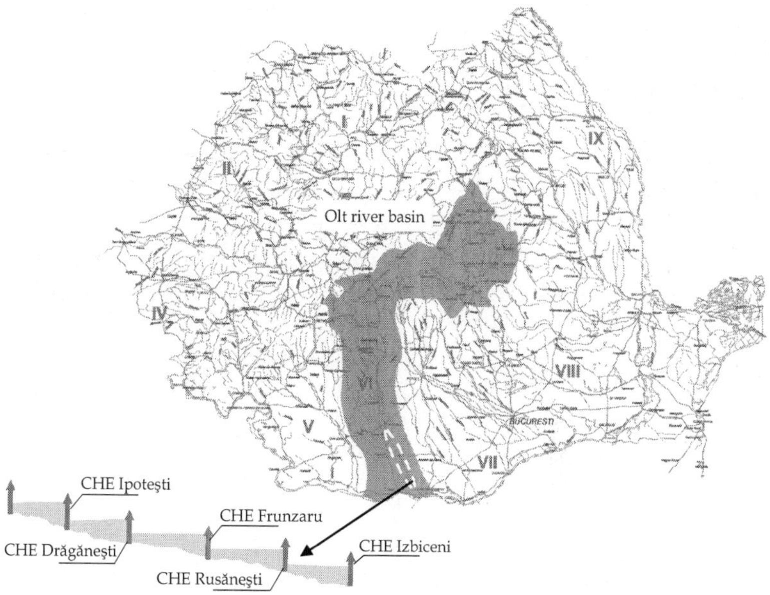

Fig. 8. Lower Olt cascade

A special particularity is that all five HPP operate both turbine and pump regime, which means important volumes of water are transported downstream and upstream in the same site. This could affect in a negative way the water quality by preserving for a longer time the accidental pollution effects.

Because water quality and environment protection are main concerns, the chemical parameters of water in Olt River, is periodically analyzed. For example, a few parameters recorded after ten years of operation and after twenty years, shows that the long time interaction between water and the equipment of HPPs, did not affect the quality of water.

Generally, all parameters remained at the same values, or even decreased, excepting sulphuretted hydrogen, but still remains in admitted values. During the analyzed period, there were no pollution accidents, neither in the hydropower plants in the Olt River and nor in its effluents.

It is also important to determine the effect of water on the equipment (corrosion). For this purpose, determinations in ten different measurement points along Lower Olt cascade were made for the level of pH indicator, chlorine content and manganese content. The results showed for pH indicator that five values are between 5.5÷6.5, and five of them are between 6.5÷7.0, all being between 6.5 ÷ 8.5, the reference limits for natural waters.

Parameter	Unit	Measured values	
		After 10 years	After 20 years
pH	u. pH	8.19-8.23	7.15
Dissolved oxygen	mg/l	4.8-5.9	5.1
CBO_5	mg/l	1.8-2.2	1.1
CCO-Mn	mg/l	5.7-6.2	2.203
Sulphuretted hydrogen	mg/l	<0.001	0.409
Phenols	µg/l	10	1.68
Phosphate	mg/l	<0.01	0.06
Chlorides	mg/l	85.3-92.1	73.4
Total hardness	°G	7.91-8.38	10.53

Table 4. Water chemical parameters in Lower Olt cascade reservoirs

The chlorine content is between 20 mg Cl/l and 106.5 mg Cl/l and the total manganese contents are up to 0.004 mg Mn/l, but mostly is zero. The total manganese content being so low the possibility of strong oxidizer appearance on the steel surface can be excluded (Bucur et al., 2010).

These results show that after twenty years of continuous, the water quality in this complex hydroelectric site was preserved from physical - chemical point of view. The dissolved oxygen quantity has a level that allows the preservation of aquatic life.

5. Conclusion

The hydraulic energy is an essential green energy source and important for the integration of the others energy sources in the energy system. However, as an essential live source - water - is used for energy generation, to conserve the green character of power generation, accompaniment measures are to be implemented on the hydropower plants sites.

The quality of water in a hydro electrical site depends on both natural factors - like temperature variations, precipitation intensity and frequency, thermal stratification reservoirs, and operational parameters and components of the hydro electrical site. The behaviour of the water in the lake is to be studied and considered in operation. To not take in consideration this behaviour, the complete eutrophication of the lake can happen, with huge consequences of the ecological system.

Another essential topic is related to the released water. The required parameters to preserve the aquatic life are to be insured; the main parameter is the DO, and 5 mg/l are needed. Aeration devices can be implemented to improve the DO content. The implementation has to consider the compromise between the positive effect of air injection and the inconvenient of energy losses due to air injection (energy needed for injection and energy losses due to flow perturbation by the air injection). Aeration devices can be implemented on new facilities or in the refurbishment process. Actual studies are done to improve the efficiency of the aeration.

The environmental impact of a hydroelectric power plant should be minim, and the water parameters should be as close as possible to natural water course values. The necessity of supervising of the water quality is a reality that should be a main concern for all hydropower users and will permit to prevent the degradation of the ecological system or to implement the needed system to improve the water quality.

6. Acknowledgment

This paper has been elaborated with the support from the Romanian National Council for Scientific Research in Education, IDEAS program, contract no. 705/2009, ID 1701, and by the Sectoral Operational Programme Human Resources Development 2007-2013 of the Romanian Ministry of Labour, Family and Social Protection through the Financial Agreement POSDRU/89/1.5/S/62557.

7. References

ASCE (American Society of Civil Engineers) standard ANSI/ASCE 2-91/1993, *Measurement of Oxygen Transfer in Clean Water - 2nd Edition*, ISBN 087262885X, 45 p.

Becke S., De Bie H. and Sweeney J. (1999). *Dynamic flow behaviour in bubble columns*, Chemical Engineering Science, V. 54, Issue 21, pp. 4929-4935

Boring, S., 2005, *Lower Saluda Site Specific DO Standard*, Joint Meeting with Fish & Wildlife RCG, Presentation, Kleinschmidt. (http://www.saludahydrorelicense.com)

Bucur D.M., Bunea F., Ciocan G.D., Băran G., Isbăşoiu E.C. (2010). *Water parameters evolution in a hydroelectric site*, Environmental Engineering and Management Journal, Vol.9, No. 11, 469 – 472

Bunea F., Houde S., Ciocan G.D., Oprina G., Băran G., Pincovschi I. (2010). *Aspects concerning the quality of aeration for environmental friendly turbines*, 25th IAHR Symposium on Hydraulic Machinery and Systems, IOP Conf. Series: Earth and Environmental Science, vol. 12, 012035

Ciocan G.D., Bunea F., Houde S., Deschênes C., (2011). *Measurement technique for air-water disperse system study*, 5TH International Conference on Energy and Environment CIEM2011, Bucharest, Romania

Connie D. DeMoyer, Erica L. Schierholz, John S. Gulliver, Steven C. Wilhelms, (2003). *Impact of bubble and free surface oxygen transfer on diffused aeration systems*, Pergamon, Water Research 37 1890–1904

Cooke D, Welch E., Peterson S., Nichols S. (2005). *Restoration and management of lakes and reservoirs*, Boca Raton Florida, Taylor&Francis Group

Dumitran G.E., Vuţă L. (2011). *Study on Lake Izvorul Muntelui rehabilitation*, Simulation Modelling Practice and Theory, vol. 19, pp 1235-1242

Ekambara K., Dhotre M.T. (2010). *CFD simulation of bubble column*, Nuclear Engineering and Design 240 963–969

Harshbarger, E.D., Herrold, B., Robbins, G., Carter, J. (1999). *Turbine Venting for Dissolved Oxygen Improvements at Bull Shoals, Norfork and Table Rock Dams*, Waterpower '99 - Hydro's Future: Technology, Markets, and Policy, CD-ROM

Harshharger, E.D., Mobley, M.H., Brock, W.G. (1995). *Aeration of hydroturbine discharges at Tims Ford Dam, San Francisco*"; ASCE, 9 pp., Waterpower '95 - Proc.of the Conf.on Hydropower, San Francisco, 1, 11-19

Katerina A. Mouza, Nikolaos A. Kazakis & Spiros V. Paras, (2004). *Bubble column Reactor design using a CFD code*, 1st International Conference "From Scientific Computing to Computational Engineering" 1st IC-SCCE, Athens, 8-10 September

partitions may be divided further into specialities. For instance, *algology* and *mycology*, the study of algae and fungi respectively, are subdivisions of botany. *Protozoology*, the study of unicellular animals, is a division of zoology whilst bacteriology and virology, the study of bacteria and viruses, are subdivisions of microbiology. Knowledge of the behaviour of micro-organisms is heavily dependent upon biochemistry although knowledge about macroscopic animals and plants may be acquired through studies of anatomy and morphology as well as structure and form.

Life does not have simple definition. It may be characterized by a list of properties which are shared by all living organisms, with the exception of the viruses, and which discriminate them from non-living matter:

Movement: It is characteristic of organisms that they, or some part of them, are capable of moving themselves. Even plants, which at first sight appear to be an exception, display movements within their cells.

Responsiveness: All organisms, including plants, react to stimulation. Such responses range from the growth of a plant towards light to the rapid withdrawal of one's hand from a hot object.

Growth: Organisms grow from within by a process which involves the intake of new materials from outside and their subsequent incorporation into the internal structure of the organisms. This is called assimilation and it necessities some kind of feeding process.

Feeding: Organisms constantly take in and assimilate materials for growth and maintenance. Animals generally feed on ready-made organic matter (*heterotrophic* nutrition) whereas plants feed on simple inorganic materials which they build-up into complex organic molecules (*autotrophic* nutrition).

Reproduction: All organisms are able to reproduce themselves. Reproduction involves the replication of the organism's genetic "blueprint" which is encoded in a nucleic acid. Generally this is *deoxyribonucleic acid* (DNA) but in some viruses it may be *ribonucleic acid* (RNA).

Release of energy: To sustain life an organism must be able to release energy in a controlled and usable form. This is achieved by breaking down *adenosine triphosphate* (ATP). The energy to generate ATP is obtained by the breakdown of food by respiration. The occurrence of ATP in living cells appears to be universal.

Excretion: The chemical reactions that take place in organisms result in the formation of toxic waste products which must be either eliminated or stored in a harmless form.

1.3 Global impact of waterborne disease

Throughout the world, many people do not have access to safe drinking water. As a consequence, there is significant morbidity and mortality due to disease-causing organisms in water. It is estimated that nearly one-fourth of all hospital beds in the world are occupied by patients with complications arising from infection by waterborne organisms (Gerba, 1996). Citing the *WHO/UNICEF Global Water Supply and Sanitation Assessment 2000 Report,* it is estimated that nearly 6000 people, mostly children, die every day because of water related diseases. Even in the United States, an estimated US$20 billion per year in lost productivity has been attributed to diseases caused by waterborne pathogens (Gerba, 1996).

1.4 Microbiological aspects

Drinking water should not contain any microorganisms known as pathogens. It should be free from bacteria indicative of excremental pollution. The primary bacterial indicator recommended for this purpose is the coliform group of organisms as a whole. Although as a group they are not exclusively of faecal origin, they are universally present in large numbers in the faeces of man and other warm-blooded animals, and thus can be detected even after considerable dilution. The detection of faecal (*thermotolerant*) coliform organisms, in particular *Escherichia coli*or *E.coli* provides definite evidence of faecal pollution.

The tiny and invisible microorganisms pose a serious threat to the safety of the world's drinking water. Water when contaminated by microorganisms, particularly by the pathogenic ones, can become a growing peril with the potential to cause significant outbreaks of various types of infectious disease. The list of potentially pathogenic microorganisms transmitted by water is increasing significantly each year. Indeed the distribution of safe drinking water to the home can no longer be taken for granted, not even in the United States and Western Europe. The average person consumes about 2-4 liters of water per day through food and drinks. All these deliberations suggest an urgent need for supply of safe drinking purposes. Pilot and mini scale carbon-soil-sand filtering system has successfully produced potable drinking water in the developed countries for a century. Cost of fabrications, usefulness of local raw materials, ease of operation and maintenance, and low energy requirements make slow carbon-soil-sand filtration system a water treatment technology that is particularly well suited for developing countries.

In the midpoint of international drinking water supply and sanitation decade (1985) only 42% of the world's rural populations and 77% of the world's urban population had access to safe drinking water excluding the Peoples of Republic of China (Rotival, 1987). Extensive research has been performed on the efficiency of carbon-soil-sand filters for treatment of water in normal climates. When source waters are relatively low in temperature and turbidity, this process has been found to be effective in the removal of toxic elements, organics and bacteria (Bellamy et al., 2006, Slezak and Sims , 1984). Tropical rivers and other surface waters, which serve as receiving waters for waste discharges and runoff, also serve as water supplies for downstream communities. Low flow rates of surface waters during dry seasons will result in high levels of toxic and microbial contamination from upstream human waste, agricultural runoff and industrial processing facilities. As concentrations of faceal coliforms in untreated water supplies vary widely in developing countries (Vsscher et al., 1986), polluted source waters were simulated by maintaining a filter influent concentration of approximately 10^6 *E. coli* cells per 100 mL.

Adsorption with carbon-soil-sand filter has been one of the most useful techniques in water treatment. In the past, activated carbon was predominantly used to remove odor and colour producing molecules in water (Suffer and McGuire, 1980). We have previously reported the removal of toxic elements (heavy metals) using the same filtering system (Yusof et al., 2002 and Rahman et al., 2011).

In these papers, the competitive separation and adsorption microbiological pollutants (Coliform, total count etc.) are reported. One of the problems related to groundwater is the reddish colour caused by the presence of iron and manganese. This colour can be seen after it has been exposed to the air, the oxidation of groundwater will promote the precipitation of iron(III) and manganese ions. A significant removal of iron and manganese was reported by Ahmad et al. (2005). While a fixed bed column or continuously flow study to remove heavy metal specifically cadmium and lead by using granular activated carbon has been successfully removed up to 99 percent (Ahmad et al., 2007).

A culture medium is a solid or liquid preparation used to grow, transport and store microorganisms also based on Prescott, et. al (2005). The medium must contain all the nutrients the microorganism requires for growth. The isolations and identifications of microorganisms need special media which is essential for testing the sensitivities of water. Sources of energy, *carbon, nitrogen, phosphorus* and various minerals are required for the growth of microorganisms and the precise composition of a satisfactory medium will depend on the species one is trying to cultivate because nutritional requirements vary so greatly.

Indicator organisms are bacteria that are used as a sign of quality or hygienic status in water. The definition of indicator is the concept of the indicator organisms which is so strictly associated with particular conditions that its presence is indicative of the existence of these conditions. The minimum requirement for an indicator is that it must be a biotype that is prevalent in sewage and excreted by humans or warm-blooded animals. Historically, these conditions have been related to insanitation and public health concerns. Over the years, however, the use of indicator organisms has been extended to provide evaluation of the quality, in addition to the safety, of particular commodities. In addition, the indicator should be present in greater abundance than pathogenic bacteria, incapable of proliferation or at least not more capable than enteric bacteria, more resistant to various disinfectants than the pathogenic bacteria, and qualified by simple and rapid laboratory procedures. To ensure of value in evaluating the risk of disease as well as water quality, the indicator should satisfy the following criteria:

i. The indicator should always be present when the source of the pathogenic microorganisms of concern is present and absent in clean uncontaminated water.
ii. The indicator must present in numbers much greater than the pathogen or pathogens it is intended to indicate.
iii. The indicator should respond to natural environmental conditions and water as well as wastewater treatment processes in a manner similar to the pathogens of interest.
iv. The indicator should be easy to isolate, identify and enumerate.

1.5 Bacteria

The bacteria (blue green bacteria) formerly known as blue-green algae, constitute in the kingdom *Procaryotae.* They are single-celled organisms which use soluble food. From all organisms, this group is the most significant to the public health engineer, since biological wastewater treatment processes rely almost exclusively on the activity of bacteria. Bacteria are relatively endurant in many habitats on earth, growing in soil, acidic hot springs, water, deep in the earth's crust, in organic matter and the live bodies of plants and animals. They constitute the highest population of microorganism in wastewater.

Bacteria can be classified into two major groups: heterotrophic and autotropic depending on the source of nutrients. Heterotrophs utilize organic matter as energy as well as a carbon source for their synthesis. Whereas autotropic bacteria use oxidizing inorganic compounds for energy and carbon dioxide as a carbon source (Hammer, 2008).

Bacteria that cause bacterial infection are called pathogenic bacteria. Pathogenic bacteria are a major cause of human death and disease and cause infections such as cholera, tetanus, typhoid fever, diphtheria, syphilis, foodborne illness, leprosy and tuberculosis. A pathogenic cause for a known medical disease may only be discovered many years after, as was the case with *Helicobacter pylori* and peptic ulcer disease. Bacterial diseases are also important in farm animals such as mastitis, salmonella and anthrax as well as in agriculture, with bacteria causing leaf spot, fire blight and wilts in plants. Some of commonly found bacteria are shown in Figure 1. They are *Bacillus, Bordetella, Clostridium, Escherichia, Spirilina, Staphylococcus, Streptococcus* and *Salmonella.*

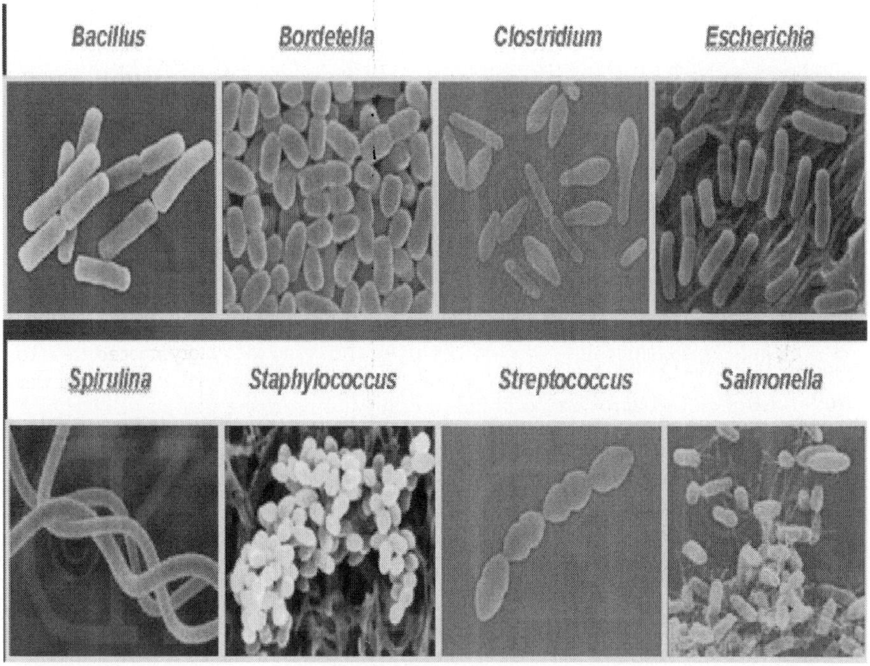

Fig. 1. Commonly found bacteria (Source http://agrobacter wikispaces com/file/list)

It is well known that soil washes into natural bodies of water, particularly after heavy rains. Many different kinds of bacteria also will be carried into water, including soil bacteria and bacteria from the faeces of animals, such as tiny invertebrates, insects, birds, and other "lower animals." In addition, water will often have its own bacterial flora, contributed by its resident species of animals. Bacterial flora from literally hundreds of different species of animals enters natural waters each week. The normal flora of man has been thoroughly studied. The stool flora of man alone may comprise over a hundred different species of bacteria. Although less well studied, other animals have their own characteristic bacterial flora. Many of the bacteria species which form the normal flora of man probably evolved from bacterial species which form the normal flora of lower animals. In the cases that have been carefully studied, it is easy to tell the difference between closely related species with a sensitive technique, such as DNA-*hybridization*. Often, however, it is not easy to tell the difference between closely related species with the simple tests samples taken from insects, birds, and mammals.

1.6 Coliform

The coliform group was the mainstay of the sanitarian's tools for detecting the presence of faecal contamination in aquatic environments. The broad general characteristics which define this group have allowed it to be one of the most useful of bacterial indicators and at the same time have been responsible for its displacement as an indicator of faecal contamination.

The coliform group is made up of bacteria with defined biochemical and growth characteristics that are used to identify bacteria that are more or less related to faecal contaminants. The total coliforms represent the whole group, and are bacteria that multiply at 37°C. The thermotolerant coliforms are bacteria that can grow at a higher temperature (44.2°C) and *E.coli* is a thermotolerant species that is specifically of faecal origin. A finding of any coliform bacteria, whether thermotolerant or not, in water leaving the treatment works requires immediate investigation and corrective action. There is no difference in the significance of total coliforms, thermotolerant coliforms and *E. coli* in water leaving a treatment works, as they all indicate inadequate treatment, and action should not be delayed pending the determination of which type of coliform has been detected. Upon detection in a distribution system, investigations must be initiated immediately to discover the source of the contamination.

Tests for detection and enumeration of indicator organisms, rather than of pathogens, are used. The cultural reactions and characteristic of this group of bacteria have been studied extensively. Coliform group density is a criterion of a degree of pollution. Membrane filter technique, which involves direct plating for detection and estimation of coliform densities, is as effective as the multiple-tube fermentation test for detecting bacteria of the coliform group. Modification of procedural details, particularly of the culture medium, has made the results comparable with dose given by the multiple-tube fermentation procedure. Although there are limitations in the application of the membrane filter technique, it is equivalent when used with strict adherence to these limitations and to the specified technical details.

Thus, two standard methods are presented for the detection of bacteria of the coliform group. It is customary to report results of the coliform test by the multiple-tube fermentation procedure as a Most Probable Number (MPN) index. This is an index of the number of coliform bacteria that, more probably than any other number, would give the results shown by the laboratory examination; it is not an actual enumeration. By contrast, direct plating methods such as membrane filter procedure permit a direct count of coliform colonies. In both procedures coliform density is reported conventionally as the (most probable number) MPN or membrane filter count per 100 mL. Use of either procedure permits appraising the sanitary quality of water and the effectiveness of treatment process. *E. coli* infection often causes severe bloody diarrhoea and abdominal cramps; sometimes the infection causes none bloody diarrhoea or no symptoms. Usually little or no fever is present, and the illness resolves in 5 to 10 days. To some people, particularly children under 5 years of age and the elderly, the infection can also cause a complication called hemolytic uremic syndrome, in which the red blood cells are destroyed and the kidneys fail. About 2%-7% of infections lead to this complication. In the United States, hemolytic uremic syndrome is the principal cause of acute kidney failure in children, and most cases of hemolytic uremic syndrome are caused by *E. coli.*

1.7 Fungi

Fungi are multi-cellular, non-photosynthetic, heterotrophic organisms. Fungi are obligate aerobes that are reproduced by a variety of methods including fission, budding, and spore formation. Their cells require only half as much nitrogen as bacteria so that in a nitrogen-deficient wastewater, they predominate over the bacteria. Fungi are plants that are unable to do photosynthesis such as yeast and moulds. Yeast is normally used for fermentation in making bread, cake and alcohol. Moulds are filament in shape that lives in acidic condition. They reduce the efficiency of secondary sedimentation tank and cause unpleasant smell and taste.

Few members of the group are readily visible when they are present. It is their fruiting body which is seen rather than the vegetative body of the organisms. Perhaps the most familiar is the mushrooms, whose Greek name *mykes* gives rise to the term applied to the scientific study of fungi and mycology.

The fungi have highly distinctive biological organization. Although some aquatic fungi and the yeasts are unicellular they are readily distinguishable from bacteria by their large cells and membrane-bound nuclei. Some aquatic fungi do show resemblances to flagellate protozoa. Fungi occupy a wide variety of habitats including the sea and fresh waters. However, the majority occupy moist habitats on land and are abundant in soil. At least 100,000 species are known. Although some produce macroscopic fruiting bodies (e.g. mushrooms) the overwhelming majority are microscopic. The fungi are heterotrophic organisms acquiring organic materials for their nutrition. Those which feed on dead organic materials are described as *saprophytic*. Saprophytes bring about the decomposition of plant and animal remains and in doing so release simpler chemical substances into the environment. In soil this is vital significance in maintaining fertility by recycling essential plants nutrients.

Fungi are present in, and have been recovered from, diverse, remote, and extreme aquatic habitats including lakes, ponds, streams, estuaries, marine environments, wastewaters, sludge, rural and urban storm-water runoff, well waters, acid mine drainage, asphalt refineries, jet fuel systems, and aquatic sediments. Fungi are widely distributed and are found wherever moisture is present. Glycogen is the primary storage polysaccharide in fungi. Most fungi use carbohydrates (preferably glucose or maltose) and nitrogenous compounds to synthesise their own amino acids and proteins. Identification of fungi which are larger than bacteria is dependent on colonial morphology on a solid medium, growth as well as reproduction morphology and for yeasts, physiological activity in laboratory cultures.

Increasing numbers of fungi usually indicate the increasing organics loadings in water or soil. Large numbers of similar fungi suggest excessive organic load while a highly diversified mycobiota indicates populations adjusted to the environmental organics. Despite their wide occurrence, little attention has been given to the presence and ecological significance of fungi in aquatic habitats. The relevance of fungi and their activities in water is emphasized by increasing knowledge of their pathogenicity for humans, animals, and plants; their role as food or energy sources; their activity in natural purification processes; and their function in sediment formation. Quantitative enumeration of fungi is not equivalent like the unicellular bacteria because a fungal colony may develop from a single cell (spore), an aggregate of cells (a cluster of spores or a single multi-celled spore), or from a mycelial or pseudo-mycelial fragment (containing more than one viable cell). It is assumed that each fungal colony developing in laboratory culture originates from a single colony-forming unit (CFU) which may or may not be a single cell.

The advantages of fungi are for food and food preparation such as edible fungi and fermentation of bread, wines and beers; for medicine e.g. Penicillin is the best known antibiotic and it is actually made from a mould; and decomposers of organic material. The disadvantages of fungi are poisonous (i.e. wild fungi can be both delicious and deadly poisonous. The high concentration of metals such as arsenic, cadmium, copper and lead in wild fungi) causing fungal diseases to plants (i.e. mildew, smuts, ruts, etc).

1.8 Algae

Algae are plant-like organisms that usually photosynthetic and aquatic but do not have true root, stem, leaf, vascular tissues and have simple reproductive structures. They range from tiny single cells to branched forms of visible length that appear as attached green slime. They may be either unicellular or multi-cellular. There are a wide variety of algal species in various shades of commonly green, brown and red. They are distributed worldwide in sea, in freshwater, waste water, marine water, and non-marine water such as mud and sand. Typical green algae are *Oocystis* and *Pediastrum*. Whereas the blue green algae that associated with polluted water are *Anacystis*, *Anabaena*, and *Aphanizomenon*.

Algae are photoautotrophic that use carbon dioxide or bicarbonate as carbon source and inorganic nutrients of nitrogen as ammonia or nitrate and phosphate. Algae have chlorophyll that can manufacture their own food through the process of photosynthesis. They conduct photosynthesis within membrane bound structure called chloroplast. In the presence of sunlight, the photosynthetic production of oxygen is greater that the amount used in respiration. At night they use up oxygen in respiration. If the daylight hours exceed the night hours by a reasonable amount, there is a net production of oxygen. They produce more oxygen than all the plants in the world.

The good of algae are: form important food source for many animals. e.g: little shrimps and huge whales; most important at the bottom of food chain with many living things depend upon them; and have an economic importance because they are a source of carotene, glycerol, and alginates and can be converted into a food source for aquaculture. While the bad of algae are: too much algae does suffocate the lake, so it will kill many fish; blue green-algae are very toxic and algal toxin can seriously affect human and animal; and toxic blue-green algal blooms cause a rash known as "swimmer's itch", while powerful neuromuscular toxins released by other cyanobacteria (blue green algae) can kill fish or animals that drink the water.

1.9 Filtration

Filtration is most often a polishing step to remove non-settleable flocs remaining after chemical coagulation and sedimentation or precipitant particles of softened water. Under certain conditions, filtration may serve as the primary turbidity removal process especially in direct filtration of raw water.

The normally used filtration process involves passing the water through a stationary bed of granular medium. Solids particles in the water are retained in the interstices of the filter media. Several modes of operation are possible in granular medium filtration. These include upflow, downflow, pressure and vacuum filtration. The most common practice is downward gravity flow filtration, with the weight of the water column above the filter providing the driving force.

The solids removal operation with granular medium filter involves several complicated processes. The most obvious processes include straining, flocculation and sedimentation. The straining process occurs at the interface between filter media and water. Initially, materials larger than the pore openings at the interface are strained. In the filtration process, conditions within pores of a filter bed promote flocculation. Flocs grow in size and become trapped in the interstices. Other processes are also important since most of the solids existence in settled water is too small to be completely removed by straining. Removal of particles and flocs in the filter media depends on transport mechanisms that carry the solids through the water to the surface of the filter grains, and on retention of the solids by the medium once contact has occurred. Transport mechanisms include gravity settling, inertial impaction, diffusion of colloid, Brownian movement and van der Waals forces (Kim and Whittle, 2006). Retention of solids once contact has occurred can be attributed primarily to electrochemical forces, van der Waals force, and physical adsorption (Yaroshevskaya, 2007).

With chemical preconditioning of the water, a well-designed and operated filter should remove virtually all solids down to the submicron size. Removal begin at the top portion of the filter. As pore opening are filled by the filtered material, increased hydraulic shear sweeps particles farther into the bed. The ideal filter media should be coarse enough for large pore openings to retain huge quantities of particles or floc however sufficiently fine to avoid passage of small floc. It should has an optimum filter depth to produce relatively long operation filter run and graded to permit effective cleaning during backwash. Dual-media filter of coarse burnt oil palm shell granules or anthracite overlaying the sand media provide higher porosity at the upper layer as well as higher filter run of more than three times than the conventional sand filter as recorded by Ahmad et al. (2009).

2. Methodology

2.1 Sampling protocols

Sampling site was whole area of Terengganu DarulIman, one of the east coast states, which was divided by 7 districts and these are Kemaman, Dungun, Marang, Kuala Terengganu, Hulu Terengganu, Setiu and Besut. Two samples were taken randomly from every single district, except Kuala Terengganu, and from there 4 samples were taken. All together 16 samples were taken in the whole Terengganu area.

Chosen places are in Kemaman, Kampong (Kpg.) Baharu and Esso petrol pump. Two samples were taken from there. Kpg. Kemenyer and Kpg. Pasir were chosen in Dungun districts and two samples were taken from there to be analyzed. Kuala Terengganu districts which consists of Tanjung, Pantai Tok Jembal, MengabangTelipot 1 and MengabangTelipot 2 were places of sampling where 4 samples were taken. Hulu Terengganu, there were two samples taken that was in Kpg. Bt. Gemuroh and Kpg. Nibong. Marang district, two samples was taken in Kpg. Lubok Perah and Wakaf Tapai. In Setiu, Kpg. Guntung Luar and Kpg. Tembila were chosen and two samples were taken there. In Besut, Kpg. Bt. Bunga and Kpg. Pasir Aka were chosen and two samples were taken there.

For the field method, refer to Figure 2 and the explanation in sampling procedure for the well.

Fig. 2. Sampling Map

2.2 Filtration system

The setup of the filtration system consisting of the carbon-soil-aquifer filtering system is shown in Figure 3. The experimental set-up consists of two identical columns, though only final setup was diagramed. Filter column height is 18 cm, external diameter is 8.5 cm, internal diameter is 7.3 cm and its constructed material is polyethylenetetrathelate (PET). Granulated activated carbon, modified activated carbon and red soil particles were prepared from locally available raw materials. Sand and silica particles were collected from locally available sources.

2.3 Sampling procedure and preserved for microbiological test

In Figure 3, firstly, the metal bucket was cleaned until it is free from earth and rubbish to avoid contamination. Next, a little methylated spirit was poured into the bucket, light it and allowed the burning alcohol to run over the walls of the bucket so as to sterilise the inner surface. After that, the bucket was lowered into the well and natural water sources, and making sure that the rope does not enter the water or the inside of the bucket, as well as the bucket should not touch the side of the well. When the bucket is full of water, it was carefully raised. Next, the string fixing the protective cover was untied and the stopper removed. Then, he sample bottle was filled with the water from the bucket. Finally, the bottle was capped and the protective sheet fixed in place with string.

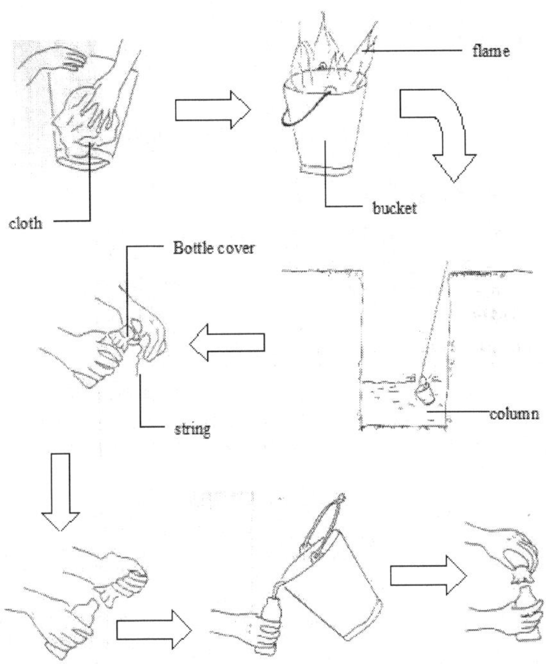

Fig. 3. Sampling method for dug well

The sequence of the components, its thickness and quantity inside the filtration system is shown in Figure 4. The Teflon plastic bottles were cleaned until free from earth and rubbish to avoid contamination. Next, a little methylated spirit was poured into the bottle, light it and allowed the burning alcohol to run over the inside wall of the bottle so as to sterilise the inner surface. When the bottle is full of water, it is carefully transferred, and then bottles are capped and protective sheet fixed in place with a piece string.

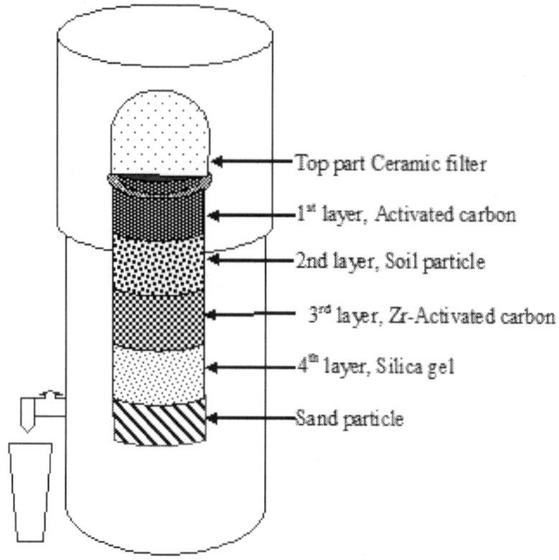

Fig. 4. Constructed filter unit (Yusof et al., 2002 and M. M. Rahman et al., 2011)

2.4 Preparation of media for coliform and total count

Endo broth: About 2.4 g of (7.5 g agar 1.25 g Yeast Extract + 2.5g Peptone) MF Endo Broth mixture was dissolved in 50 mL of distilled water containing 1.0 mL ethanol. Then this was sterilized in an autoclave at 121^{0}C, 15 psi for 20 minutes. The broth was cooled to room temperature before incubation. This medium was freshly prepared to ensure accurate results used for coliform count.

Exactly 1.25 g of tryptone was added to 50 mL of deionized water and to it was added 3.75 g agar. The mixture was then gently heated. The pH was adjusted to 8.6 to 9.0 and then 1.25 g of glucose was added and then autoclaved. Yeast and mould count was then done.

Appropriate volume of mixture was filtered through a sterile 47-mm, 0.45μm, grid membrane filter (cellulose nitrate membrane filter), under partial vacuum. The funnel was rinsed with three 20 to 30ml portions of sterile dilution water. An exact amount of 100 ml of water sample was poured through the funnel. If the water was heavily polluted, the water sample will be diluted using a dilution bottle 3 to 5 times dilution with 90 ml sterile dilution water. The filter was placed on the agar in the petri dish. Dishes were placed in close fitting box containing moistened paper towels. Incubation at 44.5± 0.5°C for 24 hours was done if using the EMB Agar medium. Duplicate plates may be incubated for other time and temperature conditions as desired. Membrane filter technique is shown in Figure 5.

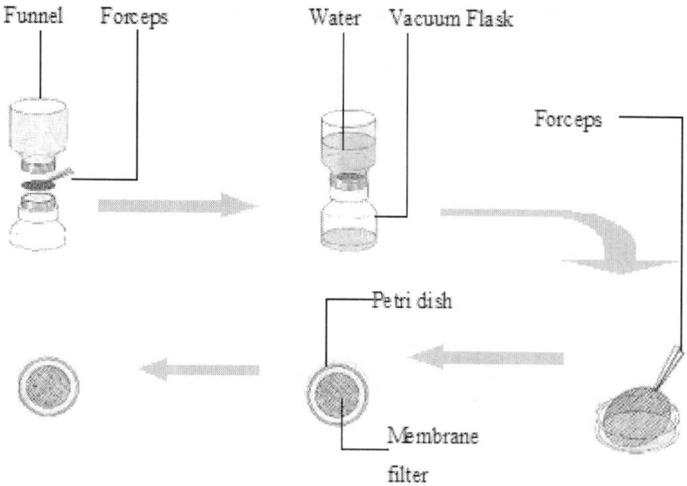

Fig. 5. Membrane filter technique

2.5 *E. coli* (faecal)

Colonies on membrane filters were counted by viable count or using a stereoscopic microscope at 10 to 15 x magnification. Preferably, the petri dish was placed on the microscope stage slanted at 45^0 and light source was adjusted vertically to the colonies. Optimal colony density per filter is 20 to 200. If colonies are small and there is no crowding, a higher limit is acceptable.

All colonies were counted on the membrane when there were 1 to 2 or fewer, colonies per square. For 3 to 10 colonies per square count 10 squares and obtain average count per square. For 10 to 20 colonies per square count 5 square and obtain average count per square. Multiply average count per square by 100 times reciprocal of the dilution to give colonies per millilitre. If there are more than 20 colonies per square, record count as >2000 times the reciprocal of the dilution. Report averaged counts as estimated colony-forming units (CFU). Make estimated counts only when there are discrete, separated colonies without spreaders

2.6 Fungi – Membrane filter method

Sample preparation: Each plate was marked with a sample number, dilution, date, and any other necessary information before examination. Duplicate plates were prepared for each volume of sample or dilution examined. All samples were thoroughly mixed or dilutions done by rapidly making about 25 complete up-and-down movements. Optionally, the use of a mechanical shaker to shake samples or dilution blanks for 15 second would be useful.

Filtration: Appropriate volumes of well shaken samples or dilution were filtered in triplicates, through membrane filter with a pore diameter of 0.45 or 0.80μm. The filters were then transferred and incubated at 15°C for 5 days in a humid atmosphere. Alternatively, incubate-ion can also be performed at 20°C for 3 days, or longer depending on the fungi present. Using a binocular dissecting microscope at a magnification of 10×, all colonies on each selected plate were then counted. If counting must be delayed temporarily, plates are held at 4°C for not longer than 24 hours. Ideal plates should have 20 to 80 colonies per filter.

2.7 Total count of bacteria – Membrane filter method

Appropriate volume of water samples were filtered through a sterile 47-mm, 0.45μm, grid membrane filter, under partial vacuum. The funnel was rinsed with three 20 to 30ml portions of sterile dilution water. The 100 ml of water sample was poured into the funnel. Dilution is required if the water was heavily polluted, and done with a dilution bottle 3 to 5 times dilution with 90 ml sterile dilution water. The filter was placed on the agar in the petri dish. The dishes were later placed in close fitting box containing moistened paper towels and incubated at 35± 0.5°C for 24 hours if the nutrient agar medium was used. Duplicate plates may be incubated for other time and temperature conditions as desired.

Counting the colonies was done on membrane filters by viable count or using a stereoscopic microscope at 10 to 15x magnification. Preferably, the petri dish was placed on a microscope stage slanted at 45° and the light source adjusted vertical to the colonies. Optimal colony density per filter is 20 to 200. If the colonies are small and there is no crowding, a higher limit is acceptable.

Counting was done on all colonies on the membrane when there are 1 to 2 or fewer, colonies per square. For 3 to 10 colonies per square count 10 squares an average count per square was obtained. For 10 to 20 colonies per square, counting was done on 5 squares and an average count per square obtained. Multiplying the average count per square by 100 times reciprocal of the dilution will give colonies per millilitre. If there are more than 20 colonies per square, the count was recorded as >2000 times the reciprocal of the dilution. The average count was reported as estimated colony-forming units. Estimated counts are made only when there are discrete, separated colonies without spreaders.

3. Discussion

Environmentally polluted water samples were taken from various sources such as rivers, columns, tube well, sea and water falls. This naturally polluted water was filtered continuously through the filtering column for 10 days. The results of coliform and TC in raw and treated shown in Table 1 and Table 2. Coliform count is the most important microbiological count for drinking water. Coliforms are quantified to assess water treatment effectiveness and the integrity of the distribution system. They are also used as a screening test for recent feacal contamination. Treatment that provides coliform-free water should also reduce the pathogens to a minimal level. A major shortcoming is that coliforms under certain circumstances may proliferate in the bio-films of water distribution systems, clouding their use as an indicator of external contamination.

From the results in Table 1 it was observed that all raw water samples contain coliform except raw water from locations 10, 13 and 16. The sample taken from location 16 was observed to contain uncountable coliforms but treated water using filter column contained no coliforms. Therefore, it could be confirmed that the filtering column has produced coliform free water and safe for drinking. From Table 2, it was found that the total count includes three types of microorganisms; bacteria, yeasts and moulds which presence in the raw water samples. Filtrate water supplies were selected at random from the stock. After 24 and 72 hours of observation the results showed that the yeasts and moulds could be distinguished from each other. The 24 hour observation could not distinguish between the yeast and mould but with the 72 hours observation yeast and mould were detected.

Sample ID	WHO Std.	Raw Water	Treated Water
1	0/100 ml	02	0
3	0/100 ml	01	0
6	0/100 ml	01	0
7	0/100 ml	20	0
8	0/100 ml	01	0
9	0/100 ml	23	0
10	0/100 ml	0	0
11	0/100 ml	11	0
12	0/100 ml	01	0
13	0/100 ml	0	0
14	0/100 ml	04	0
15	0/100 ml	25	0
16	0/100 ml	Uncountable	0

Table 1. Coliform count in the raw and treated water samples

Sample ID	WHO Std	Raw water/10 ml		Treated water/10 ml	
		24 hours	72 hours	24 hours	72 hours
1	<25/10mL	28	36/14	09	11/1
3	<25/10mL	38	40/38		
6	<25/10mL	02	4/1		
7	<25/10mL	06	20/2		
8	<25/10mL	25	30/18	13	18/12
9	<25/10mL	21	32/6		
10	<25/10mL	15	15/3		
11	<25/10mL	70	78/12	11	12
12	<25/10mL	Uncountable	/6	09	10/9
13	<25/10mL	0	13/3		
14	<25/10mL	18	22/88		
15	<25/10mL	Uncountable	/25	02	12/3
16	<25/10mL	08	8/8		
utm tap	<25/10mL			00	00

Table 2. Total count in raw and treated water samples

4. Conclusion

A newly fabricated filter unit as shown in Figure 4 has been found to be significantly improved for its removal capability of microbiological hazardous materials from the filtered water. It provides a safe drinking water in the point of view of microbiological, toxicity, softening hardness and pH value. All kinds of parameters for drinking water have fulfilled the requirements as provided by this water filter unit.

5. Acknowledgment

The authors wish to express their gratitude to the "**Ministry of Science, Technology and the Environment (MOSTE)**" (Fund no. ERGS55064) and Ministry of Higher Education (Fund no. FRGS59210) for funding this research.

6. References

Ahmad, J.; Cheng, W.H.; Wan Nik, W.B.; Noraaini, A.& Megat Mohd Noor, M.J. (2005). Study on Removal of Iron and Manganese in Groundwater by Granular Activated Carbon. *Desalination*. 182, pp. 347 - 353.

Ahmad, J.; Shiung LS.; Ali N.& Noor M.J.M.M. (2007). Assimulation Study of the Removal Efficiency of Granular Activated Carbon on Cadmium and Lead. *Desalination*. 206, pp. 9 - 16.

Ahmad, J.; Nora'aini A.;Halim A.G. & Wan Nik, W.B. (2009). Effective Depth, Initial Headloss and Backwashing Criteria as the Key Factors of Burnt Oil Palm Shell (BOPS) Granular Filtration. *Journal Desalination and Water Treatment*. 10. pp. 80 - 86.

Bellamy, W.D.; Hendricks, D.W. & Lonsdon, G.S. (2006). Slow Sand Filtration: Influences of Selected Process Variables. *J.W.W.W.A.* (77) pp. 62-66.

Gerba, C. P. (1996). Pathogens in the Environment. In: Pepper, I. L., Gerba C. P., Brusseau, M. L. (Eds.), Pollution Science Academic Press, New York, pp. 279 - 299.

Hammer M.J. (2008). *Water and Wastewater Technology*. Pearson Prentice Hall.London U. K. Sixth Edn., pp. 49 - 61.

Kim, Y.S. & Whittle, A.J. (2006). Filtration in a Porous Granular Medium: Simulation of Pore Scale Particle Deposition and Clogging. *Journal of Transport in Porous Media*. 65, pp. 53 - 87.

Rahman M.M.; YusofA.M. & Wan Nik, W.B. (2011). Increase of Purification Capacity and the Performance of Slow Filtering in the Removal of Bacteria in an Activated Carbon-Soil-Sand Filter Unit. *International Journal of Applied Science and Technology (IJAST)*. 1 (5), pp.127-130.

Prescott, Harley & Klein, "*Microbiology*" 6th Edition, International Edition (2005). McGraw Hill.

Rotival, A.H. (1987). "State of the International Drinking Water Supply and Sanitation Decade: in *Resource Mobilization for Drinking Water and Sanitation in Developing Nations*." American Society of Civil Engineers, NY.

Slezak, L.A. & Sims, R.C. (1984)."The Application and Effectiveness of Slow Sand Filtration in the United States."*J.W.W.W.A.* (76) pp. 38-43.

Suffer, I.H. & McGuire, M.J. (1980)."Activated carbon adsorption of organics from aqueous phase."Ann. Arbor., MI.; *Ann. Arbor. Science*,**1**.

Vsscher, J.T.; Paramasivam, R. & Santacruz, M. (1986). "IRC's Slow Sand Filtration Projects." *Waterlines*, (4) pp. 4-27.

Yaroshevskaya, N.V. (2007). The Method of Computing a Water-treatment Filter with the Account of Media Granulometric Composition. *Journal of Water Chemistry and Technology*. 29 (3), pp. 152 - 162.

Yusof A.M.; Rahman M.M.; Wood A.K.H.; Hamzah S. & Shamsiah A. (2002). Development and water quality parameters assessment of a filter unit using local raw materials for use in safe drinking water, *Proceedings of SKAM-15/EXPERTS 2002*, Bay view Beach Resort, Pulau Pinang, Malaysia, September 10-12, 2002.

22

Impact of Agricultural Contaminants in Surface Water Quality: A Case Study from SW China

Binghui He and Tian Guo
Southwest University
China

1. Introduction

Water is a very precious natural resource and the matter foundation that the whole lives rely on in the earth, and it is also the matter that cannot be replaced in the natural resources on the earth again. With increasingly intensive population, various community economy activity's demand for water increases, and the space-time distribution of water resource become uneven, and the supply of water resource in the whole world become tight. Soil erosion has taken away the mass of rich topsoil and makes soil become more and more thin, reduces land productive, and poses serious threat to water environment. In addition, water body pollution has negative effect on effective use of water resource and aggravates the contradiction. The management of point source pollution and circularly making use of in the developed countries currently up make more ideally, but a lot of still directly exhaust in various water body in the extensive developing countries.

Soil erosion is the reason that results in surface source pollution, and runoff is the medium of surface source pollution, while the amount of runoff and water and soil conservation are closed tightly, therefore, the basic way to resolving surface source pollution is reducing soil erosion by water and soil conservation measures. Most of analysis of runoff pollution mechanism and runoff creation, space-time distribution of runoff, and the effect analysis of different measures are all carried on runoff plots.

With the improvement of the controlling of point source pollution, water environment issues caused by the non-point source pollution are increasingly conspicuous. Especially in the developed countries, such as the America, the non-point source pollution has been major factor of the water environment pollution (Fu et al., 2011). The agriculture runoff pollution is the main fraction of surface source pollution, and agriculture surface source pollution and soil erosion is a pair of symbiosis (Zhang & Huang, 2011). Thus the most effective and the most practical approach to solve agriculture pollution problem is to implement water and soil conservation measure. With the development of small watershed comprehensive management and surface source pollution in China, the way to resolve surface source pollution by the effective small watershed comprehensive management has been a hot point in the environment science and water and soil conservation science field nowadays.

The environmental benefit of small watershed comprehensive management includes natural environment, eco-environment and social environment. The water and soil conservation works since 1980s indicate that the small watershed comprehensive management could produce obvious environmental benefit (David et al., 1998).

Water eutrophication is one of the water pollution issues that perplexed developed countries nowadays and is also the realistic issue that the developing countries facing (Leeds-Harrison et al., 1999; Heitz et al., 2000). After the water eutrophication, the excessive reproduction of algae had leaded to water hypoxia, clarity step-down, smelly, even some toxins, thus had damaged the normal function of water body. According to calculating, the eutrophication degree of most lack in china will turn worse further with the development of the industrial and agricultural. The pollutants resulting in Water eutrophication were mainly plant nutritional substances, such as nitrogen and phosphorus etc. The search indicate that it mainly exist in closed water area, such as lake, reservoir etc.

The control of runoff pollution is an important aspect in the whole water pollution controlling progress. The search and control for non-point source pollution start relatively late in china, and the search for the control way about non-point source pollution has been valued since the last few years, But the abroad has made a great deal of works and has accumulated prolific experience in this field (Chartes & Roben, 2000; Prato & Shi, 1990). For example, In 70's, the United states proposed "BMP-Best Management Practices" in the control and management fulfillment of non-point source pollution. What it points is the combination of several measures after analysis and compassion by the government or designation which is the most useful practical measure for making the identified water by perverting and cutting non-point source pollution burden. Usually it can be divided into the control of source and the runoff pollution.

To know what effects does the watershed comprehensive management had on the runoff pollution, this search adapt the combined method of runoff plot and small watershed. The small watershed model is the main demonstration for water and soil conservation - the reservoir peripheral of Gezi channel, Lianglu town, Yubei plot, Chongqing city (managed watershed) and Changxi small basin(not managed watershed), the soil in experimental runoff plot is the same as the adapted soil in small watershed. The reservoir in Gezi has played a key role in the development of agriculture economy, and the water of reservoir is the source of peripheral farmland irrigation and the headwaters that the farm tourism keeps fish again. To know the influences of different measures on runoff pollution, the search of the distribution regulation in time and space about the reservoir agriculture runoff, the estimating of amount runoff pollutant burden of reservoir stores in warehouse by the methods of synchronous monitor and field investigation to the runoff water and sediment respectively in runoff plot and basin, and calculated the decreasing runoff pollution burden after management by the compassion of the density of runoff pollutant in managed plot and not managed plot's, and then to analyze the dimension of synthesize management and measure the effect after administering.

Therefore, there are three purposes to establish this search, the first one is evaluating the control action of small watershed comprehensive management to runoff pollution; The second one is directing the measure of water and soil conservation; The third one is defining the best sloping plant model from ecology direction. Thus it can offer a certain theory basis for predicting the control action of small watershed comprehensive management to the surface source pollution in this region.

2. Research content

The soil erosion is the main reason of slope runoff pollution, the different land use and different cropping-plants influence the occurrence and development of runoff. According to the spot investigation, it indicated that the local resident plant some farm crops, such as bean, maize, sweet potato and fruit tree under the 25° sloping fields, these farm plant are been planted alone or been intercropped, and the different intercropped area proportion could had different influence on runoff. This search adapt the method of monitoring surface runoff in plot in consideration of that the natural water is the main recharge source of surface runoff, the amount of burden of different using type of land's surface runoff is relevant with this area, which also has something to do with the pollutant density of surface runoff.

This search mainly investigates the relevant intensity of surface runoff pollutant of different type of using land, thus we can use surface runoff pollutant total nitrogen and total phosphorus, Pb and Cd, chemical oxygen demand and the average density of the amount of sediment to express the discussion of different plant model's impacts on the runoff pollution. According to it, we can compare the differences of the degree of influence of different plant model's impact on runoff pollution. The study contents were as follows: 1) the influence of different plant models on runoff pollution; 2) the timely distribution of runoff pollution; 3) the spatial distribution of runoff pollutant.

2.1 Basic situation of experimental plot and research methods

The experimental plot located in Lushan village, Lianglu town, Yubei district, Chongqing city and was shallow landform, soil type is purple soil which belongs to the Jurassic formation, the soil texture was formed by the physical chemistry elegance of sandstone and shale, soil bulk density was 1.42-1.68g/cm^3, total porosity was 40.28%. And perennial mean temperature was 18.2 °C, annual yearly rainfall was 1110mm, the rainfall in May to October was about 80.62% of the whole year's, and the amount of rainfall in August was the most largest. Before the planting tillage in plot, to know what influence does different land use and slope will have on runoff pollution, we adapt the method of runoff plot to research, and the soil texture fertilizer applications was the same in each plot. The basic situation of experimental plots was as follows:

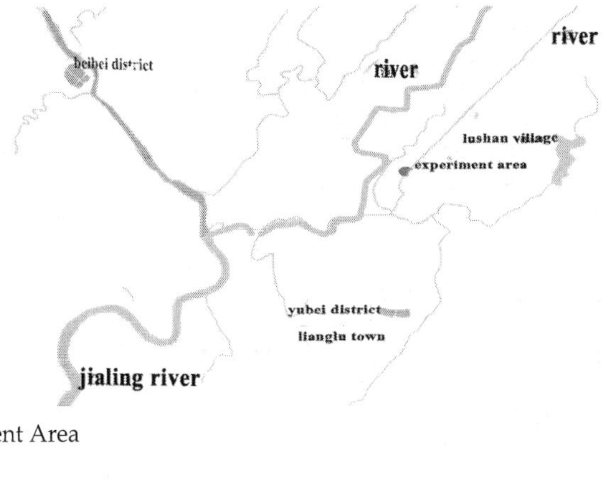

Fig. 1. Experiment Area

Plot number	Slope(°)	Aspect	Specification	Crop and planting pattern	Coverage (%)
1	25	SE	5m×20 m	bean (50%)+ sweet potato(50%)(intercropping along the slope)	20
2	25	SE	5m×20 m	seedlings of economic fruit forest(along the slope)(20cm*20cm)	5
3	25	SE	5m×20 m	seedlings of economic fruit forest (transforming slope into terrace)+ (20cm*20cm)	5
4	15	SE	5m×20 m	bean + sweet potato(25% intercropping)+ sweet potato(75%)(along the slope)	30
5	15	SE	5m×20 m	seedlings of economic fruit forest (transforming slope into terrace) (20cm*20cm)	5
6	15	SE	5m×20 m	sweet potato +maize(intercropping along the slope)	30

Table 1. Basic situation of experimental plots

Index	Plot number					
	1	2	3	4	5	6
Minimum permeability (mm/h)	3.64	3.61	3.72	3.66	3.70	3.67
Bulk density (g/cm^3)	1.47	1.48	1.45	1.34	1.44	1.30
Water content(100%)	17.5	16.9	19.5	19.8	19.5	21.2

Table 2. The physical characteristic of soil in plots Basic situation of experimental plots

Soil samples were collected from five sampling spots ("S" type) of two replicates in each plot, then combined within plots before analysis. Once after a runoff, swept the sludge of gutter into runoff pool mixed it with water, and collected three columnar samples from pool (gross 1000-3000ml). Then put mixing sample (500-2500ml) into closed container, 4ml concentrated H_2SO_4 to terminate microbial activity, and then put it in the refrigerator. The surplus water sample (500ml) was precipitated, filtered and dried, to determine sediment content.

Precipitation capacity was determined by hydrocone type journal rain gauge; Runoff capacity was calculated by SWZ type journal water column and 45° triangular weir; Sediment content was measured by oven drying method, and measure method of water sample: total nitrogen(semimicro Kjeldahl), total phosphorus(molybdo-antimony anti-colorimetry), Cd (atomic absorption spectrometry), chemical oxygen demand (potassium dichromate method).

2.2 Variation and analysis of runoff pollution

On the basis of the size of runoff pollution index content , the order of effect of runoff pollution among six cropping patterns was plot 5 > plot 3 > plot 6 > plot 4 > plot 1 > plot 2. To reduce the slope runoff pollution, the slope farmland under 25 degree should be transformed into terrace, and at the seedlings of economic fruit forest, it didn't show a strong controlling effect on runoff pollution, so it should intercrop other plants according to fruit tree's biological habit and ecological characteristics.

The species of crop and the reasonable density and its structure of ecological function should be considered when adapted to the intercropping pattern.The management effect of plot 1 and plot 2 was the most obvious in the total ratio of rainfall collection, but the management effect of plot 5 and plot 6 was the worst, the Cd content of plot 4 was the highest, and Pb content of plot 5 and plot 6 was the highest.

To decrease the slope runoff pollution, the slope farmland under 25 degree should be transformed into terrace, and at the seedlings of economic fruit forest, it didn't show a strong controlling effect on runoff pollution, so it should intercrop other plants according to fruit tree's biological habit and ecological characteristics. The species of crop and the reasonable density and its structure of ecological function should be considered when adapted to the intercropping pattern. Analyzing the each index and getting its average by the runoff samples of three times, the results were as follows:

Date	Plot number	Total nitrogen (mg/l)	Total phosphorus (mg/l)	Pb (mg/l)	Cd (mg/l)	Chemical oxygen demand (mg/l)	Sediment amount (mg/l)
8.16	1	5.9	0.68	0.1064	0.0109	14.3	6.3
	2	6.29	0.71	0.1086	0.0111	14.38	6.7
	3	4.45	0.54	0.0919	0.0099	10.32	4.7
	4	5.81	0.64	0.1092	0.0107	13.94	5.9
	5	3.95	0.49	0.0897	0.0095	9.87	4.2
	6	5.47	0.62	0.106	0.0104	13.02	5.6
8.20	1	5.82	0.61	0.1057	0.0102	12.7	5.6
	2	6.22	0.64	0.1079	0.0104	13.92	6
	3	4.4	0.47	0.0912	0.0092	9.58	4
	4	5.74	0.57	0.1087	0.01	11.72	5.2
	5	3.92	0.43	0.089	0.0088	8.88	3.5
	6	5.4	0.55	0.1055	0.0097	11.81	4.9
8.23	1	5.83	0.6	0.1056	0.0101	11.78	5.5
	2	6.23	0.63	0.1078	0.0103	12.64	5.9
	3	4.39	0.46	0.0911	0.0091	9.32	3.9
	4	5.73	0.56	0.1085	0.0099	11.08	5.1
	5	3.91	0.42	0.0889	0.0087	7.76	3.4
	6	5.39	0.54	0.1053	0.0096	10.05	4.8
the average of index of three runoff pollution	1	5.85	0.63	0.1059	0.0104	12.93	5.8
	2	6.24	0.66	0.1081	0.0106	13.65	6.2
	3	4.42	0.49	0.0914	0.0094	9.74	4.3
	4	5.76	0.59	0.1088	0.0102	12.25	5.4
	5	3.94	0.45	0.0892	0.009	8.84	3.7
	6	5.42	0.57	0.1056	0.0099	11.63	5.1

Table 3. The index value of three runoff

The general order of change rate is sediment>total nitrogen> chemical oxygen demand > total phosphorus > Pb > Cd except of a little deviation of the change of pollution index. Evaluation of different planting patterns on the environmental quality pollution is as follows:

Index	Evaluation standard				
	I	II	III	IV	V
Total nitrogen	0.5	0.5	1	2	2
Total phosphorus	0.02	0.025	0.05	0.2	0.2
Pb	0.01	0.05	0.05	0.05	0.1
Cd	0.001	0.005	0.005	0.005	0.01
Chemical oxygen demand	lower than 15	lower than 15	15	20	25

Table 4. The environmental standard of surface water (GB3838-88) (mg/L)

Index	Plot number					
	1	2	3	4	5	6
Total nitrogen	0.185	0.197	0.140	0.182	0.125	0.171
Total phosphorus	0.186	0.195	0.145	0.174	0.133	0.168
Pb	0.174	0.178	0.150	0.179	0.146	0.173
Cd	0.175	0.178	0.158	0.171	0.151	0.166
Chemical oxygen demand	0.187	0.198	0.141	0.177	0.128	0.168

Table 5. Index weight under different land utilization patterns

Index	Plot number					
	1	2	3	4	5	6
Total nitrogen	4.875	5.200	3.683	4.800	3.283	4.517
Total phosphorus	6.364	6.667	4.949	5.960	4.545	5.758
Pb	2.037	2.079	1.758	2.092	1.715	2.031
Cd	2.000	2.038	1.808	1.962	1.731	1.904
Chemical oxygen demand	0.718	0.758	0.541	0.681	0.491	0.646
Total	2.92	3.21	1.86	2.74	1.59	2.52

Table 6. Composite index of runoff pollution in plots

The total tendency of pollutant content by observing the experimental data of each plot's visually was plot 5 < plot 3 < plot 6 < plot 4 < plot 1 < plot 2, and by the calculation of pollution integrated index in each plot and the average of three runoff pollutant, runoff pollutant had a same tendency in the six cropping patterns, so it was reasonable and feasible to conduct environmental quality assessment by the organic combination of the evaluation method of environmental pollution index and the one-level fuzzy evaluation method. Compared to the expert evaluation, this method could reduce workload greatly.

There were economic fruit forest in plot 2, 3 and 5, and there were not obvious differences of the runoff pollutant content between plot 3 and plot 5 , but the runoff pollutant content of plot 3 and plot 5 was smaller than the runoff pollutant content of plot 2, which indicated that slope's effects on runoff pollution is not obvious, while it made a remarkable effect on the control action of runoff pollution by transforming slope into terrace; the effect of the economic fruit forest in plot 2 was the worst because of its stage of seedlings and the small converge that couldn't control soil erosion well.

The pollutant loss in plot 6 was smaller than the effect in plot 1, which indicated that the model of maize intercropping sweet potato was better than the model of bean intercropping sweet potato, because the underground root system and big leaf of maize could make a stronger effect on loosing soil and the rainfall interception than the root and leaf of bean during the experimental period, we can also explained this phenomenon according to soil bulk density, water content and permeability; the effect of controlling pollutant loss in plot 4 was better than that in plot 1, that was to say, planting the single sweet potato on slope land had a stronger controlling effect on runoff pollutant than the model of bean intercropping sweet potato, but which didn't match the former regulation that intercropping pattern was better than single planting pattern, that because the bean was too young in plot 1, so the rainfall interception was not strong, and for planting bean, the row spacing of sweet potato extended, the total coverage of plot 1 changed small.

2.3 The spatial distribution characteristics of runoff pollution

The runoff pollution degree of the downhill of plot 2 by the filtration of buffer plot was the smallest, and the runoff pollution degree of new citrus located in the downhill of plot 4 was the maximum. Total nitrogen content and Cd content in runoff had significant correlations with sediment, and total phosphorus content and Pb content had extremely significant correlations with sediment. And chemical oxygen demand content had the extremely significant correlations with total nitrogen, Pb and Cd, chemical oxygen demand content had significant correlations with total phosphorus and sediment, and total nitrogen content was the main affecting factor of chemical oxygen demand content. During the management of basin, we could adopt intercropping pattern to increase the vegetation coverage of initial harnessing area to come to the target of controlling runoff pollutant. Moreover, it also needed to take measures to control the quality of runoff from upslope.

The total tendency in runoff was that total nitrogen, total phosphorus, Pb and chemical oxygen demand content increased at first then decreased. Total nitrogen content came to peak during one hour to two hours, and the last content had a tendency to stability and was lower than the initial value, the time of runoff pollutant content coming into peak varied with the difference of pollutant and plot, and during the time of peak appearing, there was a big difference between each plot, the total nitrogen content of each plot was generally same at the last period of the occurrence of runoff; Total phosphorus content came to peak during 0.8 hour to two hours, the total tendency of total phosphorus content at the same time was 4>5>6>7>3>1>2, during the period of peak's appearance, there were big differences between each plot, total phosphorus content had a tendency to stability at the last time of the occurrence of runoff; Pb content came to peak during one hour to two hours, the order of Pb content of each plot after the former one hour of occurrence of runoff was 1>3>7>2>6>5>4, during the period of peak's appearance, there were big differences between each plot; Total nitrogen content of each plot was generally same at the last time of the occurrence of runoff; The order of Pb content after the one hour to three hours and a half of the occurrence of runoff was 5>6>4>7, there were no differences of Pb content between 1 plot 1, plot 2 and plot 3. And Pb content of each plot had a tendency to same at the last period runoff. Cd content of each plot came to peak during 1.8 hours to two hours, the order of Cd content after the occurrences of runoff but the former two hours before the occurrence of runoff peak was 1>3>7>2>6>5>4, Cd content of each plot was generally same after the appearance of runoff peak; Cd content of plot 4, 5, 6 and 7 changed sharply, Cd content of plot 1,2 and 3 changed gently; Chemical oxygen demand content of each plot came to peak during 1.8 hours to two hours, the order of chemical oxygen demand content in the total runoff progress was 5>6>3>7; Chemical oxygen demand content of plot 1, 2, 4 and 7 changed irregularly. Generally speaking, the change range of chemical oxygen demand content of plot 4, 5, 6 and 7 was larger than plot 1,2 and 3's. The effect of runoff management of plot 1, 2 and 3 was stronger, secondly for plot 4 and 7, the effect of runoff management of plot 5 and 6 was the worst; the time of the appearance of runoff pollution peak was fixed.

Plot number	pH	Organic matter (g/kg)	Total nitrogen (g/kg)	Total phosphorus (g/kg	Total potassium (g/kg)	Cd (mg/kg)	Pb (mg/kg)
1	7.96	22.54	1.431	0.657	10.2	0.498	27.58
2	8.19	4.69	0.786	0.716	14.6	0.4	29.64
3	6.64	22.31	1.447	0.904	12.9	0.243	24.69
4	8.17	4.87	1.426	0.992	14.8	0.552	25.01
5	8.11	22.21	1.431	0.667	12.8	0.346	29.12
6	8.14	22.19	1.429	0.662	12.7	0.345	29.12
7	6.94	4.91	0.786	0.772	15	0.377	24.68
8	8.18	4.69	1.084	0.828	13.8	0.483	26.66

Table 7. The physicochemical characteristic of soil

Plot number	Slope position	Aspect	Total nitrogen (g/kg)	Total phosphorus (g/kg)	Pb (mg/kg)	Cd (mg/kg)	Chemical oxygen demand (mg/kg)	Sediment Amount (mg/kg)
1	incomingwater	NE	3.525	0.136	0.0628	0.0086	19.567	2.9
	upslope	NE	1.872	0.088	0.0431	0.0058	10.269	1.4
	midslope	NE	1.47	0.191	0.0322	0.0027	9.157	2.6
	downslope	NE	1.311	0.152	0.0257	0.001	7.683	2.2
2	incoming water	NE	2.521	0.137	0.0555	0.0063	20.233	3.3
	upslope	NE	1.966	0.112	0.0501	0.0047	12.565	2
	midslope	NE	1.659	0.159	0.0554	0.0055	8.461	3.1
	downslope	NE	1.085	0.126	0.0252	0.0008	8.282	0.8
3	incoming water	E	2.349	0.161	0.0489	0.0057	16.243	3.1
	upslope	E	1.693	0.142	0.0432	0.0044	10.856	2.2
	midslope	E	2.344	0.199	0.0456	0.0048	15.381	2.6
	downslope	E	1.192	0.161	0.0451	0.0026	7.416	2.1
4	incoming water	SE	2.657	0.136	0.0550	0.0068	16.835	2.9
	upslope	SE	2.615	0.102	0.0441	0.006	11.235	3.8
	midslope	SE	2.057	0.208	0.0487	0.0098	26.261	3
	downslope	SE	1.878	0.341	0.0537	0.0011	12.541	2.2
5	incoming water	S	2.101	0.161	0.0612	0.0052	14.235	2.6
	upslope	S	1.898	0.159	0.0551	0.0052	8.561	2.2
	midslope	S	2.551	0.188	0.0594	0.0076	15.623	2.7
	downslope	S	3.002	0.234	0.0664	0.009	24.652	3.8
6	incoming water	SW	1.921	0.167	0.0619	0.0051	14.237	2.4
	upslope	SW	1.797	0.134	0.0431	0.0049	8.141	2
	midslope	SW	2.349	0.09	0.069	0.008	12.563	2.6
	downslope	SW	3.015	0.216	0.0584	0.0092	21.695	3.7
7	incoming water	NW	2.43	0.145	0.0432	0.0051	15.623	3.3
	upslope	NW	2.234	0.099	0.0392	0.0042	12.666	2.4
	midslope	NW	2.001	0.252	0.0414	0.0071	9.283	3.2
	downslope	NW	1.647	0.206	0.0554	0.0083	8.28	2.8
8	incoming water	NW	2.53	0.147	0.0445	0.0062	15.741	3.6
	upslope	NW	2.66	0.245	0.0524	0.0071	18.564	4.1
	midslope	NW	4.36	0.612	0.0926	0.0113	51.48	5.8
	downslope	NW	5.21	0.633	0.0942	0.0113	51.88	6.1

Table 8. The spatial distribution characteristics of runoff pollution

The management effect of plot 1 and plot 2 was the most obvious in the total ratio of rainfall collection, but the management effect of plot 5 and plot 6 was the worst, and the Cd content of plot 4 was the highest, Pb content of plot 5 and 6 plot was the highest. the runoff pollution degree of the downhill of plot 2 by the filtration of buffer plot was the smallest, and the runoff pollution degree of new citrus located in the downhill of plot 4 was the biggest. Total nitrogen content and Cd content in runoff had significant correlations with sediment, and total phosphorus content and Pb content had extremely significant correlations with sediment. And chemical oxygen demand content had the extremely significant correlations with total nitrogen, Pb and Cd, chemical oxygen demand content had significant correlations with total phosphorus and sediment, and total nitrogen content was the main affecting factor of chemical oxygen demand content. During the management of basin, we could adopt intercropping pattern to increase the vegetation coverage of initial harnessing area to come to the target of controlling runoff pollutant. Moreover, it also needed to take measures to control the quality of runoff from upslope.

2.4 The evaluation of water quality of space runoff in watershed

According to the comments on runoff quality, it indicated the runoff pollution degree of the downhill of plot 2 by the filtration of buffer plot was the smallest and the runoff pollution degree of new citrus located in the downhill of plot 4 was the biggest. The not managed slope pollution index was higher than the managed slope's, which indicated that the regulation effect was significant in Gezi channel. To analyze the sample clustering of interval runoff quality by conducting the spatial runoff pollution synthetic index using SPSS and analyze the variable cluster of runoff quality in all position of the whole reservoir, the cluster result were as Fig.2 and Fig.3.

Slope position	Plot number							
	1	2	3	4	5	6	7	8
upslope	1.09 (1-1)	1.14 (2-1)	1.13 (3-1)	1.39 (4-1)	1.29 (5-1)	1.15 (6-1)	1.20 (7-1)	1.593 (8-1)
midslope	1.25 (1-2)	1.24 (2-2)	1.54 (3-2)	1.88 (4-2)	1.64 (5-2)	1.41 (6-2)	1.70 (7-2)	6.558 (8-2)
downslope	1.05 (1-3)	0.88 (2-3)	1.09 (3-3)	2.22 (4-3)	2.00 (5-3)	1.95 (6-3)	1.56 (7-3)	7.333 (8-3)

Table 9. Average of Composite index of runoff pollution in plots on Aug 16th and Aug 20th

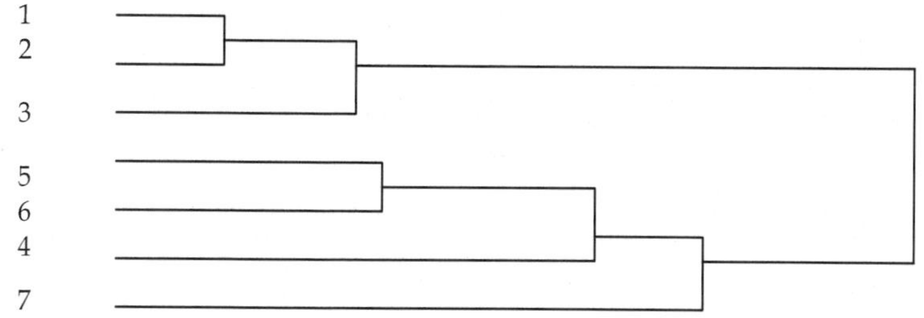

Fig. 2. Runoff water quality clustering in plots

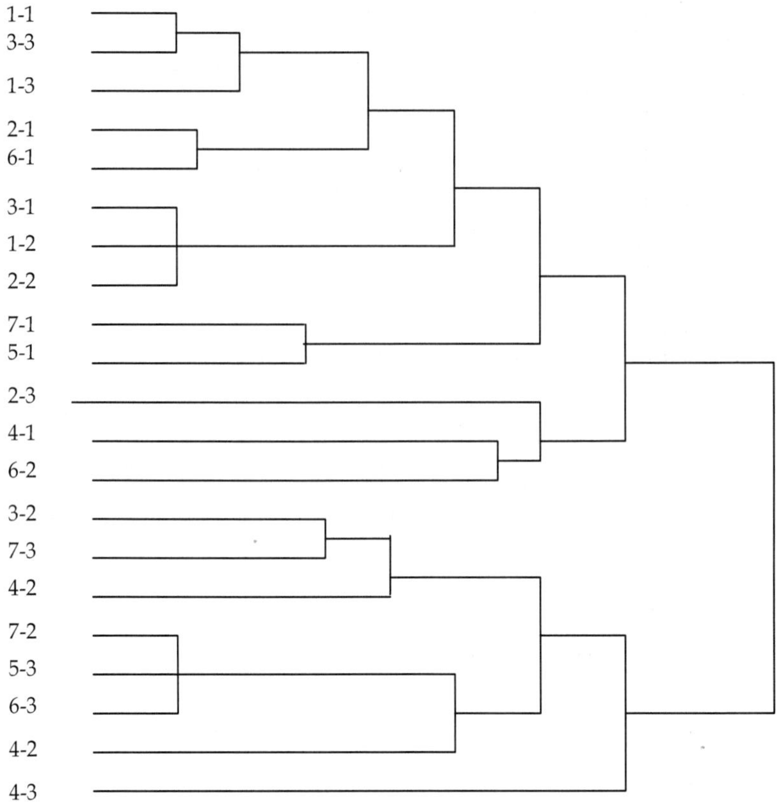

Fig. 3. Runoff water quality clustering of different position in plots

2.5 The space-time distribution characteristics of runoff pollution

The degree of runoff pollution in each plot first increased to peak then decreased gradually, and the degree of runoff pollution at the beginning was larger than in the end; the runoff pollution synthetic index of plot 1, 2 and 3 changed slowly, it appeared peak at the period 5. While the runoff pollution synthetic index of plot 5 and 6 changed sharply, the runoff pollution synthetic index of plot 4 and 7 also changed sharply.

The latter four plot appeared peak much earlier, it indicated that the runoff regulation effect of plot 1, 2 and 3 was stronger, secondly for plot 4 and 7, the runoff regulation effect of plot 5 and 6 was the worst.

The evaluation method of runoff quality is as mentioned above, the evaluation results are as follows:

Plot number	Date								
	16.50-17.20	17.20-17.40	17.40-18.00	18.00-18.30	18.30-18.50	18.50-19.20	19.20-19.50	19.50-20.20	20.20-20.50
1	0.841	0.934	0.963	1.053	1.107	0.827	0.651	0.537	0.536
2	0.786	0.856	0.911	0.973	1.027	0.803	0.657	0.603	0.551
3	0.952	1.012	1.150	1.225	1.309	0.975	0.818	0.742	0.669
4	1.444	1.738	2.125	2.154	1.959	1.730	1.332	1.074	0.976
5	1.389	1.563	1.940	2.243	2.111	1.574	1.165	0.838	1.013
6	1.293	1.422	1.727	2.115	2.019	1.312	1.041	0.903	0.887
7	1.199	1.164	1.407	1.607	1.416	1.077	0.912	0.821	0.780

Table 10. The space-time distribution characteristics of runoff pollution composite index

2.6 The timely clustering of runoff pollution composite index

The period cluster of each section: the nine periods mentioned above could be expressed by TIME1-TIME9, to cluster the runoff pollution synthetic index in each period by the Hierarchical Cluster to Q cluster, the cluster method was Euclidean distance, the cluster results were as follows:

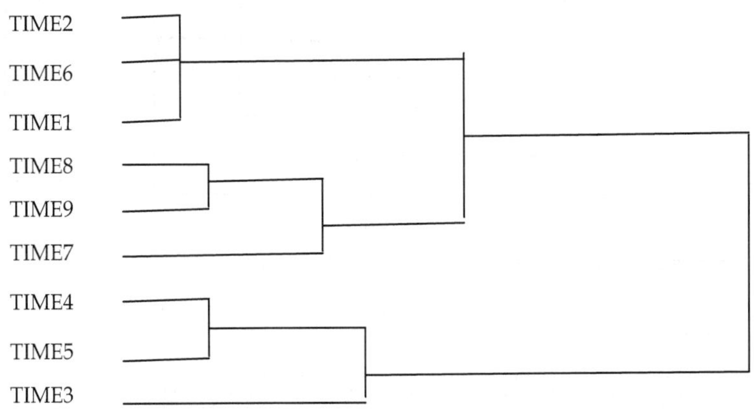

Fig. 4. The timely clustering of runoff water quality in the same plot

In the term of cluster graph, the temporal and spatial distribution characteristics of the degree of runoff pollution were as follows: the degree of runoff pollution of TIME1, TIME2 and TIME6 were the same, the degree of runoff pollution of TIME4 and TIME5 were the same, and the degree of runoff pollution of TIME8 and TIME9 were the same, the runoff pollution synthetic index of TIME4 and TIME5 were the highest, that was to say, the time of coming into peak of runoff pollution is fixed. The interval cluster results of each period were as follows:

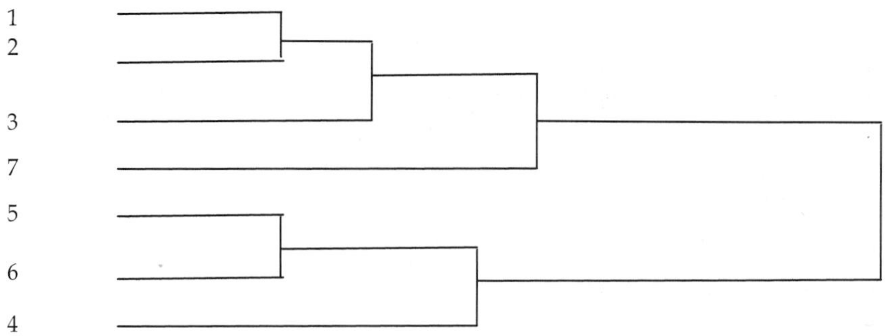

Fig. 5. Runoff water quality clustering during the same time in different plots

The cluster results were that the degree of runoff pollution of plot 1 was similar to the plot 2's and the degree of runoff pollution of plot 4 was similar to the plot 5's, the degree of runoff pollution of plot 7 was different with the other plots, it also showed the land use method and vegetation conditions influenced the development and progression of runoff pollution, which matched the chapter about the analysis of the spatial distribution characteristics of runoff pollution.

3. Conclusion

The effect of planting pattern on the runoff contamination: Adopt total nitrogen, total phosphrous, Pb, Cd, chemical oxygen demand, sedimentation and heavy metal content in surface runoff to evaluate the comprehensive effect. The order is : terrace of 15°+ economic fruit tree seedling>terrace of 25° + economic fruit tree seedling > sweet potato +maize(intercrop and along with the slope)> bean + sweet potato(25% intercropping)+ sweet potato(75%)(along with the slope)> bean(50%)+ sweet potato(50%)(intercrop and along with the slope)> economic fruit tree seedling(along with the slope).

The spatial distribution characteristic of runoff contaminate: the Gezigou reservoir small watershed is divided into 7 sections based on the land-use ways and the planting canopy percent and the origin of runoff around the watershed. In a whole, the first and the second sections are harnessed effectively, the 5th and 6th are mostly worst. The content of Cd in the 4th and the content of Pb in 5th and 6th are the highest.

The contamination degree in the second section is the least in which the lower slope is filtrated trough buffer zone, but the most contamination degree appealed in 4th in which the lower slope planted newly orange. There are remarkably correlation between total nitrogen, total phosphrous, Pb, Cd, chemical oxygen demand and sedimentation in surface runoff; there are also remarkably correlation between the content of chemical oxygen demand and total phosphrous sedimentation, moreover the content of total nitrogen is the principal factor affected chemical oxygen demand in surface runoff.

The timely distribution characteristic of runoff contaminate: In the runoff, the general trend of the content of total nitrogen, total phosphrous, Pb, Cd, chemical oxygen demand increase at first and decrease at end. The time of the contents reach the peak value different from the kinds of the contamination and the sections. At the latest time, the content of the contamination is steady to every section. The content of total nitrogen will reach the peak value within the first to second hour, at end which become steady and less than the initial value. The content of total phosphrous reach the peak valve within 0.8-2 hour; At the same time, the rank of the content of total phosphrous in every section is 4>5>6>7>3>1>2.

The content of Pb will reach the peak value within the first to second hour and at the first hour after the runoff appealed the rank of the content in every section is 1>3>7>2>6>5>4, after 1-3.5 hour is 5>6>4>7 and the difference of the content in 1, 2, 3 section is tiny; At the latest time, they are consistent. The content of Cd reach the peak value within 1.8-2 hour and at the former two hours between the runoff and peak valve appealed the rank of the content in every section is 1>3>7>2>6>5>4, the content in 4, 5, 6, 7 sections changed rapidly, but smoothly in 1, 2, 3 sections. The content of chemical oxygen demand reach the peak value within 1.8-2 hour, the content rank is 5>6>3>7 trough the whole runoff time, but the content change have not regulation in 1, 2, 4, 7 sections. In general the time of peak value of contamination is within 1.5-3.5 hours after runoff happened.

The estimation of total contamination entered into reservoir: To linearity regression between runoff and contamination of total nitrogen, total phosphrous, Pb, Cd, chemical oxygen demand according with SPSS and build a model. Based the model the contamination load of total nitrogen, total phosphrous, Pb, Cd, chemical oxygen demand in runoff in Gezigou reservoir small watershed on Aug 16th respectively 8411.69g, 1014.43g, 402.65g, 25.90g, 59829.82g, the total runoff is 6181.598m^3.

Effect analysis of comprehensive Harness: Apply with the runoff load model in Zhuchangxi small watershed to calculate the runoff load of un-harnssed Gezigou reservoir small watershed and compare with the load based on the build model on Aug 16th, results show out: the total of total nitrogen, total phosphrous and Pb are less 7.87%, 44.66%, 47.09% than un-harnessed respectively.

4. Acknowledgment

The authors are thankful to Key Laboratory of Eco-environments in Three Gorges Reservoir Region, Ministry of Education, College of Resources and Environment, Southwest University for providing the laboratory facilities to do our research work.

5. References

Chartes, A.C. & Roben, P.B. (2000).A Comparision of the Hydrologic Characteristic of Natural and Created Maimstream Flood Plain Wetlands in Pennsyvania. *Ecological Engineering,* Vol.14, pp. 221-231. ISSN 0925-8574

David, J.W. (1998). Modeling Residence Time Instreamwater Ponds. *Ecological Engineering,*Vol.10, pp. 247-262. ISSN 0925-8574

Fu, L.L.; Shuai, J.Bing.; Wang,Y.b.; Ma, H.J.& Li, J.R.(2011). Temporal Genetic Variability and Host Sources of Escherichia Coli Associated with Fecal Pollution from Domesticated Animals in the Shellfish Culture Environment of Xiangshan Bay, East China Sea. *Environmental Pollution,* Vol.159, No.10, (October 2011), pp. 2808-2814, ISSN 0269-7491

Heitz, F.L. ; Khostowpanah, S. & Nelson, J. (2000). Sizing of Surface Runoff Detention Ponds for Water Quality Improvement · *Journal of the American Water Resources Association,* Vol. 36, No. 3, pp. 541-548. ISSN1093-474X

Leeds-Harrison, P.B. ; Quinton, J.N.&Walker, M.J. (1999) Grassed Buffer Strips for the Control of Nitrate Leaching to Surface Waaters in Headwater Catchments. *Ecological Engineering,* Vol.12, pp. 299-313. ISSN 0925-8574

Prato, T. & Shi H. (1990).A Comparison of Erosion And Water Pollution Control Strategies for an Agricultural Watershed. *Water Resources Research,* Vol.26, No.2, pp. 111-114. ISSN 0043-1397

Zhang, H. &Huang, G.H.(2011). Assessment of Non-point Source Pollution Using a Spatial Multicriteria Analysis Approach. *Ecological Modelling,* Vol.222, No.2, (January 2011), pp. 313-321, ISSN 0304-3800

23

Fluxes in Suspended Sediment Concentration and Total Dissolved Solids Upstream of the Galma Dam, Zaria, Nigeria

Y.O. Yusuf, E.O. Iguisi and A.M. Falade
Department of Geography, Ahmadu Bello University, Zaria Nigeria

1. Introduction

Soils are formed from the weathering of the earth's surface and it is loose thereby subject to washing away by various erosive agents including water. Earth materials can therefore be transported in water bodies as either sheet or channel erosion. Water is the transporting medium through which sediments entering the stream from the catchment area are carried down stream. As water moves, its potential energy is transformed into kinetic energy and part of the latter is consumed for transporting the sediments. Mineral materials of many different shapes and particles sizes erode and contribute to the overall stream load. Stream load is broken into three types: dissolved load, suspended load and bed load (Ritter, 2006).

Dissolved loads are materials carried by rivers in true chemical solution. Consist mainly of materials, organic or inorganic, carried in solution by moving water. This type of load can result from mineral alteration chemical erosion or may even be the result of ground water seepage into the stream. Materials comprising the dissolved load have the smallest particle size of the three load types. The quantity of the dissolved solids in the stream is referred to as the Total Dissolved Solids (TDS) and it is important in the assessment of water quality and pollution (Leopold *et al*, 1964; Smith and Stopp, 1978 and Degens *et.al.*, 1991).

Suspended load comprises of fine sediment particles suspended and are transported through the stream. They are transported with no direct contact with the channel floor. In other words, they are too large to be dissolved, but too small to lie on the stream bed; they are rather buoyed by the flow of the stream. They account for the largest majority of the stream load in many rivers and consist of materials such as clay and silt. Suspended Sediment Concentration (SSC) is the volume of suspended sediments at successive depths along the vertical profile of a river, from the water surface to close the river bed (Colby and Hubbell, 1961). Walling and Kleo (1979) added that information on the magnitude of the suspended loads of rivers has many practical applications ranging from geomorphologic studies of denudation rates and patterns of landform development to problem of upstream soil loss and downstream channel and reservoir sedimentation.

Bed load are materials that remain in contact with the bed of the stream and moves by a combination of rolling, sliding and skipping (Knighton, 1998).

The product of the SSC and TDS with the stream discharge gives the suspended sediment discharge (load) and the dissolved sediment discharge (load) respectively. The SSC and the load computed from it is said to increase downstream, while the TDS with the load computed from it is said to decrease downstream, all things being equal (the reverse should be true for an upstream consideration) (Leopold *et al*, 1964; Smith and Stopp, 1978 and Degens *et al.*, 1991). However, since all things are not equal in nature, the study of SSC and TDS variations upstream of the Galma Dam is therefore necessary. Furthermore, an idea of the upstream fluxes of these parameters will help in the development and maintenance of the water infrastructure, which will be of great relevance to the inhabitants of Zaria and Kaduna State at large.

2. Study area

The Galma River is mainly situated in Zaria, with some portion of it extending to Kano and Katsina states. The Galma River originates from the Jos plateau in the South Western area of the Shetu hills, which is some 350km away from Zaria. The river then flows from there, Northwest towards longitude 8^{0}E and Southwest towards Zaria. It flows from Zaria to join the Kaduna River (Abdulrafiu, 1977). Galma dam is located on the Galma River, at a distance of about 10km Northeast of Zaria town (see fig.1a, 1b and 1c). The dam consists of a 550m long earthfill embankment with a maximum height of about 14 metres, and a spillway that is 91.5 metres. (NUWSRP, 2004)

The study area lies within the tropical wet and dry climatic zone, exhibiting a strong seasonality of rainfall. The area is characterized with 6 months of rain and 6 month of dryness, denoted as the Aw climatic type. The rainy season starts around May and terminates around October; on the other hand, the dry season lasts between November and April (Iguisi and Abubakar, 1998). The mean monthly temperature is about 27^{0}C, temperature varies and it is highest between the months of March and May, which represents the hot and dry period. It is lowest in December/January reaching about 22^{0}C.

The basement complex rocks primarily underlie the Zaria region. The region is an area within the Zaria plain, a dissected part of the Zaria-Kano portions, an extensive peneplain that had developed in crystalline metamorphic rock. The geology is constituted of three basic rock types: gneissic complex, metasediments and older granites. Younger granites can occur mostly as the ring complexes of Jurassic age. Alluvial deposits are quite extensive in Galma area, traversing wide open, shallow and gently sloping channels that had been cut mainly into the thick mantle of the overburden with some bedrock exposures. Shallowness of the stream channel enables the formation of extensive flood plain along the river and hence is often flooded during the rainy season (Nassef and Olugboye, 1979).

The vegetation type of the northern Guinea Savanna, which is mainly characterized by herbs and grasses, with few deciduous trees widely scattered, characterizes the area. The herbs and grasses grow very tall, in some places, along the perennial rivers and streams, riparian vegetation consisting of evergreen trees are found. The common grass and tree communities in the study area are mostly *Andropogon spp,* Mango trees, *Parkia clappertonia, Azfelia africana* and *Daniella oliveri* used for making mortar and *Acacia balanites* are found in the area.

The soil type in the area is highly leached ferruginous tropical soils that develop on weathered regolith overlain by a thin deposit of wind blown silt from the tropical continental air mass into the area (McCurry and Wright, 1970 and Torkarski, 1972). The physical conditions of the soils are generally poor. The soil aggregates are very small and unstable with tendency to compact under wet condition (Kowal and Kassam, 1978).

The rural inhabitants living around the river are mainly subsistent and peasant farmers, hence the land is subjected to intensive seasonal farming. In addition, along the Galma River and its tributaries, irrigation farming is being practised on a small scale, where farmers divert flowing water into their farmlands, or they can use machines or pumping device to supply water to their farmland. Rising population however, has led to massive deforestation in the area to create space enough for cultivation, thus posing a serious environmental threat, in terms of upstream soil loss and downstream sedimentation. The available grassland encourages grazing of cattle and other pasturing activities in the area, likewise, the river supports fishing activities, which also improves the economy of the area.

Fig. 1a. Map of Nigeria showing Kaduna State

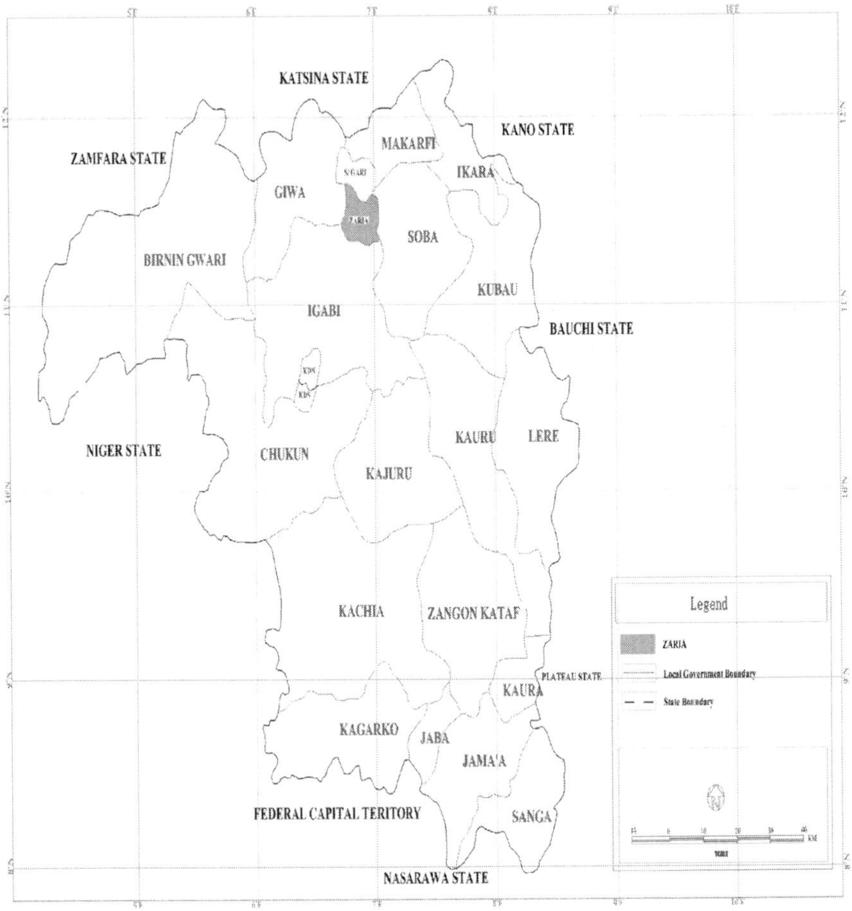

Fig. 1b. Kaduna State map showing Zaria local government

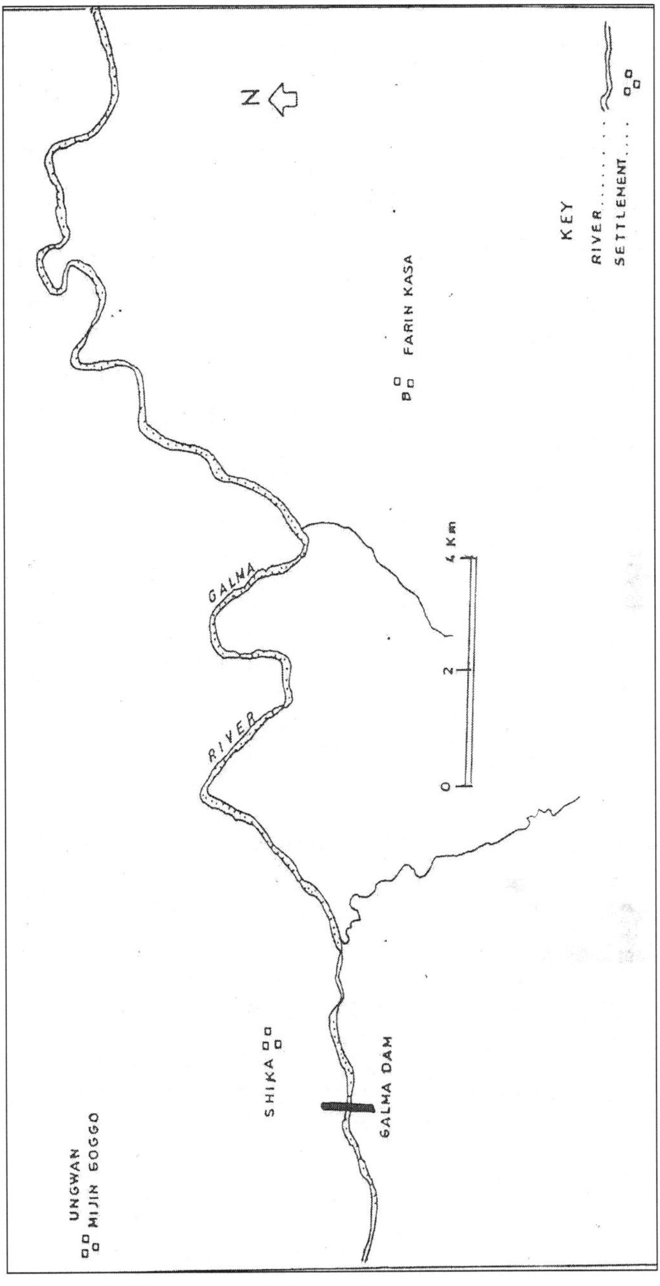

Fig. 1c. The Galma basin: Upstream of the Galma Dam

2. Methodology

2.1 Techniques of data collection

Data were collected using a systematic sampling approach, from the upstream section of the Galma dam. Starting point for the sampling was determined by moving 2km upstream away from the dam's embankment. From this starting point, 2km away from the dam's embankment, data on SSC, TDS and the stream discharge were obtained at 20 sampling points, over a distance of 100 metres each.

The velocity-area technique of discharge measurement was employed for this study, because it's relatively easier, accurate and less costly; here discharge is by definition the product of velocity and Cross Sectional Area (CSA) of flow, and this procedures evaluate these two terms for a stream section at a particular point in time. The width of the river and its depth was measured at each sampling point, the product of which gives the Cross Sectional Area. The velocity of flow of each sampling point was also read on a current metre, which is a digital equipment consisting of a graduated rod and a valeport, a propeller and the digital machine attached to it. The discharge therefore is calculated as the product of its flow velocity and Cross Sectional Area.

Sediment samples were collected at each sampling point using the USDH 48 sediment sampler, which is a depth-integrated sediment sampler. It was used because concentrations vary in a stream's cross-section and therefore, a sampling device is required to provide a representative sample of the concentration at a particular point within the channel. The sediment sample collected was then transferred into a plastic container and tightly corked. The samples were taken to the laboratory for onward analysis to determine the SSC and TDS values of the various samples.

The analysis of the sediment samples to determine their SSC and TDS values was carried out in the laboratory. The SSC was obtained by fetching 250ml of the sediment samples, which had been vigorously shaked to resuspend some of the settled particles. A membrane filter, which had been weighed, was attached to the filtration apparatus, which is equipped with a vacuum pump to make the filtration faster. The filter paper and the residue on it are reweighed. The difference between the initial weight and the new weight gives the SSC.

In getting the TDS, the aliquot from the filtration process for each sample is poured into a weighed crucible and oven-dried. On complete drying, the crucible is reweighed with the dissolved sediments that did not evaporate. The difference in the initial weight of the crucible and the weight after oven-drying gives us the TDS. However, for both the SSC and the TDS, the values gotten was multiplied by four (4), this is done to make the values up to mg/litre, since the sample taken is just 250ml.

The values of the SSC and that of the TDS multiplied by the stream discharge gives the Qs (suspended sediment discharge) and Qd (dissolved sediment discharge) respectively.

2.2 Techniques of data analysis

All generated data were logarithmically transformed. The log-transformed data were plotted against distance upstream on a line graph to reveal their upstream trends. Data on Suspended Sediment Concentration and Total Dissolved Solids were separately correlated with distance upstream to show their correlations. Also data on suspended and dissolved solids discharge were separately correlated with distance upstream to observe their correlations.

Student's t was also used to test the significance of the correlation co-efficient at the 95% significance level (α= 0.05). Mean, standard deviation, co-efficient of variation and ranges for each variable were also calculated. All analyses were carried out by the use of Microsoft excel and SPSS statistical Package.

3. Results and discussions

Table1 shows summary of the descriptive statistics data on sediment concentration (suspended sediments and dissolved solids) and discharge of the upstream section of the Galma Dam.

	N	Range	Minimum	Maximum	Sum	Mean	Standard error	Standard dev	Coefficient of variation
SSC	20	72.00	180.00	252.00	4416.00	220.80	5.09	22.76	8%
TDS	20	1280.00	120.00	1400.00	10160.00	508.00	86.48	386.75	76%
Qs	20	3.03	0.65	3.68	39.83	1.99	0.16	7.16	36%
Qd	20	15.15	0.42	15.57	88.15	4.41	0.81	3.62	82%

Table 1. Table showing Summary of Descriptive Statistics for SSC, TDS, Qs and Qd

The results from the summary statistics show that Suspended Sediment Concentration (SSC) and Suspended Discharge are moderately variable while the Total Dissolved Solids (TDS) and dissolved discharge have higher variability. SSC varied from a concentration of 180mg/l to a maximum concentration of 252 mg/l, giving a range of 72 mg/l with a standard deviation of 19.04 mg/l and a very low co-efficient of variation of 8.65%. Suspended discharge varies between 0.65g/s to 3.8g/s, with a range of 3.03g/s, mean of 1.99g/s standard deviation of 0.72g/s and co-efficient of variation of about 36%.

Total Dissolved Solids varies between 120 mg/l and 1400 mg/l, with a wide range of 1280 mg/l. TDS has a mean of 508 mg/l, standard deviation of 386.75 mg/l and co-efficient of variation of 76%. Solute discharge on the other hand ranges between 0.42g/s and 15.57g/s, giving a range of 15.15g/s mean solute discharge is 4.41g/s with a standard deviation of 3.62g/s and co-efficient of variation of 82%.

The higher mean values of TDS (and solute discharge) above SSC (and suspended discharge) indicates the dominance of chemical denudation within the basin. This confirms findings elsewhere (Dole and Stabler, 1909; Morisawa, 1968; Smith and Stopp, 1978 and Bowale, 2007)

Table 2 below shows that the correlation co-efficient of SSC with distance upstream is -0.16, while those of TDS, Qs and Qd are 0.330, -0.32 and 0.346 respectively. All the correlation coefficients are not significant at 95% significance level (α = 0.05), because the r - values are all greater than 0.05. This means that the relationship of SSC and TDS (with their respective sediment loads) with distance upstream is not well defined as to say that an increase in distance upstream will connote an increase in TDS and decrease in SSC. This is largely attributed to the fluxes experienced in the SSC and TDS as one move upstream, giving an irregular, haphazard and indiscernible pattern relating to the findings of Bowale, 2007.

Variables	Correlation with Distance (r)	t - values	Significance at 0.05 level
SSC (mg/l)	- 0.016	0.948	Insignificant
TDS (mg/l)	0.330	0.155	Insignificant
Qs (g/s)	- 0.320	0.893	Insignificant
Qd	0.346	0.135	Insignificant

Table 2. Correlation of SSC, TDS, Qs and Qd with distance upstream

4. Upstream fluxes of sediments

A more detailed and careful examination of the upstream patterns of SSC, TDS, Qs and Qd as shown in fig. 2-5 reveals that they show some pulses upstream; the pulsations are similar to waves having an amplitude and a pulse-length. SSC has four (4) pulses (fig. 2.); this is same for TDS, Qs and Qd (fig. 3-5). Nonetheless, despite showing some pulsations, the logarithmic line in figures 2-5, show that the variables increases generally upstream, except for the suspended discharge, Qs (fig. 4). The patterns exhibited by the four variables as revealed by the graphs are summarized in tables 3-6.

4.1 Fluxes of suspended sediment concentration

From fig. 2 and the summary in table 4, the first pulse of SSC started from 100m, which is the first point of measurement as well as the minimum SSC of 180mg/l, it peaked at a concentration of 232mg/l at a distance of 400m from starting point of measurement. The first pulse of SSC therefore has amplitude (range) of 52mg/l, representing 22.4% of maximum concentration, over a pulse length (distance) of 400m. The second pulse started immediately at the end of the first at 501m, got to a peak at a distance of 1000m and then declined to another minimum at a distance of 1200m from starting point. Pulse 2 had SSC amplitude of 56 mg/l, representing about 22.2% of maximum concentration and a pulse length of 700m. Pulse 3 starting from 1201m got to a maximum at 1400m and then declined to another low at 1700m, giving amplitude of 40mg/l, representing just 16.1% of maximum concentration, over a pulse length of 500m. Pulse 4 of SSC starting from 1701m reached a maximum at 2000m, the last point of measurement to give amplitude of 68mg/l, representing 27% of maximum concentration and a pulse length of 300m.

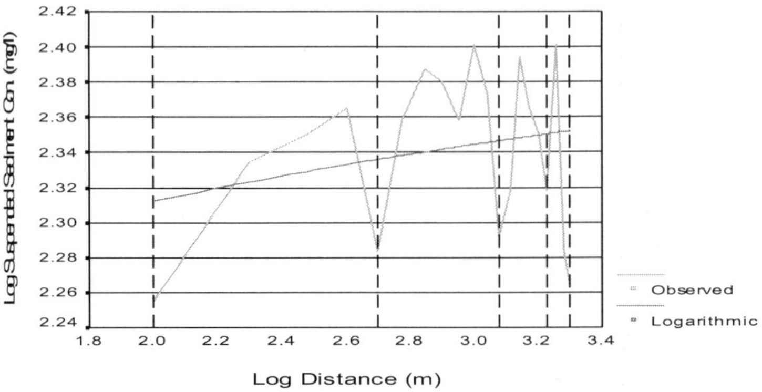

Fig. 2. Upstream Fluxes of Suspended Sediment Concentration

Pulse Parameter	Pulses			
	1	2	3	4
Distance of peak pulse concentration (m)	400	1000	1400	1800
Beginning of pulse (distance (m)	100	501	1201	1701
End of pulse (distance) (m)	500	1200	1700	2000
Pulse length (m)	400	700	500	300
Maximum concentration, (mg/l)	232	252	248	252
Minimum concentration (mg/l)	180	196	208	184
Amplitude (range) of concentration (mg/l)	52	56	40	68
Amplitude as % of maximum concentration (%)	22.4	22.2	16.1	27.0

Table 3. Upstream Fluxes (Pulses) of SSC, produced from fig. 2

4.2 Fluxes of total dissolved solids

Total Dissolved Solids, similar to the SSC also has four pulses (fig.3 and Table 4). Beginning from 100m, the first pulse of solute concentration peaked at a concentration of 200mg/l at the 100m point of measurement and thereafter declined to a minimum at 300m to give solute concentration amplitude of 80mg/l, representing about 40% of maximum concentration, over a pulse length of 300m. The 2nd pulse starting at 401m, got to a maximum concentration of 1000mg/l at 1100, thereafter, declining to a minimum about 100m from the point of maximum concentration, over a pulse length of 800m. Pulse 3 starting at 1201m, got to a maximum solute concentration at a distance of 1400m, and then declining to minimum solute concentration at 1600m. It has amplitude of 600mg/l, representing 75% of maximum solute concentration, over a pulse length of 400mg/l. The fourth pulse starting at 1601m got to a maximum solute concentration at 1800m and then declined to a minimum at a distance of 200m, the last point of measurement. Pulse 4 has an amplitude solute concentration of 1280mg/l, representing 91.4% of maximum solute concentration, over a pulse length of 400m.

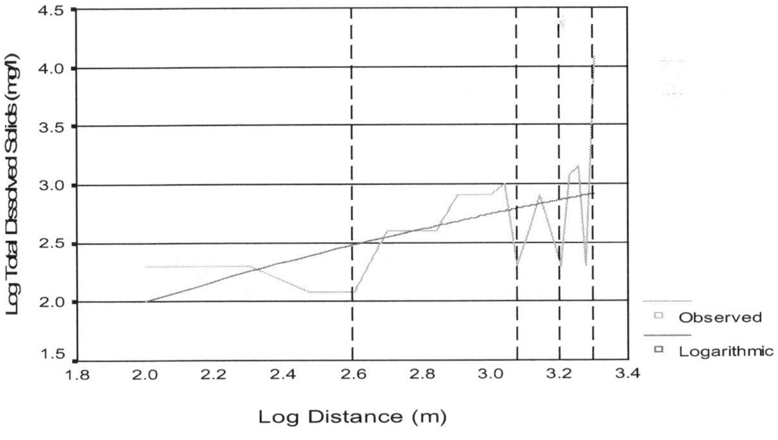

Fig. 3. Upstream Fluxes of Total Dissolved Solids

Pulse Parameter	Pulses			
	1	2	3	4
Distance of peak pulse concentration (m)	100	1100	1400	1800
Beginning of pulse (distance (m)	100	401	1201	1601
End of pulse (distance) (m)	400	1200	1600	2000
Pulse length (m)	300	800	400	400
Maximum concentration, (mg/l)	200	1000	800	1400
Minimum concentration (mg/l)	120	200	200	120
Amplitude (range) of concentration (mg/l)	80	800	600	1280
Amplitude as % of maximum concentration (%)	40	80	75	91.4

Table 4. Upstream Fluxes (Pulses) of TDS, produced from fig. 3

4.3 Fluxes of suspended sediment discharge

The suspended sediment discharge also has four pulses (fig.4 and table 5). The first pulse beginning from 100m reached the maximum suspended load of 2.39 g/s at a distance of 400m, after recording the minimum sediment discharge for the pulse at the first point of measurement of 100m, giving amplitude of 1.1g/s, representing 46% of the maximum sediment discharge over a pulse length of 400m. The second pulse beginning from 501m reached peak sediment discharge at 700m and declined to a minimum at 900m. It has an amplitude sediment discharge of 1.46, representing 50.3% of maximum sediment discharge over a pulse length of 400m. pulse 3, beginning at a distance of 901m peaked at 1000m and declining thereafter to a minimum at 1300m, with an amplitude of 1.33g/s, representing 51.8% of maximum sediment discharge, over a distance of 400m. Pulse 4, beginning from 1301m, reached peak sediment discharge at 1900m and declining to a minimum at 100m further upstream at 2000m, with amplitude of 3.03g/s, representing 82.3% of maximum sediment discharge over a pulse length of 700m.

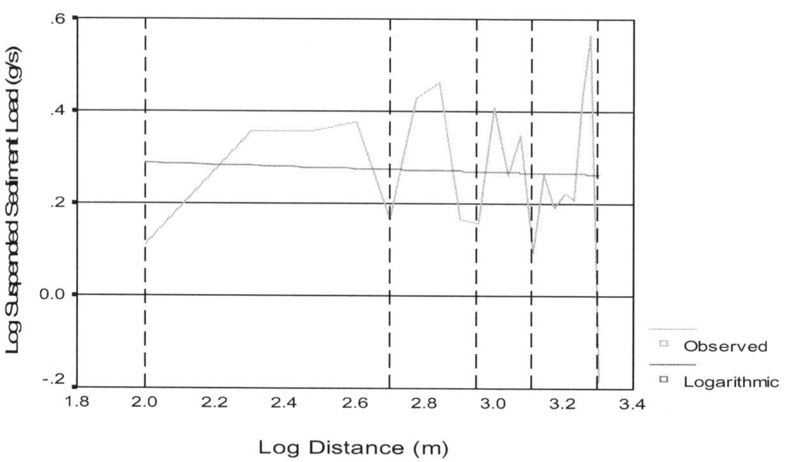

Fig. 4. Upstream Fluxes of Suspended Sediment Load

Pulse Parameter	Pulses			
	1	2	3	4
Distance of peak pulse concentration (m)	400	700	1400	1900
Beginning of pulse (distance (m)	100	501	901	1301
End of pulse (distance) (m)	500	900	1300	2000
Pulse length (m)	400	400	400	700
Maximum concentration, (mg/l)	2.39	2.90	2.57	3.68
Minimum concentration (mg/l)	1.29	1.44	1.24	0.65
Amplitude (range) of concentration (mg/l)	1.1	1.46	1.33	3.03
Amplitude as % of maximum concentration (%)	46.0	50.3	51.8	82.3

Table 5. Upstream fluxes (Pulses) of Qs produced from fig. 4

4.4 Fluxes of solute discharge (load)

The upstream fluxes of solute discharge (fig 5. and Table 6) closely replicate those of Total Dissolved Solids (fig. 3 and Table 5). Consequently, solute discharge has four pulses, same for TDS with the points of maximum discharge, discharge amplitude, pulse length and beginning and end of each pulse resembling those of TDS, except for the lower numerical values of minimum, maximum and amplitude of solute discharge.

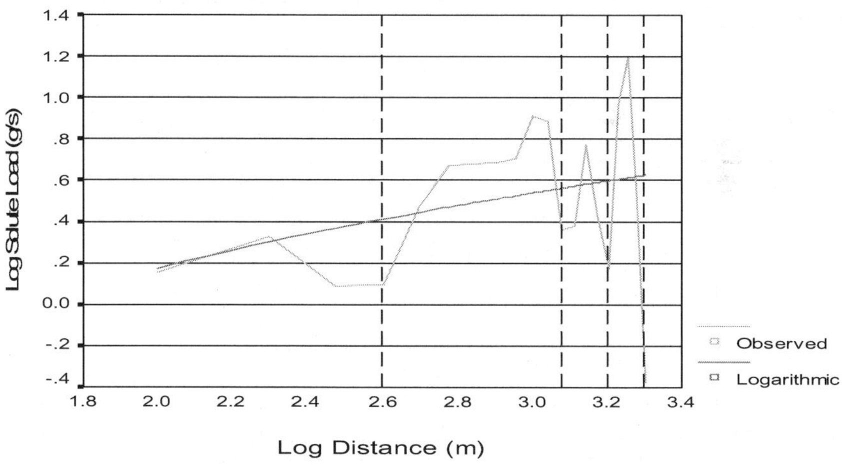

Fig. 5. Upstream Fluxes of Solute Load

Pulse Parameter	Pulses			
	1	2	3	4
Distance of peak pulse concentration (m)	200	1000	1400	1800
Beginning of pulse (distance (m)	100	401	1201	1601
End of pulse (distance) (m)	400	1200	1600	2000
Pulse length (m)	300	800	400	400
Maximum concentration, (mg/l)	2.11	8.15	5.90	15.57
Minimum concentration (mg/l)	1.22	2.28	1.48	0.42
Amplitude (range) of concentration (mg/l)	0.89	5.87	4.42	15.15
Amplitude as % of maximum concentration (%)	42.2	72	75	97.3

Table 6. Upstream Fluxes (Pulses) of Qd, produced from fig. 5

5. Comparison of sediment variables

5.1 Relationship between SSC and TDS

Suspended Sediment Concentration (SSC) in the Galma basin is lower than Total Dissolved Solids (TDS); the mean SSC is 220.8 mg/l, while mean TDS is 508 mg/l (table 2), a ratio of about 1:2.3. Looking at fig. 2 and fig. 3 in relation to each other, especially their logarithm lines, it can be observed that there is a general upstream increase in both variables, implying that a positive association exists between both variables. The test of association between both variables, shows a significant correlation between them, r=0.535, a moderately positive correlation (table 7). Therefore, generally on the average, despite the fluxes, as we move upstream, both the SSC and TDS tends to increase.

Compared Variables	Correlation values (r)	t- values	Significance at 0.05 level
SSC/ TDS	0.535	0.015	Significant
SSC/ Qs	0.321	0.168	Insignificant
TDS/ Qd	0.927	0.000	Significant
Qs/ Qd	0.317	0.174	Insignificant

Table 7. Correlations of SSC, TDS, Qs and Qd with each other

5.2 Relationship between SSC and Qs

In Table 7, the correlation between the SSC and sediment discharge (Qs) had an r-value of 0.321, which is poorly and insignificantly correlated. The correlation though positive is not strong, this can also be observed by viewing closely together the graphs of SSC and Qs (fig. 2 and 4), and they show a similar though not perfect trend upstream. Therefore, high levels of SSC correspond to high levels of suspended sediment discharge and vice versa.

However, when the logarithm line for both variables graph are viewed, there is a difference, which is reflected in the general upstream increase in concentration for the SSC, whereas there is a general upstream decrease for suspended sediment discharge, this could be a reason for the insignificant correlation between the two variables.

5.3 Relationship between TDS and Qd

TDS and Qd show a very strong, positive and significant relationship, with an r-value of 0.927 (table 6). This correlation is strongly positive indicating a perfectly direct relationship between both variables i.e. as the TDS increases, so does the solute discharge and vice-versa. This relationship is also reflected in the similar pattern and pulsations, the graphs of both variables with distance upstream gives, and in the general upstream increase as depicted by the logarithm line of both variables (fig. 3 and 5).

A downstream study of both variables by Walling and Webb (1986), however, revealed that while solute discharge increased with increase in river discharge downstream, solute concentration decreased, it can therefore be said that for an upstream scenario, the relationship will be positive, all things being equal. The decrease in the downstream case of the solute discharge can be attributed to the dilution effects of solutes, which increases as river discharge increases downstream.

5.4 Relationship between Qs and Qd

Suspended Sediment Discharge and Solute Discharge showed a poor and insignificant correlation, with an r-value of 0.317 (refer to fig. 4 and 5). At the start of measurement, especially the first and second pulse for both variables, there was an increase in suspended discharge that was associated with a decrease in the solute discharge, however, pulses 3 and 4 for both variables show a great correspondence, in that as the suspended discharge increases, the solute discharge increases and vice-versa. This kind of association is found between suspended and solute discharges due to the moderating effect of river discharge.

6. Discussion

6.1 Fluxes of SSC and Qs

The fluxes in SSC and Qs can be attributed to sand mining at irregular points along the river channel, to animals grazing around the channel banks, and the effect of cultivation near the riverbanks. Sand mining within the channel, most especially for construction works, promotes turbulence, which entrains sediments already settled on the riverbed. These sediments go into suspension to become part of suspended sediment load thereby increasing SSC. In addition, the release of sediments into the channel during mining operations enhances the concentration of suspended sediments. The entrained or released suspended sediments travel from point of release for a short distance before they settle back on the riverbed; the distance travelled by these sediments before resettling depends on the nature (weight and shape) of sediments, degree of turbulence eddies generated at the point of entrainment or release, flow characteristics and channel configuration among others. As sand mining occurs in a few irregular spots along the channel, sediment concentration and discharge show fluxes in the form of irregular pulses along the river channel. Dearing (1992) and Bowale (2007) have observed similar effect of sand mining on SSC and load in Wales and Samaru respectively.

Grazing of animals around the riverbanks has caused bank collapse in many parts of the channel by their trampling effects. Some of the bank materials slumped into the channel go into suspension and are transported in suspension before settling down on the riverbed further downstream. Likewise, human activities such as cultivation practices close to the riverbanks can lead to the intermittent addition of soil particles to the river and forthwith carried in suspension. Cultivation of lands along the river channel generally increases the level of SSC in a water body, as the soil becomes more susceptible to erosion, because of tillage.

6.2 Fluxes of TDS and Qd

The cause of the fluxes of Total Dissolved Solids and solute discharge (Qd) in the river channel is uncertain. However, fertilizer application by farmers cultivating the land along the river channel may be thought of as a possible cause. Excess chemical fertilizer not used up by plants fined their way into the river through various routes to increase the solute load of the river.

7. Conclusion

This study shows that chemical denudation seems more dominant within the basin than mechanical denudation; this is reflected in the higher mean value of dissolved solids over the suspended loads. Furthermore, sediment concentrations and discharges show wave like fluxes upstream. Using the logarithm line, the suspended sediment concentration, total dissolved solids and discharges showed an increasing trend upstream except for the suspended sediment discharge. The fluxes were attributed to certain physical and anthropogenic factors, implying that certain elements capable of creating disturbances within the river channel can lead to a diversion from the ideal patterns.

8. Recommendation

The study of sediment concentration and discharge fluxes is very important and useful in municipal water supply. This study therefore is of great importance to the Federal Ministry of Health as well as the Kaduna State Water Authority (K.S.W.A). The study was carried out in the upstream section of the Galma dam, which serves as the major reservoir for municipal water supply in Zaria and environs. The study was necessitated considering the effects of suspended and dissolved sediments on water quality and quantity in the reservoir, it also provides information on the rate at which sediment are being deposited into the reservoir.

Institution of an effective watershed management policy is recommended to help curtail the general increase for sediment released into the reservoir. The watershed management authority should be mandated to moderate or stop practices of sand mining at irregular spots on the channel; cultivation practices in the river basin; and cultivation requiring use of inorganic fertilizer should be moderated in the river basin. This will reduce the dissolved sediment sources in the basin thereby increasing the quality of water going into the reservoir.

Grazing of cattle on the riverbank should also be discouraged, to reduce sediment generated because of loosening of the soil by trampling effect of the cattle. The authority should also embark on massive afforestation programmes to reduce the rate of erosion, consequently, reducing sediment generation in the basin. Planting trees and cover crops that will protect the soil from raindrop impact and erosive work of wind and water is important. The authority should also carry out regular studies on sediment concentration and discharges to obtain data that gives an idea of the reservoir's lifespan; it can also be used to determine the cost of water treatment. This information will help the Kaduna State Water Authority, Nigeria to effectively carry out their function of municipal water supply.

9. References

Abdulrafiu, B.G. (1977). Land use changes Association with the New Galma Dam in Zaria. *Unpublished B.Sc. Dissertation*, Geography Department, Ahmadu Bello University, Zaria, Nigeria.

Bowale, A.R. (2007) "Downstream Fluxes in Suspended Sediment Concentration and Total Dissolved Solids, Upper Kubanni River Channel, Zaria, Nigeria". Paper Presented at the 49th Annual Conference of the Association of Nigerian Geographers, University of Abuja, Abuja, Nigeria.

Colby, B.R. and Hubbell, D.W. (1961). Simplified methods for computing total sediment discharge with the modified Einstein Procedure. U.S. *Geological Survey Water Supply Paper*. 1593.

Dearing, J.A. (1992). Sediment yields and sources in Welsh upland lake catchment during the past 800 years. *Earth surface Processes and Landforms*. 17, 1 - 22

Degens, E.T., Kempe, S and Richey, J.E. (1991). Summary: Biogeochemistry of major world rivers. In Degens, *et al* (eds), *Biogeochemistry of Major World Rivers*.

Dole, R.B. and Stabler, H. (1909). Denudation. *U.S. Geological Survey Water Supply Paper*, 234.

Iguisi, E.O. and Abubakar, S.M. (1998). "The Effect of Land use on Dam Siltation" Paper presented at 41st Annual Conference of Nigeria Geographical Association, University of Uyo, Akwa Ibom.

Knighton, D. (1998). *Fluvial Forms and Processes*. Arnold Hodder, Headline Group, London.

Kowal, J.M. and Kassam, A.H. (1978). Drought in the Guinea and Sudan Savanna Area of Nigeria. *Savanna*, Vol. 3. No. 1.

Leopold, L.B., Wolman, M.G. and Miller, J.P. (1964). *Fluvial Processes in Geomophology*. W.H. Freeman and Company, San Francisco.

McCurry, P.M. and Wright, J.B. (1970) Geology of Zaria. In Mortimore, M.J. (edition). *Zaria and its Region*. Occasional Paper No. 4. Department of Geography.

Morisawa, M. (1968). Streams: *Their dynamics and Morphology*. McGraw - Hill Company, New York.

Nassef, M.O. and Olugboye, M.D. (1979). The Hydrology and Hydrogeology of Galma Basin. *Federal Department of Water Resources*. File Report.

National Urban Water Sector Reform Project (NUWSRP) (March, 2004). *Report on Dam Safety Measures*. Federal Ministry of Water Resources.

Ritter, M.E. (2006). *The Physical Environment: An Introduction to Physical Geography*. Methuen & Co. Ltd. London.

Smith, D.I. and Stopp, P. (1978). *The River Basin. An Introduction to the Study of Hydrology.* Cambridge University Press, Cambridge.

Torkarski, A. (1972). Heterogeneous Terrace Arrangement of Ahmadu Bello University Zaria. *Occasional Paper,* No. 2. Dept. of Geography. A.B.U.

Walling, D.E. and Kleo, A.H.A. (1979). Sediment yield of Rivers in of areas of low precipitation: A global view in: The hydrology of areas of low precipitation, proceedings of the Canberra Symposium, 1979, *IAHS-AISH Publication.* 12, 479 - 93.

Walling, D.E. and Webb, B.W. (1986). Solute in River System in Trudgill, S.T. (ed), *Solute Processes.* Chichester. Wiley, 251 - 327.

24

An Overview of the Persistent Organic Pollutants in the Freshwater System

M. Mosharraf Hossain[1], K.M. Nazmul Islam[1] and Ismail M.M. Rahman[2*]
[1]*Institute of Forestry and Environmental Sciences, University of Chittagong, Chittagong,*
[2]*Department of Applied and Environmental Chemistry, University of Chittagong, Chittagong,*
Bangladesh

1. Introduction

Organic contamination can be viewed as the secondary and tertiary dispersions of organic compounds from various sources into the global circulation across different spheres of the environment *viz.* hydrosphere, lithosphere, atmosphere and biosphere - mediated primarily by human activities [1, 2]. Freshwater contamination occurs in diverse ways, of which, contamination by organic compounds of various kinds and traits from a wide range of sources is a major concern due to their persistent nature and harmful biological impacts, especially on human beings [3]. All over the world, supply of freshwater is shrinking in quantity and dwindling in quality [4] - forcing the modern city planners to recycle potable water from raw water coming out of waste treatment facilities [5]. Contaminants are coming into the fresh water streams from different natural and anthropogenic sources of which the anthropogenic contribution is getting larger in volume each day as do their diversity [6]. These contaminants are either attenuated naturally into less toxic or non-toxic forms [7], or they persist in the fresh water ecosystem for long time, enter the food chain through bioaccumulation and bio magnified to cause cascading effect on terrestrial and aquatic bio-diversity altogether [8, 9]. Presence of organic compounds and their harmful derivatives makes water unfit to consumption. Maintaining a supply of pure water for ever increasing population is already a daunting challenge all over the world while the organic contaminants are aggravating the challenges further [10, 11]. Hence, we need a clear understanding of the classes of organic compounds, which find their way into the fresh water system, their sources, and their transformation through physicochemical and biological processes in the fresh water system in order to control their entry and undesired transformation thereof. Moreover, we need to know the chemistry of these contaminants and their effects on the environment in general and on human health in particular while detecting their presence in fresh water in a quick and easy manner. Better understanding of these pollutants is inevitable for developing better techniques of purification of water from persistent organic contaminants. Substantial improvement has so far been achieved in all these aspects. This chapter is an effort to get a contemporary picture of our understanding

*Corresponding Author

about the organic contaminants in the freshwater system. After introducing the issue of fresh water pollution in the second section, we shall intensely focus on different aspects of contamination of fresh water by persistent organic compounds in section three. In this section, we shall describe the categories of organic compounds which contribute to fresh water contamination with their sources, fate and effects. This section will also include a review of detection techniques available for the detection of organic contaminants in the freshwater system. The section will be concluded with two sub sections - one highlighting our understanding of the natural assimilation of persistent organic contaminants through normal aquatic ecological processes and the other focusing on the techniques that we have at our discretion for the removal of such contaminants from water in order to purify water. Different conventions and protocols are in place to reduce and control the issue of organic contamination of water. Before concluding the chapter, we shall try to shed some light on the current status of such protocols and conventions regarding organic contamination of fresh water.

2. Pollution and contamination of fresh water

Contamination of fresh water is a natural process; in the passage of the normal cycling of water through the water cycle, during precipitation, different organic substances of varied origins, both soluble and insoluble, get transported into fresh water streams. Among the natural contributors of organic contaminants, volcanic eruption, forest fires, natural decomposition of organisms, are substantial sources [4]. Organic substances of biological origin also result from either excreta or wastes from of living organisms, from municipal waste decomposition, etc. Among the dissolved organic substances in water, humic and fulvic substances, polycyclic aromatic hydrocarbons (PAHs), halogenated aromatic and aliphatic hydrocarbons, phthalate esters are the major classes [12]. However, the input of organic contaminants from natural sources does not increase the way they are increasing from anthropogenic sources. Human activities are increasing the consumption of the already known organic compounds all over the world for meeting diverse needs while organic pharmaceuticals, pesticides, paints and coloring materials, cosmetics, etc with novel properties are synthesized every day for many purposes. Consequently, the quantity and diversity of anthropogenic organic contaminants that enter the fresh water system and polluting it are accumulating. According to a recent study [6] on the presence organic contaminants in untreated drinking water sources in USA, for surface sources, natural sterol - cholesterol, herbicide - metolachlor, nicotine metabolite - cotinine, natural plant sterol - β-sitosterol and caffeine metabolite - 1,7-dimethylxanthine were the five most frequently detected organic contaminants. In ground water sources, solvent - tetrachloroethylene, pharmaceutical - carbamazepine, plasticizer - bisphenol-A, caffeine metabolite - 1,7-dimethylxanthine, and fire retardant - tri (2-chloroethyl) phosphate were the five respective major contaminants. However, they could not establish any specific seasonal trends in the distribution patterns of contaminants.

3. Organic contaminants in freshwater system

In recent decades there has been escalating concern about the potential effects of the occurrence of a diverse array of organic contaminants in the fresh water system on the human and environmental health [12, 13]. Organic contaminants are mostly human-induced chemicals entering into natural fresh water through pesticide use, industrial chemicals, and as by-products of degradation of other chemicals and persist long enough in the environment to cause harmful effects. They tend to accumulate in reservoirs such as water, soil, sediments etc. From these reservoirs, they are remobilized through various processes, switch form or speciation and become available to the biological food chain. In this way, these contaminants tend to bio accumulate and bio magnify exhibiting toxicity and other related outcomes - mutagenicity, carcinogenicity and teratogenicity - resulting into chronic and acute disorders [14].

3.1 Common organic contaminants in freshwater

Organic contaminants are composed basically of hydrocarbons both from anthropogenic and natural sources. Many contaminating hydrocarbons exist in the environment but carcinogens, mutagens and teratogens among them are the most closely monitored [15]. These contaminants enter the fresh water system both from point and non-point sources. Point sources are defined in a spatially explicit manner both in terms of chemical residues from organic contaminants they contribute and by epidemiological factors thereof, like morbidity, mortality or community disruption [16]. Effluents from municipal sewage-treatment plants, industrial sources, storm sewer systems, mining and construction sites, etc. are examples of point sources. In contrast, organic contaminants from non-point-sources are diffused over broad geographical scale in a relatively uniform environmental concentration explicitly delineated into spatial or temporal patterns. Consequently, the management of non-point organic contaminants is difficult. Among non-point-sources are agricultural runoff, urban runoff and atmospheric wet and dry depositions [17]. Based on how they are formed, organic contaminants of fresh water systems can be of two categories - natural and synthetic. Organic contaminants of biological origin are natural organic contaminants of which sugars, alkaloids and terpenoids are prominent. Synthetic organic contaminants are generated through reaction among different chemical species, which subsequently get discharged into the fresh water - examples include DDT, Polychlorinated biphenyls (PCBs), Chlordane, and Dieldrin, etc. [18]. Some organic contaminants are easily broken down upon entry into the environment, but others are very persistent, popularly termed as, Persistent Organic Pollutants (POPs) which are of particular concern because of the long term risks they pose. POPs, due to their persistent nature, become widely distributed geographically to pose adverse effects to human health and the environment [2]. The *Stockholm Convention* on POPs entered into force in 2004, and it identified some POPs potential for damaging environment and human health (Table 1).

Since the mid-twentieth century, pesticides became ominous in agriculture for saving crop yields from pastes both in the field and in storage. Pesticides are vital for crop protection and human health in many parts of the world [19, 20], yet due to their detrimental effects on natural ecosystems [21-23], we are becoming increasingly worried about their short and long term harms on the environment. A number of pesticides are in use throughout the world, but the four most important groups are *insecticides, rodenticides, herbicides* and *fungicides* – each group contains a rapidly accumulating list of persistent organic compounds capable of contaminating fresh water. Upon application, a substantial amount of these chemicals found their ways into fresh water systems. Most of the pesticides tend to bioaccumulate and exert ruinous effects on environment and human health [24, 25]. Among the twelve toxic substances listed for being phased out as per the *Stockholm Convention*, nine are pesticides. The most common pesticide among the fresh water organic contaminants are DDT, Diedrin, Aldrin, Chlordane, Endrin, Mirex, Toxaphene, and Heptachlor [25, 26] - all of which has a long half-life, can break down to intermediate species and have detrimental health effects.

POPs	Class
Heptachlor	Pesticide
Endrin	Pesticide
Dieldrin	Pesticide
Aldrin	Pesticide
Chlordane	Pesticide
Hexachlorobenzene	Pesticide / industrial chemical / by-product
Dioxins	By-product
Furans	By-product
Polychlorinated biphenyls (PCBs)	Industrial chemical / by-product
DDT	Pesticide
Mirex	Pesticide
Toxaphene	Pesticide

Table 1. POPs scheduled to be phased out and eliminated under the Stockholm Convention, 2004.

3.2 Sources of persistent organic contaminants in the freshwater system

3.2.1 Industrial sources

Poly Aromatic Hydrocarbons (PAHs) are produced through burning of common fossil fuels, such as coal, petroleum and natural gas as well as from biomass fuels such as fuel wood, animal excreta etc. Naphthalene, an ingredient in dyeing industry, aluminium smelting industry and lubricant oils as well as wood procession industry, enters into water, mainly through discharges and spills during the storage, transportation and disposal of fuel oil and coal tar and incomplete combustion of organic compounds [27]. Anthracene releases from dye and pesticide manufacture, from exhaust of engines, from the incomplete combustion of organic compounds [28]. Burning of gasoline, garbage or biomass causes release of benzopyrene into the environment. It also releases from burning of tar during road construction, from wood preservative creosote and from glues used in electrical components [29]. The environmental release of benzene occurs from industries such as rubber processing, dyeing and washing, pharmaceuticals, and agrochemicals. Underground gasoline storage tanks also leak benzene into ground water [30]. Xylene is a common solvent, thinning and cleaning agent used widely in printing, rubber, wood processing, plastic and leather industries. Being liquid in form, xylene easily leaks into surface or ground water [31]. Dioxins are another class of organic contaminants released from pulp and paper mills, wood preservatives, etc. Trichloroethene is used as solvent during cleaning engine parts from grease and spillage of it to environment occurs during cleaning process or from the wastes of such cleaning facilities.

3.2.2 Agricultural and farming sources

Agrochemicals are predominantly organic in nature and diverse in classes, nature and applications. They constitute the major anthropogenic source of organic contaminants into the fresh water system [32]. Usually, agrochemicals when applied in agricultural field or agro-product storage or processing, come in direct contact with flow of fresh water and can contaminate gradually both the surface and ground sources of fresh water. As they enter through flow water, they can reach unimaginable distance from the place of their applications.

3.2.3 Natural sources

The background concentration of organic contaminants in the environment is negligible, however, a number of them exist in the environment. Some of the organic contaminants enter the environment through various natural processes [1] such as volcanic eruptions, forest fires, biological decomposition and microbial activities. PAHs may enter the fresh water system naturally from thermal geological reactions, *i.e.* volcanic eruptions, forest fires, etc. Anthracene releases from natural sources such as coal and tar, can seep into the ground and surface water from coal piles. Among seventy five dioxins, originating naturally from volcanic eruption or forest fires, 2,3,7,8- tetrachlorodibenzo-*p*-dioxin (TCDD), is the most contaminating to the environment. A deeper treatment of contaminant contribution from natural sources has been discussed elsewhere [32].

3.2.4 Domestic and municipal sources

Burning of biomass fuels, such as wood, leaves, cow dung, etc., lead to the release of organic contaminants such as PAHs, anthracene and benzopyrene, which ultimately reaches water or soil [33]. Burning of coal for cooking and home heating is another source of anthracene and benzopyrene, which find their way into the fresh water system. One of the major concerns of modern city or municipality management is the vast quantity of solid wastes produced every day in big municipal cities. The solid waste stream contains a huge amount of organic substances - human and animal excreta and wastes, vegetable and food remaining, healthcare wastes, etc. Upon their disposal into land filling or open dumping ground, the leachates thereof easily finds their way into surface or ground water [34]. Rich in organic contaminants for fresh water system, the leachates either contaminates surface water in the short run and ground water gradually, in the long run. Incineration of municipal waste also releases different organic contaminants into the air which ultimately pollute fresh water. In 37 rivers of Japan, presence of many pharmaceutical organic compounds has been found of which crotamiton, carbamazepine, ibuprofen and mefenamic acid were positively correlated to population in the respective catchment areas indicating the contribution of these compounds from sewage sources [35].

3.2.5 Atmospheric deposition

Organic contaminants from different sources easily enter to the atmosphere and subsequently adsorbed with atmospheric particles or moisture. Such deposition of contaminants from the atmosphere ultimately reaches the aquatic ecosystem. An example of atmospheric movement of organic contaminants to fresh water ecosystem has been clearly demonstrated in case of two high arctic lake systems - Lake Ellasjøen and Lake Øyangen - which are 500 km away from known point source [36]. In sediments and biota from both the lakes, high levels of POPs especially PCB and *p,p* 9-DDE, have been detected. However, the higher levels of these contaminants in Lake Ellasjøen has been linked to the higher precipitation it receives, and that it receives POPs from birds that use the lake as resting ground.

3.3 Fate of organic contaminants

Organic contaminants usually break down easily in the environment through natural assimilative processes, but there are compounds highly resistant and these needs special attention from the standpoint of pollution of fresh water systems. Popularly known as *Persistent Organic Pollutants* (POPs), they remain in the environment from many months to even years [1].

3.3.1 The media involved in the aquatic transport of organic chemicals

Some of the persistent organic contaminants, including DDT, move from the point of application through the atmosphere and translocate from relatively warm regions to get condensed at colder, higher latitudes through a process known as *global distillation effect.* This explains the deposition of organic compounds at high concentrations in the Arctic region which is free from usage of such compounds [9, 37]. Benzene is water soluble, to some extent, and can seep into groundwater as well. Among the POPs, polychlorinated biphenyles (PCBs) were wonder materials with a wide range of applications. However, they entered the fresh water system during their manufacture and use. Even after the ban on them, they are still entering into the environment from the waste products where PCBs were used in the past. In fresh surface water or ground water, most of the PCBs adhere, weakly or strongly, to suspended sediment particles and remain as such for years [38]. Sediments containing PCB settle down to the bottom of fresh water reservoirs. Such sediments become the source of PCBs and gradually but continuously release PCBs in scanty amounts for years.

3.3.2 Bio-transformation and bio-accumulation

POPs are less soluble in water and more in lipid or fat resulting in their higher accumulation into the fatty tissues of living organisms. These organic contaminants enter living organisms from water through drinking of contaminated water or through ingestion of foods either processed using such water or foods from sources, which already have accumulated these contaminants. Bioaccumulating POPs include pesticides, PCBs, dioxins, and furans [1]. Repeated intake of organic contaminants or food contaminated thereof may cause bio-magnification of these contaminants leading to the enhancement of the harm caused by these contaminants. The example of this is DDT, which shows higher concentration in fish fat tissue compared to its concentration in water where fishes grow. Different derivatives of DDT, for example, DDE or DDD forms inside the organisms and they are excreted off through excreta, but they may excrete through milk of mammals leading to transfer of them to offspring from parents. These compounds are stored most readily in fatty tissue. The tendency of these substances to persist in the environment and to be build-up in plant and animal tissues poses the greatest risk to human health and the environment. Usually, the organic contaminants function as mutagenic, carcinogenic or teratogenic agents. Benzene, in fresh water, breaks down quickly and does not build-up in plants or animals. Fat soluble dioxins bioaccumulate easily in fatty tissue as well as in skin, muscle and other organs of most of the exposed animals, including fishes grown in dioxin rich surface water sources. The risk is higher for fishes from streams where affluent from paper and pulp industry is released.

3.4 Effect of fresh water organic pollutants

DDT causes sperm decline, eggshell thinning of birds and birth defects in many animals, which have been linked to near elimination of some species of animal. Bald eagle, a carnivore, is an example of what damage a POP like DDT can do [1]. On the other hand, PCBs from fresh water bioaccumulate into fish to a concentration hundreds or thousands of times higher than their levels in water. Benzene, if ingested, is broken down into secondary metabolites in the liver and bone marrow which are suspected to have to link with liver and bone marrow tissues. Exposure to moderate or high toluene levels has potential adverse effects on their liver, kidneys, and lungs. Aquatic animals absorb only a minimal amount of the xylene available in water contaminated with it. However, methylbenzaldehyde, a breakdown product from xylene cause damage to lungs of some animals. Exposure of animals to moderate to high levels of trichloroethane (TCE) causes liver and kidney damage.

Among the organic contaminants, PCBs, DDT, DDE, DDD, naphthalene, benzopyrene, dioxins, etc., are reported or suspected carcinogens for humans. Benzopyrenes have been related to the production of metabolites doubted as a carcinogen precursor. PCBs enter the human body through eating of PCB containing meat or fish and get converted to other metabolites some of which excreted naturally, but others stay in fat and in the liver for months or years. Inclusion of PCBs into breast milk fat and their subsequent entries into babies have been reported [8, 39, 40]. In Western Japan, the consumption of rice bran oil contaminated with some thermal degradation products of PCBs, *i.e.* furans and quaterphenyls, which are more toxic than PCBs, led to a severe form of acne called chloracne followed by fatigue, nausea, and liver disorders [41, 42]. Short-term low human exposure to naphthalene in mild concentration has been linked to eye and skin irritation while at elevated exposure levels, it causes headaches, fatigue and nausea. However, ingestion of naphthalene may cause hemolytic anemia, damages to kidneys and liver. Decreasing fertility, fatal damage, lung and skin tumors are also evident in the case of chronic exposure to naphthalene. Trace amount of dioxins at parts per trillion levels may cause hormone disruption; it also causes numbness, fluctuations in liver enzymes levels, nausea, etc. Exposure to large quantity of TCE creates dizziness to senselessness and even death. Range of concentrations of different pharmaceutical organic contaminants of fresh water in water bodies around the world, their lowest predicted no-effect concentrations (PNEC), their health effects at higher doses along with natural attenuation have been summarized by Pal *et al.* [2].

In a high-latitude freshwater food web of a remote lake in the Canadian Arctic, bioaccumulation of atmospherically deposited organochlorides such as PCBs, DDT, chlordane (CHL)-related compounds and hexachlorocyclohexane (HCH)- isomers were found at ng L^{-1} concentrations in water and sediment samples, while in fish samples, 6- to 10-fold higher concentrations of these POPs compared to water samples were observed [9].

Every day, the flow of organic contaminants into the fresh water system is increasing with the increase in the industrial activities and intensification of agricultural practices alongside the expansion of big municipalities all over the world. The rapid growth of population and the changes in lifestyle is enhancing the demand for global freshwater consumption. Hence, the necessity of treating larger volume of water is increasing and the task of meeting this demand is becoming complicated in terms of technology and costlier in terms of capital investment due to increased flow of contaminants in a larger variety into the water. The challenges that the water treatment will face in near future have been discussed in a nice review by Shannon *et al.* [43].

3.5 Detection of organic contaminates in fresh water system

3.5.1 Instrumental and laboratory techniques

The ubiquitous presence of persistent organic pollutants in surface and ground water sources necessitates their detection at trace level in order to ensure safe drinking water standards for public health in one hand and to develop appropriate technology for the removal or avoidance of these contaminants, on the other hand. Significant efforts are evident to the development of novel techniques and systems for the detection of the presence of organic fresh water contaminants in sub parts per billion levels to sub parts per trillion levels along with the refinement of existing detection capabilities for these organic compounds in the environment [6]. Pressurized solvent extraction, solid phase extraction, and capillary-column gas chromatography/mass spectrometry techniques has been developed for the determination of organic contaminants [44, 45]. Gas chromatography-high resolution mass spectroscopy (GC-HRMS) has been used in Canada to detect estrogenic organic fresh water contaminants in municipal wastewater as well as in effluents from bleached kraft pulp mill at ng L^{-1} level [46]. Natural organic matter and the anthropogenic organic contaminants have distinct fluorescence signatures, which have been used to develop a technique for early detection of contamination in a groundwater based drinking water supply plant [47]. Very recently, low cost calibration free methods for determination of organic contaminants have been reported [48].

3.5.2 Bio indicators

Bio monitoring of water contamination using bio indicators is becoming a cost-effective mode of contamination detection in aquatic systems [49, 50]. Some species of aquatic biota helps the depuration of aquatic resources contaminated by organic contaminants through bioaccumulation. The presence and concentration of contaminants can therefore be bio-monitored by using these species as biological indicators. Bivalves are usually preferred as a bio indicator and Goldberg in 1975 introduced the concept of "Mussel Watch" followed by the wide use of bivalves as an ideal bio-monitor in routine monitoring for National and Regional programmes [51]. The current results from the implementation of this approach succeeded to identify hot spots of contaminants and in following up their spatial distribution in the marine environments. "Mussel Watch" approach is based simply on using the characteristic feeding habits of bivalves as filter feeders which can accumulate tremendous quantity of contaminants in their tissues; reflecting the present quality and quantity of bio-available chemicals in their surrounding waters. The approach has been used to monitor PCBs in water [52], organotin contamination of fresh water lakes [53], and for other organic contamination issues [54, 55]. Diatoms are also used in bio monitoring of water contamination [56].

3.5.3 Assimilative capacity of water and organic contaminants

Once into the fresh water system, the natural assimilative capacity of water can attenuate majority of the organic compounds in a relatively shorter period of time using natural processes such as biotransformation, photolysis, sorption, volatilization, and dispersion, or a combination thereof [57]. The completion of natural breakdown of PCBs in the water may take several years, or even decades. Benzopyrenes, once get into the fresh water system, is either photo-degraded at the surface of waters or biodegraded gradually, however, their adsorption to particulates slows down the rate of natural assimilation. After entry, xylene may remain in fresh water for months or more but it gradually breaks down into relatively less harmful or harmless organics [58]. Trichloroethane and other halogenated hydrocarbons degrade naturally by photo-oxidation and biodegradation. In surface water, natural breakdown occurs in weeks while in groundwater, the breakdown is much slower. However, the slow rate of desorption and resorption of trichloroethane in subsurface soil make their long persistence in soil from where they may find their way into ground water for a quite long time [59]. Reports on the natural attenuation of pharmaceutical organic compounds in the fresh water system have been reviewed in an excellent review by Pal *et al.* [2]. New approaches are emerging to assess the natural attenuation of organic contaminants in the freshwater system. Plume-scale electron and carbon balances have been used in UK in a Permo-Triassic Sandstone aquifer contaminated with phenolic compounds [7].

3.6 Treatment options for organic contaminants in freshwater system

The entry of organic contaminants into ground and surface fresh water source is increasing with industrialization and rapid urbanization. Accordingly, efforts towards the enhancement of treatment capabilities are increasing [60]. Bolong *et al.* [61] has reviewed the use and efficiencies of activated carbon, oxidation, activated sludge, nanofiltration and reverse osmosis membranes in removal of persistent organic contaminants. The current options for water treatment for the removal of organic contaminants are based on continued development of activated sludge method, which is basically a bio transformation method [62, 63], the membrane filtration methods [60, 64] and reverse osmosis [65-67]. Activated sludge with zeolite improves organic contaminants biodegradability during anaerobic oxidation of municipal waste water by increasing the number of heterotropic bacteria as high as 55 times compared to control activated sludge [68]. With the use of zeolite, the COD removal rate reached to about 90% while the TOC removal rates with 0.22μm and 0.45μm filter membrane reached respectively at 97% and 92%. The sludge with zeolite can depredate further organic contaminants, which was difficult to degrade using organisms. Removal of organic contaminants are also enhanced by using novel biodegradable coagulants like chitosan, poly-lactic acid derivatives, etc. [69], and adsorption into porous materials such as the activated charcoal [70], or granular carbon [71], etc. The focuses of these techniques are basically on the removal of POPs from freshwater entering into the environment through industrial or municipal effluents, since it became a binding requirement for business entities in almost all the countries of the world to release industrial effluents after making it free from such contaminants, which made the development of modern and effective removal technology an active field of research with adequate funding. However, there is a grim picture even after all these developments, as a recent report from South-East Queensland, Australia reported the presence of a total of 15 organic contaminants, including NDMA and bisphenol in potable water recycled from raw water coming off waste treatment plants with reverse osmosis equipped advanced water treatment facility [5]. On the other hand, the need for removal of organic contaminants puts tremendous pressure on the water-treatment facilities in terms of increased cost. However, some alternatives are emerging in this case as well to bring the cost of treatment down. Low-cost alternative adsorbents (LCAs), which comprise of both natural and synthetic materials, are in use for removal of POPs. LCAs have shown fast kinetics and appreciable adsorption capacities in removing organic dyes from the contaminated fresh water [72].

4. Convention/protocol on organic contaminant of fresh water

There are a number of conventions and protocols dealing with the issue of organic contaminants of the fresh water system, however, none of these protocols specifically deals only with this issue. The progress is in place, and we can expect that in the years to come, specific global convention covering the range of organic pollutants of the fresh water system will be adopted and ratified by all. Among the current conventions and/or protocols, Stockholm Convention on Persistent Organic Pollutants comprises of 173 parties, and it entered into force in 2004. The convention deals with POPs and it has identified 12 highly hazardous organic compounds, mostly agrochemicals and outlined the phasing out of these twelve gradually by ratifying parties [61]. Convention on the Protection and Use of Trans-boundary Watercourses and International Lakes, adopted at Helsinki, in 1992 by United Nations Economic Commission for Europe (UNECE), outlines the protocols for controlling trans-boundary movement of contaminants [73]. There are many country or region-specific regulations and directives related to water quality, which also covers the issue of organic contaminants of the freshwater system.

5. Conclusion

Our understanding of the extent to which organic contaminants are entering into the freshwater system is still sketchy, as do our knowledge of the subsequent chemistry of these contaminants - their fate, effect and mediation. The reasons of non-comprehensive understanding are many. The first is the impossibility, in practice, to keep a clear accounting of the entry of persistent organic contaminants into the aquatic system since the entry is taking place through plethora of means from innumerable sources. Secondly, the rapid increment of the pool of persistent organic fresh water contaminants in number and diversity; the interaction among the organic contaminants themselves, between contaminants and aquatic environment mediated by biological, physical and chemical forces leading to the formation of many intermediates and derivatives thereof. However, new information is added to our knowledge base each day, and we are becoming more capable of addressing the issue - locally and globally.

6. References

[1] T. Navratil and L. Minarik, Trace elements and contaminants, In: V. Cilek and R. H. Smith (Eds.), Encyclopedia of Life Support System, Vol. IV, pp. 1184–1213, 2005.

[2] A. Pal, K.Y.-H. Gin, A.Y.-C. Lin and M. Reinhard, "Impacts of emerging organic contaminants on freshwater resources: Review of recent occurrences, sources, fate and effects", *Science of the Total Environment*, Vol. 408, pp. 6062–6069, 2010.

[3] I. Rostami and A.L. Juhasz, "Assessment of persistent organic pollutant (POP) bioavailability and bioaccessibility for human health exposure assessment: A critical review", *Critical Reviews in Environmental Science and Technology*, Vol. 41, pp. 623–656, 2011.

[4] R.P. Schwarzenbach, T. Egli, T.B. Hofstetter, U. von Gunten and B. Wehrli, "Global water pollution and human health", *Annual Review of Environment and Resources*, Vol. 35, pp. 109–136, 2010.

[5] D.W. Hawker, J.L. Cumming, P.A. Neale, M.E. Bartkow and B.I. Escher, "A screening level fate model of organic contaminants from advanced water treatment in a potable water supply reservoir", *Water Research*, Vol. 45, pp. 768–780, 2011.

[6] M.J. Focazio, D.W. Kolpin, K.K. Barnes, E.T. Furlong, M.T. Meyer, S.D. Zaugg, L.B. Barber and M.E. Thurman, "A national reconnaissance for pharmaceuticals and other organic wastewater contaminants in the United States–II) Untreated drinking water sources", *Science of the Total Environment*, Vol. 402, pp. 201–216, 2008.

[7] S.F. Thornton, D.N. Lerner and S.A. Banwart, "Assessing the natural attenuation of organic contaminants in aquifers using plume-scale electron and carbon balances: model development with analysis of uncertainty and parameter sensitivity", *Journal of Contaminant Hydrology*, Vol. 53, pp. 199–232, 2001.

[8] N.M. Tue, A. Sudaryanto, T.B. Minh, T. Isobe, S. Takahashi, P.H. Viet and S. Tanabe, "Accumulation of polychlorinated biphenyls and brominated flame retardants in breast milk from women living in Vietnamese e-waste recycling sites", *Science of the Total Environment*, Vol. 408, pp. 2155–2162, 2010.

[9] K.A. Kidd, R.H. Hesslein, B.J. Ross, K. Koczanski, G.R. Stephens and D.C.G. Muir, "Bioaccumulation of organochlorines through a remote freshwater food web in the Canadian Arctic", *Environmental Pollution*, Vol. 102, pp. 91–103, 1998.

[10] C. Chen, X. Zhang, W. He, W. Lu and H. Han, "Comparison of seven kinds of drinking water treatment processes to enhance organic material removal: A pilot test", *Science of the Total Environment*, Vol. 382, pp. 93–102, 2007.

[11] J.D. Wilcox, J.M. Bahr, C.J. Hedman, J.D.C. Hemming, M.A.E. Barman and K.R. Bradbury, "Removal of organic wastewater contaminants in septic systems using advanced treatment technologies", *Journal of Environmental Quality*, Vol. 38, pp. 149–156, 2009.

[12] H.F. Al-Mudhaf, F.A. Alsharifi and A.S. Abu-Shady, "A survey of organic contaminants in household and bottled drinking waters in Kuwait", *Science of the Total Environment*, Vol. 407, pp. 1658–1668, 2009.

[13] R. Loos, B.M. Gawlik, G. Locoro, E. Rimaviciute, S. Contini and G. Bidoglio, "EU-wide survey of polar organic persistent pollutants in European river waters", *Environmental Pollution*, Vol. 157, pp. 561–568, 2009.

[14] K.K. Barnes, D.W. Kolpin, E.T. Furlong, S.D. Zaugg, M.T. Meyer and L.B. Barber, "A national reconnaissance of pharmaceuticals and other organic wastewater contaminants in the United States–I) Groundwater", *Science of the Total Environment*, Vol. 402, pp. 192–200, 2008.

[15] M.S. Evans, D. Muir, W.L. Lockhart, G. Stern, M. Ryan and P. Roach, "Persistent organic pollutants and metals in the freshwater biota of the Canadian Subarctic and Arctic: An overview", *Science of the Total Environment*, Vol. 351-352, pp. 94–147, 2005.

[16] A. Kronimus, J. Schwarzbauer, L. Dsikowitzky, S. Heim and R. Littke, "Anthropogenic organic contaminants in sediments of the Lippe river, Germany", *Water Research*, Vol. 38, pp. 3473–3484, 2004.

[17] I. Baranowska, H. Barchańska and A. Pyrsz, "Distribution of pesticides and heavy metals in trophic chain", *Chemosphere*, Vol. 60, pp. 1590–1599, 2005.

[18] Z.L. Zhang, H.S. Hong, J.L. Zhou, J. Huang and G. Yu, "Fate and assessment of persistent organic pollutants in water and sediment from Minjiang River Estuary, Southeast China", *Chemosphere*, Vol. 52, pp. 1423–1430, 2003.

[19] D.G. Hela, D.A. Lambropoulou, I.K. Konstantinou and T.A. Albanis, "Environmental monitoring and ecological risk assessment for pesticide contamination and effects in Lake Pamvotis, northwestern Greece", *Environmental Toxicology and Chemistry*, Vol. 24, pp. 1548–1556, 2005.

[20] P. Mineau, "A review and analysis of study endpoints relevant to the assessment of "long term" pesticide toxicity in avian and mammalian wldlife", *Ecotoxicology*, Vol. 14, pp. 775–799, 2005.

[21] C.A. Laetz, D.H. Baldwin, T.K. Collier, V. Hebert, J.D. Stark and N.L. Scholz, "The synergistic toxicity of pesticide mixtures: Implications for risk assessment and the conservation of endangered Pacific salmon", *Environmental Health Perspectives*, Vol. 117, pp. 348–353, 2008.

[22] M. Olsson, A. Bignert, J. Eckhéll and P. Jonsson, "Comparison of temporal trends (1940s-1990s) of DDT and PCB in Baltic sediment and biota in relation to eutrophication", *AMBIO: A Journal of the Human Environment*, Vol. 29, pp. 195–201, 2000.

[23] A. Baun, N. Bussarawit and N. Nyholm, "Screening of pesticide toxicity in surface water from an agricultural area at Phuket Island (Thailand)", *Environmental Pollution*, Vol. 102, pp. 185–190, 1998.

[24] W. Lopes Soares and M. Firpo de Souza Porto, "Estimating the social cost of pesticide use: An assessment from acute poisoning in Brazil", *Ecological Economics*, Vol. 68, pp. 2721–2728, 2009.

[25] L.C.M. Cuyno, G.W. Norton and A. Rola, "Economic analysis of environmental benefits of integrated pest management: A Philippine case study", *Agricultural Economics*, Vol. 25, pp. 227–233, 2001.

[26] J. Fernandez-Cornejo, S. Jans and M. Smith, "Issues in the economics of pesticide use in agriculture: A review of the empirical evidence", *Review of Agricultural Economics*, Vol. 20, pp. 462–488, 1998.

[27] J.K. Rosenfeld and R.H. Plumb, "Ground water contamination at wood treatment facilities", *Ground Water Monitoring & Remediation*, Vol. 11, pp. 133–140, 1991.

[28] D.H. Thomas and J.J. Delfino, "A gas chromatographic/chemical indicator approach to assessing ground water contamination by petroleum products", *Ground Water Monitoring & Remediation*, Vol. 11, pp. 90–100, 1991.

[29] G. Chao and H. Ting-lin, *Comparative research on petroleum composition between pro-release and post-release at the low petroleum contaminant sediments*, In: International Symposium on Water Resource and Environmental Protection (ISWREP), Xi'an, 2011.

[30] O. Adeyemi, J.O. Ajayi, A.M. Olajuyin, O.B. Oloyede, A.T. Oladiji, O.M. Oluba, I.A. Ololade and E.A. Adebayo, "Toxicological evaluation of the effect of water contaminated with lead, phenol and benzene on liver, kidney and colon of Albino rats", *Food and Chemical Toxicology*, Vol. 47, pp. 885–887, 2009.

[31] R. Kandyala, S.P.C. Raghavendra and S.T. Rajasekharan, "Xylene: An overview of its health hazards and preventive measures", *Journal of Oral and Maxillofacial Pathology : JOMFP*, Vol. 14, pp. 1–5, 2010.

[32] S. Vedal, Natural sources - wildland fires and volcanoes, In: S. Tarlo, P. Cullinan and B. Nemery (Eds.). *Occupational and Environmental Lung Diseases*, John Wiley & Sons, Ltd: Chichester, pp. 389–404, 2010.

[33] A. Asa-Awuku, A.P. Sullivan, C.J. Hennigan, R.J. Weber and A. Nenes, "Investigation of molar volume and surfactant characteristics of water-soluble organic compounds in biomass burning aerosol", *Atmospheric Chemistry and Physics*, Vol. 8, pp. 799–812, 2008.

[34] D. Barceló and M. Petrovic, *Emerging contaminants from industrial and municipal waste: Occurrence, analysis and effects*. Springer Verlag: Berlin, 2008.

[35] N. Nakada, K. Kiri, H. Shinohara, A. Harada, K. Kuroda, S. Takizawa and H. Takada, "Evaluation of pharmaceuticals and personal care products as water-soluble molecular markers of sewage", *Environmental Science & Technology*, Vol. 42, pp. 6347–6353, 2008.

[36] A. Evenset, G.N. Christensen, T. Skotvold, E. Fjeld, M. Schlabach, E. Wartena and D. Gregor, "A comparison of organic contaminants in two high Arctic lake ecosystems, Bjørnøya (Bear Island), Norway", *Science of the Total Environment*, Vol. 318, pp. 125–141, 2004.

[37] G.A. Stern, E. Braekevelt, P.A. Helm, T.F. Bidleman, P.M. Outridge, W.L. Lockhart, R. McNeeley, B. Rosenberg, M.G. Ikonomou, P. Hamilton, G.T. Tomy and P. Wilkinson, "Modern and historical fluxes of halogenated organic contaminants to a lake in the Canadian arctic, as determined from annually laminated sediment cores", *Science of the Total Environment*, Vol. 342, pp. 223–243, 2005.

[38] G. Welfinger-Smith and D.O. Carpenter, Addressing sources of PCBs and other chemical pollutants in water, In: J. M. H. Selendy (Ed.). *Water and Sanitation-Related Diseases and the Environment*, Wiley-Blackwell: Hoboken, N.J., pp. 359–384, 2011.

[39] L.-M.L. Toms, L. Hearn, K. Kennedy, F. Harden, M. Bartkow, C. Temme and J.F. Mueller, "Concentrations of polybrominated diphenyl ethers (PBDEs) in matched samples of human milk, dust and indoor air", *Environment International*, Vol. 35, pp. 864–869, 2009.

[40] P.O. Darnerud, "Toxic effects of brominated flame retardants in man and in wildlife", *Environment International*, Vol. 29, pp. 841–853, 2003.

[41] C.J. George, G.F. Bennett, D. Simoneaux and W.J. George, "Polychlorinated biphenyls a toxicological review", *Journal of Hazardous Materials*, Vol. 18, pp. 113-144, 1988.

[42] G. Ross, "The public health implications of polychlorinated biphenyls (PCBs) in the environment", *Ecotoxicology and Environmental Safety*, Vol. 59, pp. 275-291, 2004.

[43] M.A. Shannon, P.W. Bohn, M. Elimelech, J.G. Georgiadis, B.J. Marinas and A.M. Mayes, "Science and technology for water purification in the coming decades", *Nature*, Vol. 452, pp. 301-310, 2008.

[44] M.R. Burkhardt, *Determination of wastewater compounds in sediment and soil by pressurized solvent extraction, solid-phase extraction, and capillary-column gas chromatography/mass spectrometry [electronic resource]*. U.S. Dept. of the Interior, U.S. Geological Survey: Reston, Va., 2006.

[45] J.D. Cahill, E.T. Furlong, M.R. Burkhardt, D. Kolpin and L.G. Anderson, "Determination of pharmaceutical compounds in surface- and ground-water samples by solid-phase extraction and high-performance liquid chromatography-electrospray ionization mass spectrometry", *Journal of Chromatography. A*, Vol. 1041, pp. 171–180, 2004.

[46] M.P. Fernandez, M.G. Ikonomou and I. Buchanan, "An assessment of estrogenic organic contaminants in Canadian wastewaters", *Science of the Total Environment*, Vol. 373, pp. 250–269, 2007.

[47] C.A. Stedmon, B. Seredyńska-Sobecka, R. Boe-Hansen, N. Le Tallec, C.K. Waul and E. Arvin, "A potential approach for monitoring drinking water quality from groundwater systems using organic matter fluorescence as an early warning for contamination events", *Water Research*, Vol. 45, pp. 6030–6038, 2011.

[48] A. Radu, S. Anastasova, C. Fay, D. Diamond, J. Bobacka and A. Lewenstam, *Low cost, calibration-free sensors for in situ determination of natural water pollution*, In: IEEE Sensors, Kona, HI 2010.

[49] W. Liu, L.Z. Shan, Q.L. Xie and others, "Application of biomonitoring in water pollution", *Journal of Environment and Health*, Vol. 25, pp. 456–458, 2008.

[50] Q. Zhou, J. Zhang, J. Fu, J. Shi and G. Jiang, "Biomonitoring: An appealing tool for assessment of metal pollution in the aquatic ecosystem", *Analytica Chimica Acta*, Vol. 606, pp. 135–150, 2008.

[51] E.D. Goldberg, V.T. Bowen, J.W. Farrington, G. Harvey, J.H. Martin, P.L. Parker, R.W. Risebrough, W. Robertson, E. Schneider and E. Gamble, "The mussel watch", *Environmental Conservation*, Vol. 5, pp. 101–125, 1978.

[52] A.A. Raeside, S.M. O'Rourke and K.G. Drouillard, "Determination of in situ polychlorinated biphenyl elimination rate coefficients in the freshwater mussel biomonitor Elliptio complanata deployed in the huron–erie corridor, southeast Michigan, USA, and southwest Ontario, Canada", *Environmental Toxicology and Chemistry*, Vol. 28, pp. 434–445, 2009.

[53] J. Yang, H. Harino, H. Liu and N. Miyazaki, "Monitoring the organotin contamination in the Taihu Lake of China by bivalve mussel *Anodonta woodiana*", *Bulletin of Environmental Contamination and Toxicology*, Vol. 81, pp. 164–168, 2008.

[54] S. Mishra, R.K. Mishra, B.K. Sahu, L. Nayak and Y. Senga, "Differential growth of the freshwater mussel, *Lamellidens marginalis* in relation to certain drugs", *Environmental Toxicology*, Vol. 23, pp. 379–386, 2008.

[55] H. Liu, J. Yang and J. Gan, "Trace element accumulation in Bivalve Mussels *Anodonta woodiana* from Taihu Lake, China", *Archives of Environmental Contamination and Toxicology*, Vol. 59, pp. 593–601, 2010.

[56] A. Beyene, T. Addis, D. Kifle, W. Legesse, H. Kloos and L. Triest, "Comparative study of diatoms and macroinvertebrates as indicators of severe water pollution: Case study of the Kebena and Akaki rivers in Addis Ababa, Ethiopia", *Ecological Indicators*, Vol. 9, pp. 381–392, 2009.

[57] C.J. Gurr and M. Reinhard, "Harnessing natural attenuation of pharmaceuticals and hormones in rivers", *Environmental Science & Technology*, Vol. 40, pp. 2872–2876, 2006.

[58] J.P. Barker, G.C. Patrick and D. Major, "Natural attenuation of aromatic hydrocarbons in a shallow sand aquifer", *Ground Water Monitoring & Remediation*, Vol. 7, pp. 64–71, 1987.

[59] N.S. Thom, A.R. Agg and A.L. Downing, "The breakdown of synthetic organic compounds in biological processes", *Proceedings of the Royal Society of London. Series B. Biological Sciences*, Vol. 189, pp. 347–357, 1975.

[60] N. Tadkaew, M. Sivakumar, S.J. Khan, J.A. McDonald and L.D. Nghiem, "Effect of mixed liquor pH on the removal of trace organic contaminants in a membrane bioreactor", *Bioresource Technology*, Vol. 101, pp. 1494–1500, 2010.

[61] N. Bolong, A.F. Ismail, M.R. Salim and T. Matsuura, "A review of the effects of emerging contaminants in wastewater and options for their removal", *Desalination*, Vol. 239, pp. 229–246, 2009.

[62] G.M. Shaul, T.J. Holdsworth, C.R. Dempsey and K.A. Dostal, "Fate of water soluble azo dyes in the activated sludge process", *Chemosphere*, Vol. 22, pp. 107–119, 1991.

[63] M. Petrovi, S. Gonzalez and D. Barceló, "Analysis and removal of emerging contaminants in wastewater and drinking water", *TrAC Trends in Analytical Chemistry*, Vol. 22, pp. 685–696, 2003.

[64] S. Gonzalez, M. Petrovic and D. Barcelo, "Removal of a broad range of surfactants from municipal wastewater–Comparison between membrane bioreactor and conventional activated sludge treatment", *Chemosphere*, Vol. 67, pp. 335–343, 2007.

[65] T.A. Peters, "Purification of landfill leachate with reverse osmosis and nanofiltration", *Desalination*, Vol. 119, pp. 289–293, 1998.

[66] C.Y. Tang, Q.S. Fu, A.P. Robertson, C.S. Criddle and J.O. Leckie, "Use of reverse osmosis membranes to remove perfluorooctane sulfonate (PFOS) from semiconductor wastewater", *Environmental Science & Technology*, Vol. 40, pp. 7343–7349, 2006.

[67] J. Radjenovic, A. Bagastyo, R.A. Rozendal, Y. Mu, J. Keller and K. Rabaey, "Electrochemical oxidation of trace organic contaminants in reverse osmosis concentrate using RuO_2/IrO_2-coated titanium anodes", *Water Research*, Vol. 45, pp. 1579–1586, 2011.

[68] G.-W. Cheng, W.-Y. Wei, Z.-C. Wu, F.-J. Zhang and X.-F. Huang, *A study on degradation of organic contaminants of the A/O process of denitrogenation enhanced by zeolite and chemical phosphorus removal*, In: International Conference on Energy and Environment Technology, Guilin, Guangxi 2009.

[69] S. Zodi, J.-N. Louvet, C. Michon, O. Potier, M.-N. Pons, F. Lapicque and J.-P. Leclerc, "Electrocoagulation as a tertiary treatment for paper mill wastewater: Removal of non-biodegradable organic pollution and arsenic", *Separation and Purification Technology*, Vol. 81, pp. 62–68, 2011.

[70] D. Stalter, A. Magdeburg, M. Wagner and J. Oehlmann, "Ozonation and activated carbon treatment of sewage effluents: Removal of endocrine activity and cytotoxicity", *Water Research*, Vol. 45, pp. 1015–1024, 2011.

[71] D. Mohan, K.P. Singh and V.K. Singh, "Wastewater treatment using low cost activated carbons derived from agricultural byproducts–A case study", *Journal of Hazardous Materials*, Vol. 152, pp. 1045–1053, 2008.

[72] V.K. Gupta and Suhas, "Application of low-cost adsorbents for dye removal - A review", *Journal of Environmental Management*, Vol. 90, pp. 2313–2342, 2009.

[73] S.M.A. Salman, "The Helsinki Rules, the UN watercourses convention and the Berlin rules: Perspectives on international water law", *Water Resources*, Vol. 23, pp. 625–640, 2007.

25

Rainwater Harvesting Systems in Australia

M. van der Sterren[1,2], A. Rahman[1] and G.R. Dennis[1]
[1]*University of Western Sydney,*
[2]*NSW Office of Environment and Heritage,*
Australia

1. Introduction

The Australian continent has an extremely variable climate, as a result of the different oceanic currents and atmospheric variation. Australia has regular cycles of droughts and floods resulting in highly variable storage volumes in its major dams. The population in Australia is nearly 23 million (Australian Bureau of Statistics (ABS), 2011) of which the majority lives in the South-East coast of Australia. The largest cities (see Figure 1) are Sydney (4.58 million people.), Melbourne (4.08 million people), Brisbane (1.07 million people) and Canberra (358,600 people) (ABS, 2011). The water supply storage for these cities is located in the nearby mountain ranges and brought to the metropolitan areas through large distribution water pipes. The urban fringe areas, rural locations and the outback have limited reticulated water supply and often rely on capturing roof water, farm dam water and bore water for their water supply. The roof water in these regions provides the principal potable water supply, whilst farm dam and bore water are often used to meet non-potable requirements and for livestock (ABS, 2010). Historically, this has been different in the urban areas where potable and non-potable supply demands are met with a reticulated water supply.

A shift has occurred in the Australian Water industry as a result of population growth, the worst drought in living memory (Horstman, 2007) and a desire to become more sustainable. Total Water Cycle Management has gained momentum in Australia and new property developments must consider all aspects of the water cycle, including water supply, waste water treatment, stormwater control and water quality control of all discharges and supplies (Argue, 2004; Argue & Pezzaniti, 2009; Barton & Argue, 2009; Hardy, 2009; Hardy et al., 2003; Wong, 2006b, c; Wong et al., 2008; Wong & Brown, 2009). Rainwater tanks are being installed in urban areas, resulting in an increase resilience of the cities to droughts and a reduction of mains water demand. These rainwater tank installations are encouraged in various Development Control Plans (DCPs), through state legislation, such as the NSW Building and Sustainability Index (BASIX Sustainability Unit, 2009), and by providing rebates (Blacktown City Council, 2006; Blue Mountains City Council, 2005; Gardiner & Hardy, 2005; Ku-ring-gai Council, 2005; Penrith City Council, 2010). The reasons for installing a rainwater tank in Australia include reducing mains water costs, helping the environment, irrigating the garden and because it was mandatory when the house was built (ABS, 2010; Blackburn et al., 2010; White, 2010).

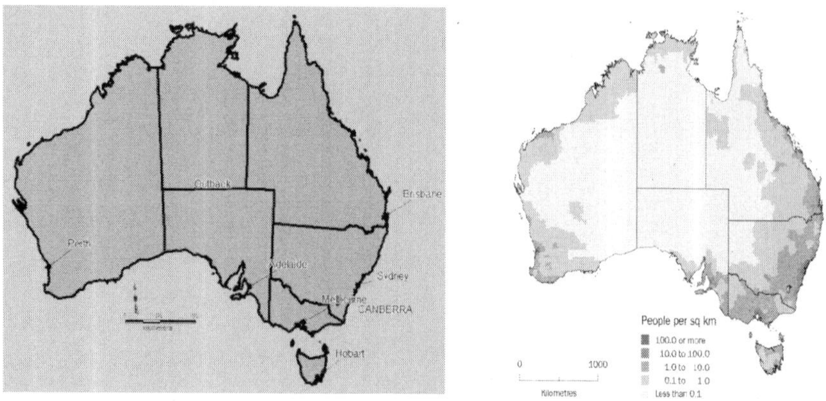

(a) (b)

Fig. 1. Australia with the location of major urban centres (a) and Population density (b) (ABS, 2011)

2. Legislation

2.1 Australian guidelines

The water quality requirements in Australia have been developed using a fit-for-purpose approach. The guidelines for water quality in Australia are supported by the Department of Sustainability, Environment, Water, Population and Communities of the Australian Commonwealth Government. The Natural Resource Management Ministerial Council (NRMMC) released the guidelines and, in some cases, with input from the National Health and Medical Research Council (NHMRC) and the Australian Health Ministers Conference. These National Councils update and publish a number of guidelines applicable to rainwater tank installation in Australia. These guidelines are discussed below and are dependent on the point of discharge, users, use and general application of the water in the rainwater tank. The guidelines are summarised below:

- Australian and New Zealand guidelines for fresh and marine water quality (AFWG) (Australian and New Zealand Environment and Conservation Council & Agriculture and Resource Management Council of Australian and New Zealand (ANZECC & ARMCANZ), 2000a, b);
- Australian drinking water guidelines (ADWG) (NHMRC & NRMMC, 2004);
- Guidelines for urban stormwater management (ANZECC & ARMCANZ, 2000c); and
- Australian guidelines for water recycling: managing health and environmental risks - Phase 2 (NRMMC et al., 2009c), including
    - Australian Guidelines for Water Recycling: Stormwater Harvesting and Reuse (NRMMC et al., 2009a) or Recycled Water Guidelines (RWG); and
    - Australian Guidelines for Water Recycling: Augmentation of Drinking Water Supplies (NRMMC et al., 2009b).

2.2 Stormwater harvesting and reuse

Stormwater harvesting and reuse guideline have been prepared as part of the National Water Quality Management Strategy (NWQMS) by NRMMC, Environment Protection and Heritage Council (EPHC) and the NHMRC (2009a). The guideline uses a risk-based approach to stormwater harvesting and can be adopted for any stormwater harvesting technique. The management and treatment techniques for the harvested stormwater depend on the type of catchment area (roof, drainage system, driveway) and its intended end-use (potable, non-potable or irrigation only). The guideline has a specific section for the management of harvested roof water and identifies these systems as low risk. In particular, the guideline indicated that roof water harvesting for residential dwellings should meet the criteria and management techniques specified in the Australian Standards (AS) (Standards Australia, 2008) and *Guidance on use of rainwater tanks* (enHealth Council, 2004). For non-residential buildings or for a larger user group (nursing homes), the stormwater harvesting guideline (NRMMC et al., 2009a) is recommended to be applied and a risk assessment is to be conducted for the proposed end-uses of the rainwater harvesting system

The end-uses for harvested roof water are also dependent on other guidelines and policies. Most council Development Control Plans (DCP) specify that water from a rainwater tank is not to be used for drinking (Blue Mountains City Council, 2005; Hawkesbury City Council, 2000; Upper Parramatta River Catchment Trust (UPRCT) et al., 2005) even though the New South Wales (NSW) Department of Health does not prohibit its use. The NSW Department of Health does recommend using the water only for non-potable purposes when town water is available (enHealth Council, 2004; Environmental Health Branch, 2008). In contrast to this, it was concluded by Heyworth, Maynard & Cuncliffe (1998, p. 9) that 'little is known about the associated health risks' in regards to using water from a rainwater tank. It should be noted that a large number of Australians still rely on their rainwater tank as their sole supply of drinking water. They are advised by various government bodies to refer to Cuncliffe (1998) and enHealth (2004) to determine how to maintain a rainwater tank and how to improve the quality of drinking water from their rainwater tank.

In addition to these guidelines, the Australian Standards (AS) specify how these systems should be constructed. The AS 3500 (Standards Australia, 2006) specifies connections, backflows prevention systems and roof gutter design. The handbook (Standards Australia, 2008) shows how the systems can be designed and constructed to ensure a low risk, high quality fit-for-purpose water supply. In Australia, all designs and constructions of roof water harvesting and drainage systems need to adhere to minimum standard specified in the Australian Standards. Stormwater harvesting can also be used augment drinking water supplies and the guideline divides the augmentation of drinking water supplies in direct augmentation and indirect augmentation (NRMMC et al., 2009b). Indirect augmentation includes storages, such as dams or rivers prior to treatment and distribution. Direct augmentation includes storages such as rainwater tanks and recycled water treatment plants and requires the treatment and adequate management of the system to protect users.

2.3 Australian Drinking Water Guidelines

The Australian Drinking Water Guidelines (NHMRC & NRMMC, 2004) give the minimum standards to which water supply authorities need to treat drinking water to protect the users, especially those who are very young, elderly or immuno-compromised. Rainwater tanks on private property are not required to meet these guidelines for low risk use. A risk assessment approach is recommended for larger harvesting systems to ensure the risk associated with the use of the rainwater is managed (NRMMC et al., 2009a). Adopted management techniques can include fit-for-purpose use of the harvested water or treating the harvested water to the ADWG (NHMRC & NRMMC, 2004).

2.4 Fresh and marine water quality guidelines

The AFWG (ANZECC & ARMCANZ, 2000a, b) are used to determine objectives for in stream water quality of fresh and marine waters from high conservation areas to heavily modified systems. The overflow from harvesting systems is commonly directly diverted to the receiving waters through a drainage system. So although the AFWG (ANZECC & ARMCANZ, 2000a, b) are ambient in-stream water quality objectives, it is important to understand the impact that rainwater tank overflows might have on the receiving waters. Mixing of flows will occur within the drainage system thereby affecting the final concentrations of the water quality parameters.

2.5 Stormwater management

The current guideline for stormwater quality, *Australian Guidelines for Urban Stormwater Management* (ANZECC & ARMCANZ, 2000c) is based on a risk management approach, which is appropriate for catchment wide management, but difficult to implement on a lot scale. Council-wide assessments are made by local councils to develop a stormwater management plan. These plans have been developed by some councils and Catchment Management Authorities to identify management scenarios and pollutant hot spots, and are developed to reduce the pollutant levels in the run-off discharges from the catchment. Some councils have implemented water quality targets or water quality objectives for new developments, based on published water quality targets (see Table 1).

Pollutant	**Percentage Detained**
Suspended Solids	80% retention of average load
Total Phosphorus	45% retention of average annual load
Total Nitrogen	45% retention of average annual load
Litter	Retention of litter (> 50 mm) up to the 3-month ARI peak flow
Coarse Sediment	Retention of sediment (> 0.125 mm) up to the 3-month ARI peak flow
Oil and Grease	No visible oils up to the 3 month ARI peak flows

Table 1. Guidelines for stormwater pollutant discharges (data from Wong, 2006, p. 1-6).

Water quality and quantity control for redevelopment in many council areas is required to include an On Site Detention system (OSD) and a trash rack in front of an orifice screen. This is often combined with an in-line commercial pollutant filter. The trash screen is a mesh screen bolted to the OSD pit wall to remove litter, vegetation and sediment (Nicholas, 1995).This screen removes the gross pollutants and may also trap other pollutants (Goyen et al., 2002; UPRCT2005). In most cases, a commercially available pollutant filter is added to the design to reduce the contaminants below the threshold values shown in Table 1. OSD systems by themselves may not be able meet the standards shown in Table 1. Brisbane City Council (2008) developed a framework for water quality requirements that can be specifically applied to a site, based on the urban stormwater point of discharge to the catchment (for example a specific stormwater outlet or receiving stream). Other councils such as Blacktown City Council (BCC) (2006) and Blue Mountains City Council (BMCC) (2005) require the designs to protect specific receiving waters and these councils refer designs to the respective Catchment Management Authorities (CMA).

Often council requirements do not take into consideration any improvement in discharge water quality as a result of rainwater tank on-site. A number of contrasting results have been reported in regard to the water quality of rainwater tank. Evans et al. (2006, p. 37) reviewed the current literature on water quality of rainwater tank and found that 'a clear consensus on the quality and health risks associated with rainwater has not been reached' for water use. Furthermore, research has also indicated that the internal processes in the rainwater tank, such as sedimentation and micro-layer flocculation, can improve the water quality in the tank (Spinks et al., 2003).

3. Construction and typical Australian rainwater tanks

The application and size of rainwater tanks in Australia is highly varied. The tanks come in all shapes and sizes, but are also made from different materials. Australian rainwater tanks are often constructed above ground (see Figure 2(a)), partially in-ground (see Figure 2(b)), in-ground (see Figure 2(c)) or under house and the tanks can be made of polypropylene (see Figure 2(a)), concrete (see Figure 2(b)), and coated corrugated iron (Zincalume®, Colorbond®) (see Figure 2(e)). The size of the tank is dependent on roof area, water demand, local rainfall characteristics, and availability of mains water including required security of supply. Each of these characteristics has a different effect on the size of the tank to meet demand (Barry & Coombes, 2006, 2007; Coombes, 2002; Standards Australia, 2008). The size of the tank also has an impact on the materials used as the forces associated with water and ground pressure have a direct impact on the total stress on the tank wall (Standards Australia, 2008). The structural integrity of a large tank can be difficult to maintain for the polypropylene tanks (PVC) and therefore often concrete is used for larger tanks. The smaller tanks are often chosen for urban areas and need to be able to be installed in confined spaces, such as small side setbacks or little alcoves. The shape, size and general visual appeal of small tanks are therefore very different than the larger concrete tanks (Standards Australia, 2008). Finally, the cost of the tank material can often also become a deciding factor in the selection of size and material.

All rainwater tanks are required to have as an absolute minimum an inlet, an outlet to meet demand and an overflow point (see Figure 3). Rainwater tanks can also be fitted with a pump for ease of use (Standards Australia, 2008), a top-up system for areas where mains water is available (Barry & Coombes, 2006, 2007; Standards Australia, 2008), a first flush (Cuncliffe, 1998; enHealth Council, 2004; Standards Australia, 2008) and a water treatment system (Föster, 1996; Standards Australia, 2008). Furthermore in some urban areas stormwater is also detained in the tank, adding a third outlet in the tank to separate detention[1] and retention[2] of run-off from the roof. These additional items are dependent on

[1]Detention of run-off is to keep the water on the site for a period of time and slowly release the flow to the downstream system.

[2]Retention of run-off is to keep the water on the site for an unknown period of time and reuse the captured volume on the site.

the local legislation, but a first flush is recommended to be installed everywhere to improve the water quality on the tank.

Fig. 2. Rainwater tank site locations and materials

In a rainwater tank study in Western Sydney, it was identified that topping-up of a rainwater tank with mains water can cause a change in the harvested water quality (van der Sterren et al., 2010a; 2009; 2010b). A larger value of hardness and conductivity was observed when mains water was added to the tank. The tank had reached a low volume as a result of high levels of water use and top-up of the water in the tank was needed. Mains water in the area has a typical value between 35 and 62 mg L^{-1} $CaCO_3$, which increased the hardness of water within the rainwater tank after, which created similar concentration in the tank as in the mains water supply (Sydney Water, 2007). The results were significantly higher than the average conductivity in the rainwater tank and the equilibrium was not re-established until 3 months after the mains water was added (see Figure 4).

A first-flush device diverts the first part of the run-off (e.g. first 5 minutes)and it is allowed to drain overland. Griffin et al. (1980) showed that the first 30% runoff generated by a storm has a higher contaminant concentration and contains nearly 70% of the total pollutant load. Bucheli et al. (1998) and Föster (1999) found that the run-off from the first 2 mm of rain contains most of the total pollutants on a number of different roof types. The remaining flows for the roofs have a lower contaminant concentration and are treated by the natural processes within the rainwater tank itself. Goonetilleke et al. (2005, p. 33) disagrees with

[3]Reproduced with permission from Rocla Pty Ltd (www.rocla.com.au)

these findings as the first-flush is 'never precisely defined' and argued that the pollutant load, rather than the concentration, should be the governing factor for the design.

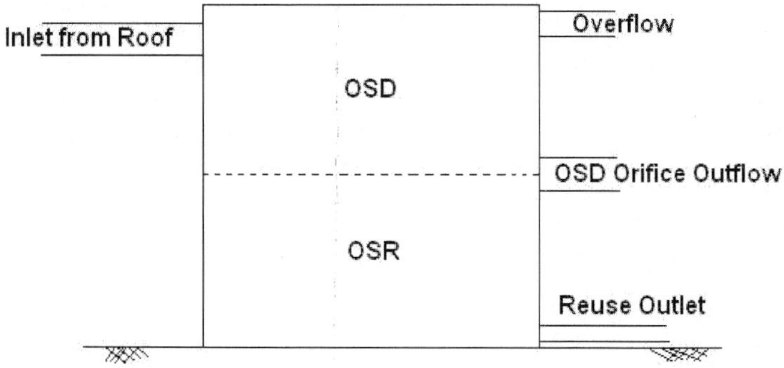

Fig. 3. Rainwater tank with inlet, overflow, outlet and optional third outlet for detention control.

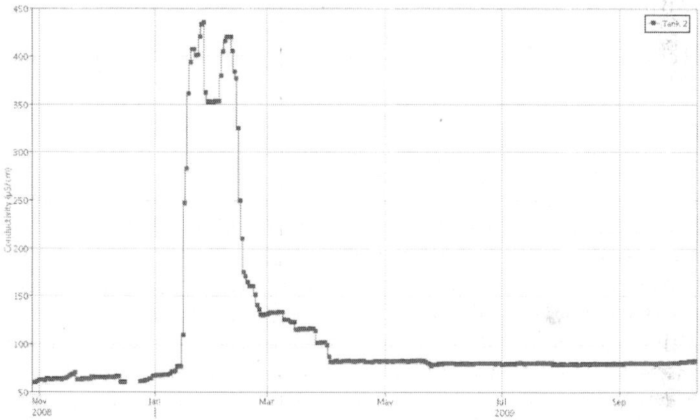

Fig. 4. Conductivity measurements for a rainwater tank in Western Sydney showing the effect of the mains water added to the tank.

Brodie (2007) indicated that the volume required for an effective first-flush is hard to quantify when duration of the storm event is similar to the time of concentration (t_c) of the roof catchment and this can cause difficulties in designing a first-flush. Recently a new technique was suggested by Bach et al. (2010) to determine the first-flush based on statistical analysis of the concentration of pollutants in the first-flush versus the background concentrations rather than a set amount of rainfall as the criteria. This could provide a volume-based design approach suitable for industry application. The removal of pollutants by collecting the first-flush and separating it from the water in the rainwater tank can result in better water quality in the tank. Mendez et al. (2011), however, did not find a significant difference in the water quality in rainwater tank with or without a first-flush. A recent study in Western Sydney showed elevated levels of microbes, nutrients and heavy metals in the first-flush, which were significantly higher than the samples of the harvested water (van der Sterren et al., 2009; 2010b). In addition, the *E.coli* and *Enterococcus* enumeration exceeded secondary contact guidelines, which indicates that pathogens can be present in the first-flush. The data collected in the Western Sydney study supports the emptying of the first-flush after a storm event, which often is not conducted on a regular basis. This defeats the purpose of the system, highlighting the importance of having a first-flush device that can operate without human intervention, especially to prevent the inhalation of pathogens when cleaning is conducted. Recently some councils (e.g. Penrith City Council, 2008, 2010) have added the requirement to connect the overflow from the rainwater tank directly to the street drainage; recent research has found that the overflow quality is not significantly cleaner than the harvested tank water due to settlement of pollutants in the rainwater tank (van der Sterren et al., 2010a; 2009; 2010b). The impact of the overflow quality and comparison of overflow water quality to tank water quality is included in Section 4.2.

4. Water quality of harvested water in Australia

4.1 Processes within the rainwater tank

During storage of water in a rainwater tank, various chemical and biological processes take place. Well known processes are those of settlement and re-suspension of particles (Davis & Cornwell, 2006), conservation of energy (Moeller et al., 1984; Young et al., 2004) and conservation of mass (Moeller et al., 1984). Each of these processes plays a vital role in improving the water quality in a raïnwater tank and in turn ensures that the water in the tank ages and is purified through natural processes.

Settlement and the monolayer/multilayers at the air/liquid interface are two critical processes contributing to water quality in rainwater tanks (Spinks et al., 2003). The settlement could significantly reduce Total Solid (TS) concentrations in the overflow and tank water quality, whilst a biofilm and/or chemical gradients could potentially be critical contributors to overflow contamination. A number of pollutants have been shown to adhere to Suspended Solids (SS) and TS. The principal law that governs the settling of particles is known as Stoke's Law and is a function of particle size and gravitation forces through the medium (Davis & Cornwell, 2006). The particle density and radius vary within the solution and are dependent on the pollution run-off characteristics. In Australia, the average particle size of SS is larger than 125 μm (Melbourne Water, 2005). Miguntanna et al. (2010a; 2010b) found that the 75 to 150 μm particle range had the strongest influence on the Total Nitrogen (TN) build-up, whilst the 1 to 150 μm particle range had the strongest influence on the Total Phosphorous (TP) build-up. Brodie (2007), on the other hand, investigated sediment size and found that over 50% was smaller than 63 μm in roof run-off. The particle density varies from 1600 kg m^{-3} for sand, 1720 kg m^{-3} for coarse materials and 1210 kg m^{-3} for organic fine sediment (Verkerk et al., 1992).

Previous research has found that the water quality at the surface of the water in a rainwater tank is significantly different from the water quality at the bottom of the rainwater tank (Coombes, 2002). This difference has been attributed to stratification. Stratification occurs in lakes and wetlands (Doods, 2002) and could occur in rainwater harvesting systems, which results in chemical and physical gradients in the water column. In addition, a biofilm can form on the air/liquid interface and the solid/liquid interface (Hermansson & Dahlbäck, 1983; Hermansson et al., 1987; Marshall, 1980; Parker & Barson, 1970; Percival et al., 2000; Sigee, 2005; Spinks et al., 2003). Stratification occurs in a body of water, due to warm water being less dense than cooler water. The mixed layer of warm water is the epilimnion and floats on the cooler hypolimnion (Davis & Cornwell, 2006; Doods, 2002). De-stratification will occur if the temperature of the total water body is 3.9°C (Doods, 2002), or when storms and wind mix the water. The mixing by wind is minimal in a rainwater tank, but mixing by inflow could occur depending on the storm intensity and depth of water in the tank. Australia also has reasonably high temperatures. Cool temperatures resulting in rainwater tank water cooling to 3.9°C are likely to only occur at high elevations or during extreme winters in Western Sydney or in the high altitude areas of New South Wales and Victoria, for example in the Blue Mountains and the Snowy Mountains. The epilimnion promotes Brownian motion, due to the higher temperatures and is likely to be influenced by inflows causing turbulence. In addition, the movement between layers is mainly molecular diffusion, thereby creating chemical and physical gradients (Doods, 2002). These result in higher chemical and biological activity, and higher concentrations in the upper layers of the water column in comparison to the lower layers. The higher concentrations and warmer temperatures in the epilimnion also support the growth of micro-organisms, and therefore biofilm growth on the air/liquid interface is likely to occur (Parker & Barson, 1970).

The biofilm surface layer may be between 0.1-10 µm, but divers significantly from the remainder of the water body (Characklis & Cooksey, 1983; Sigee, 2005). Biofilms are defined as "microbial cells, attached to a substratum and immobilised in a three dimensional matrix of extra cellular polymers enabling the formation of an independent functioning ecosystem homeostatically regulated" (Percival et al., 2000, p. 61). These biofilms consist of 50% to 90% of Extracellular Polymeric Substances (EPS) and can provide a number of benefits, including (but not limited to) the absorption of nutrients, microbes and heavy metals, protection of temperature changes, desiccation and mediation by protozoa (Ancion et al., 2010; Bester et al., 2010; Characklis & Cooksey, 1983; Percival et al., 2000; Sigee, 2005). The biofilms have a continuous growth cycle and have been shown to protect natural systems against pollution, but are very susceptible to the flow and turbulence of the liquid (Blanchard & Syzdek, 1970; Percival et al., 2000). Biofilm formation was shown to be favoured during non-optimal microbial growing conditions (Landini, 2009) and growth often start off as small micro-colonies (Walker et al., 1995) attached to the tank, pipe walls or sediments (Characklis & Cooksey, 1983; Characklis & Marschall, 1990). The biofilms can form on the air/liquid and solid/liquid interfaces (Marshall, 1980), but the formation is also dependent on van der Waals, electrostatic and steric forces (Van Houdt & Michiels, 2010).

It is expected that these biofilms exist on the air/liquid interface of the rainwater tank and the solid/liquid interface within the rainwater tank, pipes and in bioretention systems (Blanchard & Syzdek, 1970; Characklis & Marschall, 1990; Flemming, 1993; Hermansson & Dahlbäck, 1983; Sawyer et al., 2010; Spiers et al., 2003; Spinks et al., 2003). The velocity of flow within the rainwater tank itself is slow, even when it is raining and when the tank is utilised. It is, therefore, expected that the detachment of biofilms and microbes is minimal inside the tank (Spinks et al., 2003). The biofilms formed on the air/liquid interface provide the bacteria access to oxygen from the atmosphere and nutrients from the liquid (Spiers et al., 2003); however, when the tank is full, it is expected that the air/liquid interface biofilm is removed from the tank through the overflow and thereby increasing the pollutants in the overflow. This allows the water to age and therefore ensures a better water quality within the tank. This however, does not take into consideration the impact of this biofilm in discharges on the receiving waters. The air/liquid interface was tested in a recent study by the authors in Western Sydney to determine if a difference was present in water samples from the bottom of tank and the overflow, thereby investigating if notable stratification occurs in the rainwater tanks.

A sample of the water at the top of two rainwater tanks were taken in the Western Sydney study and tested for *E. coli*, TC and TTC, *Enterococcus sp.* and HPC (van der Sterren, 2011; van der Sterren et al., 2010a). All microbiological results for top of the tanks for the same week as the tanks and overflow samples were higher than results from the overflow and tank samples. The water surface was determined to contain an average of 143 cfu 100 mL^{-1} *Enterococci spp.*, which was significantly higher than the 7 cfu 100 mL^{-1} recorded in the tap sample. All these results showed higher readings in all microbes test in the overflow than in the tank samples, which has been theorised by Coombes et al. (2002; 2000b) to support a microlayer. This is further supported by theory of biofilms (Percival et al., 2000). It is considered critical to test the overflow during rainwater tank water quality studies to determine the long-term effect on overflow water quality, because of the biofilm growth on the air/liquid interface or the effect of chemical gradients and investigate their effect on the receiving waters.

4.2 Water quality within the rainwater tank

The heavy metal results of the tank samples can vary significantly between the sites and can be above the ADWG (NHMRC & NRMMC, 2004). Aluminium concentrations were found to be above the ADWG ((NHMRC & NRMMC, 2004)) for galvanised roofs and tanks and below the ADWG for concrete tanks in a yearlong water quality study in Western Sydney (van der Sterren, 2011; van der Sterren et al., 2010b) and this supported the findings of Berdhal et al. (2008), Föster (1996, 1999) and Magyar et al. (2007; 2008).

The aluminium concentrations in water at sites with galvanized roofs and tanks are expected to be higher than those from those sites with plastic or concrete systems (BlueScope Steel, 2010; Wight et al., 2000), because of the zinc or zinc-aluminium alloy protective layers. The aluminium concentration could also be linked to location and usage of the tanks, as a high water use household is more likely to expose the inside of the tank to atmosphere due to the higher rate of drawdown on the tank. This drawdown could expose the inside of the tank more frequently to oxygen, thereby likely causing an increase in the rate of corrosion (Askeland, 1998; Berdhal et al., 2008). The aluminium concentration determined in the Western Sydney research study were lower than those of Sorenson et al. (1974) in the United States of America (USA), but similar results were found by Herngren et al. (2005) on streets in Brisbane.

Roofs made of Zincalume® can also show elevated levels on zinc in the tank water. The coating on Zincalume® is 55% aluminium, 43.5% zinc and 1.3 % silicon (BlueScope Steel, 2010). The zinc concentrations of tanks samples on sites with galvanised materials were significantly higher ($p<0.05$) than those sites without these materials and were above the recommended guidelines of 3 mg L^{-1} (NHMRC & NRMMC, 2004). The higher levels of zinc in tank samples are likely due to the protective layer that contains zinc, which can cause the higher zinc concentrations in these tanks. The risk of zinc pollution can be significantly reduced using a concrete or plastic tank (enHealth Council, 2004; NRMMC et al., 2009a). The zinc concentrations in the Western Sydney study were similar to the roof run-off results found by Thomas and Greene (1993), and Herngren et al. (2005). The Western Sydney study does not support the finding by Duncan (1999) who obtained a zinc concentration of 10.2 mg L^{-1} (±5 mg L^{-1}), which is higher than the Western Sydney results (μ_{all}=2.63 mg L^{-1}).

Rainwater tank catchment roof areas should not contain lead flashing, especially if the harvested water is to be used for consumption purposes (Cuncliffe, 1998; enHealth Council, 2004; NRMMC et al., 2009a). Sites containing some lead flashing on the rainwater catchment area show elevated levels of lead (Magyar et al., 2007; 2008; O'Connor et al., 2009; van der Sterren, 2011; van der Sterren et al., 2010b). Sites without lead flashing, but with Poly Vinyl Chloride (PVC) pipes can also show elevated levels of lead concentrations (Magyar et al., 2007; 2008; O'Connor et al., 2009; van der Sterren, 2011; van der Sterren et al., 2010b). Lead concentrations are of particular concern, when the residents use their tank for all in-house purposes, including consumption. Magyar et al. (2007; 2008) and O'Connor et al. (2009) found similar lead values in rainwater tank water in Victoria as van der Sterren (2011) found in Western Sydney. Duncan (1999) also showed lead concentration in roof run-off to be on average 0.054 mg L^{-1} (± 5.01 mg L^{-1}). The results collected in the Western Sydney study (μ_{Tall}= 0.009 mg L^{-1}) were similar to other studies in Australia (Magyar et al., 2007; 2008; O'Connor et al., 2009) and are therefore, not considered unusually high.

The copper concentrations for all tanks tested in the Western Sydney study ($\mu_{T\text{-}all}$=0.221 mg L^{-1}) were well below the ADWG (NHMRC & NRMMC, 2004), but all tank samples had a hardness falling into the soft range, which is likely to result in corrosion of copper pipe lines. Sites containing concrete tiles or a concrete tank can be beneficial for the users, as the higher hardness reduces copper corrosion. A higher hardness would be beneficial to all sites in the Western Sydney study, especially because the ADWG (NHMRC & NRMMC, 2004) suggest that a good hardness is between 60 and 200 mg L^{-1} CaCO3, which was not met by any of the samples on any of the sites.

Total Nitrogen is an important pollutant for receiving water bodies, but in respect to drinking water, the ADWG (NHMRC & NRMMC, 2004) have made recommendations in regards to nitrate, nitrite and ammonium. The tank samples in the Western Sydney study indicated nitrate to be as low as 0.001 mg/L in the tank water (van der Sterren, 2011; van der Sterren et al., 2010b). This is significantly different to the roof catchment study conducted by Evans et al. (2006), who found nitrate concentrations between 0.31 and 4.63 mg L^{-1}. The results are similar to the findings by Nicholson et al. (2010), whom studied roof catchment in the United States of America (USA). The long-term recommended trigger value for Total Phosphorous (TP) in collected roof run-off used for irrigation is 0.05 mg L^{-1}. This can be increased to 0.2 mg L^{-1} if algal blooms in their irrigation system are acceptable (NRMMC et al., 2009a). The RWG (NRMMC et al., 2009a) indicates that it is not practical to reduce the TP concentrations in roof water for domestic applications. The average TP concentrations in the tap samples in the Western Sydney study were 0.18 ± 0.20 mg L^{-1}, which is above this guideline. The results were similar to those found by Duncan (1999) (0.15 ± 1.95 mg L^{-1}) and slightly below the roof run-off TP concentrations found by Miguntanna (2009) (1.85 mg L^{-1}).

The microbiological contamination in tanks used for drinking water must have minimal pathogenic contaminants (enHealth Council, 2004). The ADWG (NHMRC & NRMMC, 2004) states that *E. coli* concentrations should not be detected and the enHealth guidelines indicate that the microbial quality of rainwater tank water is 'not as good as urban water supplies' (enHealth Council, 2004, p. 2). *E. coli* was detected above 1 cfu 100 mL^{-1} for at least 75% of the tank samples in the Western Sydney study. High enumeration of *E. coli* was attributed to low water levels and long antecedent dry periods, resulting in an increased concentration of faecal contamination levels in the roof run-off and tank water. Lower water levels reduce the effect of dilution and the longer antecedent dry period increases the likelihood and build-up of faecal contamination from wildlife deposited onto the roof. In addition, the low water levels can potentially cause significant mixing of the water column as a result of a rainfall event. It is possible that a biofilm or the air/liquid interface are broken up and mixed throughout the small water column during low water levels and a rainfall event.

The mixing of the water column could potentially increase faecal contamination at the bottom of the tank. It has been suggested that a rainwater tank should not be grossly over-designed to ensure that the biofilm is removed from the tank on a regular basis (van Olmen, 2009). This could be done by ensuring that the roof size and usage of the water is sufficient for the site, thereby finding a balance between the formation and cleaning out of the biofilm or air/liquid interface. The results of the enumeration of *E. coli* in the Western Sydney study showed lower concentrations than reported by McCarthy et al. (2008) in stormwater run-off (50 to 34,770 cfu 100 mL^{-1}). The difference is attributed to the different surface areas contributing to the stormwater run-off tested by McCarthy et al. (2008). Roofs are considered to have lower concentrations of *E. coli* than other impervious areas. The enHealth guidelines (2004, p. 12) also indicate that there is 'no measurable difference in rates of gastrointestinal illness in children who drank rainwater [sic. harvested roofwater] compared to those who drank mains water', but also indicates that those who are immunocompromised are at a much greater risk.

The turbidity in tank water varies with location and some sites show a greater variation in turbidity due to the location and size of the tank (Thomas & Greene, 1993). The Western Sydney study showed exceedance of the turbidity guidelines from 2% to 33% of the samples taken from the tanks. This is in contrast with the results of Mobbs (1998), where the rainwater tank did not exceed the ADWG (NHMRC & NRMMC, 2004) at all throughout a year of monthly sampling. It should be considered that testing was conducted on a weekly basis for the Western Sydney study, instead of monthly as in Mobbs (1998) and as settlement of solids is often associated with turbidity, this could give different results than monthly grab samples. The increases in turbidity from the tank samples can be attributed to the tanks not being used. When users in the Western Sydney study were on extended leave, the water quality test results showed outliers and when irrigation was not required, an increase in turbidity levels was noted. The second attribute influencing the turbidity levels is the impact of rain on the volume remaining in the tank. Han and Mun (2008) suggest using a 3 m deep tank to reduce the impact of rainfall on the mixing in the tank. Smaller tanks having a high water use and drawdown can result in low water levels on a regular basis. During rainfall events, the rain is likely to disturb the sediment in the bottom of the tank if the water levels are low. Other contributors to elevated turbidity levels have been the dust storm of 23 September 2009 in Sydney and the impact of runoff containing leaf litter (see Section 4.3).

The Dissolved Oxygen in the Western Sydney study was in the acceptable range for the ADWG (NHMRC & NRMMC, 2004), but for one tank in the study, the DO was mainly below the recommended guideline of 85% saturation (HNMRC & NRMMC, 2004). This could be an indication of more oxygen consuming processes taking place in the tank, for example, corrosion and microbial growth, but Chemical Oxygen Demand and Biological Oxygen Demand were not tested for in the Western Sydney study and therefore this hypothesis cannot be confirmed.

The pH was expected to be approximately 6.5 for the tank water (Coombes et al., 2000b; Duncan, 1999; Herngren et al., 2005; Thomas & Greene, 1993) and the average results for the sites containing galvanised materials in the Western Sydney study were not statistically different from this pH ($\mu \neq 6.5$, $p > 0.05$). The sites containing concrete were slightly higher than this, but still within the ADWG (2004) (HNMRC & NRMMC, 2004). The effect of materials on the pH of the tank samples was also presented by Thomas and Greene (1993) and Mendez et al. (2011). The main cause for the pH increase in concrete roofs was attributed to efflorescence (Berdhal et al., 2008). Calcium hydroxide reacts with carbon dioxide in the air producing calcium carbonate, which is transported into the tank during a rain event. The calcium carbonate in the tank dissolves and neutralises the pH. The results for pH found in the Western Sydney study were higher than Duncan (1999) (5.7 ±1.1), but similar to Thomas and Greene (1993) (6.8 to 7.0). Duncan (1999) analysed values from around the world, which could have brought the mean result down due to acid rain in the Northern Hemisphere. Bridgman et al. (1989; 1988) indicated that severity of acid rain in Australia and New Zealand is significantly lower than other parts of the world, which could explain the higher readings in the Western Sydney study and by Thomas and Greene (1993) in comparison to Duncan (1999).

The conductivity concentrations were expected to be between 15 and 297 $\mu S\ cm^{-1}$ (Camp Scott Furphy Pty Ltd, 1991; Herngren et al., 2005; Thomas & Greene, 1993). The recorded results from the Western Sydney study fell mostly within this range and any outliers were attributed to extremely low water level. Thomas and Greene (1993) discussed that the higher conductivity was likely to be the result of concrete on their sites, which is supported by the results from the Western Sydney Study.

In summary, the materials of the rainwater tank and roof largely govern the water quality within the rainwater tank. The galvanised steel (and Zincalume® or Colourbond®) roofs and rainwater tank are likely to add significant concentrations of aluminium and zinc to the water supply, as the aluminium and zinc coating is the sacrificial layer. The polypropylene rainwater tank, PVC piping and lead flashing increase the lead concentrations within the rainwater tank, whilst concrete increases the hardness and conductivity in the rainwater tank. A lower hardness can also increase the scaling and therefore the concentration of copper in the rainwater tank and plumbing.

4.3 Water quality of the overflow

A very limited amount of data is available for overflow water quality in the literature. Most rainwater tank water quality testing has focused on the potable or non-potable quality of the harvested tank water. In regards to stormwater management, the overflow quality is of a much greater concern, as it is often directly connected to the receiving drainage system. A study on rainwater tank water quality and quantity discharges has been conducted in Australia, in which the first 2500 mL of overflow was tested (van der Sterren, 2011; van der Sterren et al., 2010a). The parameters analysed were DO, pH, turbidity, conductivity, temperature, TN, TP, aluminium, copper, lead, zinc, *E.coli*, *Enteroccocus spp.*, TC and TTC, which give an indication of the potential water quality of overflows from rainwater tanks. This section summarises the results of the recent study in Western Sydney(van der Sterren, 2011; van der Sterren et al., 2010a) and contrasts the results to the tank samples taken at the same time from the same sites. The overflows tested in the Western Sydney study had significantly greater pollutant loadings than the tank sample itself, which is hypothesised to be the result of the epilimnion characteristics of the air/liquid interface, the potential of biofilm growth or the microlayer on the air/liquid interface, as discussed in Section 4.1.1

Statistically the mean values and distributions of conductivity, copper hardness, lead, pH and turbidity were found to be similar ($\mu_T \neq \mu_O$, $p>0.05$) between the tank and overflow samples, but aluminium and zinc standard deviations of the tank and overflow samples were significantly different ($\sigma_T \neq \sigma_O$, $p<0.05$). The concentrations for aluminium ($min_{O\text{-}all}$=0.020 mg L^{-1}) were all above 0.8 µg L^{-1}, which is above the guideline for freshwater of 55 µg L^{-1} (ANZECC & ARMCANZ, 2000a, b). Sites containing galvanised roofs recorded higher readings for aluminium than those without. The majority of the lead results for the overflow (min_{Oall}= 0.001 mg L^{-1}) were above the AFWG (ANZECC & ARMCANZ, 2000a, b) guideline of 34 µg L^{-1}. A greater variation was observed on sites that contained lead flashing on the roof and conventional PVC downpipes.

The zinc concentrations for the overflow samples had higher concentrations for two sites as a result of the galvanised roofs and tank. All recorded zinc ($min_{O\text{-}all}$=0.050 mg L^{-1}) results were above the AFWG (ANZECC & ARMCANZ, 2000a, b) of 8 µg L^{-1}. The RWG (NRMMC et al., 2009a) indicate that the zinc concentrations in water stored from zinc coated roofs exceeded the long and short term trigger values in the AFWG (ANZECC & ARMCANZ, 2000a, b) and can be toxic to species. The RWG (NRMMC et al., 2009a) recommend low irrigation rates to protect sensitive plants and to minimise contamination of the soils. This would mean that the overflow discharge should be treated using best management stormwater treatment techniques prior to discharging it into the existing drainage system.

The copper concentrations in the overflow also varied, but showed a narrower distributions and similar results between overflows. A comparison of the mean copper concentrations indicated that at 95% confidence interval of the mean values (μ_{all}= 0.081 mg L^{-1}) were similar in the Western Sydney study, but all results were above the AFWG (ANZECC & ARMCANZ, 2000a, b). According to the RWG (NRMMC et al., 2009a) copper concentrations greater than 0.2 mg L^{-1} can be toxic to plants, which occurred occasionally in the overflow samples, thereby posing a potential threat to sensitive plants growing in the overflow discharge path. The high values for the copper concentration occurred on the same day as those outliers for lead and zinc concentrations in the results. It is hypothesised in the Western Sydney study (van der Sterren, 2011) that the humidity and warmth of summer increases the degradation, as identified by Berdahl et al. (2008), thereby increasing the pollutant wash-off.

In addition, a high rate of drawdown can also increase the corrosion, because of wetting, drying and exposure to oxygen on the inside of the tank. It should be noted that the hardness of overflows tested in the Western Sydney study were below the AFWG (ANZECC & ARMCANZ, 2000a, b) criteria of 30 mg L^{-1} CaCO3, which allows the heavy metals guidelines to be applied to the test results. All the hardness results were also below the recommended ambient in-stream guideline of lowland rivers (ANZECC & ARMCANZ, 2000a, b). Sites containing concrete tanks and tiled roofs showed higher hardness concentrations in the overflow, just like the tank samples in the Western Sydney study.

The pH values of the overflows can have an impact on receiving waters, however, dilution of the overflows through the stormwater system, which is mostly concrete, could increase the pH and reduce the risk to the receiving waters. The AFWG (ANZECC & ARMCANZ, 2000a, b) requires the pH to be in between 6.5 and 9.0 for inland rivers, but results indicate average higher values than the minimum guideline for sites containing concrete. This is supported by the findings from Thomas and Greene (1993), whom also found elevated pH results when concrete was used as a catchment area. The conductivity of the overflows is recommended to be below 300 µS cm^{-1} (ANZECC & ARMCANZ, 2000a, b). The conductivity results in the overflow samples indicate that the conductivity was higher in those samples from sites with concrete than those without concrete.

E. coli, TTC, TC and *Enterococci spp.* counts can vary significantly between different sources and values above the guideline can be expected in rainwater tanks (Evans et al., 2006; Richardson et al., 2009). The *E.coli* enumeration from the overflows in the Western Sydney study showed a high variation. The enumeration of *E.coli* exceeded the primary contact standard for faecal coliforms (150 cfu 100 mL^{-1}) and one sample exceeded the secondary contact guideline (150-1000 cfu 100 mL^{-1}) (ANZECC & ARMCANZ, 2000a, b). The results of *Enterococcus spp.* enumeration also exceeded the secondary contact guidelines in the AFWG (ANZECC & ARMCANZ, 2000a, b).

The variation in the enumeration of the *E. coli* and *Enterococcus spp.* may be due to discharge volume, time elapsed between different storm events, and environmental condition, and the total indicator organisms in the run-off entering the tank during an event. The higher readings were found to occur in summer, which was similar to the findings by van Olmen (2009), who indicated that higher temperatures in the water may potentially increase the enumeration of indicator organisms.

The microbiological and nutrient analysis showed that an epilimnion, microlayer or biofilm could exist. As for the microbial and nutrient concentrations, the tank samples were significantly lower than the overflow samples. The *E. coli* and E*nterococcus sp.* counts show significantly higher counts in the overflows and were considerably higher than the outliers for the tank samples. In addition, the tank outliers were all within the variation of the overflows and occur after a significant rainfall event or during low water levels. This further supports the possibility of mixing of the water column and biological gradients discharging through the overflow. It should be noted that 50% of the overflows are above the secondary contact guideline of 230 *Enterococcus sp* cfu 100 mL^{-1} (ANZECC & ARMCANZ, 2000a, b).

All mean overflow temperatures were within the range of 15°C to 23°C with some outliers in summer and some lower temperatures in winter, which were slightly higher than the tank samples. Thermal pollution is, therefore, considered not to be a fundamental issue for the overflow from rainwater tank; however, if all urban areas are considered, thermal pollution of receiving waters may be significant. The heat exchange laws support the growth of biofilms or microlayer at the air/liquid interface, as the direct contact between the water surface and the air is likely to warm up the water at the surface more than the main water body within the tank (Doods, 2002). There is direct contact between the air and liquid at the top of the tank, whilst the main body of the tank has indirect contact with the air. More heat is transferred through air/liquid interface than through the air-solid-liquid interface (tank wall). Furthermore the first 2500 mL of overflow was collected, which is a smaller volume than the tank and therefore after collection, more than likely to have a greater variation in temperature until removed from site as the thermal exchange laws are dependent on mass and thermal conductivity (Askeland, 1998; Young et al., 2004). The temperature and volume in the collection bottle would also affect the DO, as the re-aeration rate of oxygen is dependent on temperature, velocity and water depth (Davis & Cornwell, 2006).

The DO concentrations were highly variable and mostly above the guideline (ANZECC & ARMCANZ, 2000a, b). Some of the results were below the 7.6 mg L^{-1} minimum limit. This indicates that DO in the overflow should be increased prior to discharge into receiving rivers, which often occurs as a result of flow velocity and turbulence through the drainage system. The top of the tank had a higher DO content than the bottom because of the re-aeration is dependent on the depth which causes a temperature gradient (Davis & Cornwell, 2006; Doods, 2002). This higher DO concentration at the air/liquid interface and in the overflow, as a result of the gradients, combined with increases in water temperature could potentially sustain the microbes, thereby resulting in potential higher counts. In addition, when overflows occur, the rainfall coming into the rainwater tank also contains microbes, further increasing the potential survival of indicator organisms and therefore pathogens in the rainwater tank.

The TP concentrations ($\mu_{o\text{-}all}$=0.976 mg L^{-1}) however, were all above the AFWG (ANZECC & ARMCANZ, 2000a, b) (0.050 mg L^{-1}) and RWG (NRMMC et al., 2009a). Outliers were attributed to the eucalyptus blossom and other plants in the area. According to the RWG (NRMMC et al., 2009a), the short-term impact on soils for phosphorus contamination is low. It is suggested that phosphorous levels be reduced by using bio-retention and filtration systems to minimise the effect of phosphorus contamination on receiving waters. The overflow from these rainwater tanks should, therefore, be directed to a filtration or retention system to minimise wash-off of TP to the receiving water bodies. All of the nitrate concentrations were well below the AFWG (ANZECC & ARMCANZ, 2000a, b) (600 µg L^{-1}) and the RWG (NRMMC et al., 2009a) (30 mg L^{-1}). This further supports the hypothesis that there is a microlayer or biofilm at the air liquid interface, as these layers are often formed as survival mechanisms in low nutrient environments (Landini, 2009).

Biofilm formation is promoted by non-optimal growing temperature or limited nutrients (Bester et al., 2010; Blanchard & Syzdek, 1970; Landini, 2009). The rainwater tank has been shown to be low in nutrients and is influenced by the similar chemical gradients as a lake. The epilimnion has a limited source of DO and temperature (Chapra, 1997), thereby most likely promoting the formation of a biofilm at the air/liquid interface. The atmosphere provides another additional source reducing the strain on the microbes and allowing the formation of new cells (Bester et al., 2010; Spiers et al., 2003). The chemical and biological gradients (Chapra, 1997), the possible biofilm, as well as the pollutant run-off from the roof water can cause the overflow to have higher pollutants than the bottom of the rainwater tank. This indicates that the rainwater tank, especially larger rainwater tank, can have stratification and therefore should not be modelled as completely mixed tanks (or continuously stirred tanks). On the other hand, when the rainwater tank has a low volume of water, the rainfall can have a significant stirring effect on the minimal volume thereby creating a mixed tank (Chapra, 1997). There is a further need to examine the overflow from rainwater tank and their effects on the stormwater quality to understand its impact. Furthermore, detailed analysis is also required on the stratification effects in the rainwater tank and the potential of biofilm growth. These effects are considered to age and clean the water, but could have a detrimental effect on the outflow water quality of the site. It clearly highlights the need for more than one source control and using a treatment train approach, including UV treatments to control micro-organisms. Furthermore, recent research in Europe indicated that the water quality of rainwater tank improves if an overflow occurs on a regular basis, therefore indicating that a rainwater tank should not be over-designed (van Olmen, 2009).

4.4 Impact on quality from climatic events

Different climatic events can have impacts on the water quality of the rainwater tank. A dust storm covered Sydney and surrounds in a haze on the 23 September 2009 and deposited large amounts of fine particles onto all surfaces, which includes all roofs (Leys et al., 2009). A low intensity rainfall event occurred on the 24 September 2009 and the research conducted in Western Sydney showed high turbidity results (14.8 NTU) in the first-flush after this event. The increase of turbidity levels in the tested tanks were not detected until 6 October 2009, after a more intense rainfall event as well as settling time in the rainwater tank. The overflows also showed higher than average turbidity levels on 6 October 2009. In addition to the elevated turbidity levels, copper, TN, conductivity and hardness also showed significantly high values as a result of the storm event of 2 October 2009. Testing of the deposited dust was conducted by Radhi et al. (2010), who found that the dust contained salt particles and could be traced back to Lake Eyre Basin. The dust storm event is therefore the most likely contributor to these elevated levels. This clearly shows the needs for cleaning and maintenance of the roof after an event like this. Heavy rainfall, is the primary cause of pollutant run-off from the roof and was therefore examined here. All high values were compared to rainfall events and antecedent periods. The tank samples were more likely to show elevated levels when antecedent periods were longer, whilst overflows were more likely to show high values when there is a prolonged period between overflow events (van der Sterren, 2011).

Elevated levels and extremes were more likely to occur in the tank water quality when the antecedent rainfall event exceeded two days. The overflow, on the other hand, was more likely to contain elevated levels when the time to previous overflow exceeded 8.5 days. This is of course a guideline as events such as the dust storm or the intensity of the rainfall event can alter the high values and concentrations after a smaller antecedent period. The focus was on long term sampling (i.e. a year). The benefit of the long term testing is that the seasonal changes as well as long-term (in other words, year) impact can be examined. The antecedent time period can affect the build-up on roofs and wash-off into tanks, as high values are more likely to occur during long antecedent periods. This was previously shown to be important in pollutant build-up by other studies (Egodawatta & Goonetilleke, 2008; Egodawatta et al., 2009; Miguntanna et al., 2010b; Sartor & Boyd, 1972). Overflows are also more likely to contain extreme values when time intervals between overflows are longer.

5. Government subsidies and financial viability

The Australian, State and local governments have a number of subsidies and grants that are provided to citizens to install a rainwater tank on their property. This section looks at the financial viability of rainwater harvesting system and these government subsidies for rainwater tanks. The amount of rebate varies from state to state and also on the size of the rainwater tank and where the water is utilised. For New South Wales, the maximum rebate is $1500. For Australian Capital Territory and Victoria, the maximum rebate is $1000.

A number of studies have examined the financial viability of rainwater harvesting system. These studies have looked at the financial viability of rainwater tanks by themselves, or rainwater tanks as part of an integrated catchment management method. The financial benefit of rainwater tanks as a single entity has been shown to be minimal; however, commonly reduction of cost associated with a reduction in on site detention volume and environmental benefits are often not included in these studies.

A cost benefit analysis of the financial viability of rainwater tanks is dependent on the cost of water, any subsidies and at what time the net present value is computed. In addition, the design and country of origin are also important variables in conducting a life cycle or cost - benefit analysis. Roebuck and Ashley (2006) examined the life cycle costing of a rainwater tank for a school building in United Kingdom (UK) with a project life of 65 years. The long-term savings were estimated to be £18,370 opposed to £122,230 estimated by the tank supplier. They found that capital cost, maintenance cost and mains top-up cost were 31%, 26% and 40% respectively of the total cost and argued that many of the methods of rainwater tank analysis overestimated the cost savings of rainwater tanks. Domenech and Sauri (2011) examined the financial benefit of rainwater tanks in single and multi-unit buildings in Barcelona, Spain. They found minimum pay back periods of 33 to 43 years for a single-family household and 61 years for a 20 m^3 tank for multi-unit buildings. Grant and Hallmann (2003) conducted the financial viability of a 600 and a 2250 L rainwater tank in Australia. They found that neither tanks would create a positive return over 30 years using the 2003 water pricing. Life cycle costing of rainwater tanks for multi-storied building in Sydney using a 75 kL tank and for toilet, laundry and outdoor use, the cost benefit ration has been found to range from 0.64 to 1.15 for discount rates of 0% to 7.5% (Rahman et al., 2010)

Although both Spain (Barcelona) and Sydney are considered a mediteranian climate, there is a significant difference between the benefits of rainwater tanks and the pay back periods. The UK life cycle costing can expected to be very different, as the chance of rain and annual volume in the UK is higher than in Spain and Australia. These rainwater tanks are designed according to the climate and therefore affect the size, cost and use of the rainwater tank. It is therefore difficult to compare Australian data with other countries, but overall, there is a significant pay back period of a single rainwater tank and the cost-benefit analysis should take all aspects of the benefits and the cost into account.

When the financial viability of rainwater tanks are considered as part of an integrated water cycle management plan, the rainwater tanks are shown to have significant benefits to authorities. The demand on dam and mains water is significantly reduced when rainwater tanks are implemented throughout the catchment. Various studies have shown that the cost reduction by implementing rainwater tanks throughout the catchment is significant

(Coombes & Kuczera, 2003; Coombes et al., 2000a; Davis & Birch, 2008; Lucas, 2009; Lucas et al., 2009). This clearly indicates that any cost-benefit analysis should not only investigate the cost and benefits of a single rainwater tank, but the cost and benefits of the whole system, including water mains upgrade, cost to the environment and reduction in cost of treating and managing stormwater.

6. Conclusion

Rainwater tanks are popular in Australia as a source of alternative water supply and a means of stormwater management. From a literature review and findings from a field study conducted in Western Sydney in Australia, it has been found that materials of the rainwater tanks and roof largely govern the water quality within the rainwater tanks. The overflow and first flush water quality from the rainwater tanks exceeded water quality guidelines, highlighting the needs for further treatment before discharging the water to urban stormwater systems. The water quality in the rainwater tanks does not meet the drinking water guidelines on many instances, especially during low water levels and the days immediately after a rainfall event. Under these circumstances, the tank water should be disinfected before drinking. Rainwater tanks are found to be not financially viable at lot scale given the current water price, but if all the associated benefits of rainwater tanks at catchment scales are accounted for their financial viability is likely to increase significantly.

7. Acknowledgements

The work presented in this Chapter would not have been completed without the assistance of Jonathan Barnes, Solomon Donald and Ian Turnbull. Thanks also to Surendra Shrestha, John Bavor, Jeff Scott, Sharon Armstrong, Rhonda Gibbons, Bert Aarts, Wayne Higgenbotham, Paul Roddy, Heidi Fitzpatrick from the University of Western Sydney. Thanks also to Turnbull Electrical Contracting, who contributed time and resources for the installations of the systems on site and the funding from the UWS office of Disabilities.

8. References

Ancion, P.-Y., Lear, G., & Lewis, G. D. (2010). Three common metal contaminants of urban runoff (Zn, Cu & Pb) accumulate in freshwater biofilm and modify embedded bacterial communities. *Environmental Pollution*, 158(8), 2738-2745.

Askeland, D. R. (1998). *The science and engineering of materials*. Cheltenham: Stanley Thornes Ltd.

Australian and New Zealand Environment and Conservation Council, & Agriculture and Resource Management Council of Australian and New Zealand (ANZECC & ARMCANZ) (2000a). Australian and New Zealand guidelines for fresh and marine water quality - volume 1. *National Water Quality Management Strategy*. Artarmon: Australian Water Association.

Australian and New Zealand Environment and Conservation Council, & Agriculture and Resource Management Council of Australian and New Zealand (ANZECC & ARMCANZ) (2000b). Australian and New Zealand guidelines for fresh and marine water quality - volume 2 - aquatic ecosystems - rationale and background information. Artarmon: Australian Water Association.

Australian and New Zealand Environment and Conservation Council, & Agriculture and Resource Management Council of Australian and New Zealand (ANZECC & ARMCANZ) (2000c). Australian guidelines for urban stormwater management. *National Water Quality Management Strategy*. Artarmon: Australian Water Association.

Australian Bureau of Statistics (2010). Australia's Environment: Issues and Trends. Canberra: Australian Bureau of Statistics.

Australian Bureau of Statistics (ABS) (2011). Regional Population Growth, Australia. Canberra: Australian Bureau of Statistics.

Bach, P. M., MacCarthy, T. D., & Deletic, A. (2010). Redefining the stormwater first flush phenomenon. *Water Research*, 44(8), 2487-2498.

Barry, M. E., & Coombes, P. J. (2006). Optimisation of mains trickle topup supply to rainwater tanks in an urban setting. *Australian Journal of Water Resources*, 10(3), 269-275.

Barry, M. E., & Coombes, P. J. (2007). Optimisation of mains trickle topup volumes and rates supplying rainwater tanks in the Australian urban setting. *Rainwater and Urban Design Conference*. Sydney: The Institution of Engineers Australia.

BASIX Sustainability Unit (2009). Building Sustainability Index. Sydney: New South Wales Department of Planning.

Berdhal, P., Akbari, H., Levinson, R., & Miller, W. A. (2008). Weathering of roofing materials - an overview. *Construction and Building Materials*, 22(4), 423-433.

Bester, E., Kroukamp, O., Edwards, E. A., & Wolfaardt, G. M. (2010). Biofilm form and function: carbon activity affects biofilm architecture, metabolic activity and planktonic cell yield. *Journal of Applied Microbiology*, 110(2), 387-398.

Blackburn, N., Morison, P., & Brown, R. (2010). A review of factors indicating likelihood of and motivations for household rainwater tank adoption. *Stormwater 2010: National Conference of the Stormwater Industry Association*. Sydney, pp. 1-10.

Blacktown City Council (2006). Blacktown development control plan 2006. Blacktown: Blacktown City Council.

Blanchard, D. C., & Syzdek, L. (1970). Mechanisms for the water-to-air transfer and concentration of bacteria. *Science, New Series*, 170(3958), 626-628.

Blue Mountains City Council (2005). Better Living Development Control Plan. Katoomba: Blue Mountains City Council.

BlueScope Steel (2010). Benefits of ZINCALUME® steel.

Bridgman, H. A. (1989). Acid rain studies in Australia and New Zealand. *Archives of Environmental Contamination and Toxicology*, 18(3), 137-146.

Bridgman, H. A., Rothwell, R., Tio, P.-H., & Pang-Way, C. (1988). The Hunter region (Australia) acid rain project. *Bulletin of the American Meteorological Society*, 69(3), 266-271.

Brisbane City Council (2008). Subdivision and development guidelines - Part C water quality management guidelines. Brisbane: Brisbane City Council.

Brodie, I. M. (2007). Investigation of stormwater particles generated from common urban surfaces. *Faculty of Engineering and Surveying*: University of Southern Queensland.

Bucheli, T. D., Müller, S. R., Heberle, S., & Schwarzenbach, R. P. (1998). Occurrence and behaviour of pesticides in rainwater, roof runoff, and artificial stormwater infiltration. *Environmental Science and Technology*, 32(22), 3457-3464.

Camp Scott Furphy Pty Ltd (1991). *Urban runoff study*: Nepean-Hawkesbury Catchment Management Council.

Chapra, S. C. (1997). *Surface water-quality modeling*: WCB/McGraw-Hill.

Characklis, W. G., & Cooksey, K. E. (1983). Biofilms and microbial fouling. *Advances in Applied Microbiology*, 29, 93-138.

Characklis, W. G., & Marschall, K. C., eds. (1990). *Biofilms*. New York: John Wiley & Sons.

Coombes, P. J. (2002). Rainwater tanks revisited: new opportunities for Urban Water Cycle Management. Newcastle: University of Newcastle.

Coombes, P. J., & Kuczera, G. (2003). A sensitivity analysis of an investment model used to determine the economic benefits of rainwater tanks. *28th International Hydrology and Water Resources Symposium*. Wollongong, Australia: Institution of Engineers Australia, pp. 243-250.

Coombes, P. J., Kuczera, G., & Kalma, J. D. (2000a). Economic benefits arising from use of Water Sensitive Urban Development source control measures. *Hydro 2000 - 3rd International Hydrology and Water Resources Symposium of the Institution of Engineers Australia*. Perth, Western Australia: The Institution of Engineers Australia, pp. 152-157.

Coombes, P. J., Kuczera, G., & Kalma, J. D. (2000b). Rainwater quality from roofs, tanks and hot water systems at Figtree Place. *Hydro 2000 - 3rd International Hydrology and Water Resources Symposium of the Institution of Engineers Australia*. Perth, Western Australia: The Institution of Engineers Australia, pp. 1042-1047.

Cuncliffe, D. A. (1998). Guidance on the use rainwater tanks. *National Environmental Health Forum Monographs, Water Series Number 3*. Rundle Mall: Department of Human Services.

Davis, B. S., & Birch, G. F. (2008). Catchment-wide assessment of the cost-effectiveness of stormwater remediation measures in urban areas. *Environmental Science & Policy*, 12(1), 84-91.

Davis, M. L., & Cornwell, D. A. (2006). *Introduction to environmental engineering*. New York: McGraw-Hill.

Domènech, L., & Saurí, D. (2011). A comparative appraisal of the use of rainwater harvesting in single and multifamily buildings of the Metropolitan Area of Barcelona (Spain): social experience, drinking water savings and economic costs. *Journal of Cleaner Production*, 19(1), 598-608.

Doods, W. (2002). Freshwater ecology - concepts and environmental applications. Elsevier.

Duncan, H. P. (1999). Urban stormwater quality: A statistical overview. Cooperative Research Centre for Catchment Hydrology.

Egodawatta, P., & Goonetilleke, A. (2008). Modelling pollutant build-up and wash-off in urban road and roof surfaces. *31st Hydrology and Water Resources Symposium and 4th International Conference on Water Resources and Environmental Research: Water Down Under 2008*. Adelaide, Australia, pp. 418-427.

Egodawatta, P., Thomas, E., & Goonetilleke, A. (2009). Understanding the physical processes of pollutant build-up and wash-off on roof surfaces. *Science of the total Environment*, 407(6), 1834-1841.

enHealth Council (2004). Guidance on use of rainwater tanks. In: Australian Government, ed: Australian Government Department of Health and Ageing, pp. 1-72.

Environmental Health Branch (2008). Rainwater tanks where a public water supply is available - use of. North Sydney: Department of Health, New South Wales Government.

Evans, C. A., Coombes, P. J., & Dunstan, R. H. (2006). Wind, rain and bacteria: The effect of weather on the microbial composition of roof-harvested rainwater. *Water Research*, 40(1), 37-44.

Flemming, H.-C. (1993). Biofilms and environmental protection. *Water Science and Technology*, 27(7-8), 1-10.

Föster, J. (1996). Patterns of roof runoff contamination and their potential implications on practice and regulations of treatment and local infiltration. *Water Science and Technology*, 33(6), 39-48.

Föster, J. (1999). Variability of roof runoff quality. *Water Science and Technology*, 39(5), 137-144.

Goonetilleke, A., Thomas, E., Ginn, S., & Gilbert, D. (2005). Understanding the role of land use in urban stormwater quality management. *Journal of Environmental Management*, 74(1), 31-42.

Goyen, A. G., Lees, S. J., & Phillips, B. C. (2002). Analysis of allotment based storage, infiltration and reuse drainage strategies to minimize urbanization effects. *9th International Conference on Urban Drainage, Global Solutions for Urban Drainage:9ICUD 2002*. Portland, Oregon USA: American Society of Civil Engineers, pp. 1-17.

Grant, T., & Hallmann, M. (2003). Urban domestic rainwater tanks: Life cycle assessment. *Water*, 30(5), 36-41.

Griffin, D. M. J., Randall, C., & Grizzard, T. J. (1980). Efficient design of stormwater holding basins used for water quality protection. *Water Research*, 14(10), 1540-1554.

Han, M. Y., & Mun, J. S. (2008). Particle behaviour consideration to maximize the settling capacity of rainwater storage tanks. *Water Science and Technology*, 56(11), 73-79.

Hawkesbury City Council (2000). Development control plan 2000. Windsor: Hawkesbury City Council.

Hermansson, M., & Dahlbäck, B. (1983). Bacterial activity at the air/water Interface. *Microbial Ecology*, 9(4), 317-328.

Hermansson, M., Jones, G. W., & Kjelleberg, S. (1987). Frequency of antibiotic and heavy metal resistance, pigmentation and plasmids in bacteria of the marine air-water interface. *Applied and Environmental Microbiology*, 53(10), 2338-2342.

Herngren, L., Goonetilleke, A., & Ayoko, G. A. (2005). Understanding heavy metal and suspended solids relationships in urban stormwater using simulated rainfall. *Journal of Environmental Management*, 76(2), 149-158.

Heyworth, J. S., Maynard, E. J., & Cuncliffe, D. A. (1998). Who drinks what? Potable water use in South Australia. *Water (Melbourne)*, 25(1), 9-13.

Horstman, M. (2007). 1000 year drought. In: Hiscox, M., ed. *Catalyst*. Australia: Australian Broadcasting Company, p. 9:38.

Landini, P. (2009). Cross-talk mechanisms in biofilm formation and responses to environmental and physiological stress in *Escherichia coli*. *Research in Microbiology*, 160(4), 259-266.

Leys, J., Heidenreich, S., & Case, M. (2009). DustWatch report for week ending 21 September 2009. 1-6.

Lucas, S. A. (2009). Mains water savings and stormwater management benefits from large architecturally - designed under-floor rainwater storages. *32nd Hydrology and Water Resources Symposium*. Newcastle: Engineers Australia, pp. 628-640.

Lucas, S. A., Coombes, P. J., & Sharma, A. K. (2009). Residential diurnal water use patterns and peak demands - implications for integrated water infrastructure planning. *32nd Hydrology and Water Resources Symposium*. Newcastle Australia: Engineers Australia, pp. 1081-1091.

Magyar, M. I., Mitchell, V. G., Ladson, A. R., & Diaper, C. (2007). An investigation of rainwater tanks quality and sediment dynamics. *Water Science and Technology*, 56(9), 21-28.

Magyar, M. I., Mitchell, V. G., Ladson, A. R., & Diaper, C. (2008). Lead and other heavy metals: common contaminants of rainwater tanks in Melbourne. *31st Hydrology and Water Resources Symposium and 4th International Conference on Water Resources and Environmental Research: Water Down Under 2008*. Adelaide, Australia: Institute of Engineers Australia, pp. 409-417.

Marshall, K. C. (1980). Microorganisms and interfaces. *BioScience*, 30(4), 246-249.

McCarthy, D. T., Deletic, A., Mitchell, V. G., Fletcher, T. D., & Diaper, C. (2008). Uncertainties in stormwater *E.coli* levels. *Water Research*, 42(-), 1812-1824.

Melbourne Water (2005). *Water Sensitive Urban Design engineering procedures: stormwater*. Melbourne Australia: CSIRO Publishing.

Mendez, C. B., Klenzendorf, J. B., Afshar, B. R., Simmons, M. T., Barrett, M., Kinney, K. A., & Kirisits, M. J. (2011). The effect of roofing material on the quality of harvested rainwater *Water Research*, Article in Press, 1-11.

Miguntanna, N. (2009). Determining a set of surrogate parameters to evaluate urban stormwater quality. *Faculty of Built Environment and Engineering*: Queensland University of Technology.

Miguntanna, N., Egodawatta, P., Kokot, S., & Goonetilleke, A. (2010a). Determination of a set of surrogate parameters to assess urban stormwater quality. *Science of the total Environment*, 408(24), 6251-6259.

Miguntanna, N., Goonetilleke, A., Egodawatta, P., & Kokot, S. (2010b). Understanding nutrient build-up on road surfaces. *Journal of Environmental Sciences*, 22(6), 806 - 812.

Mobbs, M. (1998). *Sustainable house*. Marrickville, Australia: CHOICE Books.

Moeller, T., Bailar, J. C., Jr., Kleinberg, J., Guss, C. O., Castellion, M. E., & Metz, C. (1984). *Chemistry with inorganic qualitative analysis*. Orlando, Florida: Academic Press Inc.

National Health and Medical Research Council & Natural Resource Management Ministerial Council (NHMRC & NRMMC) (2004). National water quality management strategy - Australian Drinking Water Guidelines. Australian Government.

Natural Resource Management Ministerial Council, Environment Protection and Heritage Council, & National Health and Medical Research Council (NRMMC / EPHC & HNMRC) (2009a). National water quality management strategy - Australian guidelines for water recycling: Stormwater harvesting and reuse. Australian Government.

Natural Resource Management Ministerial Council, Environment Protection and Heritage Council, & National Health and Medical Research Council (NRMMC / EPHC & HNMRC) (2009b). National water quality management strategy - Australian Guidelines for Water Recylcing (Phase 2): Augmentation of drinking water supplies. Australian Government.

Natural Resource Management Ministerial Council, Environment Protection and Heritage Council, & National Health and Medical Research Council (NRMMC / EPHC & HNMRC) (2009c). National water quality management strategy - Overview of the Australian guidelines for water recycling: Managing health and environmental risks 2006. Australian Government.

Nicholas, D. I. (1995). On-site Stormwater Detention: improved implementation techniques for runoff quantity and quality management in Sydney. *Water Science and Technology*, 32(1), 85-91.

Nicholson, N., Clark, S. E., Long, B. V., Siu, C. Y. S., Spicher, J., & Steele, K. A. (2010). Roof runoff water quality - a comparison of traditional roofing materials. *World Environmental and Water Resources Congress 2010: Challenges of Change*. Providence, Rhode Island: American Society of Civil Engineers, pp. 3349-3355.

O'Connor, J. B., Mitchell, V. G., Magyar, M. I., Ladson, A. R., & Diaper, C. (2009). Where is the lead in rainwater tanks coming from? *32nd Hydrology and Water Resources Symposium*. Newcastle, Australia: Institute of Engineers Australia, pp. 653-660.

Parker, B., & Barson, G. (1970). Biological and chemical significance of surface microlayers in aquatic ecosystems. *BioScience*, 20(2), 87-93.

Penrith City Council (2008). *Exhibition of stage 1 planning documents*. Penrith, Australia: Penrith City Council.

Penrith City Council (2010). Penrith City Council Development Control Plan. Penrith.

Percival, S. L., J. T. Walker, et al. (2000). Microbiological aspects of biofilms and drinking water. Vreeland, R. H. (ed). Boca Raton: CRC Press.

Radhi, M., Box, M. A., Box, G. P., & Cohen, D. D. (2010). Size-resolved chemical composition of the September 2009 Sydney dust storm. *Air Quality and Climate Change*, 44(3), 25-30.

Rahman, A., Dbais, J., & Imteaz, M. A. (2010). Sustainability of RWHSs in Multistorey Residential Buildings. *American Journal of Engineering and Applied Sciences*, 1(3), 889-898.

Richardson, H. Y., Nichols, G., Lane, C., Lake, I. R., & Hunter, P. R. (2009). Microbiological surveillance of private water supplies in England - the impact of environmental and climate factors on water quality. *Water Research*, 43(8), 2159-2168.

Roebuck, R. M., & Ashley, R. (2006). Predicting the hydraulic and life-cycle cost performance of rainwater harvesting systems using a computer based modeling tool. *4th International Conference on Water Sensitive Urban Design*. Melbourne: Engiener Australian, pp. 699-709.

Sartor, J. D., & Boyd, G. B. (1972). Water pollution aspects of street surface contaminants. In: monitoring, O. o. r. a., ed. Washington: US Environmental Protection Agency, p. 236.

Sawyer, C. B., Hayes, J. C., & English, W. R. (2010). Characterization of *Escherichia coli* for sediment basin systems at construction sites. *World Environmental and Water Resources Congress 2010: Challenges of Change*. Providence, Rhode Island: American Society of Civil Engineers, pp. 3198-3208.

Sigee, D. C. (2005). *Freshwater microbiology - biodiversity and dynamic interactions of microorganisms in the aquatic environment*. Chichester: John Wiley & Sons.

Sorenson, J. R. J., Campbell, I. R., Tepper, L. B., & Lingg, R. D. (1974). Aluminium in the environment and human health. 8, 3-95.

Spiers, A. J., Bohannon, J., Gehrig, S. M., & Rainey, P. B. (2003). Biofilm formation at the air-liquid interface by the *Pseudomonas fluroscens* SBW25 wrinkly spreader requires an acetylated form of cellulose. *Molecular Microbiology*, 50(1), 15-27.

Spinks, A. T., Coombes, P. J., Dunstan, R. H., & Kuczera, G. (2003). Water quality treatment processes in domestic rainwater harvesting systems. *28th International Hydrology and Water Resources Symposium*. Wollongong, Australia: Institution of Engineers Australia, pp. 227-234.

Standards Australia (2006). Plumbing and drainage part 3: Stormwater drainage. In: Australian and New Zealand Standards, ed. Sydney: Australian Standards.

Standards Australia (2008). HB 230-2008 Rainwater tank design and installation handbook. In: Australian and New Zealand Standards, ed. Sydney, Australia: Standards Australia.

Sydney Water (2007). Water analysis - Typical water analysis for Sydney Water's drinking water supply. Sydney: Sydney Water.

Thomas, P. R., & Greene, G. R. (1993). Rainwater quality from different roof catchments. *Water Science and Technology*, 28(3-5), 291-299.

Upper Parramatta River Catchment Trust (UPRCT), Cardno Willing (NSW) Pty Ltd, Haddad Khalil Mance Arraj Partners, & Brown Consulting (NSW) Pty Ltd (2005). On-site stormwater detention handbook. Upper Parramatta River Catchment Trust.

van der Sterren, M. (2011). Assessment of the impact of rainwater tanks on urban run-off quantity and quality characteristics (under review). *School of Engineering*. Penrith: University of Western Sydney, p. 604.

van der Sterren, M., Dennis, G. R., Chuck, J., & Rahman, A. (2010a). Rainwater tank water quality testing in Western Sydney, Australia. *World Environmental and Water Resources Congress - Challenges of Change*. Providence, USA, pp. 4048-4058.

van der Sterren, M., Rahman, A., & Dennis, G. R. (2009). A case study of rainwater tank quality testing in Western Sydney, Australia. *The 7th International Symposium on Ecohydraulics and 8th International Conference on Hydroinformatics* Concepcíon, Chile.

van der Sterren, M., Rahman, A., & Ryan, G. (2010b). Investigation of water quality and quantity of five rainwater tanks in Western Sydney, Australia. *World Environmental and Water Resources Congress - Challenges of Change*. Providence, USA, pp. 3933-3941.

Van Houdt, R., & Michiels, C. W. (2010). Biofilm formation and the food industry, a focus on the bacterial outer surface. *Journal of Applied Microbiology*, 109(4), 1117-1131.

van Olmen, R. (2009). Gebruik van hemelwater in woningen - economische en ecologische analyse. Belgium: Universiteit Hasselt.

Verkerk, G., Broens, J. B., Kranendonk, W., van der Puijl, F. J., Sikkema, J. L., & Stam, C. W. (1992). *BINAS*. Groningen: Wolters-Noordhoff.

Walker, J. T., Mackerness, C. W., Rogers, J., & Keevil, C. W. (1995). Heterogeneous mosaic biofilm - a haven for waterborne pathogens. In: Lappin-Scott, H. M., & Costerton, J. W., eds. *Microbial Biofilms*. Cambridge: Cambridge University Press.

White, I. (2010). Rainwater harvesting: theorising and modelling issues that influence household adoption. *Water Science and Technology*, 62(2), 370-377.

Wight, I., Huddle, L., & Mullens, R. (2000). Why any old galvanised iron won't do... *The Heritage Advisor*.

Wong, T. H. F., ed. (2006). *Australian runoff quality - A guide to Water Sensitive Urban Design*. Crows Nest: Engineers Australia

Young, H. D., Freedman, R. A., & Ford, L. A. (2004). *Sears and Zemansky's university physics with modern physics*. Sydney: Pearson Education.